Fluid Mechanics

Fluid Mechanics: A Problem-Solving Approach offers a unique textbook intended for both undergraduate and graduate mechanical, chemical, and civil engineering students.

This book works through the comprehensive coverage of fluid mechanics with a gradual introduction of theory in a straightforward, practical manner. The book contains many worked-out examples and illustrations which will help the readers to easily understand the necessary concepts and solution techniques.

This versatile instructional resource contains many interesting elements like:

- The chapter on pipe flows includes a detailed listing of entrance length and flow friction correlations.
- The chapter on oceanic and river waves is included with discussion on various wave types.
- The chapter on laminar flows offers exact solutions for cannonical flow cases like Stoke's problem, Hagen-Poisuille flow, flow between journal and bearing, etc.
- The chapter on turbulent flows discusses the turbulence length scales, the turbulence spectrum, Reynolds averaging and turbulence modeling, and simulations approaches.
- The boundary layer theory includes the Thwaites integral method which is not covered in many fluid mechanics textbooks.
- The book contains many end-of-chapter problems, and a Solutions Manual is available for instructor use.

This eBook+ version includes the following enhancements:

- 3 videos are placed throughout the text to help apply real-world examples to concepts of Newtonian vs. Non-Newtonian fluids, vortices, and additional information on surface tension.
- Pop-up explanations of selected concepts as interactive flashcards in each chapter.
- Quizzes within chapters to help readers refresh their knowledge.

Fluid Mechanics

A Problem-Solving Approach

Naseem Uddin

CRC Press is an imprint of the
Taylor & Francis Group, an **informa** business

First edition published 2023
by CRC Press
6000 Broken Sound Parkway NW, Suite 300, Boca Raton, FL 33487-2742

and by CRC Press
4 Park Square, Milton Park, Abingdon, Oxon, OX14 4RN

CRC Press is an imprint of Taylor & Francis Group, LLC

ISBN: 978-1-032-32453-1 (hbk)
ISBN: 978-1-032-32454-8 (pbk)
ISBN: 978-1-003-31511-7 (ebk)
ISBN: 978-1-032-32456-2 (ebk+)

DOI: 10.1201/9781003315117

Typeset in Nimbus font
by KnowledgeWorks Global Ltd.

Access the Support Material at: www.routledge.com/9781032324531

Dedication

This book is dedicated to my parents.

Contents

Nomenclature

$\vec{a}, a$	Acceleration and its magnitude, m^2
a	Speed of sound, m/s; radius of coiled tube, m; Amplitude of a wave
A	Area, m^2; cross-sectional area, m^2
b	Width or dimensions of lateral side, m; turbomachinery blade width, m; jet's half width
B	Coefficient in power law of viscosity; the matrix
c	Specific heat for incompressible substance, kJ/kg · K; wave celerity, m/s; chord length of an airfoil, m
c_c	Coefficient of contraction
c_d	Discharge coefficient
c_p	Constant-pressure specific heat, kJ/kg · K
c_v	Constant-volume specific heat, kJ/ kg · K; coefficient of velocity
C	Chezy coefficient, $m^{1/2}$/s
C_D	Drag coefficient; local drag coefficient
$C_f, C_{f,x}$	Fanning friction factor or skin friction coefficient; local skin friction coefficient
C_F	Surface-averaged skin friction coefficient
C_H	Head coefficient
C_L	Lift coefficient
C_P	Power coefficient
C_p	Pressure coefficient
C_Q	Flow coefficient
C_W	Drag coefficient due to water for ships
D, d	Diameter, m
D_h, d_H	Hydraulic diameter, m
d_{mol}	Molecule diameter, m
E	Effectiveness; strength of source and sink; energy
f	Darcy friction factor; frequency, cycles/s; Blasius boundary layer dependent similarity variable
f_b	Body force per unit mass in general
$\mathbf{f}$	Body force field
f_s	Surface force per unit mass
f_x, f_y, f_z	Body force per unit mass in x, y, and z directions
$\mathbf{F}, F$	Force and its magnitude, N
F_D	Drag force, N
F_L	Lift force, N
$\mathbf{F}_i$	Body force
g	Acceleration due to gravity, m/s^2

$h, \Delta h$	Manometer level difference, vertical height or depth, m
h_f	Head loss due to friction, m
H	Depth; head, m
$\mathbf{H}$	Moment of momentum
I	Moment of inertia, kg m^2
$I_{xx,cen}$	Moment of the area that goes through the centroid of the immersed surface
$\hat{i}$	Unit vector in x-direction
$\hat{j}$	Unit vector in y-direction
J	Flux of jet's momentum per unit time per unit span
k	Wave number; turbulent kinetic energy, m^2/s^2; specific heat ratio
$\hat{k}$	Unit vector in z-direction
k_B	Boltzmann constant
K	Minor loss coefficient; flow coefficient
L_e	Hydrodynamic entry length, m
$m, \dot{m}$	Mass, kg; and mass flow rate, kg/s; radii ratio
M_{mol}	Molecular mass, kg/kmol
n	Number of parameters in Buckingham pi theorem; Manning coefficient; turbulent velocity profile index
N_s	Specific speed; physical quantities in Reynolds Transport Theorem
P	Pressure, kPa
P_{abs}	Absolute pressure, kPa
P_{atm}	Atmospheric pressure, kPa
P_g	Gage pressure, N/m^2 or kPa
P_{thermo}	Thermodynamic pressure
P_v	Saturation pressure or vapor pressure, kPa
$\tilde{P}$	Average pressure
Q	Volume flow rate, m^3/s
$\dot{Q}$	Heat transfer rate, kW
r	Radial coordinate, radius, m
$\mathbf{r}$	Force arm
R	Gas constant, kJ/kg·K; resistance, ohm; radius of curvature of the coil
R_H	Radius of runner from axis till hub, m
R_o	Radius till guide vanes from the axis of turbines, m
R_T	Radius from axis till tip of the blade, m
R^t	Time correlation function
R_u	Universal gas constant, kJ/kmol K
s	Displacement, m; entropy kJ/kg·K
S	Slope in open-channel flow; Sutherland's law constant; jet spreading rate; shear correlation
S_w	Wetted surface area of the ship
t	Time, s
T	Temperature; torque
Tu	Turbulence intensity
$\mathbf{u}$	Internal energy

u	Velocity component in x-direction, m/s
$\widetilde{u}$	Velocity of fluid particle associated with kinetic energy
u_τ	Friction velocity in turbulent boundary layer, m/s
u_r	Cylindrical velocity component in r-direction, m/s
u_θ	Cylindrical velocity component in θ-direction, m/s
u_z	Cylindrical velocity component in z-direction, m/s
$\bar{u}$	Mean velocity
U_o	Velocity of wall, blade velocity
U_∞	Free stream velocity
v	Velocity component in y-direction, m/s
V	Absolute fluid velocity from stationary observer perspective
V_a	Axial velocity
V_o	Inlet speed of water entering through guide vanes
$\vec{V}$, $\mathbf{V}$	Velocity vector
$\vee$, $\forall$	Volume
w	Velocity component in z-direction, m/s
W	Weight, N; relative fluid velocity
W_∞	Freestream velocity in vertical direction
$\dot{W}$	Rate of work, kW
$\dot{W}_{flow}$	Rate of flow-work, kW
x	Cartesian x coordinate, m
y	Cartesian y coordinate, m
$y_{1/2}$	Wake half width
z	Cartesian z coordinate, m

Greek and other symbols

α	Angle; angle of attack, angle of the guide vanes; absolute velocity angles
β	Coefficient of volume expansion, K^{-1}, Mach cone angle, relative velocity angles; diameter ratio; turbulent equilibrium layer parameter; relative velocity angles
δ	Boundary layer thickness, m; the zone of viscosity
δ^*	Displacement thickness
δ_s	Shear thickness
ε	Mean surface roughness, m
ε_{ij}	Strain rate tensor, s^{-1}
$\varepsilon_{xx}, \varepsilon_{yy}, \varepsilon_{zz}$	Strain rates in x, y and z direction and orientation, s^{-1}
ℓ	Chord length of an airfoil, m: characteristic length, duct length, m
ψ	Stream function, kg/m·s
γ	Specific weight, N/m^3
Γ	Circulation or vortex strength, m^2/s
η	Efficiency; physical quantities per unit mass; similarity variable; Kolmogorov scales
κ	Bulk modulus of compressibility, kPa or atm; log law constant in turbulent boundary layer

λ	Mean free path length, *m*; wavelength, *m*; second coefficient of viscosity; tip speed ratio; Taylor length scales; dimensionless pressure gradient
λ_s	Time lapse rate
Λ	Thermodynamic pressure
μ	Absolute viscosity, $\mathrm{kg/m \cdot s}$
ν	Kinematic viscosity $\mathrm{m^2/s}$
$\omega_x, \omega_y, \omega_z$	Vorticity in Cartesian coordinates
$\omega_r, \omega_\theta, \omega_z$	Vorticity in cylindrical coordinates
ϕ	Potential function
Π	Pi group
ρ	Density, $\mathrm{kg/m^3}$
σ	Normal stress, $\mathrm{N/m^2}$
σ_{ij}	Stress tensor, $\mathrm{N/m^2}$
σ_e	Surface tension, N/m
τ	Shear stress, $\mathrm{N/m^2}$
τ_{ij}	Viscous stress tensor, $\mathrm{N/m^2}$
τ	Torque and its magnitude, N-m
Θ	div $\vec{V}$
θ	Angle or angular coordinate; boundary layer momentum thickness, *m*; pitch angle of a turbomachincry blade; turning or deflection angle of oblique shock
ζ_s	Distance from the center of buoyancy to the meta-center, *m*, ft
υ_o	Suction/blowing velocity
ϑ	Volume
$\hbar$	Channel depth

Acronyms

API	American Petroleum Institute
cb	Center of buoyancy
cg	Center of gravity
CS	Control surface
CV	Control volume
DNS	Direct Numerical Simulation
EGL	Energy grade line, *m*
F-L-T	Force, length, time system
CFD	Computational Fluid Dynamics
GB	Distance from center of buoyancy to center of gravity
iwg	Inches of water
KE	Kinetic energy, kJ
KH	Kelvin Helmholtz
LES	Large Eddy Simulation
M-L-T	Mass, length, time system
NPSH	Net positive suction head coefficient
PE	Potential energy, kJ

SG	Specific gravity
WP	Wetted perimeter, m

Dimensionless Numbers

Bo	Bond number
De	Dean number
Eo	Eötvös number
Eu	Euler number
Fr	Froude number
Kn	Knudsen number
La	Laval number
M	Mach number
Mo	Morton number
Re	Reynolds number
Sc	Scruton number
St	Strouhal number
We	Weber number
Wo	Womersley number
W_D	Dynamic vortex number
σ_v	Cavitation number
σ_c	Cavitation index
$\sigma_{v,pump}$	Thoma cavitation number

Subscripts

$(.)_{actual}$, $(.)_{act}$, $(.)_a$	Actual
$(.)_b$	Bulk
$(.)_{cen}$	Centroid
$(.)_{crit}$	Critical
$(.)_D$	Diameter
$(.)_{dis}$	Displaced
$(.)_e$	Exit
$(.)_{ideal}$	Ideal
$(.)_i$	Inner
$(.)_{in}$	Input
$(.)_{lam}$	Laminar
$(.)_L$	Length of plate
$(.)_{LM}$	Least moment
$(.)_m$	Mean
$(.)_{max}$	Maximum
$(.)_{sm}$	Submerged
$(.)_o$	Outer; stagnation quantities; wall values
$(.)_{open}$	Open system
$(.)_r$	Radial quantities
$(.)_R$	Resultant
$(.)_{sd}$	Standard deviation

$(.)_{steady}$, $(.)_{s}$	Steady
$(.)_{T}$, $(.)_{Total}$	Total
$(.)_{theo}$	Theoretical
$(.)_{turb}$	Turbulent
$(.)_{w}$	Water; wall
$(.)_{\delta^*}$	Based on displacement thickness
$(.)_{\tau}$	Based on shear stress
$(.)_{\theta}$	Based on momentum thickness; tangential quantities

Superscripts

$(.)^*$	Dimensionless quantity
$(.)^+$	Dimensionless turbulence quantities
$(.)'$	Fluctuating turbulence quantities

Conversion Factors

DIMENSION	METRIC	METRIC/ENGLISH
Length	1 m = 100 cm = 1000 mm = $10^6 \mu m$ 1 km = 1000 m	1 m = 39.370 in = 3.2808 ft = 1.0926 yd 1 ft = 12 in = 0.3048 m 1 mile = 5280 ft = 1.6093 km 1 in = 2.54 cm
Area	1 $m^2 = 10^4$ $cm^2 = 10^6$ $mm^2 = 10^{-6}$ km^2	1 m^2 = 1550 in^2 = 10.764 ft^2 1 ft^2 = 144 in^2 = 0.0929 m^2
Velocity	1 m/s = 3.60 km/h	1 m/s = 3.2808 ft/s = 2.237 mi/h 1 mi/h = 1.46667 ft/s 1 mi/h = 1.6093 km/h
Acceleration	1 m/s^2 = 100 cm/s^2	1 m/s^2 = 3.2808 ft/s^2 1 ft/s^2 = 0.30488 m/s^2
Mass	1 kg = 1000 g 1 metric ton = 1000 kg	1 kg = 2.2046226 lb_m 1 lb_m = 0.45359237 kg 1 ounce = 28.3495 g 1 slug = 32.174 lb_m = 14.5939 kg 1 short ton = 2000 lb_m = 907.1847 kg
Density	1 g/cm^3 = 1 kg/L = 1000 kg/m^3	$1g/cm^3 = 62.428lb_m/ft^3$ = 0.036127 lb_m/in 1 lb_m / in^3 = 1728 lb_m/ft^3 1 kg/m^3 = 0.062428 lb_m/ft^3
Force	1 N = 1 kg · m/s^2 = 10^5 dyne 1 kg_f = 9.80665 N	1 N = 0.22481 lb_f 1 lb_f = 32.174 lb_m · ft/s^2 = 4.44822 N $1lb_f = 1slug \cdot ft/s^2$

DIMENSION	METRIC	METRIC/ENGLISH
Pressure or stress,	1 Pa = 1 N/m^2 1 kPa = 10^3 Pa = 10^{-3} MPa	1 Pa = 1.4504 $\times 10^{-4}$ psi = 0.020886 lb$_f$/ft^2
	1 atm = 101.325 kPa = 1.01325 bar	1 psi = 144 lb$_f$/ft^2 = 6.894757 kPa
Pressure as head	= 760 mm Hg at 0°C = 1.03323 kg/cm^2 1 mm Hg = 0.1333 kPa	1 atm = 14.696 psi = 29.92 inches Hg at 30° F 1 inch Hg = 13.60 inches H_2O = 3.387 kPa
Energy, heat, work,	1 kJ = 1000 J = 1000 N · m = 1 kPa · m^3	1 kJ = 0.94782 Btu
and specific energy	1 kJ/kg = 1000 m^2/s^2 1 kWh = 3600 kJ	1 Btu = 1.055056 kJ = 5.40395 psia · ft^3 = 778.169 lb$_f$ · ft $1 Btu/lb_m = 25.037 ft^2/s^2 =$ 778.169 lb$_f$ · ft 1 kWh = 3412.14 Btu
Power	1 W = 1 J/s	1 kW = 3412.14 Btu/h = 1.341 hp
	1 kW = 1000 W = 1 kJ/s	= 737.56 lb$_f$ · fts
	1 hp = 745.7 W	1 hp = 550 lb$_f$ · ft/s = 0.7068 Btu/s = 42.41 Btu/min = 2544.5 Btu/h = 0.74570 kW 1 Btu/h = 1.055056 kJ/h
Specific heat	1 kJ/kg · °C = 1 kJ/kg · K = 1 J/g · °C	1 Btu/lb$_m$ · °F = 4.1868 kJ/kg · °C 1 Btu/lbmol · °R = 4.1868 kJ/kmol · K 1 kJ/kg · °C = 0.23885 Btu/lb$_m$ · °F = 0.23885 Btu/lb$_m$ · °R
Specific volume	1 m^3/kg = 1000 L/ kg = 1000 cm^3/g	1 m^3/kg = 16.02 ft^3/lb$_m$ 1 ft^3/lb$_m$ = 0.062428 m^3/kg

DIMENSION	METRIC	METRIC/ENGLISH
Temperature	$T(K) = T°C + 273.15$ $\Delta T(K) = \Delta T\,(°C)$	$T(°R) = T°F + 459.67 = 1.8T(K)$ $T(°F) = 1.8T(°C) + 32$ $\Delta T(°F) = \Delta T(°R)$ $= 1.8\Delta T(K)$
Viscosity, dynamic	1 kg/m · s = 1 N · s/m^2 = 1 Pa · s = 10 poise	1 kg/m · s = 2419.1 lb$_m$/ft · h = 0.020886 lb$_f$ · s/ft^2 = 0.67197 lb$_m$/ft · s
Viscosity, kinematic	1 m^2/s = 10^4 cm^2/s 1 stoke = 1 cm^2/s = 10^{-4} m^2/s	1 m^2/s = 10.764 ft^2/s = 3.875 ×10^4 ft^2/h 1 m^2/s = 10.764 ft^2/s
Volume	1 m^3 = 1000 L = 10^6 cm^3 (cc)	1 m^3 = 6.1024 ×10^4 in^3 = 35.315 ft^3 = 264.17 gal (U.S.) 1 U.S. gallon = 231 in^3 = 3.7854 L 1 fl ounce = 29.5735 cm^3 = 0.0295735 L 1 U.S. gallon = 128 fl ounces
Volume flow rate	1 m^3/s = 60,000 L/min = 10^6 cm^3/s	1 m^3/s = 15,850 gal/min = 35.315 ft^3/s = 2118.9 ft^3/min (CFM)

Author

Dr.-Ing. Naseem Uddin, CEng MIMechE, MIEAust CPENG earned his PhD in aerospace engineering from Universität Stuttgart, Germany in 2008. He is a senior assistant professor in the Mechanical Engineering Programme, Universiti Teknologi Brunei (UTB), Brunei Darussalam. Previously, he worked as full professor at NED University of Engineering and Technology, Pakistan, from 2010–2019. Dr. Naseem is a registered chartered engineer with Engineers Australia and the Institution of Mechanical Engineers, UK. He is also listed in the National Engineering Register (NER) of Australia and recognized as a professional engineer by the board of Professional Engineers of Queensland, Australia, and by the Pakistan Engineering Council. He is a member of the American Society of Mechanical Engineers (ASME), USA. He teaches fluid mechanics to both graduate and undergraduate students at UTB, and his research interests are in the areas of heat transfer and computational turbulence.

Preface

Fluid Mechanics: A Problem-Solving Approach is written to be used as a textbook for teaching the core module of fluid mechanics in mechanical, civil, chemical, and biomedical engineering programs. The book is based on the lecture notes developed over time for teaching fluid mechanics. This book introduces fundamental concepts, flow physics, and practical applications, keeping the same readers in mind.

As a primer of fluid mechanics, the topics discussed in this book are not new. However, the topics are introduced in a gradual manner to the readers along with the necessary tools so that they can grasp the basic concepts and understand different approaches to the analysis of problems encountered in fluid mechanics.

This book has been divided into several chapters. Fluid and its properties are discussed in Chapter 1. A section on the properties of nanofluids is included where formulas for necessary thermophysical quantities are presented.

A complete chapter is dedicated to the useful dimensional analysis tool and similitude theory. Several important dimensionless parameters are discussed. In the author's viewpoint, these dimensionless parameters should be included in new textbooks on fluid mechanics to make students aware of the utility of these numbers.

Different types of flow phenomena require different approaches for mathematical modeling. The basic laws of physics, namely the continuity equation, the Navier-Stokes equations, and the energy equation, are introduced up front.

The integral and differential formulations are separated, and the inviscid or potential flow model of the flow is discussed first before the viscous fluid flow analysis. A complete chapter dedicated to the turbulent flows is introduced. The Reynolds Averaged Navier-Stokes equations are developed, along with the rudimentary concept of the turbulent spectrum, turbulent length scales, and turbulent fluctuations. Several correlations on the friction factor for pipe flows are provided in this book. The limitations of the hydraulic diameter concept are highlighted. The external flows and the boundary layer theory are presented for curved surfaces. The use of different boundary layer thicknesses is explained with examples.

Jets, wakes, and separated flows are important fluid flow phenomena. Unlike many undergraduate books, which usually do not cover jet flows, a chapter on this crucial topic is included. The ocean waves and tsunamis are discussed in this book. The channel flows, the compressible gas flows, and the turbomachines are essential topics for any undergraduate module on fluid mechanics, and thus they, too, are discussed. A reference list for the literature has been added to every chapter. The Appendix can be found online at https://routledge.com/9781032324531.

I am grateful to Prof. Dr.-Ing. Bernard Weigand, Universität Stuttgart, Germany Prof. Dr. Ramesh Singh Kuldip Singh, Dean Faculty of Engineering, Universiti Teknologi Brunei, my colleagues Dr. Denni Kurniawan (UTB) and Dr. Rama Rao Karri (UTB) for encouragement to write this book. I am grateful to my teachers who taught me fluid and gas dynamics at Universität Stuttgart, Germany. The author would like to appreciate and acknowledge the valuable comments and suggestions from Prof. Dr. Duraisamy Sambasivam Sankar and Dr. Maziyar Sabet for improvement of book. I am also thankful to my students Irfan and Razzaq for typing some of the problems and solutions. I am also greatly indebted and wish to extend my gratitude to the following organizations for the fluid mechanics-related images and videos.

1. NASA Earth Observatory

2. International Association of Hydro-Environment Engineering and Research

   `https://www.iahrmedialibrary.net/`

Finally, special thanks must go to my family for their continued patience and help. I hope that the readers find this book a refreshing and straightforward approach.

N. Uddin
Bandar Seri Begawan, Brunei

1 Introduction

In the beginning, we first present the basic definition of substance called fluid. We will define the Newtonian and the non-Newtonian fluids; we will discuss the properties of the fluid like density, viscosity, and surface tension; and we will also discuss the properties of nanofluids.

Learning outcomes: After finishing this chapter, you should be able to:

- Define the substance fluid.
- Understand the concept of continuum hypothesis.
- Identify the Newtonian and non-Newtonian fluids.
- Understand the need to separate analysis for compressible and incompressible flows.
- Calculate surface tension forces.

In the dictionary, fluid is defined as a substance with no fixed shape that easily yields to external pressure. In literature and newspapers, the word *fluid* is used not only for matter but also for situations. However, in technical terms, a fluid is a substance that does not have shear stress in a state of rest. Conversely speaking, fluid is a substance that deforms continuously under the action of shear force.

We know that matter has three states: solid, liquid, and gas (Figure 1.1), and out of these three states, the liquid and gas are classified as fluid. We know that when force is applied to solid matter, it goes into deformation that can be elastic, or in the plastic state. If the force is continuously applied or increased, the solid will be permanently deformed or fractured. On the contrary, in case of fluids, the forces applied will cause fluid movement, and thus, fluid deformation is fluid flow. We know that matter is composed of atoms and molecules, and we know that liquid is different from gas. A gas can expand and fill up the space available, and no free surface is formed. However, in case of liquids, an interface separating liquid and vapor is created, which we call a free surface. Also, the liquids of two different densities do not mix, and an interface is formed between them. Figure 1.2 shows an example of liquids at nominal temperature (left) and at very high temperature (right).

1.1 CONTINUUM HYPOTHESIS

The molecular structure of fluids is essential in distinguishing one fluid from another. Still, in engineering applications, this description is sometimes too cumbersome and complex to handle. We invoke the hypothesis that a fluid is composed of a bundle of molecules instead of regular atoms and molecules. These bundles of molecules are

DOI: 10.1201/9781003315117-1

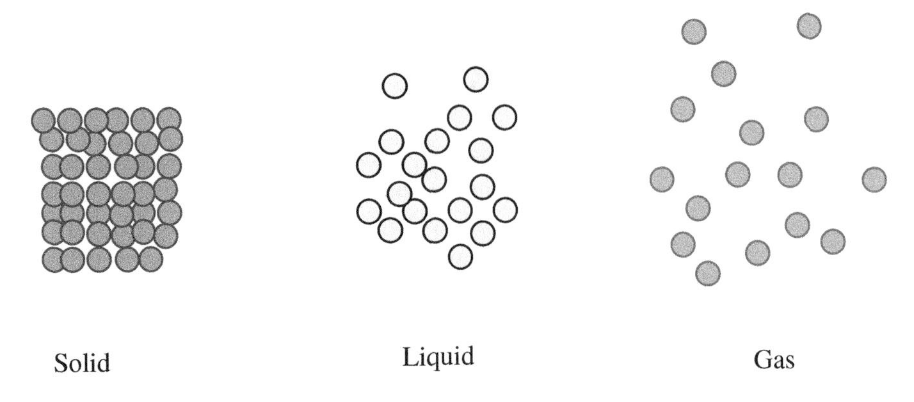

Figure 1.1 The states of matter.

Figure 1.2 Water is crucial for human life. The solid rocks converted into liquid lava.

called fluid particles in the fluid mechanics literature. In cases where the continuum hypothesis is invalid, which mostly happens in the case of gases, the principle of statistical thermodynamics and kinetic molecular theory can be used for the analysis. But apart from that, as long as we are within bounds of the continuum hypothesis, differential equations can be formed for the fluid description.

The validity of the continuum hypothesis can be confirmed by calculating a number called Knudsen number (Kn), defined as:

$$Kn = \frac{\lambda_{mfp}}{\ell}$$

where ℓ is the characteristic length and λ_{mfp} is the mean free path of the gas molecules. For a Boltzmann gas, the mean free path may be readily calculated as

$$Kn = \frac{k_B T}{\sqrt{2}\pi d_{mol}^2 PL}$$

where k_B is the Boltzmann constant (1.380649×10^{-23} J/K, in SI units), T is the absolute temperature, d_{mol} is the particle hard-shell diameter, and P is the total pressure.

The mean free path of a gas can be estimated as

$$\lambda_{mfp} = \frac{1}{\pi\sqrt{2}}\left(\frac{M_{mol}}{\rho d_{mol}^2}\right)$$

where d_{mol} is the molecule diameter, M_{mol} is molecular mass, and ρ is the density of the gas.

A crude approximation for the mean free path length (about $\pm 30\%$) in meters, which ignores the influence of molecular sizes is

$$\lambda_{mfp} = \frac{3 \times 10^{-4} T}{p}$$

where T is the temperature in Kelvin and p is pressure in bars (Böswirth, 1993).

If we start taking gas molecules out of a container, the point is reached where the pressure is less than the atmospheric pressure; such a state is called vacuum. Also, at greater distances from the Earth, around the elevation of 100 km, the molecules are very far apart. The gases in that region are considered rarefied gases, as their density is very low. The distance of 100 km above the Earth is regarded as the starting of the space zone. For very high vacuum and rarefied gases, the continuum hypothesis is invalid. The Federation Aeronautique Internationale (FAI) has introduced the idea of the Karman line (or von Karman line) as a boundary between the upper atmosphere and space.

$$Kn < 0.01 \;\; Continuum$$
$$0.01 < Kn < 0.1 \; Slip\ Flow$$
$$0.1 < Kn < 10 \;\; Transitional\ Flow$$
$$Kn > 10 \; Free\ Molecular\ Flow$$

Table 1.1
Atomic and Molecular Diameter for Some Gases.

Gas	Diameter (A)	Diameter ($\times 10^{-10}$ m)
H_2	2.73	2.73
He	2.18	2.18
N_2	3.74	3.74
O_2	3.57	3.57
CO_2	4.56	4.56
Ar	3.62	3.62
Air	3.7	3.7

Example 1.1

Example A small capillary-like tube (diameter 0.2 mm) is being used to measure the pressure experienced by an airplane. Calculate the mean free path of air at 10,000 m elevation. The diameter of the air molecule is 3.7×10^{-10} m and average mass of an air molecule is 4.8×10^{-26} kg. May we still consider the fluid as a continuum?

Solution At an altitude of 10,000 m, the density of air is 0.4136 kg/m^3.

$$\lambda_{mfp} = \frac{1}{\pi\sqrt{2}}\left(\frac{M_{mol}}{\rho d_{mol}^2}\right)$$

$$\lambda_{mfp} = 0.225\left(\frac{4.8 \times 10^{-26}}{0.4136 \times (3.7 \times 10^{-10})^2}\right) = 0.1907\ \mu m$$

$$Kn = \frac{\lambda_{mfp}}{d_{capillary}} = \frac{0.1907\ \mu m}{0.2\ mm} = 0.0009535$$

As Knudsen number is less than 0.01 we can treat the flow as a continuum.

Figure 1.3 shows different states of matter in the thermodynamic surface. Note that fluid above the critical point is called supercritical fluid. The symbols ∀ and V are used to represent fluid volume in this book.

1.2 FLUID PROPERTIES

Fluids are known for different properties, such as viscosity, density, surface tension, and boiling point. To describe the fluid, it is necessary to understand several properties that define its physical state. Qualitatively, the properties can be defined in the System International (metric system) or in the British system, which have been commonly used in the English-speaking world since the second industrial revolution.

Anything that describes the system is its property, so properties are also called state variables. All the fluid properties, including viscosity, density, surface tension, specific heat, and thermal conductivity, are called physical properties of the matter. There are some other properties called thermodynamic properties that help in identifying the state of a system. Knowing the state of a system is a starting step for the solution of a problem. Thermodynamically speaking, states are determined by the properties of the system. Properties help in the identification of the state of the matter in the system. In thermodynamics, we are primarily interested in directly measurable properties like pressure, temperature, and volume. These variables are directly measurable and give us a lot of information about the state of the matter. However, certain other properties are of equal importance and are used in the thermodynamic analysis of the fluid flow. These properties are enthalpy and entropy, and they are not directly measurable; instead, they are calculated using the first and second laws of thermodynamics.

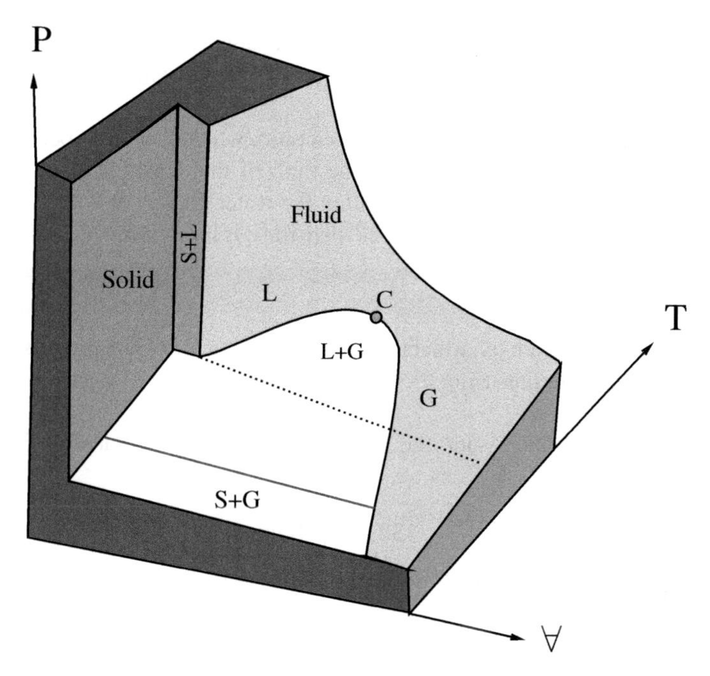

Figure 1.3 Thermodynamic surface in pressure, volume, and temperature (P, ∀, T) coordinates.

In thermodynamics, we classify the properties as either intensive or extensive. Intensive properties are independent of the system's mass, meaning that if you take a sample of the fluid, then the property value will not change with the mass. Examples of such quantities are density, pressure, and temperature. These quantities are generally represented by lowercase letters in the literature, except for pressure and temperature, which are represented by capital letters. Extensive quantities are also represented on a per unit mass basis, and we call them specific properties. Some examples are specific volume, which is the reciprocal of density, and specific total energy. On the other hand, the extensive properties depend on the size and total mass of the system. Examples of such properties are volume, momentum, etc. These quantities are typically represented by an uppercase letter in the literature (except for mass).

1.3 DENSITY

Density is defined as a mass per unit volume. Often in fluid mechanics, we use the reciprocal of density, which is called specific volume. The density of most gases is a function of temperature and pressure. On the other hand, liquids and solids exhibit usually negligible variation in density as the temperature or pressure changes. Liquids that we use in engineering or industrial applications are considered as incompressible fluids, i.e. their densities do not change with pressure and temperature. The density of water exhibits only a 2 to 3% change when the temperature is changed from 20°C to 100°C.

In liquids, the densities are also stated as a ratio, which is called specific gravity (SG) or relative density. It is defined as the ratio of the density of a substance to the density of the standard liquid (most often the water density at 4°C is taken as a reference). Specific gravity is a ratio of the densities; it is a unitless quantity.

Specific Gravity

Baumé Scale: French pharmacist Antoine Baumé in 1768 proposed to report density of liquids using the following definitions:

For liquids lighter than water, the specific gravity (SG) is related to Baumé Scale as:

$$SG = \frac{140}{130 + \text{deg Baumé}}$$

For liquids heavier than water, the SG is defined as:

$$SG = \frac{145}{145 - \text{deg Baumé}}$$

This scale is variously mentioned as degrees Baumé, B°, Bé° or simply Baumé. This scale is often used in the sugar and pharmaceutical industries.

The American Petroleum Institute (API) scale:

$$SG = \frac{141.15}{131.5 + \text{deg API}}$$

The above equation is called American Petroleum Institute (API) scale. The use of the API scale is recommended by the American National Standards Institute (ANSI). If in an industrial literature the specific gravity of liquid is reported as 11.5 API at 60°F/60°F, then this is

$$SG_{60°F/60°F} = \frac{141.15}{131.5 + 11.5} = 0.9870$$

Note that density is expressed as kg/m^3 in SI (Système Internationale) units. In the British Imperial measurement system and in the United States customary measurement system, mass (m) is measured in slugs, hence density units are $slugs/ft^3$. The mass is also reported in the units of pound-mass (lb_m), which is related with slugs as:

$$1 \text{ slug} = 1 \text{ lb}_m \cdot 32.174$$

The force of one pound-force (lb_f) is defined as the force required to move a mass of one slug with an acceleration of 1 ft/s^2. The force of one pound-force is related to pound-mass as

$$1lb_f = 32.174lb_m ft/s^2$$

1.4 COMPRESSIBILITY

The bulk modulus of compressibility or the bulk modulus of elasticity (κ) of the fluid is defined as:

$$\kappa = \rho\left(\frac{\partial P}{\partial \rho}\right)_T$$

Note that $\kappa \approx \left(\frac{\Delta P}{\Delta\rho/\rho}\right)$ for constant temperature conditions. This indicates that the bulk modulus of compressibility has units of pressure. In the SI system, κ is usually expressed in units of Pascal, and in the FPS system, κ is represented in units of psi. A large value of the bulk modulus of compressibility means that a large change in pressure is required to bring about a small change in the volume. Typically liquids have a large volume of the above models of compressibility.

A considerable value of bulk modulus of compressibility shows that a substantial variation in pressure is required to produce a modest fractional variation in volume. Therefore a fluid with a large bulk modulus of compressibility value is incompressible. This is common for liquids and explains why liquids are usually treated as being incompressible. Let's say you want to compress water up to 1%. To accomplish this, the pressure of water at standard atmospheric conditions must be increased up to 212.78 bars, which corresponds to a coefficient of compressibility value equal to 15,960,000 torrs. The speed of sound, or acoustic speed (a), in liquids is related to a bulk modulus of compressibility as:

$$a = \sqrt{\left(\frac{\kappa}{\rho_o}\right)}$$

In fluid mechanics, a dimensionless parameter called Mach number is used to do a quick check on incompressibility assumption. The above number is defined as:

$$M = \frac{u}{a} = \frac{u}{\sqrt{kRT}}$$

where R is the universal gas constant, T is the absolute temperature, and k is the ratio of specific heats, $k = c_p/c_v$. A quick check on the compressibility of the gases can be judged based on the Mach number value. If $M < 0.3$, then such flow can be considered incompressible.

Figure 1.4 shows the variation in bulk modulus of compressibility with temperature for saturated water. As the water is heated, the bulk modulus of compressibility will be reduced.

1.5 COEFFICIENT OF VOLUME EXPANSION

The density of a fluid depends more on temperature than it does on pressure. The alteration of density with temperature is responsible for various natural phenomena

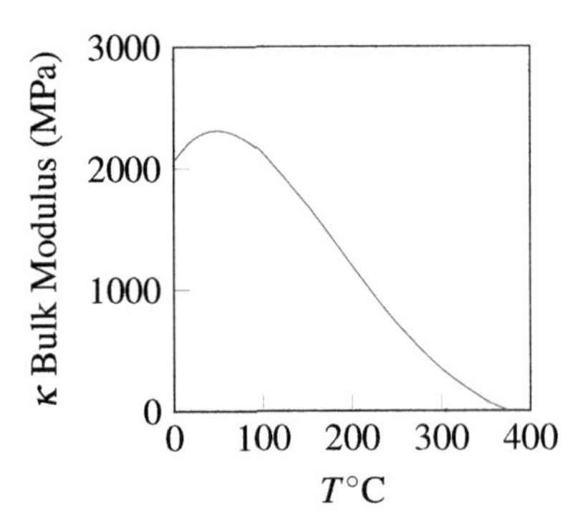

Figure 1.4 Bulk modulus of saturated water.

such as cyclones, currents in oceans, an ascent of plumes in stacks, the rising of hot air balloons, etc.

$$\beta = \left(\frac{\partial V}{\partial T}\right)_P = \frac{-1}{\rho}\left(\frac{\partial \rho}{\partial T}\right)_P$$

With constant pressure conditions, we have:

$$\beta \approx \frac{\Delta V/V}{\Delta T} = \frac{-\Delta\rho/\rho}{\Delta T}$$

With constant temperature condition:

$$\beta \approx \left.\frac{-\Delta P}{\Delta V/V}\right|_T = -\left.\frac{\Delta P}{\Delta\rho/\rho}\right|_T$$

Example 1.2

Example Find the percentage of volume variation in 10 m^3 of water at 15°C, if the pressure on the water is increased to 12 MPa from atmospheric pressure.

Solution The atmospheric pressure is 101 kPa and β for water is 2.14 GPa.

$$\Delta V/V = -\frac{\Delta P}{\beta} = -\frac{(12E6 - 101E3)}{214E7} = -0.00556$$

The percentage change in volume is just 0.556%, which can safely be ignored.

1.6 SPECIFIC HEAT

Fluids are capable of storing thermal energy, and this property is reflected by the value of specific heat, defined as:

$$c_p = \left(\frac{\partial h}{\partial T}\right)_p$$

where c_p is specific heat at constant pressure, h is specific enthalpy, p is pressure, and T is absolute temperature.

$$c_v = \left(\frac{\partial u}{\partial T}\right)_v$$

where c_v is specific heat at constant volume, u is internal energy per unit mass, and v is volume.

Specific heat is a function of temperature, and often we assume that specific heats are constant and the gas is called calorically perfect gas. If c_p and c_v are constant, the ratio of specific heats is also constant and defined as:

$$k = \frac{c_p}{c_v}$$

1.7 VISCOSITY

Liquid molecules are closely spaced, with strong cohesive forces between molecules. When the temperature rises, the level of cohesive forces is reduced, with a comparable reduction in resistance to motion. Viscosity is a property of fluid that quantifies the measure by which a particular fluid resists deformation when subjected to shear force. Viscosity depends on the intermolecular attraction between the fluid molecules. In the case of gases, the molecules are apart and have a weak intermolecular attraction, but as the gas is heated the molecular collision increases and this in turn gives rise to an increase in the viscosity of gases.

In literature, μ is called the coefficient of viscosity or absolute viscosity or dynamic viscosity. Dividing this coefficient with density, we get the kinematic viscosity.

$$\nu = \frac{\mu}{\rho}$$

In the SI system the absolute viscosity is reported in units of Pa·s. However, there are some other units of viscosity that are used in industrial literature. The dynamic viscosity is often reported in Poise, a unit named after French scientist Jean Louis Poiseuille and represented by *P*.

$$1\,P = 10^{-1}\ \text{Pa}\cdot\text{s}$$

The stoke is a unit of kinematic viscosity defined as 1 cm^2/s. It is named after English scientist George Gabriel Stokes.

$$1\ stoke = 10^{-4}\ m^2/s$$

$$1\ stoke = 1.076 \times 10^{-3}\ ft^2/s$$

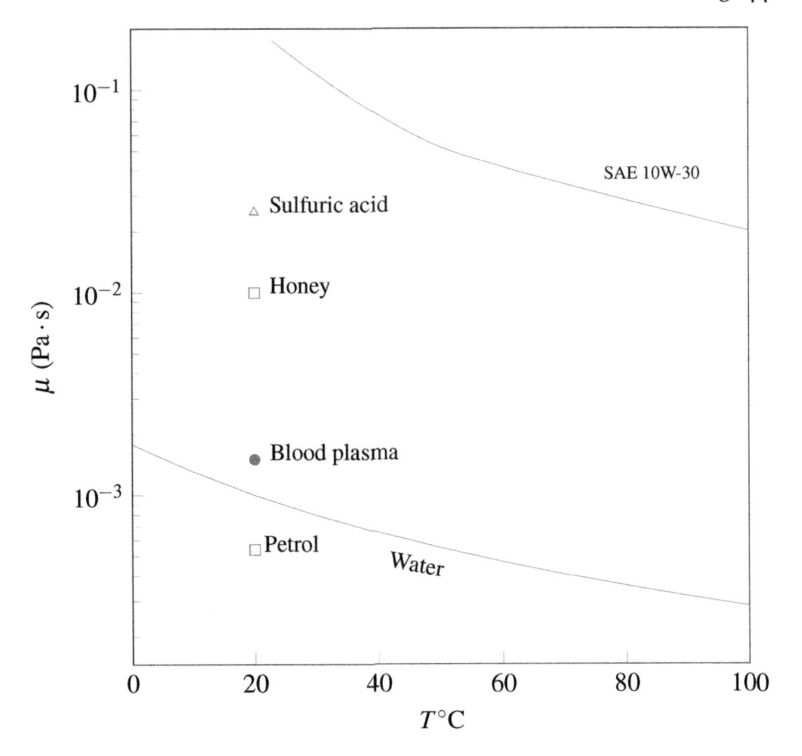

Figure 1.5 Viscosity of some fluids in the liquid state.

The viscosity of liquids decreases with an increase in temperature, whereas for gases, an increase in temperature causes an increase in viscosity.

Andrade proposed the equation for liquid viscosity as:

$$\mu = A \cdot \exp\left[\frac{E}{R.T}\right]$$

Fulcher proposed the following equation for the liquid viscosity:

$$\log \mu = A + \frac{B}{(T - T_o)}$$

The variation in liquid viscosity with temperature is shown in Figure 1.5, and the variation in gas viscosity with temperature is shown in Figure 1.6.

Sutherland's viscosity law comes from a kinetic theory by Sutherland (1893). It is based on an idealized intermolecular force potential. The Sutherland's law with two coefficients has the following form:

$$\mu = \frac{C_1 T^{3/2}}{T + C_2}$$

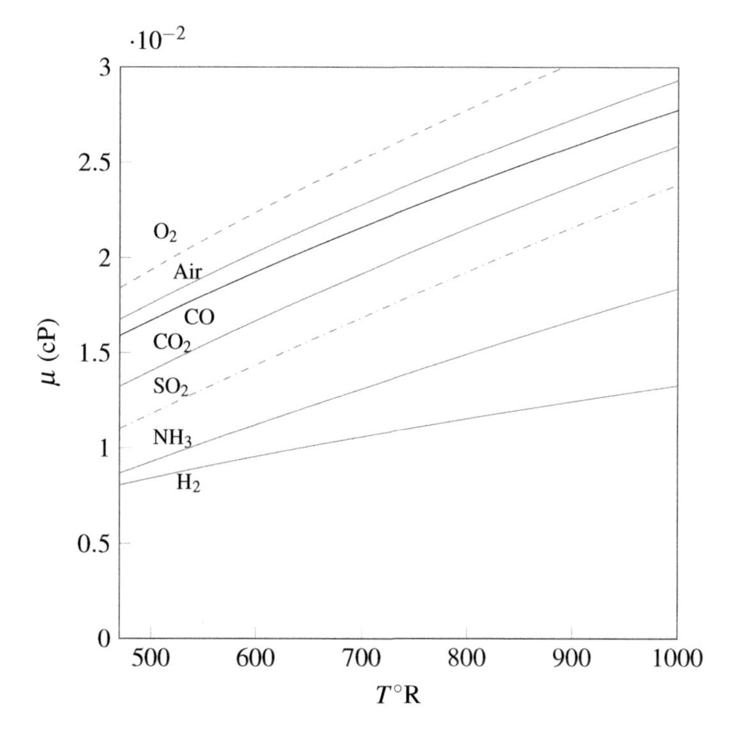

Figure 1.6 Viscosity of some fluids in gas state using three coefficients Sutherland's equation.

where C_1 and C_2 are empirical constants, and T is the absolute temperature. For air at moderate temperatures and pressures, the values are:

$$C_1 = 1.458 \times 10^{-6} \frac{kg}{s \cdot m \cdot K^{1/2}}$$

$$C_2 = 110.4K$$

Sutherland's law with three coefficients has the form

$$\mu = \mu_o \left(\frac{T}{T_o} \right)^{3/2} \left(\frac{T_o + S}{T + S} \right)$$

where $\mu =$ the viscosity in kg/m · s, T is the static temperature in K.

For fast calculations, the following equation can be used:

$$\frac{\mu}{\mu_o} = \left(\frac{T}{T_o} \right)^m$$

Figure 1.7 shows the comparison between different models. The simple relation eq. 1.2 is also very close to three coefficients Sutherland correlation.

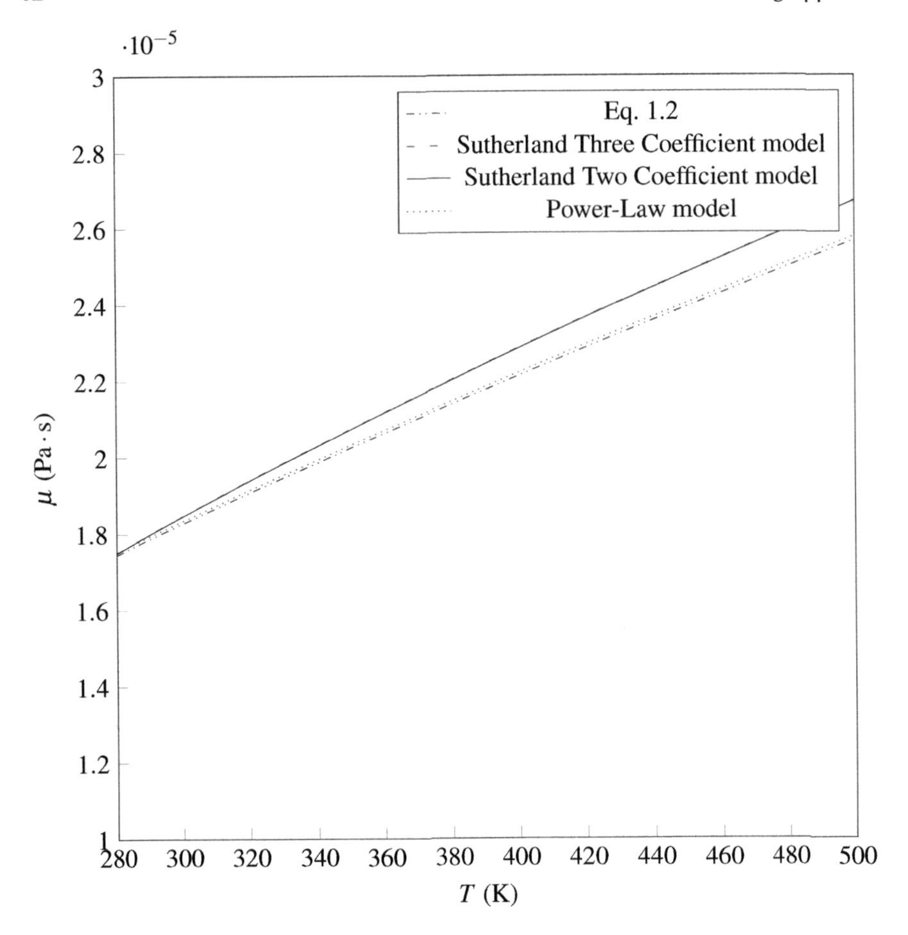

Figure 1.7 Comparison of viscosity of air as predicted by simple, two coefficient correlation, three coefficient correlation, and power-law model.

Power-law model. Another common approximation for the viscosity of dilute gases is the power-law form. For dilute gases at moderate temperatures, this form is considered to be slightly less accurate than Sutherland's law. A power-law viscosity law with two coefficients has the form

$$\mu = BT^n$$

where μ is the viscosity in kg/m-s, T is the static temperature in K, and B is a dimensional coefficient. For air at moderate temperatures and pressures, $B = 4.093 \times 10^{-7}$ and $n = 2/3$.

For the problems that involve heat transfer or shockwaves, we can define the viscosity as a function of temperature for improved accuracy.

1.8 NEWTONIAN AND NON-NEWTONIAN FLUIDS

When force acts on the body of a fluid, it may generate either normal stress or shear stress. When the force is applied perpendicular to the area, we are applying the pressure to the fluid. However, when the force is applied at an angle, a component to force parallel to the fluid would generate the shear stress inside the fluid body. In fluids, the shear stress is present due to the momentum exchange between the molecules of the fluid. The fluid layers close to the surface experience more shear stress compared to the fluid layers that are away from the surface. This molecular transport of momentum will create a force. We know that force per unit area is called shear stress. It means that the fluid layers have a distribution of shear stress that is not uniform, and this would lead to the movement of fluid layers at different rates or create a gradient of velocity. Different fluids behave differently under the action of the shear force. Some fluids exhibit a direct relationship between the applied shear stress and the velocity gradient; these are classified as Newtonian fluids. For such fluids, we can write:

$$\tau = \mu \frac{du}{dy} \tag{1.1}$$

This equation is called **Newton's law of viscosity**.

> There is no shear stress in static fluid, and only normal stresses are present.

The fluids in which the shear stress is not linearly proportional to the rate of shear strain are labeled non-Newtonian fluids. Figure 1.8 shows some common non-Newtonian fluids, and Figure 1.9 shows the general behavior of different fluids.

i Fluids in which viscosity increases as stress is applied are called **dilatant** fluids. Solutions containing a high concentration of powder usually exhibit such

Figure 1.8 Paint and ketchup are examples of non-Newtonian fluids.

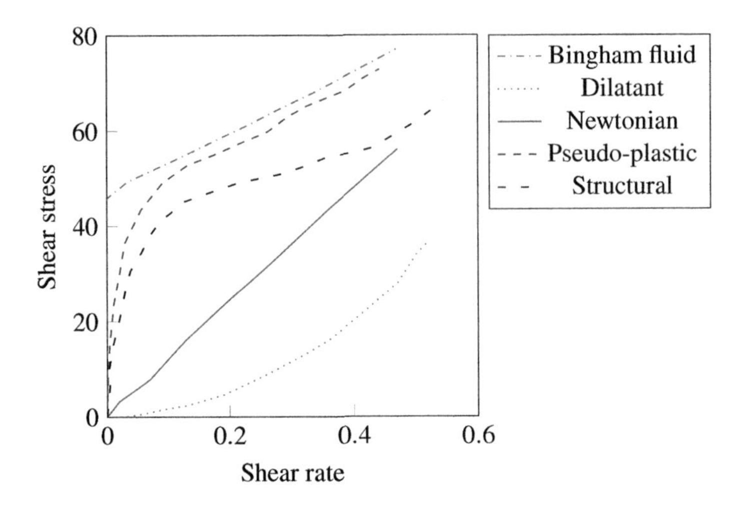

Figure 1.9 Shear stress vs. shear rate for various fluids.

behavior, and they can be modeled as:

$$\tau = K\left(\frac{dV}{dy}\right)^n$$

where $n > 1$.

ii Suspensions are examples of **pseudoplastic** fluids like starch, mayonnaise, etc.

$$\tau = K\left(\frac{dV}{dy}\right)^n$$

where $n < 1$.

iii **Bingham plastic fluids** are fluids that behave like solids until initial yield stress is exceeded, for example chocolate, toothpaste, sewage sludge, mud slurry, etc.

$$\tau = \tau_o + \mu_o\left(\frac{dV}{dy}\right)$$

Figure 1.10 shows typical Bingham plastic fluid behavior. Note that when the shear force is applied, the fluid would not immediately exhibit deformation until certain a level of stress is induced.

iv Some fluids exhibit both elastic and viscous behavior and are called **viscoelastic** fluids, for example flour dough.

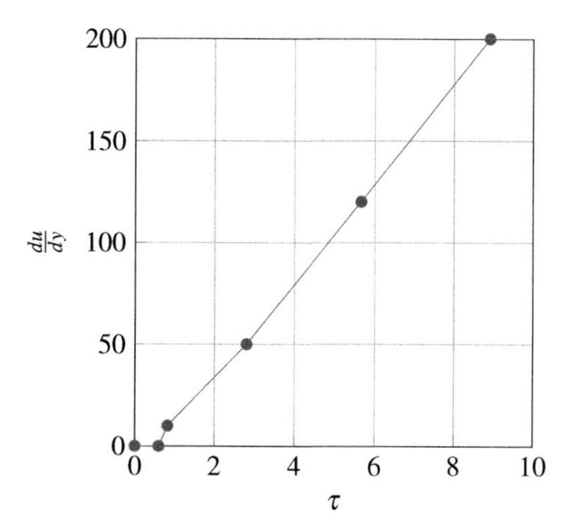

Figure 1.10 Typical Bingham fluid behavior.

Example 1.3

Example The velocity of pseudoplastic fluid follows the relation:

$$\frac{u(r)}{U_o} = \left[1 - \left(\frac{r}{r_o}\right)^{\left(\frac{m+1}{m}\right)}\right]$$

where U_o and r_o are the centerline velocity and radius of the pipe. Find the shear stress at the wall of the pipe.

Solution We can calculate the shear stress using the following relation:

$$\tau_{rz} = \mu_o \left(\frac{\partial u}{\partial r}\right)^m$$

$$\tau_{rz} = \mu_o \left(- \frac{U_o \left(\frac{r}{r_o}\right)^{\frac{m+1}{m}} (m+1)}{m \cdot r} \right)^m$$

$$\tau_{rz} = \frac{\mu_o}{r^m} \left(\frac{m+1}{m}\right)^m \left[U_o^m \left(\frac{r}{r_o}\right)^{m+1}\right]$$

$$\tau_{rz}\big|_{r=r_o} = \frac{\mu_o U_o^m}{r_o^m} \left(\frac{m+1}{m}\right)^m$$

Blood is a specialized bodily fluid in animals that delivers necessary substances such as nutrients and oxygen to the cells, and transports metabolic waste products away from those same cells. In vertebrates, blood is composed of blood cells suspended in a liquid called blood plasma. Plasma, which constitutes 55% of blood fluid, is mostly water (92% by volume), and contains dissipated proteins, glucose, mineral ions, hormones, carbon dioxide (plasma being the main medium for excretory product transportation), platelets, and blood cells themselves. Albumin is the main protein in plasma, and it functions to regulate the colloidal osmotic pressure of blood. The blood cells are mainly red blood cells (also called RBCs or erythrocytes) and white blood cells, including leukocytes and platelets. Blood viscosity is $\mu = 0.0345$ P.

Example 1.4

Example A cylinder of diameter 7 in., length 5 in. and weight 20 lb slides in a lubricated passage which is like a pipe, as shown in Figure 1.11. The gap between the cylinder and passage is 0.001 in. The cylinder decelerates at a rate of 3 ft/s^2 when speed is 20 ft/s; find the oil's viscosity.

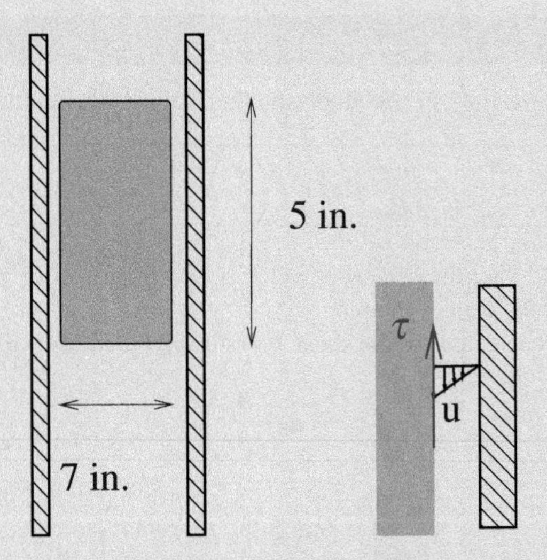

Figure 1.11 The flow through a passage and the zoom-in view of the gap.

Solution The gap between the cylinder and passage is very narrow and we can assume that the velocity distribution is linear.

$$\tau = \mu \left(\frac{\Delta u}{\Delta y} \right)$$

$$\tau = \mu \left(\frac{u}{0.001/12} \right) = 12000\, \mu u \, \frac{lb}{ft^2}$$

$$F_\tau = 12000\ \mu u\ \frac{lb}{ft^2} \times A\ ft^2 = 12000\ \mu u \times (\pi D \ell)$$

$$F_\tau = 12000\ \mu u \times \left(\pi \times \frac{7}{12} \times \frac{5}{12}\right) = 2916.66\ \pi\ \mu u$$

From Newton's second law of motion:

$$\downarrow + \quad F_y = ma_y$$

$$W - F_\tau = ma_y$$

$$20 - 2916.66\ \pi\ \mu u = \left(\frac{20\ lb}{32.2}\right) a_y$$

At $u = 20$ ft/s the cylinder was decelerating at -3 ft/s^2:

$$20 - 2916.66\ \pi\ \mu(20\ ft/s) = \left(\frac{20\ lb}{32.2}\right)(-3\ ft/s^2)$$

$$\mu = 1.19 \times 10^{-4} \frac{lb.s}{ft^2}$$

Example 1.5

Example An unknown fluid is exhibiting the following shear stress with shear strain behavior.

Shear Stress (Pa)	Shear Rate (1/s)	Shear Stress (Pa)	Shear Rate (1/s)
21.73	1.64	171.90	15.68
43.18	3.53	193.36	18.07
64.63	5.40	214.81	20.06
86.09	7.58	236.27	22.09
107.54	9.77	257.72	24.13
129.00	11.76	279.17	26.26
150.45	13.82	300.63	28.25

Identify the fluid type.

Solution We plot the data to understand the behavior of the unknown fluid.

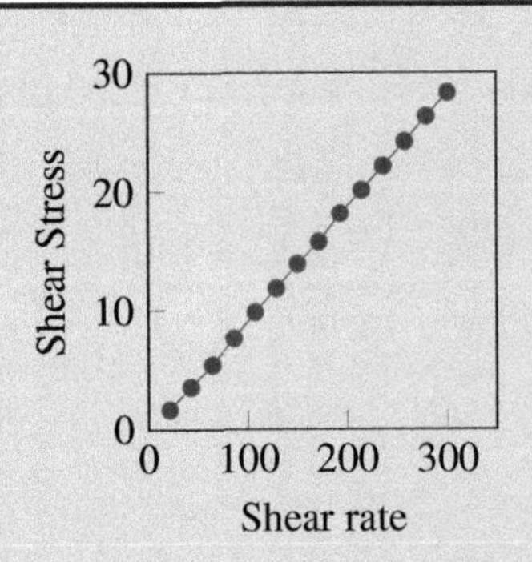

Figure 1.12 The linear relationship shows that the fluid is Newtonian.

As can be seen in Figure 1.12, the shear stress distribution is linear with the velocity gradient; this fluid is behaving as a Newtonian fluid.

1.9 SURFACE ENERGY AND SURFACE TENSION

Surface energy measures the amount of work done to stretch the surface per unit area. The magnitude of this force per unit length of the surface is called surface tension. The units of surface tension are N/m (or lbf/ft), whereas the units of surface energy are Joules per unit area. Dimensionally, surface tension and surface energy are equivalent as:

$$\frac{W}{A} = \frac{F \cdot s}{A} = \frac{F}{\ell}$$

where F is force, s is displacement, A is area, and ℓ has dimensions of length. Therefore, both the quantities are the same.

Surface free energy can also be defined a measure of the excess energy present at the surface of a material compared to its bulk. In the case of liquids, liquid surface tension and surface energy are equivalent. A surface with low surface energy will cause poor wetting, resulting in a high contact angle. This is because the surface cannot form strong bonds, so there is a bit of energetic reward for the liquid to break bulk-bonding in favor of interacting with the surface. An example of typical surfaces with low surface energy are hydrocarbons, which are held together with weak van der Waals forces.

The surface tension acts along the circumference, and the pressure acts on the area; horizontal force balances for the droplet as shown in Figure 1.13.

$$\sigma_e \ell_p \approx \Delta p A_c$$

where ℓ_p is the length of the perimeter and A_c is the cross-sectional area.

$$\sigma_e (\pi D) \approx \Delta p \left(\frac{\pi}{4} D^2\right)$$

Table 1.2
Surface Tension/Energy of Some Liquids

Liquids	10^{-3}(N/m)	Liquids	10^{-3}(N/m)
Glycerin	63	Water:	
SAE 30 oil	35	0°C	76
Mercury	440	20°C	73
Ethyl alcohol	23	100°C	59
Blood 37°C	58		
Gasoline	22	Benzene-air 20°C	28.88
Ammonia	21	Soap solution	25
Kerosene	28	Salt at 1,000°C	98

Table 1.3
Static Contact Angle for Some Liquid–Solid Pairs in Air

Solid	Liquid	Contact Angle (degrees)
Glass	Water	0
Glass	Mercury	128–148
Glass	Hydrogen	0
Glass	Nitrogen	0
Glass	Oxygen	0
Steel	Water	70–90
Steel	Hydrogen	0
Steel	Nitrogen	0
Steel	Oxygen	0
Paraffin	Hydrogen	106
Aluminum	Nitrogen	7
Platinum	Oxygen	105

Source: With kind permission from CRC Press: *Handbook of Fluid Dynamics* 1st Edition by Richard W. Johnson (Editor)

$$\sigma_e \approx \Delta p \left(\frac{D}{4}\right)$$

$$\Delta p \equiv \frac{4\sigma_e}{D}$$

For a droplet we have

$$p_i - p_o \equiv \frac{4\sigma_e}{D}$$

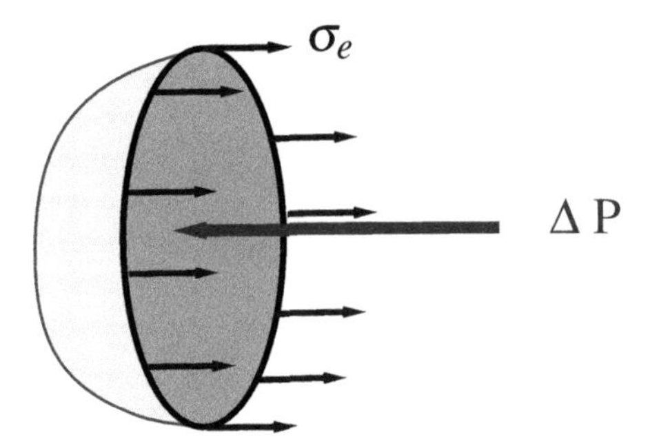

Figure 1.13 Surface tension in a droplet.

where p_i and p_o are droplet internal and outer pressures. The above relation is called the Young-Laplace law.

Note that in the case of a droplet, only one liquid-gas interface is present at which surface tension forces are important. However, if we consider a spherical bubble, there are two liquid-gas interfaces that are present where surface tension is important; one is inside the bubble in contact with gas or vapor, and one surface is outside the bubble. For the spherical bubble, the balance of forces leads us to a result:

$$p_i \equiv \frac{8\sigma_e}{D} + p_o = \frac{4\sigma_e}{r} + p_o$$

Example 1.6

Example Two bubbles of soap solution having radii $r_1 = 1$ cm, $r_2 = 2$ cm have merged and formed a bigger bubble of radius r_3. Find the radius r_3. The outside pressure is 1 atm. The soap solution has a surface energy of 0.025 N/m.

Solution We can use the surface tension relation we developed for the spherical bubble. From the internal pressure relation, we have:

$$p_i \equiv \frac{8\sigma_e}{D} + p_o = \frac{4\sigma_e}{r} + p_o$$

The masses of bubbles are m_1 and m_2 and when the bubbles coalesce they formed the bubble of mass m_3:

$$m_3 = m_1 + m_2$$

We know that density times volume is mass, so:

$$\rho_3 \forall_3 = \rho_1 \forall_1 + \rho_2 \forall_2$$

$$\rho_3 \left(\frac{4}{3}\pi r_3{}^3\right) = \rho_1 \left(\frac{4}{3}\pi r_1{}^3\right) + \rho_2 \left(\frac{4}{3}\pi r_2{}^3\right)$$

$$\left(\frac{p_3}{RT_3}\right)\left(\frac{4}{3}\pi {r_3}^3\right) = \left(\frac{p_1}{RT_1}\right)\left(\frac{4}{3}\pi {r_1}^3\right) + \left(\frac{p_2}{RT_2}\right)\left(\frac{4}{3}\pi {r_2}^3\right)$$

Assuming the density of the bubble is the same as that of vapor, and substituting density from the ideal gas equation, we have:

$$\left(\frac{4\sigma_e}{r_3} + p_o\right)\left(\frac{4}{3}\pi {r_3}^3\right)\frac{1}{R} = \frac{1}{R}\left(\frac{4}{3}\pi\right)\left[{r_1}^3\left(\frac{4\sigma_e}{r_1} + p_o\right) + {r_2}^3\left(\frac{4\sigma_e}{r_2} + p_o\right)\right]$$

$$\left(\frac{4\sigma_e}{r_3} + p_o\right)\left({r_3}^3\right) = \left[{r_1}^3\left(\frac{4\sigma_e}{r_1} + p_o\right) + {r_2}^3\left(\frac{4\sigma_e}{r_2} + p_o\right)\right]$$

$$4\sigma_e {r_3}^2 + p_o {r_3}^3 = 4\sigma_e({r_1}^2 + {r_2}^2) + p_o({r_1}^3 + {r_2}^3)$$

$$4(0.025){r_3}^2 + (1\times 10^5){r_3}^3$$
$$= 4(0.025)\left[(0.01)^2 + (0.02)^2\right] + (1\times 10^5)\left[(0.01)^3 + (0.02)^3\right]$$

$$0.1{r_3}^2 + 10^5{r_3}^3 = \left(\frac{0.1}{2000}\right) + 0.9 = 0.90005$$

$$0.1{r_3}^2 + 10^5{r_3}^3 \approx 0.9$$

This equation can be solved numerically, which gives us $r_3 = 0.0208$ m.

Example 1.7

Example Liquid eye droplets of density ρ are dropped into the eye using an eye dropper. The radius at the dropper outlet is r and a drop of size R has been formed. The weight of the liquid drop is W. Find the vertical force due to the surface tension on the drop. Also, find the minimum radius of the drop when it detaches from the dropper.

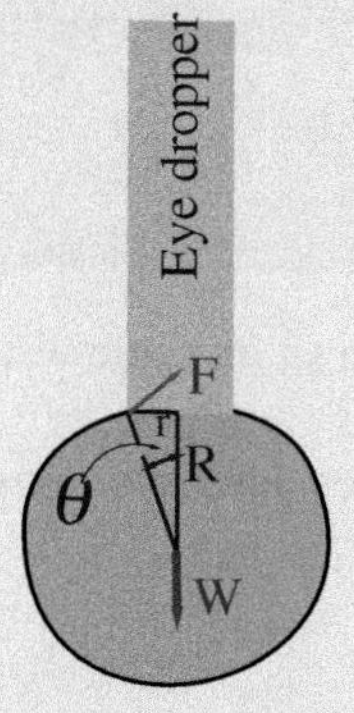

Figure 1.14 The forces in the eyedropper.

Solution We know that surface tension will play an important role in this problem. Figure 1.14 shows the forces important in the case of a liquid drop that is coming out of an eyedropper. The surface tension is acting circumferentially at the dropper's outlet:

$$F_\sigma = \sigma_e(2\pi r)\sin(\theta)$$

The force angle with the outlet is θ so:

$$F_\sigma = \sigma_e(2\pi r)\left(\frac{r}{R}\right)$$

We now replace $\sin(\theta)$ with (r/R):

$$F_\sigma = \sigma_e(2\pi)\left(\frac{r^2}{R}\right)$$

This is the force due to surface tension that is acting on the dropper outlet. Now, for the liquid to drop, this has to be equal to the weight of the spherical drop:

$$W = F_\sigma$$

$$\left(\frac{4}{3}\pi R^3\right)\rho g = \sigma_e(2\pi)\left(\frac{r^2}{R}\right)$$

$$R = \sqrt[4]{\frac{3}{2}\sigma_e\left(\frac{r^2}{\rho g}\right)}$$

Figure 1.15 Some insects can sit on static water due to surface tension.

Due to the property of surface tension, some insects can even sit on the surface of water (see Figure 1.15).

1.9.1 JURIN'S LAW

In case of capillary rise, the surface will form a meniscus, which is a curve in the upper surface of a liquid close to a solid surface, as shown in Figure 1.16.

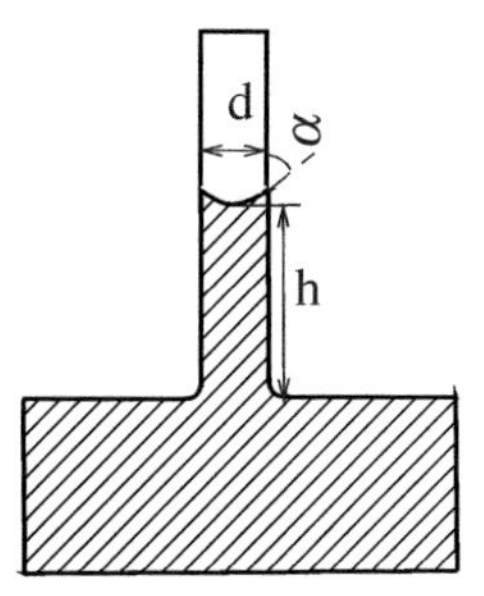

Figure 1.16 The rise of liquid inside a column and formation of the meniscus.

Figure 1.16 shows the formation of the meniscus inside a vertical column of internal diameter d. The liquid level height is h, and the meniscus is curved due to adhesive forces and surface tension.

The balance of forces is

$$F_1 = 2 \cdot \pi \cdot r \cdot \sigma_e \cdot \cos\alpha$$

$$F_2 = h \cdot \pi r^2 \cdot \rho g$$

$$F_1 = F_2$$

$$h = \frac{4\sigma_e \cdot \cos\alpha}{\rho \cdot g \cdot d}$$

where α is the contact angle of the liquid on the wall of the tube.

This relation is called Jurin's law, which is named after James Jurin, who formulated this relation between 1718 and 1719.

Example 1.8

Example A clean glass tube will be used in a manometer to measure the pressure of glycerin. If the capillary rise is to be limited to 1 mm, specific gravity of glycerin = 1.263, surface tension of glycerin = 0.063 N/m, and the contact angle is 46°, find the diameter of the glass tube in mm.

Solution From Jurin's law we have:

The density of glycerin is

$$\rho_{glycerin} = SG \times \rho_{water} = 1.263 \times 1000 = 1263 kg/m^3$$

$$\gamma_f = \rho_{glycerin} \cdot g = 12390.03 N/m^3$$

We take the contact angle as zero.

$$h = \frac{4\sigma_e \cdot \cos\alpha}{\rho \cdot g \cdot d}$$

We rearrange, and the diameter is

$$d = \frac{40.063 \cdot \cos(46^\circ)}{12390.03 \times 1 \times 10^{-3}}$$

$$d = 14.13\ mm$$

Example 1.9

Example To what height above the reservoir level will water rise in a glass tube, such as that shown in Figure 1.16, if the reservoir is maintained at 25°C and the inside diameter of the tube is 2 mm?

Solution From the Appendix[a] we know that water viscosity at 25°C is $\mu = 8.91 \times 10^{-4}$ Pa · s, and surface tension $\sigma_e = 7.20 \times 10^{-2}$ N/m, $\gamma = 9.77$ kN/m^3. We take cos $\alpha = 1$.

$$h = \frac{4\sigma_e \cos\alpha}{\gamma \cdot d} \approx \frac{4\sigma_e}{\gamma \cdot d}$$

$$h = \frac{4 \times 7.20 \times 10^{-2}}{9770 \times 2 \times 10^{-3}} = 0.01473 \text{ m} = 14.73 \text{ mm}$$

[a] Appendix can be found online at https://routledge.com/9781032324531

If a tube of diameter D is inserted into a liquid bath and, after that, another liquid is poured over it in such a manner that two immiscible liquids having different properties then an interface will be formed inside the tube. The schematic diagram of this problem is shown in Figure 1.17. The expression for the height of liquid derived before is no longer valid. We will now develop the expression for this case.

The force balance gives:

$$\sum F = \rho_2 g \Delta h \left(\frac{\pi D^2}{4}\right) - \rho_1 g \Delta h \left(\frac{\pi D^2}{4}\right) + [\pi D \cos\alpha]\, \sigma_e$$

Rearranging:

$$\Delta h = \frac{-4\sigma_e \cos(\alpha)}{(\rho_2 - \rho_1) \cdot g \cdot D}$$

where σ_e is the surface tension of fluid 1.

Fluids in contact with the vertical surfaces will form a liquid free surface attached with the wall, as shown in Figure 1.18.

The liquid layer that rises and attaches itself to the surface will exert a small hydrostatic pressure on the wall, as shown in Figure 1.18. The balance of surface tension and pressure forces is

$$\rho \cdot g \cdot y(x) = \frac{\sigma_e}{R(x)}$$

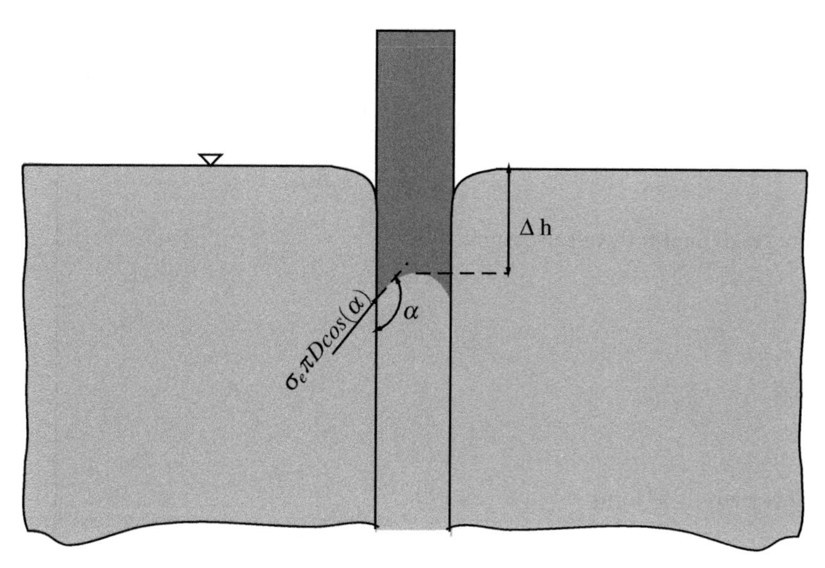

Figure 1.17 Two immiscible liquids in the same tube.

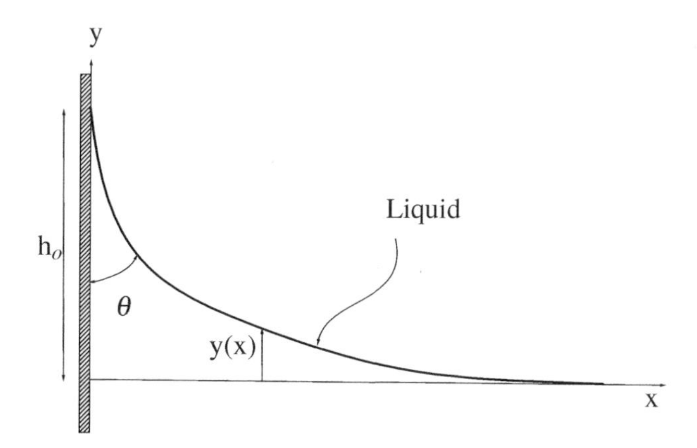

Figure 1.18 Liquid free surface near a wall.

where $R(x)$ is the radius of curvature of the liquid surface, and $y(x)$ is the surface position with respect to x. The radius of curvature is defined as:

$$R(x) = \frac{\sqrt[3]{1+\left(\frac{dy}{dx}\right)^2}}{\frac{d^2y}{dx^2}}$$

$$\rho \cdot g \cdot y(x) = \frac{\sigma_e \left(\frac{d^2y}{dx^2}\right)}{\sqrt[3]{1+\left(\frac{dy}{dx}\right)^2}}$$

For small angles θ we can assume that

$$\left(\frac{dy}{dx}\right)^2 \ll 1$$

and

$$1+\left(\frac{dy}{dx}\right)^2 \approx 1$$

This leads us to form

$$\rho \cdot g \cdot y(x) = \sigma_e \left(\frac{d^2y}{dx^2}\right)$$

If we take

$$\lambda^2 = \frac{\sigma_e}{\gamma}$$

we can formulate a differential equation:

$$y(x) = \lambda^2 \left(\frac{d^2y}{dx^2}\right)$$

$$y'' - \frac{y}{\lambda^2} = 0$$

The solution of this differential equation is

$$y = \lambda \cot(\theta) \cdot \exp\left(\frac{-x}{\lambda}\right)$$

$$y(x) = \cot(\theta) \cdot \left(\sqrt{\frac{\sigma_e}{\gamma}}\right) \cdot \exp\left(\frac{-x}{\sqrt{\sigma_e/\gamma}}\right)$$

At $y = 0$ the h_o is the length of attachment with the surface and

$$h_o = \cot(\theta) \cdot \left(\sqrt{\frac{\sigma_e}{\gamma}}\right)$$

1.10 NANOFLUIDS

The discovery of nanofluids is attributed to Argonne National Laboratory, USA. The first-ever report on nanofluids was published by Choi (1995). Nanofluids contain suspended nanoparticles that have high thermal conductivity. They are colloidal suspensions engineered by including the nano-sized particles in some base fluid. They are classified into distinct categories depending on dispersed particles: metal, metal oxide, metal hybrid, and carbon-based. The most common choice of host fluid or base fluid is water, ethylene glycol, oils, and methanol. Suspensions based on hybrid nanofluids are also investigated, which use a combination of different nanoparticles. Nanofluids quickly gained popularity in the heat exchanger industry, as they give better thermal performance than conventional fluids. Because of the growing importance of nanofluids, the properties' correlations are collected in this book.

The flow and thermal characteristics of a nanofluid are strongly influenced by the properties of the nanofluid. A great amount of research has been done investigating properties (see Ding et al., 2006; Eastman et al., 1999; Jang and Choi, 2006; Koo and Kleinstreuer, 2004; Sahoo et al., 2009). The Al_2O_3-water nanofluid, with nanoparticle concentrations of less than 10%, behaves as a Newtonian fluid for the temperature range of 0–90°C (Sahoo et al., 2009).

The relationship proposed by Brinkman (1952) can be used to calculate the effective viscosity of the nanofluid.

$$\mu_{nf} = \frac{\mu_f}{(1-\varphi)^{2.5}}$$

where φ is the nanoparticle concentration and subscripts nf and f refer to the nanofluid and the base fluid properties, respectively.

The effective density and specific heat capacity of the nanofluid can be modeled as per the proposal of Xuan and Li (2003):

$$(c_p)_{nf} = (1-\varphi)(c_p)_f + \varphi(c_p)_p$$

$$\rho_{nf} = (1-\varphi)\rho_f + \varphi\rho_p$$

The thermal conductivity of nanofluid can be modeled by use of Maxwell theory, which was proposed in 1873. For the two-component mixture of the spherical-particle suspension, the Maxwell theory proposed the following model:

$$\frac{k_{nf}}{k_f} = \frac{(k_p + 2k_f) - 2\varphi(k_f - k_p)}{(k_p + 2k_f) + \varphi(k_f - k_p)}$$

The thermal and momentum diffusivities of a nanofluid can be calculated as:

$$\frac{\alpha_{nf}}{\alpha_f} = \frac{(k_{nf}/k_f)}{(1-\varphi) + \varphi\left[(\rho c_p)_p/(\rho c_p)_f\right]}$$

$$\frac{\nu_{nf}}{\nu_f} = \frac{1}{(1-\varphi)^{2.5}\left[(1-\varphi) + \varphi(\rho_p/\rho_f)\right]}$$

Table 1.4 lists the properties of two base fluids and some nanoparticles. Table 1.5 lists the properties of a nanofluid formed by adding Al_2O_3 particles in water. The concentration of nanoparticles will change the properties of the mixture.

Table 1.4
Thermophysical Properties of Base Fluids and Nanoparticles, a Calculation Based on Pak and Cho (1998) and Brinkman (1952) Models

	ρ (kg/m^3)	c_p (J/kg · K)	k (W/m · K)
Water (base fluid)	997	4179	0.613
Ethylene glycol (base fluid)	1115	2428	0.253
Copper (Cu)	8933	385	401
Silver (Ag)	10,500	235	429
Alumina (Al_2O_3)	3970	765	40
Titanium oxide (TiO_2)	4250	686.2	8.9538

Table 1.5
Thermophysical Properties of the Water-Al_2O_3 Nanofluid for Different Volumetric Concentrations

φ%	ρ_{nf} kg/m^3	k_{nf} W/m · K	μ_{nf} Pa · s	$c_{p,nf}$ J/kg · K
0	996.24	0.616	0.000821	4179
0.2	1002.24	0.675	0.000825	4172.4
0.7	1017.30	0.822	0.000835	4158.9
1.5	1041.30	1.056	0.000851	4129.5
3	1086.40	1.497	0.000882	4080

1.11 AN OVERVIEW OF FLUID ANALYSIS TYPES

In fluid mechanics, we divide the study into real fluids and ideal fluids. Also, we can assume certain constraints on flow field.

1.11.1 VISCOUS VS. INVISCID FLOW

Flow can be viscous or inviscid. To understand this approximation, we first need to understand the concept of vorticity. Vorticity is a measure of rotation of fluid particle and it arises due to the presence of viscous forces in fluid. The shear forces acting on a fluid particle tend to stop the particle, and in doing so the particle is deformed. This causes a rotation in fluid particles and it is strongest near the walls. If we consider the flow as viscous then we need to identify the type of flow as well, which can be either **laminar** or **turbulent flows**. A laminar flow is a flow in lamina, i.e. layers. Turbulent flow is a highly diffusive and fast-moving flow. Both of these flows occur commonly in nature.

1.11.2 STEADY VS. UNSTEADY FLOW

We can declare the flow as steady by considering the following derivatives of zero at a specific location in the flow field:

$$\frac{du}{dt}=0,\frac{dv}{dt}=0,\frac{dw}{dt}=0,\frac{dp}{dt}=0,\frac{dT}{dt}=0,\frac{d\rho}{dt}=0$$

where u, v and w are three velocity components in Cartesian coordinates. The p, ρ and T are the pressure, density, and temperature, respectively. In this book, we will deal mostly with isothermal flows, so setting the time-derivative of temperature to zero is not needed.

If any of these derivatives are not zero in the flow, the flow field cannot be declared as steady.

1.11.3 UNIFORM FLOW

A uniform flow is the flow in which the following derivatives are zero:

$$\frac{du}{dx}=0,\frac{dv}{dx}=0,\frac{dw}{dx}=0,\frac{dp}{dx}=0$$

where x is a coordinate in flow direction. The flow of wind over a flat surface, cars, etc, is often considered as uniform flow in fluid mechanics.

1.11.4 WALL-BOUNDED VS. FREE-SHEAR FLOW

If the flow is bounded by walls, then such flow is called wall-bounded flow. This happens in the case of flow through a pipe, tube, or duct, for example. This flow is also called internal flow in the literature. If the flow is over the surface, as in submarines, boats, pillars, then such flow is called external flow. Free shear flows are the flow which are not confined by the surfaces or walls like in jets, wakes and separated flows.

1.11.5 ONE-, TWO-, AND THREE-DIMENSIONAL FLOW

All flows have three-dimensional velocity fields. However, in some cases, the variation of velocity in other dimensions is not significant; such flows can be approximated as one-dimensional flows. Flow over a flat plate is a kind of two-dimensional flow. The flow through a square duct and in the pipe bends is an example of a three-dimensional flow. It is essential to understand that the flows near walls cannot be treated as one-dimensional. Note that one-dimensional flow assumption is used in isentropic flows as they do not have friction present. The two-dimensional analysis is suitable when the model or prototype is sufficiently long and slender so that the end effects are negligible and the approach flow is uniform. Some flows are symmetric along the axis of the pipe or an object if flow is external. This approximation is suitable for the flow over a bullet. A bullet moving through the air can be considered an example of an axisymmetric flow. However, experiments reveal that there are secondary shear-layer vortices and strong secondary flow on conical bodies. Therefore, the axisymmetric flow approximation may be valid only at the beginning of the laminar boundary layer flow. In fluid mechanics literature, the bodies that generate a lot of aerodynamic drag are called bluff, or blunt, bodies compared with the bodies that create less aerodynamic drag, which are called streamlined bodies.

1.11.6 COMPRESSIBLE VS. INCOMPRESSIBLE FLOW

In an analysis of liquid flows, the density variation is small, and thus density can be considered constant throughout the flow field. The density will be a significant quantity in gas flows, and assuming the density constant is only valid if flow speed is less than 30% of the speed of sound in that medium.

Pioneers of Fluid Mechanics

CLAUDE-LOUIS NAVIER (February 10, 1785–August 21, 1836) Claude-Louis Navier was a French engineer and a physicist who developed the equation of motion of viscous flows.

SIR GEORGE GABRIEL STOKES (August 13, 1819–February 1, 1903) Stokes was an Irish English physicist and mathematician who proposed the constitutive relations related to stresses with velocity gradients.

OSBORNE REYNOLDS (August 23, 1842–February 21, 1912) Osborne Reynolds was an Irish scientist who identified two kinds of flows. One he called layered flow and another one he called sinuous flow now called turbulent flow. Reynolds worked also on heat transfer between solids and fluids.

VINCENC STROUHAL (April 10, 1850–January 26, 1922) Strouhal's significant contribution to the fundamentals of fluid mechanics was his discovery in 1878 of the Strouhal number (St).

CARL WILHELM OSEEN (April 17, 1879–November 7, 1944) Oseen's contributions to fluid mechanics were the Oseen equations, which describe the viscous and incompressible fluid flow at small Reynolds numbers.

MORITZ WEBER (1871–1951) Weber created the Weber number (We), a dimensionless number in fluid mechanics that is often useful in analyzing fluid flows. There is an interface between two different fluids, especially for multiphase flows with strongly curved surfaces.

LUDWIG PRANDTL (February 4, 1875–August 15, 1953) Prandtl introduced the boundary layer theory in 1905 to understand the viscous fluid flow near a solid boundary.

ANDREY KOLMOGOROV (April 25, 1903–October 20, 1987) Kolmogorov's greatest contribution to the field of fluid mechanics was the turbulence spectrum. The turbulence energy travels from large scales to small scales, and he also discovered that the energy spectrum of turbulence follows the slope of $-5/3$.

BRIAN SPALDING (January 9, 1923–November 27, 2016) Dudley Brian Spalding was Professor of Heat Transfer and Head of the Computational Fluid Dynamics Unit at Imperial College, London. He was among the pioneers of computational fluid dynamics (CFD) and an internationally recognized contributor to heat transfer, fluid mechanics, and combustion.

PROBLEMS

1P-1 The equation of state for water is approximated by the following relation:

$$\frac{p}{p_o} \approx 3001\left(\frac{\rho}{\rho_o}\right)^7 - 3000$$

where $p_o = 1$ atm and $\rho_o = 998$ kg/m^3. Find the bulk modulus of water (κ).

1P-2 Viscosity of water is approximated by the following relation:

$$\frac{\mu}{\mu_o} = \left(\frac{T}{T_o}\right)^{3/2}\left(\frac{T_o + S}{T + S}\right)$$

Find the viscosity of steam at $T = 1200°$C.

1P-3 A substance exhibited the following trend for shear stress and deformation. The data were gathered under constant temperature conditions. Classify the substance.

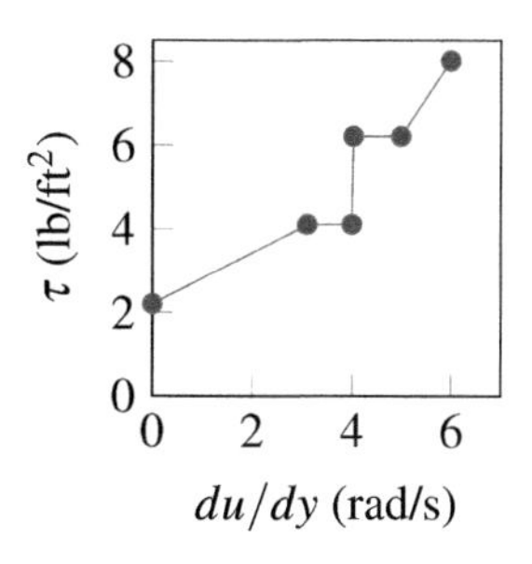

Figure 1.19 Figure of problem 1P-3.

1P-4 A substance exhibited the following trend for shear stress and deformation. The data were gathered under constant temperature conditions. Classify the substance.

Table 1.6
Commonly Used Properties of Air and Water

Property	SI	USCS & English
Air at 20°C (68°F) and 1 atm		
Specific gas constant	$R_{air} = 0.2870$ kJ/kg · K $R_{air} = 287.0$ m^2/s^2 · K	$R_{air} = 0.06855$ Btu/lb$_m$ °R $R_{air} = 53.34$ ft · lb$_f$/lb$_m$ °R $R_{air} = 1717$ ft · lb$_f$/slug °R
Specific heat ratio	$k = c_p/c_v = 1.40$	$k = c_p/c_v = 1.40$
Specific heats	$c_p = 1.007$ kJ/kg · K $c_p = 1007$ m^2/s^2· K $c_v = 0.7200$ kJ/kg · K $c_v = 720.0$ m^2/s^2 · K	$c_p = 0.2404$ Btu/lb$_m$°R $c_p = 187.1$ ft · lb$_f$/lb$_m$ °R $c_v = 6019$ ft^2/s^2°R $c_v = 0.1719$ Btu/lb$_m$°R $c_v = 133.8$ ft · lb$_f$/lb$_m$°R $c_v = 4304$ ft^2/s^2 · °R
Speed of sound	$a = 343.2$ m/s $= 1236$km/h	$a = 1126$ ft/s $= 767.7$ mi/h
Density	$\rho = 1.204$ kg/m^3	$\rho = 0.07518$ lbm/ft^3 $\rho = 0.00237$ slug/ft^3
Abs. viscosity	$\mu = 1.825 \times 10^{-5}$ kg/m · s	$\mu = 1.227 \times 10^{-5}$ lb$_m$/ft · s
Kinematic viscosity	$\nu = 1.516 \times 10^{-5}$ m^2/s	$\nu = 1.632 \times 10^{-4}$ ft^2/s
Water at 20°C (68°F) and 1 atm		
Specific heat	$c = 4.182$ kJ/kg · K $c = 4182$ m^2/s^2 · K	$c = 0.9989$ Btu/lb$_m$ · °R $c = 777.3$ ft · lb$_f$//lb$_m$ · °R $c = 25,009$ ft^2/s^2 · °R
Density	$\rho = 998.0$ kg/m^3	$\rho = 62.30$ lb$_m$/ft^3 $\rho = 1.936$ slug/ft^3
Abs. viscosity	$\mu = 1.002 \times 10^{-3}$ kg/m · s	$\mu = 6.733 \times 10^{-4}lb_m$/ft · s
Kinematic viscosity	$\nu = 1.004 \times 10^{-6}$ m^2/s	$\nu = 1.081 \times 10^{-5}$ft^2/s

1P-5 Compute an approximate distance d for mercury in a glass capillary tube. The surface tension for mercury and air here is 0.512 N/m and angle is 40°. The specific gravity of mercury is 13.6.

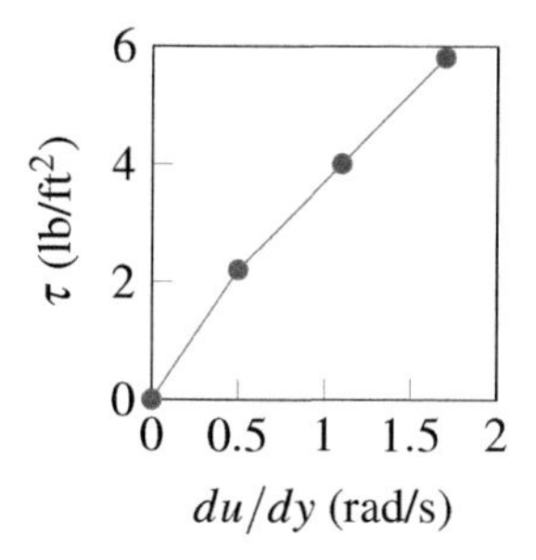

Figure 1.20 Figure of problem 1P-4.

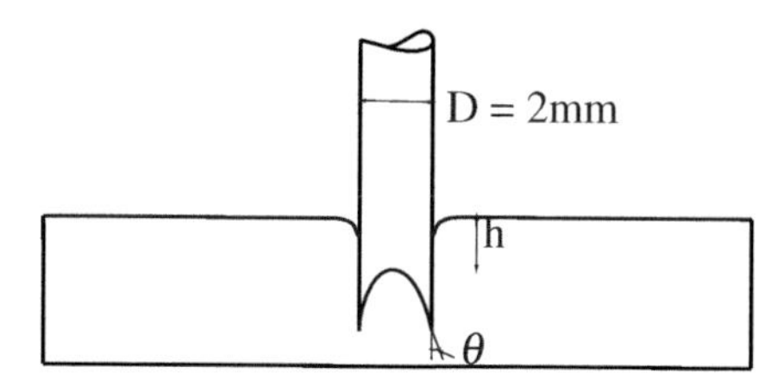

Figure 1.21 Figure of problem 1P-5.

1P-6 Estimate the height to which water at 25°C will rise in the capillary tube of diameter 5 mm.

1P-7 Water at 30°C rises in a clean glass of a 0.1-mm diameter tube. Take the water-glass angle (α) as 0 degree with the normal. How much would be the rise of water in the glass column?

1P-8 A rigid steel container is filled up with a liquid at 15 atm. The volume of the liquid is 1.231 liters at a pressure of 30 atm and 1.232 at 15 atm, respectively. Find the modulus of elasticity of the liquid for the range of pressures given in the isothermal case. So what is the compressibility coefficient?

1P-9 A dirigible with a volume of $\forall = 80{,}000\ \text{m}^3$ contains helium under conditions of standard atmospheric (pressure 101 kPa, temperature 15°C). Find the weight of the dirigible if the lift force acting on it is 6×10^6 N.

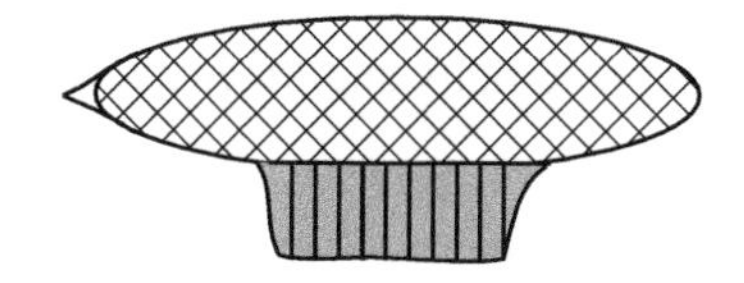

Figure 1.22 Figure of problem 1P-9.

1P-10 The liquid rises into an annular gap (formed by two concentric tubes) when the tubes are immersed into water. The inner and outer radii of the tubes are R_o and R_i. Find the expression for the rise of the water into the annular gap, if the water's surface tension is σ_e.

REFERENCES

W. Sutherland, The viscosity of gases and molecular force, Philosophical Magazine, S. 5, vol. 36, pp. 507–531, 1893.

U. S. Stephen and J. A. Choi, Eastman, Enhancing Thermal Conductivity of Fluids with Nanoparticles, SME International Mechanical Engineering Congress and Exposition, November 12–17, San Francisco, CA, 1995.

H. C. Brinkman, The viscosity of concentrated suspensions and solutions, J. Chem. Phys., vol. 20, no. 4, pp. 571, 1952.

B. C. Pak and Y. I. Cho, Hydrodynamic and heat transfer study of dispersed fluids with submicron metallic oxide particles, Exp. Heat Transf., vol. 11, no. 2, pp. 151–170, 1998.

Y. Xuan and Q. Li, Investigation on convective heat transfer and flow features of nano-fluids, ASME J. Heat Transf., vol. 125, no. 1, pp. 151–155, 2003.

Y. Ding, H. Alias, D. Wen, and R. A. Williams, Heat transfer of aqueous suspensions of carbon nanotubes (CNT nanofluids), Int. J. Heat Mass Transf., vol. 49, no. 1–2, pp. 240–250, 2006.

J. A. Eastman et al., Novel thermal properties of nanostructured materials, J. Metastable Nanocryst. Mater., vol. 2–6, pp. 629–634, 1999.

S. P. Jang and S. U. S. Choi, Cooling performance of a micro channel heat sink with nanofluids, Appl. Therm. Eng., vol. 26, no. 17-18, pp. 2457–2463, 2006.

J. Koo and C. Kleinstreuer, A new thermal conductivity model for nanofluids, J. Nanopart. Res., vol. 6, no. 6, pp. 577–588, 2004.

B. C. Sahoo, R. S. Vajjha, R. Ganguli, G. A. Chukwu, and D. K. Das, Determination of rheological behavior of aluminum oxide nanofluid and development of new viscosity correlations, Pet. Sci. Technol., vol. 27, no. 15, pp. 1757–1770, 2009.

J. C. Maxwell, A Treatise on Electricity and Magnetism. Oxford: Clarendon Press, 1873.

L. Böswirth, Technische Strömungslehre, Viewegs Fachbcher der Technik, 1993.

G. S. Fulcher and J. Amer. Cerarn. Soc. vol. 8, p. 339, 1925.

2 Pressure and Stationary Fluid

In this chapter we will discuss the pressure distribution inside static fluid. We will discuss atmospheric pressure and the different units that are used to report pressure. Also, we will learn about buoyant forces.

Learning outcomes: After finishing this chapter, you should be able to:

- Determine the pressure distribution inside a static fluid.
- Report the pressure in different units.
- Understand buoyant force.

2.1 PRESSURE IN STATIONARY FLUID

Pressure is independent of direction and fluid in a state of rest. Let us consider a very small fluid body that looks like a prism in the Cartesian coordinates system, as shown in Figure 2.1. The pressure over each face of the prism is considered constant. The prism is assumed to be placed in a gravitational field, and the body forces act on it. In fluid mechanics literature, the body forces are exerted due to field effects, i.e., the magnetic field or the gravitational field, which do not require physical contact with the control volume system. For the sake of analysis, the body forces per unit mass due to gravitational field are symbolized as f_x, f_y, and f_z.

As no external force is applied on the prism, the balance of forces can be written as follows

$$\sum F_x = p_x A_x - p_i A_i cos(\beta) + f_x \rho \left(\frac{A_x x_o}{3} \right) = 0$$

where

p_x is the average of pressure in the x-plane,

A_x is the area in the x-plane $= A_i cos(\beta)$,

p_i is the average pressure on the inclined surface

A_i is the area of the inclined surface

β is the angle between normal to the x-plane area and inclined area

DOI: 10.1201/9781003315117-2

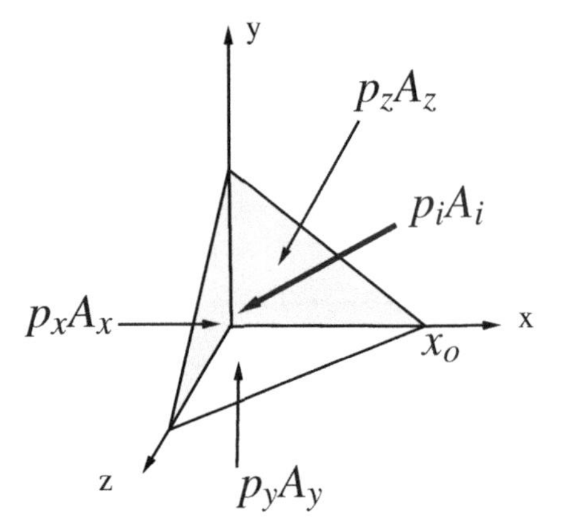

Figure 2.1 The pressure forces on a fluid prism body at rest.

x_o is the intercept of prism volume on the x-axis

f_x is body force per unit mass

The quantity $A_x x_o/3$ represents the volume of the prismatic control mass:

$$\sum F_x = A_x\left[p_x - p_i + f_x\rho\left(\frac{x_o}{3}\right)\right] = 0$$

Since the area is not zero, the natural conclusion is that

$$\left[p_x - p_i + f_x\rho\left(\frac{x_o}{3}\right)\right] = 0$$

For an infinitesimally small prism, x_o tends to zero and we reach the condition:

$$[p_x - p_i] = 0$$

The pressure on the inclined surface is also the same as the pressure in the x-direction. This condition is valid as long as fluid is in a state of rest.

Hence, it is proved that pressure is the same in all directions in static fluid.

2.2 HYDROSTATICS

In case of liquids placed in the gravitational field with y as vertical direction normal to the earth, then body force (m · g) per unit mass is

$$f_y = -\frac{m \cdot g}{m} = -g \tag{2.1}$$

where g is the local gravitational acceleration. Also

$$f_x = 0 \tag{2.2}$$

In static fluid, the gradient of pressure field is balanced by the body forces, so:

$$\frac{1}{\rho}\nabla p(x,y,z) = \mathbf{f} \tag{2.3}$$

where $\mathbf{f}$ is the body force field.

The pressure gradient is proportional to the force field vector. Note that pressure is a scalar quantity, but the gradient of pressure will be a vector quantity.

$$\frac{1}{\rho}\frac{\partial p}{\partial x} = f_x = 0$$

$$\frac{1}{\rho}\frac{\partial p}{\partial y} = f_y = -g$$

$$\frac{1}{\rho}\frac{\partial p}{\partial z} = f_z = 0$$

From the above equation we found that

$$p = -\rho g y + const \tag{2.4}$$

This shows that in hydrostatic liquids the pressure is decreasing linearly as height increases.

Pascal's law: From eq. 2.4 it is clear that in the case of rest, the pressure in the fluid is independent of the shape of the container. So whatever the shape of the vessel is, the wall of the vessel will experience the same pressure distributions. The law is named after the French philosopher Blaise Pascal (1623–1662).

For most engineering applications, the variation in g is negligible, so our main concern is with the possible variation in fluid density, which can be characterized as a function of depth and temperature. Using eq. 2.4 we get:

$$\frac{\partial^2 p}{\partial x \partial y} = -g\frac{\partial \rho}{\partial x}$$

As the pressure is changing only vertically, we know that

$$\frac{\partial p}{\partial x} = 0,$$

also, as density is independent of x-direction, we get:

$$\frac{\partial^2 p}{\partial x \partial y} = -g\frac{\partial \rho}{\partial x} = 0$$

This shows that both pressure and density are functions of y only. A fluid with constant density is called an incompressible fluid. A gas in which density changes

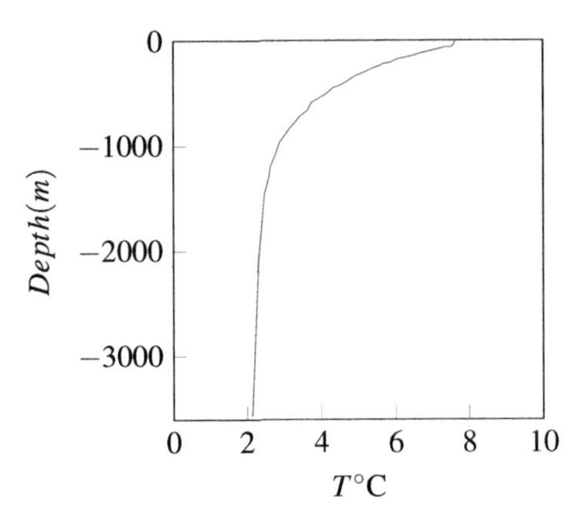

Figure 2.2 Typical temperature profile of the ocean.

due to temperature and pressure change is called compressible fluid. Under normal pressure and temperature conditions, liquid densities are not changing much. However, due to certain other factors like salt concentration, etc., in some cases, the density changes may occur with depth but independent of pressure. Such liquid are usually giving stratified layers of the fluid. In such liquids, the density is varying with vertical direction only, so

$$\rho = \rho(y).$$

Baroclinic fluid: A fluid in which the density varies with both depth and horizontal position is called baroclinic fluid.
Barotropic fluid: A fluid in which the density may vary with depth but not with horizontal distance is called barotropic fluid. In this fluid, the density is a function of the pressure only, or vice versa.

A typical temperature profile of the ocean is shown in Figure 2.2. The regions or layers of large temperature variation are called thermoclines. Thermoclines form in oceans due to convection in the ocean that is generated through wind or temperature differences at the water's surface.

Example 2.1

Example The density of the salt solution in a solar pond varies as a function of depth z, such that

$$\rho(z) = \rho_o \sqrt{1 + \tan^2\left(\frac{\pi}{4}\frac{z}{H}\right)}$$

where ρ_o is the density at the surface, which is 1,040 kg/m^3, and where $H =$ 10 m, which is the depth of the solar pond (see Figure 2.3). Find the pressure variation as a function of z.

Solution We convert the density function into a series to simplify the integration:

$$\rho(z) = 1040 + 3.20z^2 + 0.00822z^4 + O\left(z^6\right)$$

We ignore the higher order terms and plot the function and series expansion to see the influence of truncation.

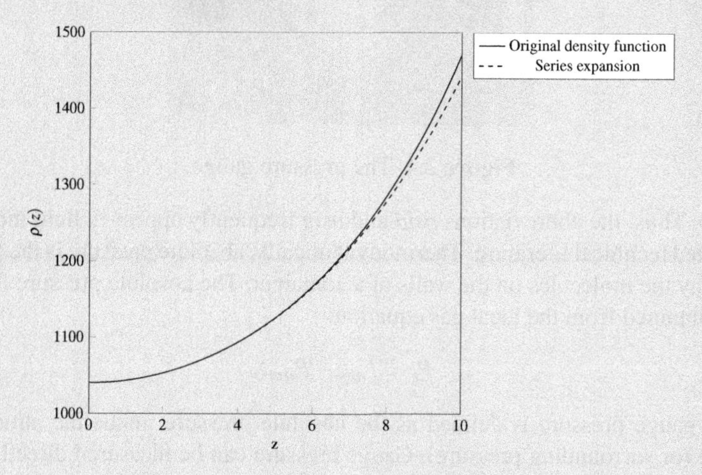

Figure 2.3 Density inside the solar pond.

Since the truncation has little influence till $z = 10$, we can safely use the truncated series as representation of density. We now integrate this function to get pressure:

$$dp = \rho(z) \cdot g \cdot dz$$

$$p = \int_0^z \rho(z) \cdot g \cdot dz$$

$$p(z) = 10202.40z + 10.47z^3 + 0.01614z^5$$

2.3 PRESSURE UNITS

Pressure is defined as force per unit area exerted in a direction normal to that area. The pressure is expressed as newtons per square meter (N/m^2) in SI units. In the British units, it is described as pound-force per square foot (lb_f/ft^2), and it can also be expressed as pound-force per square inch (lb_f/in^2), or psi. The pressure units can also appear with the alphabet, like *abs* for absolute pressure, and *g* for the gauge

Figure 2.4 The pressure gauge.

pressure. Thus, the abbreviations *psia* and *psig* frequently appear in fluid mechanics and related technical literature. Thermodynamically, absolute pressure is the pressure exerted by the molecules on the walls of a container. The absolute pressure for gases can be obtained from the ideal gas equation:

$$P_g = P_{abs} - P_{atm}$$

The gauge pressure is defined as the absolute pressure minus the atmospheric pressure (or surrounding pressure). Gauge pressure can be measured directly with a pressure gauge, which is shown in Figure 2.4.

The atmospheric pressure is 14.7 psi, 2116.22 lb_f/ft^2, or 101,325 Pa. The gauge pressure is negative if the pressure of the system is less than atmospheric pressure. If the pressure is less then the atmospheric pressure then this state is called a vacuum. In industrial and laboratory applications, varying levels of vacuum are required, and the requirements dictate that pressure can be conducted at a pressure level that is above the perfect vacuum state. Some vacuum applications are metallic coating, nuclear fuel refinement, and sperm and biological sample storage. Chemical processing industries like food processing, pharmacology, agriculture, and textiles also have operations that required a vacuum state.

Vacuum drying is a mass transfer process in which the moisture present on a solid surface can be removed under vacuum. Figure 2.5 shows different pressure levels. Note that vacuum is a pressure below atmospheric pressure.

The height of a pyramid: In 1798, Napoleon attacked Egypt and occupied it. He brought many scientists from France to help in his grandiose plan to modernize and colonize Egypt. During that time, one of his scientists (known as savants) measured the height of the great pyramid for the very first time. Nicholas Conté used his latest invention, a new type of barometer, which he had brought with him from France. Through this invention, he measured the height of the Great Pyramid of Giza, which he calculated to be 428 ft, or 130.45 m.

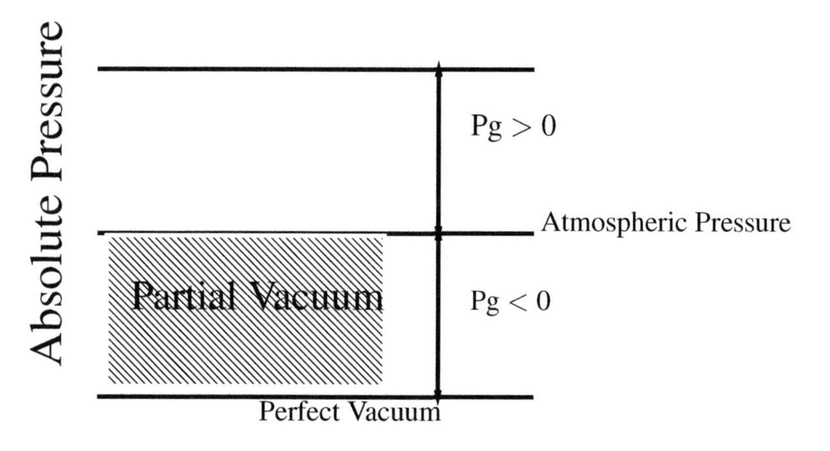

Figure 2.5 Different pressure levels in fluids.

Atmospheric Pressure

$$1\ atm = 101325\ Pa$$

$$1\ torr = 133.322\ Pa$$

$$1\ bar = 99999.72\ \text{Pa}$$

$$1\ lb_f/in^2 = 1\ psi = 6894.76\ Pa$$

$$760\ mmHg = 1\ atm$$

$$1\text{kg}_f/cm^2 = 98.0665\ kPa\ = 1\ \text{at (Technical atmosphere)}$$

$$1\ at = 0.981\ bar$$

Example 2.2

Example A barometric pressure of $\Delta\ h = 24.4$ in. (619.76 mm) Hg corresponds to how much atmospheric pressure in psia and pascals?

Solution

$$p = \gamma \cdot \Delta h = 847 \frac{lb_f}{ft^3} \times \left(\frac{24.4}{12}\right) ft \times \left(\frac{1\ ft^2}{144\ in^2}\right) = 11.95\ psia$$

$$p = \gamma \cdot \Delta h = 133\ E3 \frac{N}{m^3} \times (0.61976)\, m = 82.42\ kPa$$

Example 2.3

Example If the mercury column of 1,000 mm is replaced by SAE30 oil for the same pressure condition then how high would the oil column be? Take $\gamma_{Hg} = 133\ \text{kN/m}^3$ and $\gamma_{oil} = 8.95\ \text{kN/m}^3$.
Solution As the pressure is related to column height, we know that

$$P = \gamma\,\Delta\text{h}$$

We can compare the pressure on mercury and oil together as:

$$\gamma_{oil}\,\Delta\text{h}_{oil} = \gamma_{Hg}\,\Delta\text{h}_{Hg}$$

$$\Delta\text{h}_{oil} = \left(\frac{\gamma_{Hg}}{\gamma_{oil}}\right)\Delta\text{h}_{Hg} = \left(\frac{133}{8.95}\right)1{,}000 = 14.86\,m$$

2.4 MANOMETRY

A standard technique for pressure measurements is to use the manometer. This device comes in many configurations, for example, an inclined tube manometer, a U-tube manometer, and a piezometer tube. The fluid inside the tube remains static for a while. Once the pressure is balanced, we can evaluate the pressure difference using Pascal's law. If the fluid inside the manometer is a homogeneous fluid, then we can estimate the pressure with elevation as

$$p = \gamma h + p_o$$

where p_o is the reference pressure and h is the vertical distance.

Figure 2.6 shows a manometer. Here, the line RQ is an isobaric plane. Therefore, the pressure from station 1 in pipe till isobaric plane should be the same as the pressure from station 2 in pipe till isobaric plane RQ. Here Δh and y are the vertical distances from the pipe to the manometer.

$$p_1 + \rho_A \cdot g \cdot (\Delta h + y) = p_2 + \rho_A g \cdot y + \rho_B \cdot g \cdot \Delta h$$

$$p_1 - p_2 = (\rho_B - \rho_A) \cdot g \cdot \Delta h$$

$$\left(\frac{p_1 - p_2}{g\rho_A}\right) = \left[\left(\frac{\rho_B}{\rho_A}\right) - 1\right] \cdot \Delta h$$

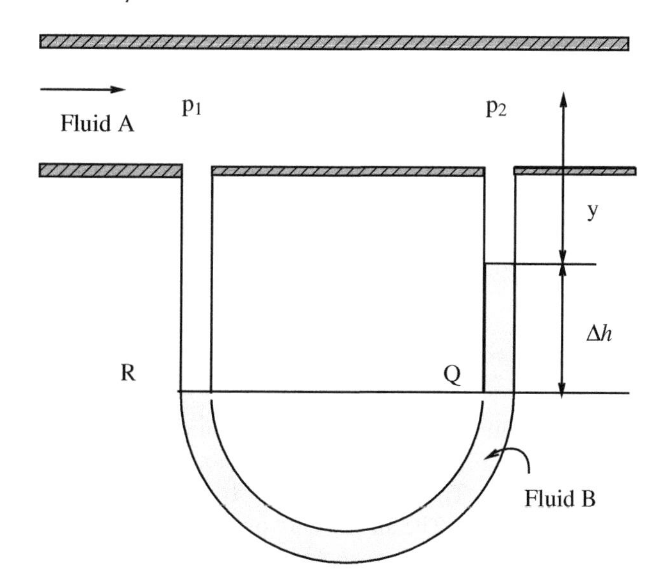

Figure 2.6 The manometer measuring the pressure difference as fluid A flows through the pipe.

Example 2.4

Example An inclined manometer measuring the pressure at bulb B is connected with a vessel having a manometer liquid of density $\rho_2 = 10,020$ kg/m^3 (see Figure 2.7). Over the manometer liquid, there is another lighter density liquid with $\rho_1 = 670\ \mathrm{kg/m^3}$. The thickness of manometer liquids is $h_1 = 20$ cm and $h_2 = 90$ cm, respectively. Find the angle θ of the inclined manometer if the atmospheric pressure is 1 atm. The length of fluid ℓ that rises into the tube is 0.8 m.

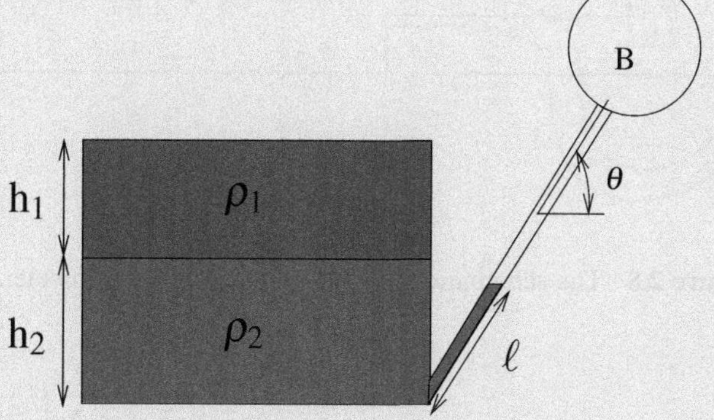

Figure 2.7 The schematic diagram of an inclined manometer.

Solution We balance the pressure at the base of the vessel as

$$p_{atm} + 9.81 \cdot (\rho_1 \cdot h_1 + \rho_2 \cdot h_2) = p_B + \ell \cdot 9.81 \cdot \rho_2 \cdot \sin(\theta)$$

$$\theta = \sin^{-1}\left[\frac{(p_{atm} + 9.81 \cdot (\rho_1 \cdot h_1 + \rho_2 \cdot h_2) - p_B)}{\ell \cdot 9.81 \cdot \rho_2}\right]$$

$$\theta = 0.883\ rad$$

Example 2.5

Example A pressurized tank contains an organic fluid of specific weight $\gamma =$ 50 lb_f/ft^3, as shown in Figure 2.8. A U-tube manometer is used to measure the pressure of the tank. The vertical distance from the free surface to the base of the tank is $z = 5$ ft. The air pressure over the liquid is maintained at 2 psia. The specific gravity of the manometer liquid is 3.0. Find the pressure at the base of the tank and the value of the manometer differential reading h.

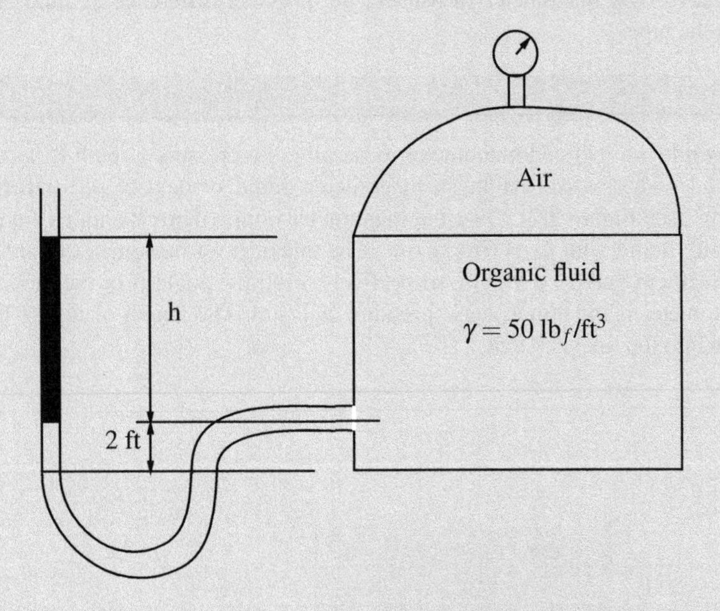

Figure 2.8 The schematic diagram of the U-tube manometer.

Solution

$$p_{base} = \gamma_{organic\ fluid}\, z + p_{air} = 50 lb_f/ft^3 \times 5 ft + 2 psia \times 144 \frac{in^2}{ft^2} = 538\ lb_f/ft^2$$

Since the pressure at the isobaric plane is the same, we can equate the pressures as:

$$P_{base} = P_{isobaric\ plane} = \gamma_{organic} \times 2ft + SG \times \gamma_w \times h$$

where SG is the specific gravity of the organic fluid.

$$P_{base} = 50lb_f/ft^3 \times 2ft + 3 \times 62.4 \times h$$

$$538\ lb_f/ft^2 = 50lb_f/ft^3 \times 2ft + 3 \times 62.4 \times h$$

$$h = 2.33\ ft$$

2.5 ATMOSPHERIC AIR PRESSURE

In many applications, the density will vary with the altitude. This is true for air as well. At sea level, air can be treated as an ideal gas, and the density can be obtained from the ideal gas relation:

$$p = \rho RT \tag{2.5}$$

where R is a gas constant and T is the absolute temperature. The ideal gas model is valid as long as the gases are not close to critical point and they are in a superheated state.

In the US standard atmosphere, the temperature decreases linearly with altitude up to an elevation of 10 km. This variation of temperature can be modeled as any equation of a straight line as:

$$T = T_o - \lambda_s \cdot y \tag{2.6}$$

where T_o is the temperature at sea level and λ_s is the slope of the line, called the temperature lapse rate. If p_o is pressure at sea level, then from Pascal's law and ideal gas relation we have

$$dp = -\rho g dy = -\frac{pg}{RT}dy = -\frac{gp}{R(T_o - \lambda_s \cdot y)}dy$$

$$\frac{dp}{p} = -\frac{g}{R(T_o - \lambda_s \cdot y)}dy$$

$$\int_{p_o}^{p} \frac{dp}{p} = -\int_0^y \frac{g}{R(T_o - \lambda_s \cdot y)}dy$$

$$ln\left(\frac{p}{p_o}\right) = ln\left(1 - \frac{\lambda_s y}{T_o}\right)^{g/\lambda_s \cdot R}$$

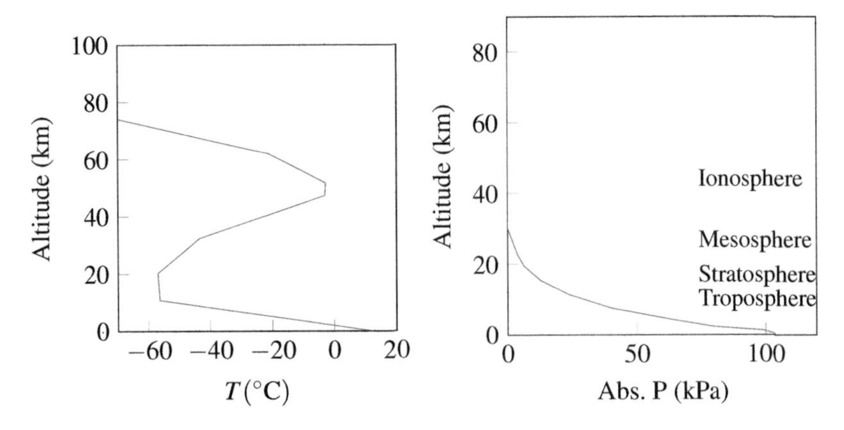

Figure 2.9 Temperature and pressure variation with altitude in US standard atmosphere.

Figure 2.10 A blimp can go to an altitude of 1,000 to 7,000 ft (305 to 2,135 m) and a hot air balloon can go as high as 3,000 ft (914 m).

$$ln\left(\frac{p}{p_o}\right) = \frac{g}{\lambda_s \cdot R} ln\left(1 - \frac{\lambda_s y}{T_o}\right)$$

Figure 2.9 shows the standard atmospheric pressure variation with elevation. A hot air balloon can reach up to 3,000 ft, whereas, a blimp can reach as high as 7,000 ft (Figure 2.10).

Example 2.6

Example According to the standard model of the atmosphere adopted by the International Civil Aviation Organization (ICAO), the temperature lapse rate of 6.5°C/km from the surface to the tropopause (at a height of 11 km) can be calculated. Show that this corresponds to law:

$$\left(\frac{p}{p_o}\right)=\left(\frac{\rho}{\rho_o}\right)^{1.235}$$

Also, find the pressure at a height of 3,000 m in the atmosphere taking datum conditions at 15°C and 101 kN/m^2.

Solution We consider the index as n and let the law be:

$$\left(\frac{p}{p_o}\right)=\left(\frac{\rho}{\rho_o}\right)^{n}$$

From ideal gas law we replace density and the equation will take the form:

$$\left(\frac{p}{p_o}\right)=\left(\frac{p/RT}{p_o/RT_o}\right)^{n}=\left(\frac{T_o p}{p_o T}\right)^{n}$$

$$\left(\frac{T_o}{T}\right)^{n}=\left(\frac{p}{p_o}\right)\left(\frac{p_o}{p}\right)^{n}=\left(\frac{p_o}{p}\right)^{n-1}$$

$$\left(\frac{T_o}{T}\right)=\left(\frac{p_o}{p}\right)^{\frac{n-1}{n}}$$

$$\left(\frac{T}{T_o}\right)=\left(\frac{p}{p_o}\right)^{\frac{n-1}{n}}$$

In the atmosphere, both pressure and temperature vary with altitude (h). In the atmosphere, the pressure varies as:

$$p=\rho g h$$

Differentiating it with altitude and replacing density with the ideal gas law gives:

$$\frac{dp}{dh}=\rho g=\left(\frac{p}{RT}\right)g$$

$$\frac{dT}{dh}=\left(\frac{n-1}{n}\right)\left(\frac{p^{-1}}{p_o{}^{(n-1)}}\right)^{1/n}\left(\frac{T_o}{T}\right)\left(\frac{pg}{R}\right)$$

For air, $R=287\ \text{kJ/kg}\cdot\text{K}$ and lapse rate dT/dh is 6.5/1,000, so we have:

$$\frac{-6.5}{1000}=\left(\frac{n-1}{n}\right)\left(\frac{9.81}{287}\right)$$

This gives $n = 1.235$.
At height of 3,000 m,

$$T = 288 - 19.5 = 268.5K$$

$$p = p_o\left(\frac{T}{T_o}\right)^{\frac{n}{n-1}}$$

$$p = 101\left(\frac{268.5K}{288}\right)^{\frac{1.235}{1.235-1}} = 69.9kN/m^2 Ans.$$

Table 2.1
Conversion of Some Common Pressure Units

Units	Pascal	Torr	mm water gauge
1 bar	10^5	750.06	10,197.2
1 atm	1.01325×10^5	760	10,332.3
1 mm WS	9.80665	0.073556	-
1 psi	6894.74	51.7148	703.068
1 psf	47.8802	0.35913	48.824
in. Hg	3386.39	25.4	345.316
in. WG	249.09	1.86832	25.4

The water gauge unit is usually in centimeters of water or millimeters of water (in German mm Wassersäule), and is frequently used to measure the pressure inside human or animal bodies, like the central venous pressure, the intracranial pressure, and the cerebrospinal fluid. It is also used for reporting pressures encountered in ventilation systems and water supply networks. It is also a commonly used unit to report pressure in the speech sciences.

Conversion of some common pressure units are listed in Table 2.1.

Example 2.7

Example Natural gas is being supplied to a facility through a pipe at a supply pressure that is reported as 8 iwg. How much is the supplied gas pressure?
Solution The gas supplied pressure is reported in water head units. Note that the inches of water is a non-SI unit for pressure that is frequently used in industry. The inches of water gauge is sometimes written as iwg or in.w.g.

Here,

$$h = 8\ inch = 0.2032\ m$$

We know that

$$p = \rho g h = \rho_w g (0.2032\ m)$$

Taking water density as 1,000 kg/m^3 we have the supply line pressure as

$$p = 1{,}000 \times 9.81 \times (0.2032\ m) = 1{,}993.39\ Pa = 1.99\ kPa$$

2.6 STATIC LIQUID FORCE ON AN INCLINED SURFACE

Figure 2.11 shows the forces that arise due to the pressure that acts on an irregularly shaped, submerged, inclined surface. We analyze this force by taking an

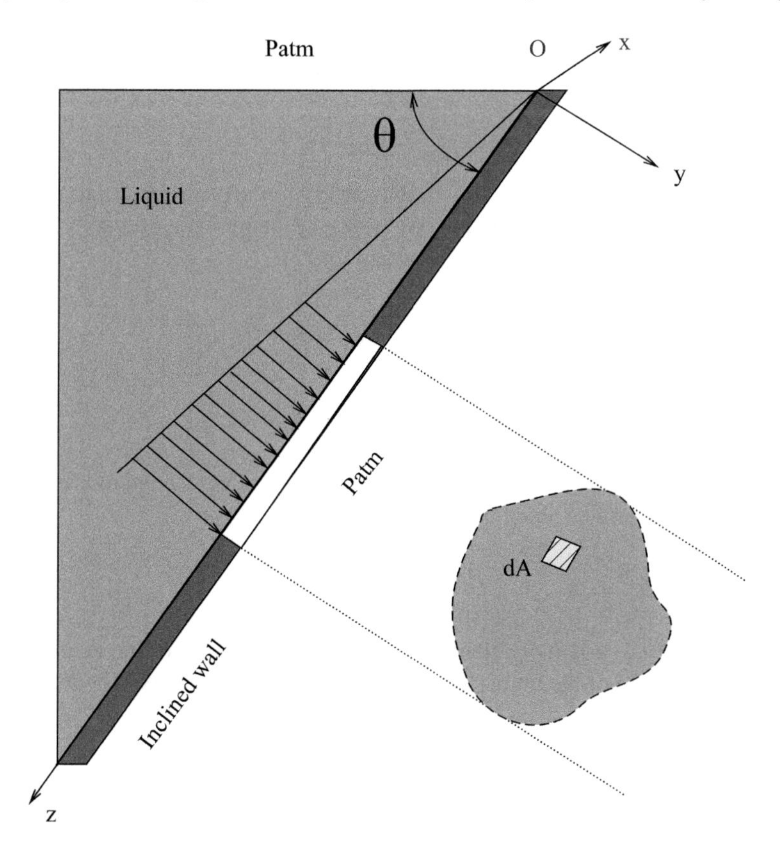

Figure 2.11 The pressure force acting on an irregularly shaped, submerged, inclined surface.

infinitesimally small area, dA. The force acting on this area is dF.

$$dF = pdA$$

The vertical hydrostatic pressure acting on the small area is the sine component

$$p = \rho gz\sin(\theta)$$

The force element that acts on the area dA is

$$dF = \rho gz\sin(\theta)\cdot dA$$

The integration gives us the total force exerted by liquid on the submerged surface

$$F = \rho g\sin(\theta)\int z\cdot dA$$

The centroid of the plane area can be estimated as

$$z_{cen} = \frac{1}{A}\int zdA$$

$$F = z_{cen}\cdot\rho g\sin(\theta)\cdot A$$

Here z_{cen} is the distance from the liquid surface to the centroid of the submerged surface as measured in the z-direction. We now take the moment of this force around position O to find the position of the resultant force

$$M_o = \int z\cdot p\cdot dA = \int z\left[\rho gz\sin(\theta)\right]dA$$

$$M_o = \left[\rho g\sin(\theta)\right]\int z^2 dA$$

The integral in the above equation can be identified as the second moment of area of the submerged surface about the x-axis.

$$I_{xx} = \int z^2 dA$$

Instead of using I_{xx}, it is more advantageous to use a moment of the area that goes through the centroid of the immersed surface, using the parallel axis theorem:

$$I_{xx,O} = I_{xx,cen} + A\cdot z_{cen}^2$$

where z_{cen} is the distance, in the z-direction, to the centroid of the submerged surface's portion as measured from the liquid's free surface.

$$z_R = \frac{\gamma\sin(\theta)\left(I_{xx,cen} + Az_{cen}^2\right)}{\gamma\sin(\theta)Az_{cen}}$$

where γ is the specific weight of the liquid in units of N/m^3 or lb$_f$/m^3 or similar units.

$$z_R = z_{cen} + \frac{I_{xx,cen}}{A \cdot z_{cen}}$$

This equation gives the location of that force that is acting vertically on the submerged surface from the free surface of the liquid.

We can also formulate an equation for the lateral force that is acting on the submerged surface as

$$x_R = x_{cen} + \frac{I_{xz,cen}}{z_{cen} \cdot A}$$

Center of pressure: The point of application of pressure distributed over a surface.
Center of gravity: The point where weight is acting. Center of gravity may also be located outside the body.
Centroid: The geometric center of a plane surface located at the arithmetic mean position.

Figure 2.12 shows a water dam. The hydrostatic pressure is greatest at the base of the dam.

Figure 2.13 shows $I_{xx,cen}$ for a few common areas.

Figure 2.12 Water reservoir and dams have walls that can withstand hydrostatic pressure.

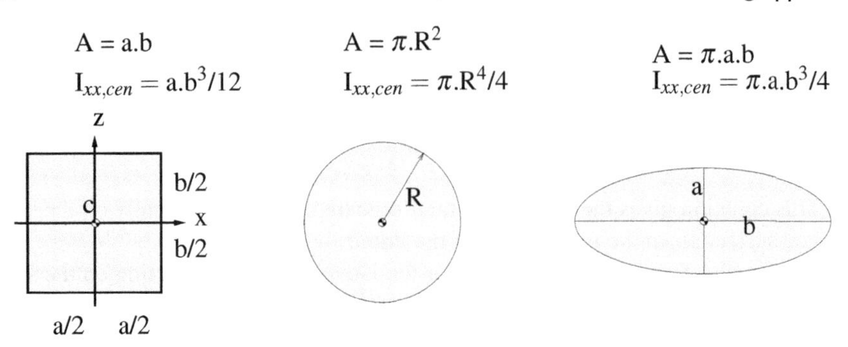

Figure 2.13 The centroid and the centroidal moment of inertia for three common areas.

Example 2.8

Example: Water is being poured into and drained from a vessel such that the level of the water is maintained constant, as shown in Figure 2.14. How much is the pressure force on the inclined wall at an angle of $\theta = 30$ degrees. The length of the inclined surface is 0.5 m and the width is 0.6 m. The depth of fluid from the free surface to the centroid of the area y_1 is 0.7 m.

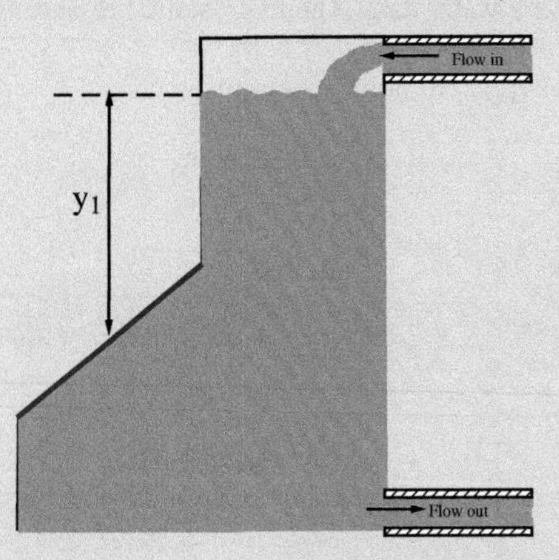

Figure 2.14 The schematic diagram of Example 2.8.

Solution The inclined wall is a rectangular area

$$A = a \cdot b = 0.5 \times 0.6 = 0.3\ m^2$$

The moment of the area is

$$I_{xxc} = \frac{a \cdot b^3}{12} = 0.009\ m^4$$

The centroid of the plane area can be estimated as

$$z_{cen} = \frac{y_1}{\sin(\theta)} = \frac{0.7}{\sin(30)} = 1.4\,m$$

The location of force that is acting vertically on the submerged surface from the free surface of the liquid.

$$z_R = z_{cen} + \frac{I_{xxc}}{A} = 1.43\,m$$

The force acting on the submerged surface is

$$F = \gamma \cdot z_{cen}(A\sin(\theta))$$

$$F = 1000 \cdot 9.81 \cdot 1.4 \cdot 0.3 \cdot \sin(30) = 2060.1\,N$$

2.7 NORMAL STRESSES IN STATIC FLUID

The normal stresses in flow are $\sigma_{xx}, \sigma_{yy}, \sigma_{zz}$, which act normally to the surface. The sum of all normal stresses is also the trace of stress field, and it is independent of direction and should be a scalar term. Therefore the normal stresses summation at a point is considered to be equal to the thermodynamic pressure of the system.

$$p = -\tilde{\sigma} = \frac{-1}{3}\left[\sigma_{xx} + \sigma_{yy} + \sigma_{zz}\right]$$

This relationship is valid for both liquids and gases as long as states are not close to a critical state.

The types of forces that do not require a physical contact and generate due to some field are called the body forces. Examples of such forces are the forces that arise due to gravitational and magnetic fields. The magnetic field around the Earth is influenced by the solar wind effect. Earth is not a perfectly spherical planet, nor is its density uniform. This asymmetric distribution generates a gravitational pull that varies from place to place. In the solar system, Jupiter's gravitational pull is highest, at around 24.79 m/s^2.

2.8 BOUYANCY FORCE IN FLUID

Eureka! (I found it!) Many books on writings of Archimedes (c. 287–212 BC), a scientist who lived in the city of Syracuse in Sicily, have mentioned the story that King Hiero II of Syracuse (ruled from 270 to 215 BC) had ordered a crown to be made from pure gold ($\rho = 19,300$ kg/m^3). However, when the crown was delivered, King Hiero got suspicious, and he gave the crown to Archimedes to investigate the purity of gold in the crown. While Archimedes laid himself in the bathtub with the crown pondering over the issue, he realized that the weight of water displaced should be equal to the crown's weight. Hence, he could measure the weight of the counterfeit crown with the original weight of pure gold and estimate the non-gold metal's weight. Archimedes was so delighted that he came out on the street saying Eureka! (I found it). Later Archimedes wrote:

> Any solid lighter than a fluid will, if placed in the fluid, be so far immersed that the weight of the solid will be equal to the weight of the fluid displaced.
>
> -On Floating Bodies, Book 1, Proposition 5, by Archimedes

Mathematically, the weight of water spilled out of the bathtub is equal to the buoyant force on the object:

$$W_{solid} = W_{liquid}$$
$$m_{solid} \cdot g = m_{liquid} \cdot g$$
$$W_{solid} = m_{solid} \cdot g = \rho_{liquid} \cdot V_{liquid} \cdot g$$
$$W_{crown} = \rho_{water} \cdot \underbrace{V_{water}}_{measurable} \cdot g$$

Archimedes, through his experiments, found that

$$W_{crown} \ll W_{pure\ gold}$$

and hence bad luck for the goldsmith.

Dimensionally, the product of the terms $\rho_{liquid} \cdot V_{liquid} \cdot g$ is a force, and this force was not named until around 1570s, when the word *buoyancy* was adapted in English from either Spanish or Dutch languages. Scientists have given this force a name. In fluid mechanics literature, this force is now called the Buoyant force. The phenomenon that an upward force act on object that is in proportion to its volume is called buoyancy.

Buoyancy force is a function of geometry. It acts on humans and objects in the air. However, as the weight of the human body is more than the buoyancy force acting on it, it is normally ignored in the physical analysis. Also, the buoyancy force functions to the density of the surrounding fluid, and as air density is very small ($\rho = 1.22$ kg/m^3), the buoyancy force exerted by air is negligible. On the other hand, helium balloons rise due to buoyancy, until they reach an altitude where the air density equals the helium density in the balloon.

The principle of buoyancy is also used in a device called a hydrometer, which measures the density or specific gravity of liquids. A hydrometer consists of a glass bulb partly filled with a lead shot and a graduated stem. Once placed in liquid, the hydrometer will float, depending on the density of the liquid. Part of the graduated stem will be outside of the liquid. The stem provides the direct measurement of the specific weight or gravity of the liquid.

For a ready reference, we have listed the volumes of some common objects in Table 2.2.

Table 2.2
Volume of Some Common Shapes

Name	Shape	Volume
Cube	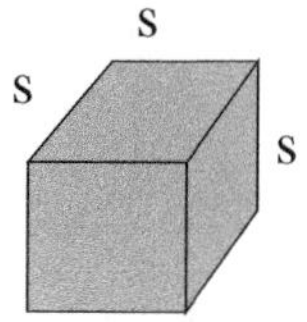 s s s	$V_{cube} = s^3$ $c_{g,cube} = s/2$
Sphere	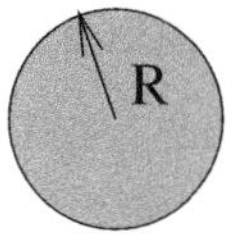 R	$V_{sphere} = \frac{4\pi}{3} R^3$ $c_{g,sphere} = R$
Ellipsoid	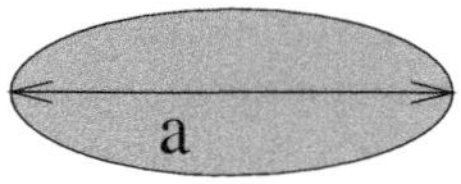 a b, c widths	$V_{ellipsoid} = \frac{4\pi}{3} a.b.c$
Prism	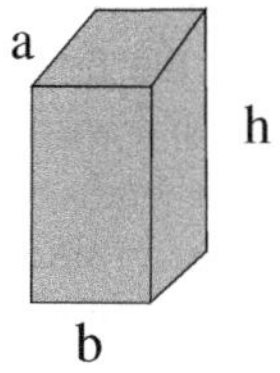 a h b	$V_{prism} = a.b.h$ $c_{g,prism} = h/2$
Cone	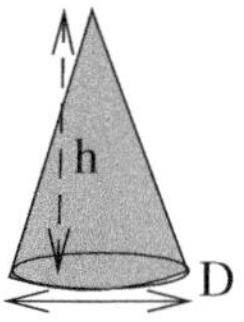 h D	$V_{cone} = \frac{\pi.h.D^2}{12}$ $c_{g,cone} = h/4$

Figure 2.15 Container ships have stability issues in high tides.

2.9 STABILITY OF FLOATING OBJECTS

The stability condition for floating bodies is different from that for fully submerged bodies. A container ships' stability is a very important issue in sea transportation. Due to ocean waves of varied types and the size of container ships, stability issues cannot be ignored. They increase the risk associated with sea transportation and associated insurance costs (see Figure 2.15).

Figure 2.16 shows an object with weight W floating over a liquid partially submerged into it. A vertical line, the objects' vertical axis, runs through these points. Figure 2.16(a) shows the floating body at its equilibrium orientation, and the center of gravity (cg) is above the center of buoyancy (cb). The center of gravity is the average location of the weight of an object. Figure 2.16(b) shows that if the body is rotated slightly and submerged more into the liquid for some reason, the center of buoyancy would shift to a new location. The buoyant force and the weight generate a righting couple that tends to return the body to its original orientation. Thus, the body is stable. This is also the case in Figure 2.16(c). The metacenter (Mc) is the intersection of the vertical axis of a body when in its equilibrium position and the vertical line through the new center of buoyancy (cb) when the object is rotated. A floating object is stable as long as its center of gravity is below the metacenter. GB is the distance from cg to cb.

The distance from the center of buoyancy to the metacenter is called ζ_s and is calculated from

$$\zeta_s = \frac{I_{LM}}{\forall_{dis}}$$

where I_{LM} is the area of least moment of inertia of a horizontal section of an object taken at the surface of the fluid; $\forall$ is the volume of liquid displaced by an object.

The criterion for stability is

$$\frac{I_{LM}}{\forall_{dis}} > GB \quad (stable)$$

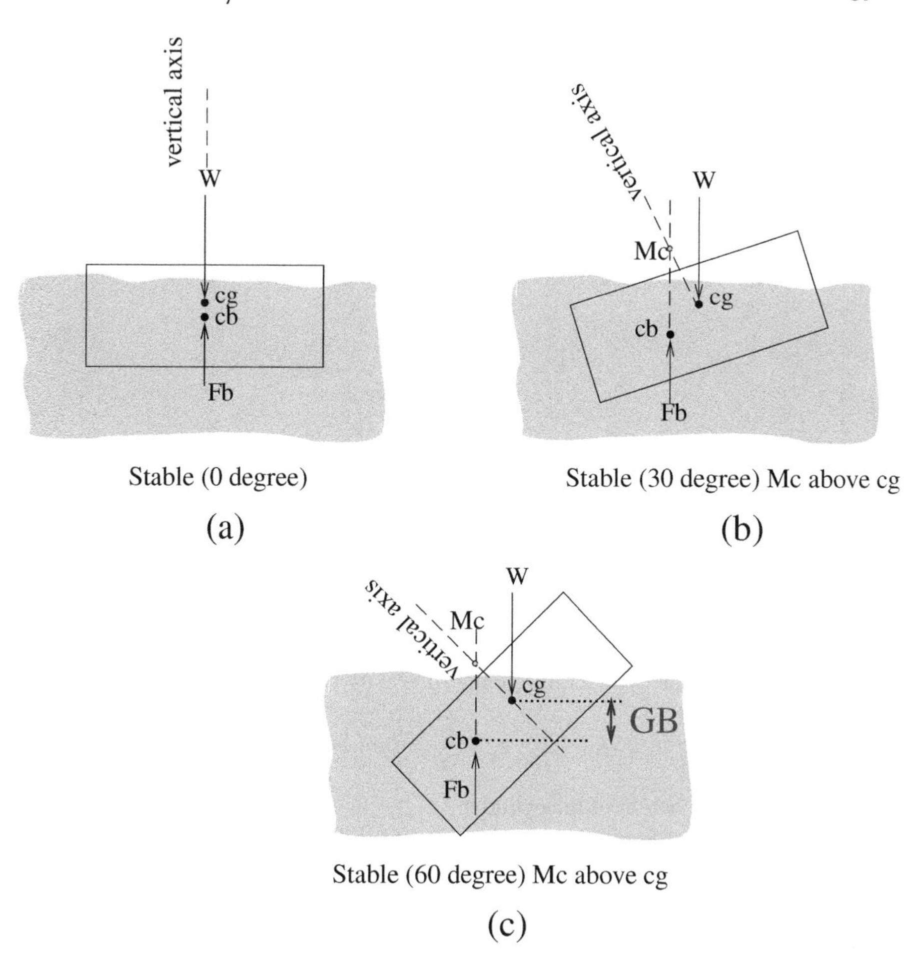

Figure 2.16 Location of center of buoyancy and center of gravity.

$$\frac{I_{LM}}{\forall_{dis}} = GB \quad (neutral)$$

$$\frac{I_{LM}}{\forall_{dis}} < GB \quad (unstable)$$

Example 2.9

Example A styrofoam cylinder of diameter 5 m and total height of 4 m is floating in seawater vertically, as shown in Figure 2.17. The length of cylinder inside the water is 3.3 m. The density of styrofoam and seawater is 1,050 kg/m^3 and 1,025 kg/m^3, respectively. Find the metacenter location with reference to the center of buoyancy, if a gust of wind tilted the cylinder. Also, find the weight and the buoyant force acting on this cylinder.

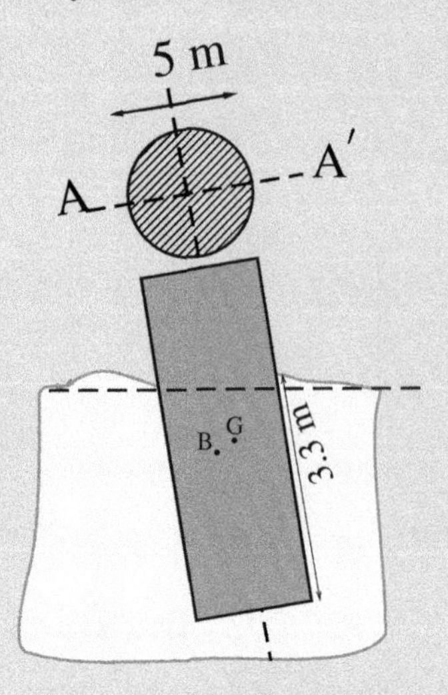

Figure 2.17 The schematic diagram of Example 2.9.

Solution For the cylinder, the center of gravity is located at

$$c_g = \frac{h}{2} = 2m$$

where h is the height of the cylinder.

The center of buoyancy is approximated to be the same as the center of gravity of the submerged volume of cylinder

$$c_b = \frac{h_{sm}}{2} = \frac{3.3}{2} = 1.65m$$

where h_{sm} is the height of submerged volume. Therefore, the distance GB is

$$GB = c_g - c_b = 2 - 1.65 = 0.35m$$

The moment of inertia is

$$I_{LM} = \frac{\pi}{64} \cdot d^4 = 30.664$$

and the submerged volume of cylinder is

$$\forall_{sm} = \frac{\pi}{4} \cdot d^2 \cdot h_{sm} = 64.76$$

Note that this submerged volume is less than the total geometric volume of the cylinder.

$$\frac{I_{LM}}{\forall_{sm}} = 0.473m$$

As 0.473 > 0.35, this shows that the floating cylinder is in a state of stable equilibrium.

Using the density of styrofoam and total volume of the object, we compute the weight of an object:

$$W = \rho_{st} \cdot g \cdot \forall_T = 1050 kg/m^3 \times 9.81 m/s^2 \times 78.5 m^3 = 808.5 kN$$

Using the density of seawater and the submerged volume of the object we compute the buoyant force on an object:

$$F_b = \rho_w \cdot g \cdot \forall_{sm} = 1025 kg/m^3 \times 9.81 m/s^2 \times 64.76 m^3 = 651.2 kN$$

PROBLEMS

2P-1 A spherical-shaped weather balloon was released by a meteorological department into the atmosphere. The diameter of the balloon is 1.5 m and the total mass is 1.2 kg. If the temperature lapse rate in the atmosphere is 0.0065 K $\cdot$ m^{-1} determine how high the balloon will rise above sea level. Atmospheric temperature and pressure at sea level are 15°C and 101 kPa respectively. Take $R = 287$ J/kg$\cdot$K for air.

2P-2 A submarine can go up to 300 m deep in sea. The sea water density is 1,023 kg/m^3. The submarine can be approximated as an ellipsoid of length $a = 175$ m, and widths $b = c = 30$ m. The mass is 17,000 tons. Find the net force this submarine will experience.

$$V_{ellipsoid} = \frac{4\pi}{3} a.b.c$$

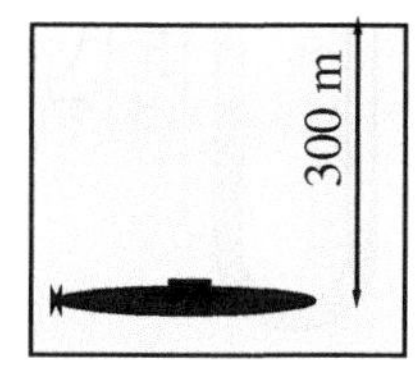

Figure 2.18 Figure of problem 2P-2.

2P-3 Human blood pressure is normally reported as 120/80, where 120 is the maximum systolic pressure in mm of Hg and 80 is the minimum diastolic pressure in mm of Hg. These units are still used in electronic blood pressure measurement apparatus, as traditionally blood pressure was measured by a mercury manometer. Convert these pressures into units of Pascal and bar. The specific weight of mercury is 133 kN/m^3.

2P-4 A hydrometer weighs 20 mN and has a cylindrical stem of 3 mm diameter. Find the level difference h when a hydrometer is placed in fluids of two different specific gravities or relative densities, as shown in Figure 2.19.

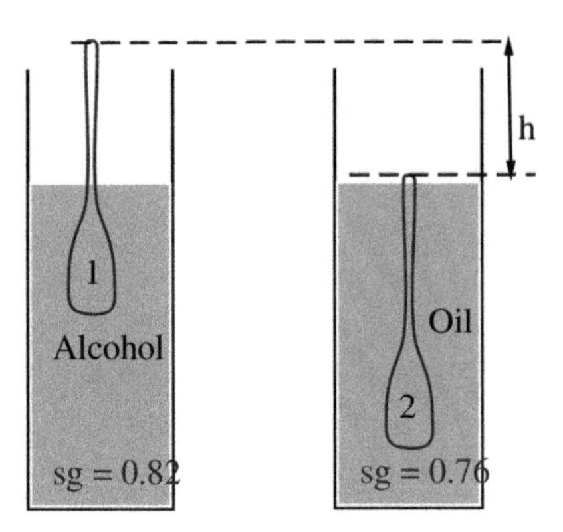

Figure 2.19 Figure of problem 2P-4.

2P-5 Bulb A and bulb B contain water under the pressure of 3.29 bar and 1.24 bar, respectively, as shown in Figure 2.20. Find the deflection of the mercury, h, in the differential gauge.

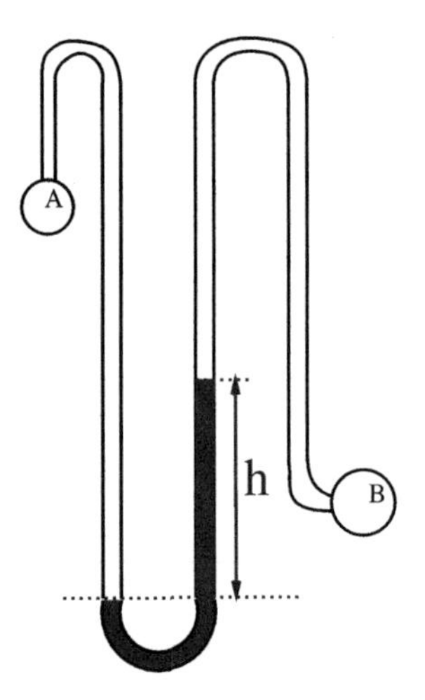

Figure 2.20 Figure of problem 2P-5.

2P-6 A rectangular gate of length $\ell = 2$ m and width $w = 2$ m is connected at A and is held in place by a block positioned at the base. The block is exerting a horizontal force of $R_x = 60$ kN. The weight of the gate is 5 kN. Find the angle θ of the depth of the water $h = 2$ m. Assume that there is no vertical force exerted by the gate.

2P-7 A tank of rectangular cross-section is filled up with water up to a height of 3.5 m. At a position of 0.6 m from the base of the tank, a U-tube manometer is connected to the tank, which is showing the manometer reading 0.25 m below the connection. Find the manometer difference if specific gravity of manometer fluid is 1.7.

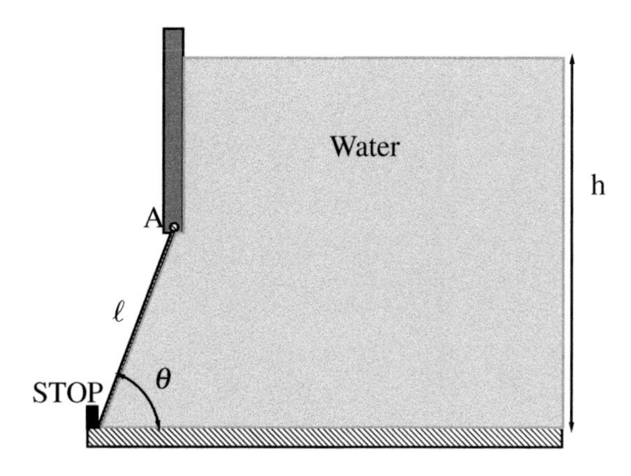

Figure 2.21 Figure of problem 2P-6.

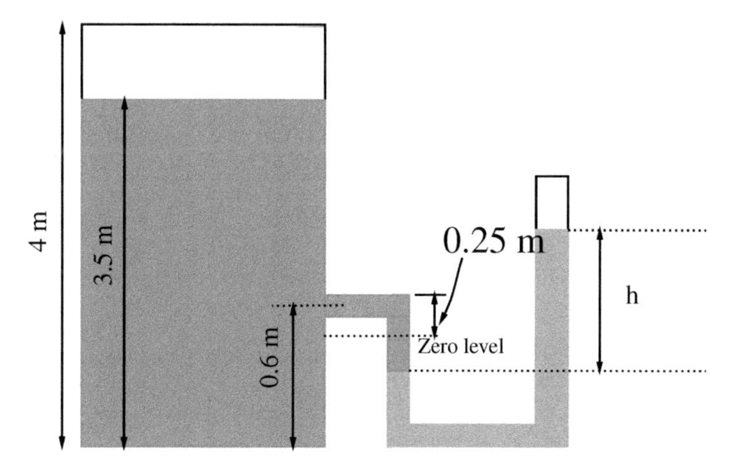

Figure 2.22 Figure of problem 2P-7.

2P-8 Find the atmospheric density and pressure at a height of 1,000 m. The pressure is 760 mm of mercury and the temperature is 25° at sea level. Assume a uniform decrease of temperature at rate of 4.94° per 1,000 m.

REFERENCES

R. L. Mott, Applied Fluid Mechanics, Prentice Hall, Upper Saddle River, NJ, USA, 2006

R. H. Sabersky, E. G. Hauptman and A. J. Acosta, Fluid Flow: A First Course in Fluid Mechanics, Prentice Hall PTR, USA, 1989.

W. S. Janna, Introduction to Fluid Mechanics, Sixth Edition, 6th Edition, CRC Press, USA, 2020.

3 Kinematics of Fluid Particle

In classical mechanics, the kinematics of the moving body are described by the displacement, velocity, and acceleration. We will introduce the concept of Lagrangian and Eulerian descriptions of the flow field. The flow trajectories known as pathlines and the streaklines will be introduced. The mathematical description of the deformation of a fluid particle is presented.

Learning outcomes: After finishing this chapter, you should be able to:

- Understand the significance of Eulerian and Lagrangian descriptions of the flow field.
- Understand the pathline and streakline used for describing the fluid particles trajectories.
- Derive mathematical expressions for normal and shear strains.

3.1 LAGRANGIAN AND EULERIAN DESCRIPTIONS OF FLOW FIELD

A particle's motion can be described by giving the three velocity components, density, temperature, pressure, at every point in space as a function of time. In classical dynamics, we have two kinds of approaches to describe the motion for a moving particle, whether solid or fluid. One approach is called the Lagrangian description, or material description, named after **Joseph Louis Lagrange** (1736–1836). The another approach is called the Eulerian description, or spatial description, named after **Leonard Euler** (1707–1783). In the Lagrangian description, the observer is moving with the particle, whereas in the Eulerian approach, the observer is stationary and watching a station. These two approaches do not depend on the flow steadiness or unsteadiness. Both approaches are used in fluid dynamics, depending on the nature of the problem. For a simplified example, you can relate these two approaches to the golf tournament. If you are following your favorite player, then such an approach is called the Lagrangian approach. If you view the whole flow field then such an approach is the Eulerian approach. Also, say you want to measure the temperature of a moving river. If you were sitting in a boat and the boat was moving at the river speed and you placed the probe in the water then you have recorded the temperature of a moving stream; this is Lagrangian approach. However, if you have recorded the temperature just at a single location then you have measured the temperature of several fluid particles that have touched the probe or thermometer; this is the Eulerian approach.

In this book, the Eulerian description is used to describe the motion.

DOI: 10.1201/9781003315117-3

The difficulty with the Lagrangian approach is mainly due to the description of the three displacement components in x, y, z direction, which shall be used in the velocity description. Since the observer is moving with a particle, the displacements cannot be fixed but must continuously vary in such a manner as to always be able to locate a particle. In this description the displacements are a function of the time, hence velocities are expressed as:

$$\vec{V} = f[x(t), y(t), z(t), t]$$

Particle tracking velocimetry (PTV) is a Lagrangian approach toward the measurements. In this technique, the particles are seeded into flow and are illuminated by a laser sheet. Each of these particles are then tracked for several frames. In contrast to PTV, with the particle image velocimetry (PIV) technique, we measure the velocity of the flow fluid as it passes the observation point, which is fixed in space. Thus this measurement technique is based on the Eulerian approach.

Steady flow: A flow is steady if in the Eulerian description the following derivatives are zero or independent of time:

$$\frac{\partial u}{\partial t}, \frac{\partial v}{\partial t}, \frac{\partial w}{\partial t}, \frac{\partial p}{\partial t}, \frac{\partial \rho}{\partial t}, \frac{\partial T}{\partial t}$$

where u, v and w are three velocity components and ρ, p, and T are density, pressure, and temperature, respectively.

Example 3.1

Example Which of the following velocities are one-, two-, and three-dimensional? Also identify the steady and unsteady flows

(i) $\vec{V} = b \cdot x\,\hat{i} + [a \cdot \exp(b \cdot x)]\,\hat{j}$

(ii) $\vec{V} = b \cdot x\,\hat{i} + [a \cdot y]\,\hat{j}$

(iii) $\vec{V} = b \cdot x\,\hat{i} + \exp(c \cdot y)\hat{j} + d \cdot z \cdot t\,\hat{k}$

(iv) $\vec{V} = 2 \cdot x\,\hat{i} + a \cdot y\exp(c \cdot t)\hat{j} + d \cdot z\,\hat{k}$

where a, b, c, d are constants.

Solution

(i) $\vec{V} = b \cdot x\,\hat{i} + a \cdot \exp(b \cdot x)\,\hat{j}$

The velocity vector is a function of x coordinate, so this flow is one-dimensional. The velocity vector is independent of time.

(ii) $\vec{V} = b \cdot x\,\hat{i} + a \cdot y\,\hat{j}$

The velocity vector is a function of x and y coordinates, so this flow is two-dimensional. The velocity vector is independent of time.

(iii) $\vec{V} = b \cdot x\,\hat{i} + \exp(c \cdot y)\hat{j} + d \cdot z \cdot t\,\hat{k}$

The velocity vector is a function of x, y, and z coordinates, so this flow is three-dimensional. The velocity vector is dependent on time, so flow is unsteady.

(iv) $\vec{V} = 2 \cdot x\,\hat{i} + a \cdot y\,\exp(c \cdot t)\hat{j} + d \cdot z\,\hat{k}$

The velocity vector is a function of x, y, and z coordinates, so this flow is three-dimensional. The velocity vector is dependent on time, so flow is unsteady.

Example 3.2

Example The velocity field of a flow is given by $\vec{V} = (4y + 10)\hat{i} + (2x - 7)\hat{j} + 3z\hat{k}$ ft/s, where x, y, and z are in feet. Determine the fluid speed at the origin ($x = y = z = 0$) and on the y axis ($x = z = 0$)

Solution The velocity components are:

$$u = 4y + 10$$

$$v = 2x - 7$$

$$w = 3z$$

The magnitude of resultant velocity:

$$\vec{V} = \sqrt{(u^2 + v^2 + w^2)}$$

$$\vec{V} = \sqrt{16y^2 + x^2 - 14x + 49 + 9z^2}$$

At origin ($x = y = z = 0$), we have $\vec{V} = 7$.

At y-axis ($x = z = 0$), we have:

$$\vec{V} = \sqrt{16y^2 + 80y + 149}$$

3.2 ACCELERATION IN FLUID

In mechanics, the variables that describe the motion of a particle are called kinematics. The variables like displacement, velocity, and acceleration describe the motion of the fluid particle in fluid mechanics. In Cartesian coordinates, if u, v, and w are the component of velocities in x, y, and z directions, respectively, then the velocity vector $\vec{V}$ is defined as $\vec{V} = u\hat{i} + v\hat{j} + w\hat{k}$.

The acceleration is defined as:

$$a_x = \lim_{\Delta t \to 0} \frac{u(x+u.\Delta t, y+v.\Delta t, z+w.\Delta t, t+\Delta t) - u(x,y,z,t)}{\Delta t}$$

$$a_x = \frac{\partial u}{\partial t} + u\frac{\partial u}{\partial x} + v\frac{\partial u}{\partial y} + w\frac{\partial u}{\partial z}$$

and similarly we can define:

$$a_y = \frac{\partial v}{\partial t} + u\frac{\partial v}{\partial x} + v\frac{\partial v}{\partial y} + w\frac{\partial v}{\partial z}$$

$$a_z = \frac{\partial w}{\partial t} + u\frac{\partial w}{\partial x} + v\frac{\partial w}{\partial y} + w\frac{\partial w}{\partial z}$$

in vector notation we may write:

$$\vec{a} = \underbrace{\frac{\partial \vec{V}}{\partial t}}_{1} + \underbrace{\vec{V}.\nabla\vec{V}}_{2}$$

The term 1 in above equation is called the temporal acceleration, and the term 2 is called the convective acceleration, or advective acceleration. The acceleration a thus has the cumulative influence of both space and time change of velocity in it and it is customary to write it as follows:

$$a_x = \frac{Du}{Dt} = \frac{\partial u}{\partial t} + u\frac{\partial u}{\partial x} + v\frac{\partial u}{\partial y} + w\frac{\partial u}{\partial z}$$

The operator D/Dt is called the material derivative or substantial derivative, or the derivative following the particle. In algebraic notation the material derivative can be written as:

$$\frac{Dq}{Dt} = \frac{\partial q}{\partial t} + \begin{bmatrix} u \\ v \\ w \end{bmatrix} \cdot \begin{bmatrix} \frac{\partial q}{\partial x} \\ \frac{\partial q}{\partial y} \\ \frac{\partial q}{\partial z} \end{bmatrix}$$

where q is any flow quantity that can vary in space and time. Also, we can write:

$$\frac{Dq}{Dt} = \frac{\partial q}{\partial t} + \sum_{j=1}^{3} u_j \frac{\partial q}{\partial x_j}$$

where u_j is velocity with $j = 1,2,3$.

Example 3.3

Example A velocity field in fluid is given by:

$$\vec{V} = 3xy^3\hat{\imath} + 3xy\hat{\jmath} + (3zy + 4t)\hat{k}$$

Find acceleration in x, y and z directions at location $(x,y,z) = (1,1,1)$ and at time $t = 1$ s.

Solution The velocity components are:

$$u = 3xy^3$$

$$v = 3xy$$

$$w = (3zy + 4t)$$

Acceleration components are:

$$a_x = \frac{\partial u}{\partial t} + u\frac{\partial u}{\partial x} + v\frac{\partial u}{\partial y} + w\frac{\partial u}{\partial z} = 9xy^6 + 27x^2y^3$$

$$a_y = \frac{\partial v}{\partial t} + u\frac{\partial v}{\partial x} + v\frac{\partial v}{\partial y} + w\frac{\partial v}{\partial z} = 9xy^4 + 9x^2y$$

$$a_z = \frac{\partial w}{\partial t} + u\frac{\partial w}{\partial x} + v\frac{\partial w}{\partial y} + w\frac{\partial w}{\partial z} = 4 + 9xyz + [3(3yz + 4t)]y$$

At location $(x,y,z) = (1,1,1)$ and time $t = 1$ s, we have $a_x = 36\ m/s^2$, $a_y = 18\ m/s^2$, $a_z = 34\ m/s^2$.

Acceleration in cylindrical coordinates

$$a_r = \frac{\partial u_r}{\partial t} + u_r\frac{\partial u_r}{\partial r} + \frac{u_\theta}{r}\frac{\partial u_r}{\partial \theta} - \frac{u_\theta^2}{r} + u_z\frac{\partial u_r}{\partial z}$$

$$a_\theta = \frac{\partial u_\theta}{\partial t} + u_r\frac{\partial u_\theta}{\partial r} + \frac{u_\theta}{r}\frac{\partial u_\theta}{\partial \theta} + \frac{u_r u_\theta}{r} + u_z\frac{\partial u_\theta}{\partial z}$$

$$a_z = \frac{\partial u_z}{\partial t} + u_r\frac{\partial u_z}{\partial r} + \frac{u_\theta}{r}\frac{\partial u_z}{\partial \theta} + u_z\frac{\partial u_z}{\partial z}$$

3.3 DEFORMATION OF FLUID PARTICLE

A fluid particle will be distorted as the flow moves, so much so that a particle will be elongated and rotated, and then it will break down and a new particle will form. This type of fluid-particle movement is very complex. A simplified mathematical model accounting only fluid particle translation, rotation, and dilatation is presented.

Figure 3.1 shows the deformation of a fluid particle. Initially, the particle was square-shaped. The rotation is measured as the movement of the diagonal.

Figure 3.2 shows the simplified model of fluid particle movements and deformation. The mathematical description of the fluid particle deformation, translation, and rotation are shown here.

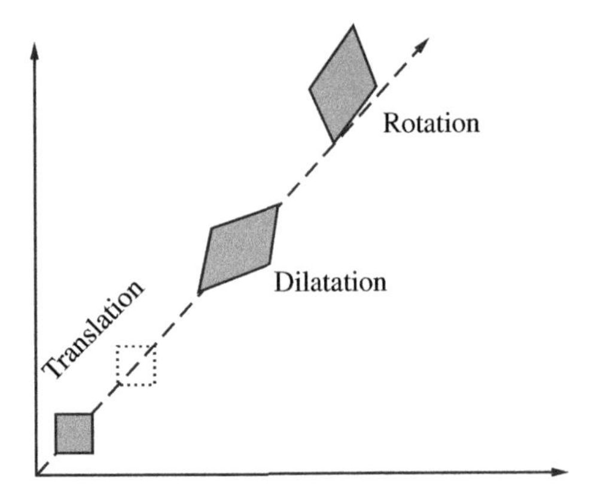

Figure 3.1 Deformation of a fluid particle.

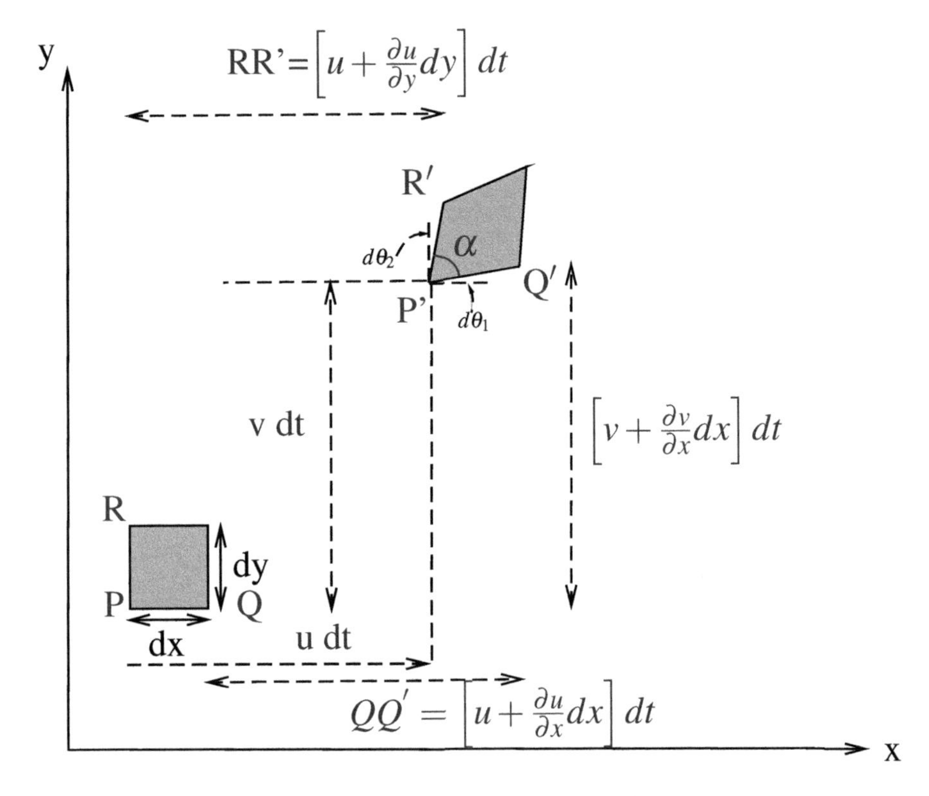

Figure 3.2 The deformation and movement of a fluid particle.

$$Rotation\ of\ diagonal = d\Omega_z$$

$$d\Omega_z = \underbrace{d\theta_1 + \frac{1}{2}\alpha}_{New\ diagonal\ rotation} - \underbrace{45}_{Original\ angle} \tag{3.1}$$

Further,

$$d\theta_1 + \alpha - d\theta_2 = 90°C \tag{3.2}$$

$$Rotation\ of\ diagonal = d\Omega_z = \frac{1}{2}(d\theta_1 + d\theta_2)$$

$$d\theta_1 = \lim_{dt \to 0}\left(tan^{-1}\frac{(vdt + \frac{\partial v}{\partial x}dxdt) - vdt}{(udt + \frac{\partial u}{\partial x}dxdt + dx) - udt}\right) = \frac{\partial v}{\partial x}dt \tag{3.3}$$

$$-d\theta_2 = \lim_{dt \to 0}\left(tan^{-1}\frac{(\frac{\partial u}{\partial y}dydt)}{vdt + \frac{\partial v}{\partial y}dydt}\right) = \frac{\partial u}{\partial y}dt \tag{3.4}$$

$$\frac{d\Omega_z}{dt} = \frac{1}{2}\left(\frac{\partial v}{\partial x} - \frac{\partial u}{\partial y}\right) \tag{3.5}$$

$$\frac{d\Omega_x}{dt} = \frac{1}{2}\left(\frac{\partial w}{\partial y} - \frac{\partial v}{\partial z}\right) \tag{3.6}$$

$$\frac{d\Omega_y}{dt} = \frac{1}{2}\left(\frac{\partial u}{\partial z} - \frac{\partial w}{\partial x}\right) \tag{3.7}$$

Vorticity defined as twice of rate of rotational of the diagonal:

$$\omega_x = \left(\frac{\partial w}{\partial y} - \frac{\partial v}{\partial z}\right) \tag{3.8}$$

$$\omega_y = \left(\frac{\partial u}{\partial z} - \frac{\partial w}{\partial x}\right) \tag{3.9}$$

$$\omega_z = \left(\frac{\partial v}{\partial x} - \frac{\partial u}{\partial y}\right) \tag{3.10}$$

3.3.1 SHEAR STRAIN

Average decrease of angle between two lines that are initially perpendicular in the unstrained state gives the shear strain:

$$\varepsilon_{xy} = \frac{1}{2}(d\theta_1 - d\theta_2) = \frac{1}{2}\left(\frac{\partial v}{\partial x} + \frac{\partial u}{\partial y}\right) \tag{3.11}$$

Similarly, we can write for shear strain in other directions as:

$$\varepsilon_{yz} = \frac{1}{2}\left(\frac{\partial w}{\partial y} + \frac{\partial v}{\partial z}\right) \tag{3.12}$$

$$\varepsilon_{zx} = \frac{1}{2}\left(\frac{\partial u}{\partial z} + \frac{\partial w}{\partial x}\right) \tag{3.13}$$

3.3.2 EXTENSIONAL STRAIN

The extensional strain is defined as the fractional increase in length of the horizontal side of the fluid particle.

$$\varepsilon_{xx}dt = \frac{(udt + \frac{\partial u}{\partial x}dxdt) - dx}{dx} = \frac{\partial u}{\partial x}dt \tag{3.14}$$

$$\varepsilon_{xx} = \frac{\partial u}{\partial x} \tag{3.15}$$

$$\varepsilon_{yy} = \frac{\partial v}{\partial y} \tag{3.16}$$

$$\varepsilon_{zz} = \frac{\partial w}{\partial z} \tag{3.17}$$

The strain rate tensor will have six components:

Cartesian coordinates

$$\varepsilon_{ij} = \begin{bmatrix} \varepsilon_{xx} & \varepsilon_{xy} & \varepsilon_{xz} \\ \varepsilon_{yx} & \varepsilon_{yy} & \varepsilon_{yz} \\ \varepsilon_{zx} & \varepsilon_{zy} & \varepsilon_{zz} \end{bmatrix}$$

Cylindrical coordinates

$$\varepsilon_{ij} = \begin{bmatrix} \varepsilon_{rr} & \varepsilon_{r\theta} & \varepsilon_{rz} \\ \varepsilon_{\theta r} & \varepsilon_{\theta\theta} & \varepsilon_{\theta z} \\ \varepsilon_{zr} & \varepsilon_{z\theta} & \varepsilon_{zz} \end{bmatrix}$$

Example 3.4

Example For the velocity field $\vec{V} = \left[3 \cdot x \cdot y^2,\ 5 \cdot t \cdot z^2,\ -4 \cdot x \cdot z\right]$ find the vorticity vector at the point $(1, -1, 1)$ at $t = 2$.

Solution

$\vec{V} = 3 \cdot x \cdot y^2\, \hat{\imath} + \ 5 \cdot t \cdot z^2\, \hat{\jmath} + \ -4 \cdot x \cdot z\, \hat{k}$

The components of velocity are:

$$u = 3\,x\,y^2;$$
$$v = 5\,t\,z^2;$$
$$w = \ -4\,x\,z;$$

$$\omega = \frac{1}{2}\left(\nabla \times \vec{V}\right)$$

$$\omega_x = \left(\frac{\partial w}{\partial y} - \frac{\partial v}{\partial z}\right) = -10tz$$

$$\omega_y = \left(\frac{\partial u}{\partial z} - \frac{\partial w}{\partial x}\right) = 4.0z$$

$$\omega_z = \left(\frac{\partial v}{\partial x} - \frac{\partial u}{\partial y}\right) = -6.0xy$$

$$\omega = -10tz\,\hat{i} + 4z\,\hat{j} - 6xy\,\hat{k}$$

At point (1,-1,1) and $t = 2$: $\omega = -20\hat{i} + 4\hat{j} + 6\hat{k}$

Strain rates in cylindrical coordinates

$$\varepsilon_{rr} = \frac{\partial u_r}{\partial r}$$

$$\varepsilon_{\theta\theta} = \frac{1}{r}\left(\frac{\partial u_\theta}{\partial \theta}\right) + \left(\frac{u_r}{r}\right)$$

$$\varepsilon_{zz} = \frac{\partial u_z}{\partial z}$$

$$\varepsilon_{r\theta} = \frac{1}{2}\left[\frac{1}{r}\left(\frac{\partial u_r}{\partial \theta}\right) + \left(\frac{\partial u_\theta}{\partial r}\right) - \left(\frac{u_\theta}{r}\right)\right]$$

$$\varepsilon_{\theta z} = \frac{1}{2}\left[\frac{1}{r}\left(\frac{\partial u_z}{\partial \theta}\right) + \left(\frac{\partial u_\theta}{\partial z}\right)\right]$$

$$\varepsilon_{rz} = \frac{1}{2}\left[\left(\frac{\partial u_r}{\partial z}\right) + \left(\frac{\partial u_z}{\partial r}\right)\right]$$

Properties of strain rate tensor: The following invariants are independent of direction or choice of axes:

$$I_1 = \varepsilon_{xx} + \varepsilon_{yy} + \varepsilon_{zz}$$

$$I_2 = \left(\varepsilon_{xx}\varepsilon_{yy} + \varepsilon_{xx}\varepsilon_{zz} + \varepsilon_{yy}\varepsilon_{zz}\right) - \left(\varepsilon_{xy}^2 + \varepsilon_{yz}^2 + \varepsilon_{zx}^2\right)$$

$$I_3 = \begin{vmatrix} \varepsilon_{xx} & \varepsilon_{xy} & \varepsilon_{xz} \\ \varepsilon_{yx} & \varepsilon_{yy} & \varepsilon_{yz} \\ \varepsilon_{zx} & \varepsilon_{zy} & \varepsilon_{zz} \end{vmatrix}$$

Example 3.5

Example A velocity field in fluid is given by:

$$\vec{V} = 3xy^3\hat{i} + 3xy\hat{j} + (3zy + 4t)\hat{k}$$

Find

(i) magnitude and direction of u, v, and w velocity components

(ii) vorticity vector

Solution (i) The velocity vector in Cartesian coordinates is:

$$\vec{V} = u\hat{i} + v\hat{j} + w\hat{k}$$

so

$$u = 3xy^3$$
$$v = 3xy$$
$$w = (3zy + 4t)$$

(ii) The vorticity vector is:

$$Vorticity = \omega = \nabla \times \vec{V} = \begin{vmatrix} \hat{i} & \hat{j} & \hat{k} \\ \frac{\partial}{\partial x} & \frac{\partial}{\partial y} & \frac{\partial}{\partial z} \\ u & v & w \end{vmatrix}$$

$$= 3\hat{i} - 30\hat{k}$$

The rotational velocity vector is:

$$V_{rot} = \frac{1}{2}(\nabla \times \vec{V}) = \left(\frac{3}{2}\right)\hat{i} - 15\hat{k}$$

Vorticity vectors in cylindrical and spherical coordinates

Cylindrical coordinates (r, θ, z):

$$\omega_r = \left(\frac{1}{r}\frac{\partial u_z}{\partial \theta} - \frac{\partial u_\theta}{\partial z}\right)$$

$$\omega_\theta = \left(\frac{\partial u_r}{\partial z} - \frac{\partial u_z}{\partial r}\right)$$

$$\omega_z = \left(\frac{\partial u_\theta}{\partial r} - \frac{1}{r}\frac{\partial u_r}{\partial \theta} + \frac{u_\theta}{r}\right)$$

Spherical coordinates (r, θ, φ):

$$\omega_r = \left(\frac{1}{r}\frac{\partial u_\theta}{\partial \varphi} - \frac{1}{r\sin\varphi}\frac{\partial u_\varphi}{\partial \theta} - \frac{u_\theta}{r}\cot\varphi\right)$$

$$\omega_\theta = \left(\frac{\partial u_\varphi}{\partial r} + \frac{u_\varphi}{r} - \frac{1}{r}\frac{\partial u_r}{\partial \varphi}\right)$$

$$\omega_\varphi = \left(\frac{1}{r\sin\varphi}\frac{\partial u_r}{\partial \theta} - \frac{\partial u_\theta}{\partial r} - \frac{u_\theta}{r}\right)$$

3.4 MOVEMENT OF FLUID PARTICLE

We can describe the movement of a fluid particle or many particles by drawing the lines for particle movements. In fluid mechanics literature, they are correctly called pathline or streakline, however, in some CFD software and visualization packages the same is mistakenly described as streamlines. Note that pathlines and streaklines can be visualized experimentally, but the streamlines are hypothetical lines and should not be used for viscous flows. Figure 3.3 shows pathlines of three fluid particles that are moving in a sinusoidal manner. Note that the pathlines may intersect each other.

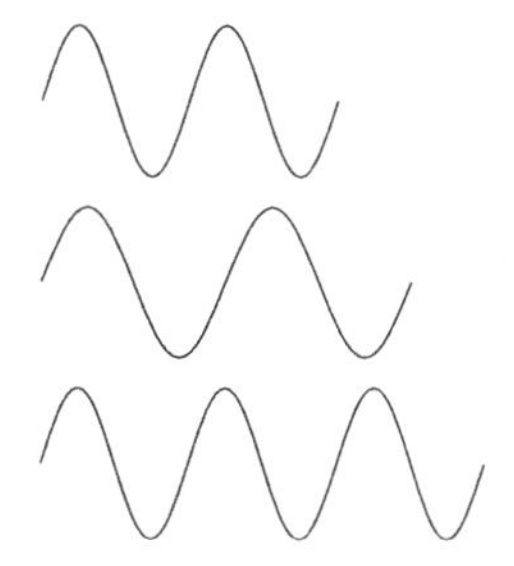

Figure 3.3 The pathlines of three different particles in an oscillating flow.

3.4.1 PATHLINES

Pathline: A path or trajectory of an individual fluid particle is called the pathline.

Calculation procedure for pathline

i Make $u = u(x,t)$ and $v = v(y,t)$ from experimental data

ii Find x and y displacements

iii Find t from equation of x and substitute it in y

iv The equation $y = f(x)$ only is eq. of pathline

Example 3.6

Example: A two-dimensional flow field is described by the following velocity components:

$$u = 3x(t+1)$$

$$v = 3y(t-1)$$

Determine the trajectory of the fluid particle that passes through the point (x_p, y_p) at $t = 0$.

Solution As u and v are the velocities in x and y direction respectively, we may write:

$$u = \frac{dx}{dt} = 3x(t+1)$$

$$dx = 3x(t+1)dt$$

We substitute $\zeta = t+1$ and RHS will be integrated as:

$$RHS = 3\int \zeta d\zeta = \frac{3}{2}(t+1)^2$$

$$\ln x = \frac{3}{2}(t+1)^2 + \ln C_1$$

$$\ln x = \frac{3}{2}(t+1)^2 + \ln C_1$$

$$x = \ln\left\{C_1 \exp\left[\frac{3}{2}(t+1)^2\right]\right\}$$

$$x = C_1 \exp\left[\frac{3}{2}(t+1)^2\right]$$

Similarly,

$$v = \frac{dy}{dt} = 3y(t-1)$$

$$dy = 3y(t-1)dt$$

$$\int \frac{dy}{y} = \int 3(t-1)dt$$

$$y = C_2 \exp\left[\frac{3}{2}(t-1)^2\right]$$

The trajectory of the fluid particle that passes through the point (x_p, y_p) at $t = 0$ will be:

$$C_1 = \frac{x_p}{\exp\left[\frac{3}{2}(0+1)^2\right]} = \frac{x_p}{e^{3/2}}$$

$$C_2 = \frac{y_p}{\exp\left[\frac{3}{2}(0-1)^2\right]} = \frac{y_p}{e^{3/2}}$$

Finally the x and y are:

$$x = \frac{x_p}{e^{3/2}} \exp\left[\frac{3}{2}(t+1)^2\right] = x_p e^{(t+1)^2}$$

$$y = \frac{y_p}{e^{3/2}} \exp\left[\frac{3}{2}(t-1)^2\right] = y_p e^{(t-1)^2}$$

from x equation, we find t expression:

$$t = \sqrt{\ln\left(\frac{x}{x_p}\right)} + 1$$

substituting it into y, we get:

$$y = y_p \exp\left[\left(\sqrt{\ln\left(\frac{x}{x_p}\right)} + 1\right) - 1\right]^2$$

The pathline is plotted in Figure 3.4.

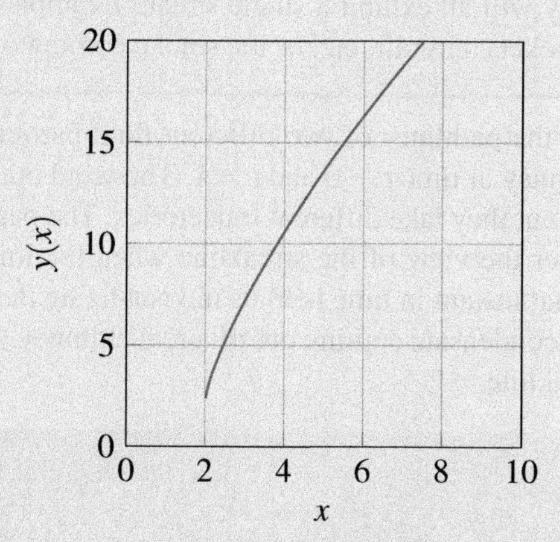

Figure 3.4 Pathline of particle passing through (2,2).

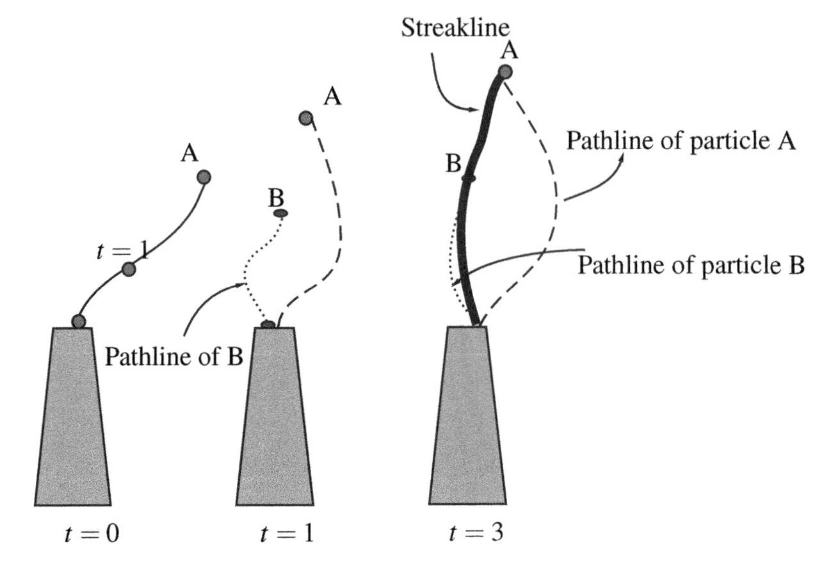

Figure 3.5 The streakline of two fluid particles A and B when they followed different pathlines, but eventually at time instant $t = 3$ they can be traced in one single line.

STREAKLINES

> **Streakline**: If many particles are issuing from one single location or loci, then in a flow they will all exhibit a single streak. Examples are chem-trails in the sky from rockets, aircraft, etc. or the smoke coming out of a chimney.

Figure 3.5 shows the pathlines of two different fluid particles A and B, as they emerge from the chimney at time $t = 0$ and $t = 1$. The wind outside is unsteady and as the particles move out they take different trajectories. The particles at time instant $t = 3$ gave an observer the view of the streakline when the line connecting all the current locations at that instant in time help us in visualizing the streakline.

Figure 3.6 shows condensate coming out of a real chimney. The condensate here is an example of streakline.

Figure 3.6 Industrial smokestack chimney with condensate flowing out.

Mathematically speaking, we search for the particles ζ that at one time instant passed through the same known point x_p, y_p.

Calculation procedure for streakline

i Make $u = u(x,t)$ and $v = v(y,t)$ from experimental data.

ii Find x and y displacements.

iii Find particle ζ from equation of x and substitute it in y.

iv The equation $y = f(x)$ only is eq. of pathline.

Example 3.7

Example A two-dimensional flow field is described by the following velocity components:

$$u = 3x(t+1)$$

$$v = 3y(t-1)$$

Determine the trajectory of the fluid particle ζ that passes through the point $(x_p, y_p) = (4,4)$ at $t = 3$ s. Plot streakline and several pathlines.

Solution We have already found the x and y expression for this flow field in a previous example

$$x = C_1 \exp\left[\frac{3}{2}(t+1)^2\right]$$

$$y = C_2 \exp\left[\frac{3}{2}(t-1)^2\right]$$

For particles ζ passing it will be:

$$x_p = C_1 \exp\left[\frac{3}{2}(\zeta+1)^2\right]$$

$$y_p = C_2 \exp\left[\frac{3}{2}(\zeta-1)^2\right]$$

$$C_1 = \frac{x_p}{\exp\left[\frac{3}{2}(\zeta+1)^2\right]}$$

$$C_2 = \frac{y_p}{\exp\left[\frac{3}{2}(\zeta-1)^2\right]}$$

Leading to x and y expressions:

$$x = \left(\frac{x_p}{\exp\left[\frac{3}{2}(\zeta+1)^2\right]}\right) \exp\left[\frac{3}{2}(t+1)^2\right]$$

$$y = \left(\frac{y_p}{\exp\left[\frac{3}{2}(\zeta - 1)^2\right]} \right) \exp\left[\frac{3}{2}(\zeta - 1)^2\right]$$

Note the parameter ζ represents different particles that are on one streakline. From x eq. we have

$$\frac{x}{x_p} = \frac{e^{3/2} e^{(t+1)^2}}{e^{3/2} e^{(\zeta+1)^2}} = \frac{e^{(t+1)^2}}{e^{(\zeta+1)^2}} = e^{(t+1)^2 - (\zeta+1)^2}$$

$$\zeta = -1 + \sqrt{(t+1)^2 - \ln\left(\frac{x}{x_p}\right)}$$

Substituting it into y, we get the equation of the streakline:

$$\frac{y}{y_p} = e^{(t-1)^2 - (\zeta-1)^2}$$

where

$$\zeta = -1 + \sqrt{(t+1)^2 - \ln\left(\frac{x}{x_p}\right)}$$

The pathlines are plotted in Figure 3.7.

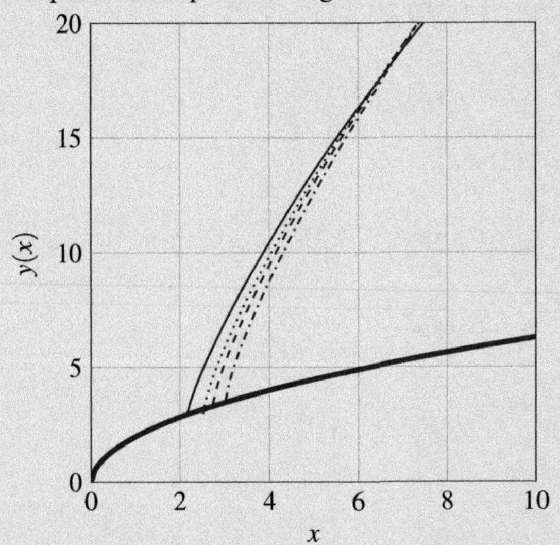

Figure 3.7 Several pathlines. A single streakline of particle passing through (4,4) at time $t = 3$ s.

PROBLEMS

3P-1 The velocity from a synthetic jet is modeled as follows:
$v(y,t) = V_o \cdot \left[1 + a\sin(\omega t)e^{(-by)}\right]$ where $V_o = 5$ m/s, $a = 0.05$ and $b = 0.3; \omega =$ 80 rad/s. Find the expression for acceleration in y direction as function of time t.

3P-2 A two-dimensional velocity field is described by

$$u(x,t) = 7x(t+1)$$

$$v(y,t) = 8y(t-1)$$

where u and v are the unsteady velocity components, u is the velocity in x direction and v is the velocity in y direction. Find the pathline if it passes through $x = 5, y = 5$ at $t = 0$.

3P-3 The velocity of a fluid flowing over the body is given by:

$$u = \frac{U_o}{3}\left[1 + \exp\left(-\frac{x^2}{L}\right)\right]$$

Find the acceleration a_x for this flow and plot the distribution.

3P-4 A viscous unsteady flow in polar coordinates is described by the velocities:

$$v_r = 0$$

$$v_\theta = \frac{\beta}{r}\left[1 - \exp\left(\frac{-r^2}{4\nu t}\right)\right]$$

Find the acceleration in r and theta coordinate.

3P-5 The velocity over the cylinder is defined as:

$$u_r = U_\infty\left[1 - \left(\frac{R}{r}\right)^2\right]\cos(\theta)$$

$$u_\theta = -U_\infty\left[1 + \left(\frac{R}{r}\right)^2\right]\sin(\theta)$$

Compute the strain rates ε_{rr}, $\varepsilon_{\theta\theta}$ and $\varepsilon_{r\theta}$ at $r = R$ and $\theta = \pi$.

3P-6 The dynamic vortex number is defined as:

$$W_D = \frac{\left|\omega \times \vec{V}\right|}{\left|\frac{\partial \vec{V}}{\partial t} + \nabla\left(\frac{\vec{V}\cdot\vec{V}}{2}\right)\right|}$$

If the velocity is defined by the vector:

$$\vec{V} = u\hat{i} + v\hat{j} + w\hat{k}$$

where $u = 3xy^3$, $v = 3xy$, and $w = (3zy + 4t)$, find the dynamic vortex number at $x = 1$, $y = 2$, $z = 1$, and $t = 1$.

3P-7 A fluid flows past a sphere with an upstream velocity as shown in Figure 3.8. It is found that the speed of the fluid along the sphere is

$$V = \frac{3}{2}U_\infty \sin(\theta)$$

Find the streamwise and normal components of acceleration at point S if the radius b of the sphere is 10 cm.

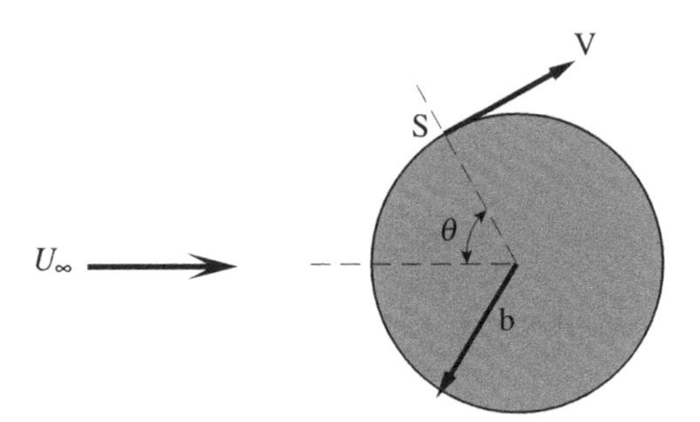

Figure 3.8 Figure of problem 3P-7.

3P-8 A velocity field is described by the vector:

$$\vec{V} = 0.3xt\ \hat{i} + 3\ \hat{j}$$

Find and plot the equations of pathline and streaklines, if they originate from $(x, y) = (0, 0)$.

3P-9 The velocity field over a cylinder is described by relation:

$$\vec{V} = B\cos(\theta)\left[1 - \left(\frac{r_o}{r}\right)^2\right]\hat{e}_r$$

$$-B\sin(\theta)\left[1 + \left(\frac{r_o}{r}\right)^2\right]\hat{e}_\theta$$

where B is 10 m/s. Find acceleration a_r and find its maximum value.

3P-10 A shock wave is a very thin wave in a supersonic gas flow across which the properties of the fluid (velocity, density, pressure etc.) will change abruptly from state

(1) to state (2) as shown in Figure 3.9 $U_1 = 560$ m/s, $U_2 = 217$ m/s. If the shockwave thickness $\ell = 2.5\mu$ m, then calculate the deceleration of the gas when it flows through the shock wave. How many "gs" does this represent?

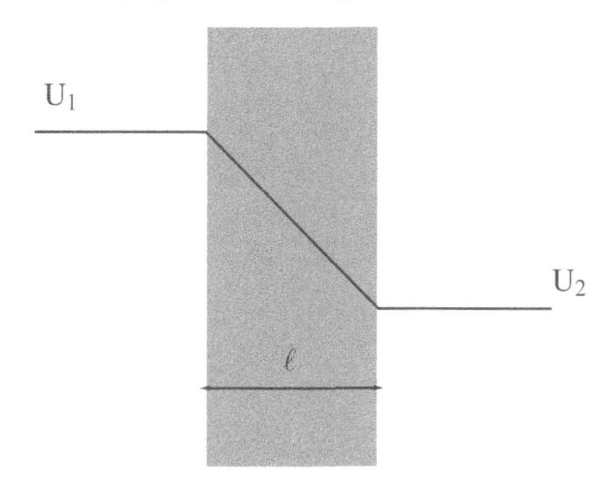

Figure 3.9 Figure of problem 3P-10.

3P-11 An organic fluid flows through the nozzle at a flow rate of 2 ft^3/s. Find the time taken by a fluid particle on the x axis at the center of a circular nozzle as it pass through it, from $x = 0$ to $\ell = 6$ in. The radius of nozzle at inlet is $r_1 = 3$ in. and at the outlet is $r_2 = 0.4$ in.

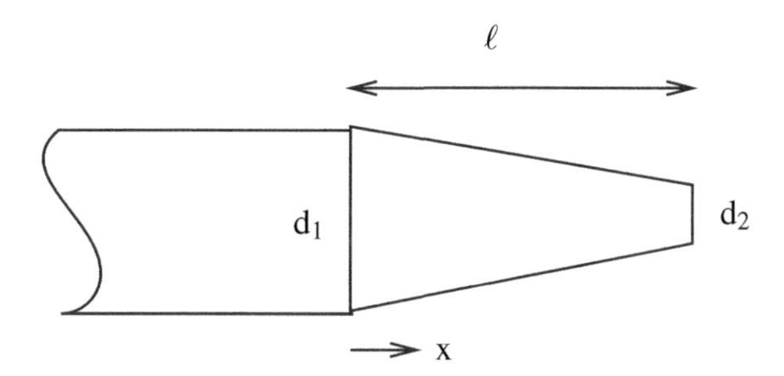

Figure 3.10 Figure of problem 3P-12.

3P-12 An organic fluid flows through the nozzle at flow rate of 2 ft^3/s. The length of nozzle is $\ell = 6$ in. The radius of the nozzle at inlet is $r_1 = 3$ in. and at the outlet is $r_2 = 0.4$ in. Determine the velocity and acceleration of a particle on the x axis at $x = 3$ in.

REFERENCE

G. Currie, Fundamental of of Fluids Mechanics, Third Edition, Marcel Dekker, Inc. USA, 2003.

4 Differential Formulation of Conservation Laws

This chapter is devoted to the derivation of the basic laws of conservation of mass and momentum in differential form. The law of conservation of mass is called the continuity equation in fluid mechanics literature, whereas the law of conservation of momentum is called the Navier-Stokes equation. The derivation of differential form of energy equation is not provided, as the focus of this book is not on convective heat transfer.

Learning outcomes: After finishing this chapter, you should be able to:

- Drive the differential formulation for mass conservation equation, which is also called the continuity equation.
- Learn about the Stokes constitutive relations for viscous flow.
- Drive the differential formulation for momentum conservation equation, also called Navier-Stokes equation.

In this chapter, we would recall fundamental laws of physics: the law of conservation of mass and the law of conservation of momentum for fluid flow. To develop the governing equations from these laws, we have been assuming that the continuum hypothesis is valid.

4.1 CONTINUITY EQUATION

In this section we will develop the differential formulation of the continuity equation in Cartesian coordinates. Mass cannot be created or destroyed. This law for an infinitesimal control volume or region results in continuity equation. The balance of mass can be described by the following equation.

$$\left\{\begin{array}{l} Net\ mass \\ flow\ rate \end{array}\right\} = \left\{\begin{array}{l}\text{Rate of mass}\\ \text{accumulation/decrease}\\ \text{in control volume}\end{array}\right\} + \left\{\begin{array}{l}\text{Net mass flow rate at}\\ \text{the surface of control}\\ \text{volume}\end{array}\right\} = 0$$

Figure 4.1 shows the mass balance for a fixed-volume element in two-dimensions (z plane). The control volume selected has unit depth. The velocity vector is $\vec{V} = u\hat{i} + v\hat{j} + w\hat{k}$. Note that mass is issuing at the center of this infinitesimal control volume where the origin of the coordinate system is placed, i.e. at the source of mass origin the (x, y) coordinates are (0,0).

DOI: 10.1201/9781003315117-4

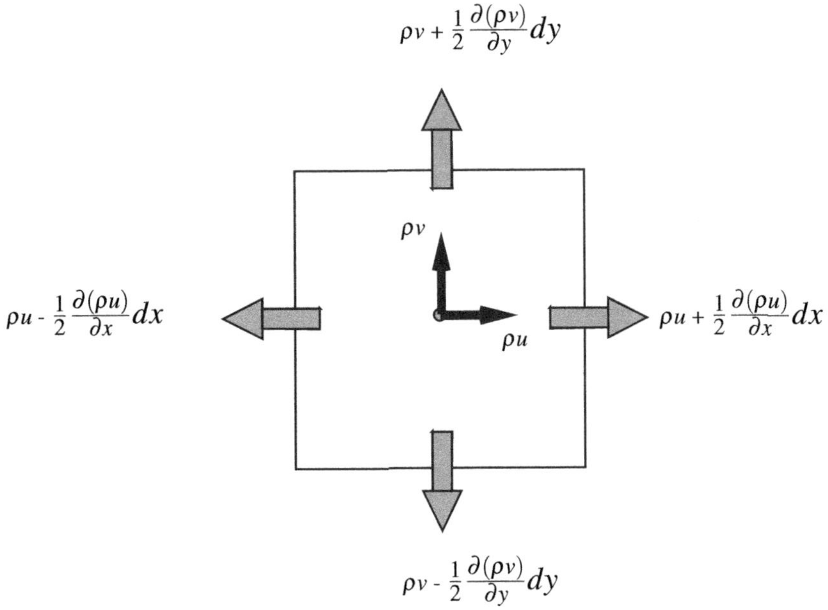

Figure 4.1 The two-dimensional control volume representing the mass balance with a unit depth.

Velocity in x direction is $u = u(x, y, z, t)$, velocity in y direction is $v = v(x, y, z, t)$ and velocity in z direction is $w = w(x, y, z, t)$. The density ρ is varying in the control volume $\rho(x, y, z, t)$. The dimensions of this two-dimensional infinitesimal control volume are dx and dy. We will balance the mass flux, so dimensionally we can formulate the mass flux as mass flow rate ($\dot{m}$) equals velocity normal to surface times the area.

$$\frac{\dot{m}_x}{A} = \rho u, \quad \frac{\dot{m}_y}{A} = \rho v, \quad \frac{\dot{m}_z}{A} = \rho w$$

Using Taylor series expansion (neglecting higher order terms), we get the net mass out-flow rate as:

$$\begin{aligned} \text{Net mass flow through control surface} = &\left\{\rho u + \frac{1}{2}\frac{\partial \rho u}{\partial x}dx\right\}dy - \left\{\rho u - \frac{1}{2}\frac{\partial \rho u}{\partial x}dx\right\}dy \\ &+ \left\{\rho v + \frac{1}{2}\frac{\partial \rho v}{\partial y}dy\right\}dx - \left\{\rho v - \frac{1}{2}\frac{\partial \rho v}{\partial y}dy\right\}dx \end{aligned} \tag{4.1}$$

$$Mass\ decrease\ in\ CV = -\frac{\partial \rho}{\partial t}dxdy \tag{4.2}$$

After doing net mass balance, we get a two-dimensional form of the continuity equation:

$$\frac{\partial \rho}{\partial t}+\frac{\partial(\rho u)}{\partial x}+\frac{\partial(\rho v)}{\partial y}=0 \tag{4.3}$$

For three-dimensional flow, the equation of continuity can be obtained as:

$$\frac{\partial \rho}{\partial t}+\frac{\partial(\rho u)}{\partial x}+\frac{\partial(\rho v)}{\partial y}+\frac{\partial(\rho w)}{\partial z}=0 \tag{4.4}$$

The continuity eq. 4.4 can be written in vector form as:

$$\frac{\partial \rho}{\partial t}+\nabla\cdot(\rho\vec{V})=0 \tag{4.5}$$

where $\vec{V}=u\hat{i}+v\hat{j}+w\hat{k}$.

4.2 THE NAVIER-STOKES EQUATIONS

The dynamic behavior of fluid motion is governed by equations called the equations of motion. The basis of these equations is Newtons second law of motion. We consider the infinitesimal mass m of fluid-particles to drive these equations. We know that the masses are related to density times the volume; therefore, we can cast the second law of motion into the variables of control volume and the density of the fluid particle. When the equations of motion of a fluid particle are written, the reference frame coordinate system is called the inertial reference frame, which must not be accelerating or rotating. The Eulerian formulation can describe acceleration as derived in previous chapters. We must remember that the fluid particle is a bundle of fluid molecules. The acceleration of the fluid particle is already described by the material derivative/substantial derivative or the derivative following the fluid.

4.2.1 ACCELERATION IN FLUID

Fluid acceleration is described by substantial derivative as discussed in a previous chapter:

$$\frac{Du}{Dt}=\frac{\partial u}{\partial t}+u\frac{\partial u}{\partial x}+v\frac{\partial u}{\partial y}+w\frac{\partial u}{\partial z} \tag{4.6}$$

$$\frac{Dv}{Dt}=\frac{\partial v}{\partial t}+u\frac{\partial v}{\partial x}+v\frac{\partial v}{\partial y}+w\frac{\partial v}{\partial z} \tag{4.7}$$

$$\frac{Dw}{Dt}=\frac{\partial w}{\partial t}+u\frac{\partial w}{\partial x}+v\frac{\partial w}{\partial y}+w\frac{\partial w}{\partial z} \tag{4.8}$$

4.2.2 BALANCE OF FORCES

The acceleration in the fluid is the manifestation of the imbalanced forces acting on the fluid body. There are two kind of forces that we are interested in. In mechanics, we divide the forces into, namely:

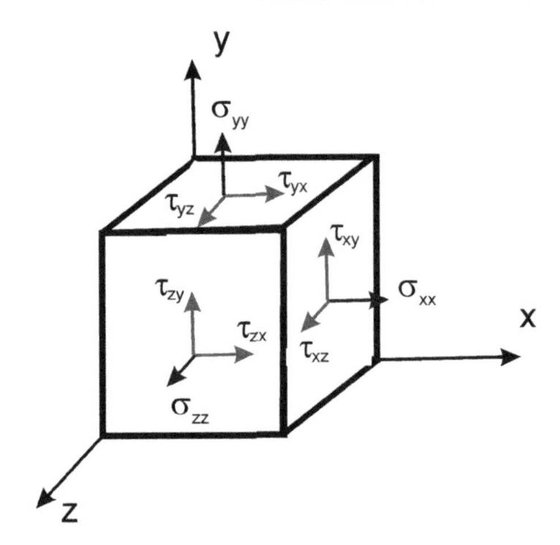

Figure 4.2 The three-dimensional control volume used for force balance.

Body forces f_b

Surface forces f_s

The forces that require physical contact with a body for action are known as surface forces, whereas those that need no physical contact with the body to manifest their influence are known as body forces. These forces arise because of some field-effect like gravitational or magnetic fields. The surface forces are due to direct action on the fluid body, and they can exhibit as pressure on the fluid, or they can arise after the external forces are applied on the fluid, like forces due to shear and normal stresses. The imbalances in forces would cause the movement of the fluid.

$$F = m\, a \tag{4.9}$$

$$F = m\, \frac{D\vec{V}}{Dt} \tag{4.10}$$

$$F = \rho\, \forall\, \frac{D\vec{V}}{Dt} \tag{4.11}$$

where $\forall$ is the control volume.

Forces due to surface stresses: The stresses in viscous flow vary from point to point and give rise to the surface forces, which act on a small fluid element and tend to accelerate it. There are two types of stresses: normal stresses (σ) and shear stresses (τ). The distribution of stresses on fluid control volume are shown in Figures 4.2 and 4.3. As a common reference point, the stresses at the center of the control volume are set equal to σ_{xx}, σ_{yy}, and τxy and so on. The normal stress forces and shear forces on each surface are reported in a right-hand coordinate system in which the outwardly directed surface normal is taken as the positive direction.

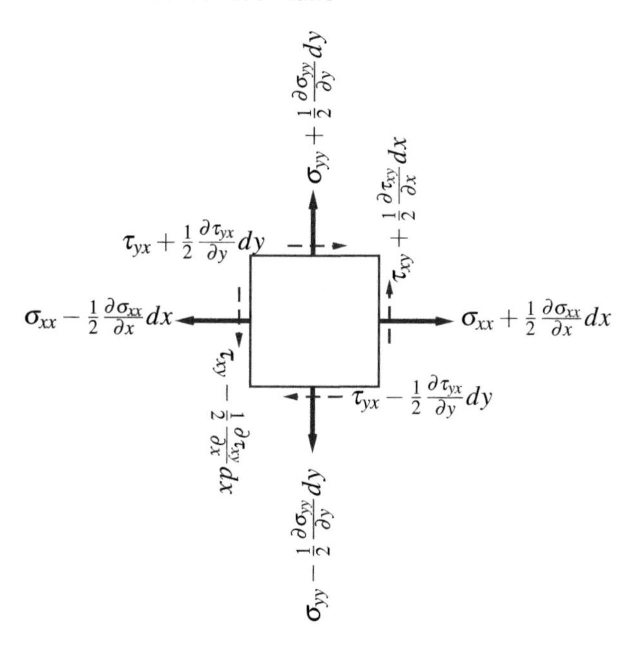

Figure 4.3 The two-dimensional control volume with normal and shear stress distribution.

The forces acting on the surface of the fluid can be obtained by considering the variation of the stresses with distance.

$$Net\ stress\ force\ in\ x\ direction = \frac{\partial(\sigma_{xx})}{\partial x}dxdy + \frac{\partial(\tau_{yx})}{\partial y}dxdy$$

This represents the viscous forces in fluid of 2D control volume. We now do the balance of forces using Newton's second law of motion and also extend it for 3D flows by adding the z-direction terms:

$$F_{viscous\ forces} + F_{pressure\ force} + F_{body\ forces} = m \cdot a$$

Substituting the forces we have force balance for x-direction:

$$\rho\frac{Du}{Dt} = \rho f_x + \frac{\partial(\sigma_{xx})}{\partial x} + \frac{\partial(\tau_{yx})}{\partial y} + \frac{\partial(\tau_{zx})}{\partial z} \tag{4.12}$$

for y direction:

$$\rho\frac{Dv}{Dt} = \rho f_y + \frac{\partial(\sigma_{yy})}{\partial y} + \frac{\partial(\tau_{xy})}{\partial x} + \frac{\partial(\tau_{zy})}{\partial z} \tag{4.13}$$

for z direction:

$$\rho\frac{Dw}{Dt} = \rho f_z + \frac{\partial(\sigma_{zz})}{\partial z} + \frac{\partial(\tau_{yz})}{\partial y} + \frac{\partial(\tau_{xz})}{\partial x} \tag{4.14}$$

4.2.3 CONSTITUTIVE RELATIONS

The relationship between stress tensor and velocity gradient (strain rate tensor) was not available when Navier developed the preceding equation. Stoke hypothesized that stresses can be related with velocity gradients. His relationship is called the constitutive relations. Stoke hypothesized that the stress tensor is linearly related with the velocity gradients, which is similar to Hookes law of elasticity for solid. This relationship is independent of the coordinate systems. Thus, his constitutive relationship is invariant to coordinates and can be used when fluid swirls, rotates, turns, etc. The principal axes of the deviatoric stress tensor and the strain rate tensor are the same. The stress tensor can be separated into two components. One component is a **hydrostatic** or **dilatational stress**, which can cause the change of the volume of the fluid; the other stress is the **deviatoric stress**, which cause to change the fluid body distortion.

Relation between stresses and velocity gradients:

$$\sigma_{xx} = \left(\lambda - \frac{2}{3}\mu\right)\Theta + 2\mu\frac{\partial u}{\partial x} + \Lambda$$

$$\sigma_{yy} = \left(\lambda - \frac{2}{3}\mu\right)\Theta + 2\mu\frac{\partial v}{\partial y} + \Lambda$$

$$\sigma_{zz} = \left(\lambda - \frac{2}{3}\mu\right)\Theta + 2\mu\frac{\partial w}{\partial z} + \Lambda$$

where λ is second viscosity coefficient, which is important for compressible flows.

$$\Theta = \frac{\partial u}{\partial x} + \frac{\partial v}{\partial y} + \frac{\partial w}{\partial z}$$

$$\tau_{xy} = \mu\left(\frac{\partial v}{\partial x} + \frac{\partial u}{\partial y}\right)$$

$$\tau_{xz} = \mu\left(\frac{\partial w}{\partial x} + \frac{\partial u}{\partial z}\right)$$

$$\tau_{yz} = \mu\left(\frac{\partial w}{\partial y} + \frac{\partial v}{\partial z}\right)$$

For simple flows, Λ= - Thermodynamic pressure = - P_{thermo}

$$\text{Average pressure} = \tilde{P} = -\frac{1}{3}(\sigma_{xx} + \sigma_{yy} + \sigma_{zz})$$

It is related to thermodynamic pressure as:

$$\widetilde{P} = -\lambda\Theta + P_{thermo} \tag{4.15}$$

$$\sigma_{xx} = -\widetilde{P} - \frac{2}{3}\mu\Theta + 2\mu\frac{\partial u}{\partial x}$$

$$\sigma_{yy} = -\widetilde{P} - \frac{2}{3}\mu\Theta + 2\mu\frac{\partial v}{\partial y}$$

$$\sigma_{zz} = -\widetilde{P} - \frac{2}{3}\mu\Theta + 2\mu\frac{\partial w}{\partial z}$$

For constant density flows Θ is zero and we may have the reduced form of these equations. Thermodynamic pressure $\widetilde{P}$ is written as P further on. Also, as only the thermodynamic pressure is relevant for incompressible flows, we can cast the normal stress-strain relationship as:

$$\sigma_{xx} = -P + 2\mu\frac{\partial u}{\partial x}$$

$$\sigma_{yy} = -P + 2\mu\frac{\partial v}{\partial y}$$

$$\sigma_{zz} = -P + 2\mu\frac{\partial w}{\partial z}$$

4.2.4 DIFFERENTIAL FORMULATION

We now insert the Stoke's constitutive relations into the Navier equation.

$$\frac{Du}{Dt} = \frac{\partial u}{\partial t} + u\frac{\partial u}{\partial x} + v\frac{\partial u}{\partial y} + w\frac{\partial u}{\partial z} = -\frac{1}{\rho}\frac{\partial P}{\partial x} + \nu\nabla^2 u + f_x \tag{4.16}$$

$$\frac{Dv}{Dt} = \frac{\partial v}{\partial t} + u\frac{\partial v}{\partial x} + v\frac{\partial v}{\partial y} + w\frac{\partial v}{\partial z} = -\frac{1}{\rho}\frac{\partial P}{\partial y} + \nu\nabla^2 v + f_y \tag{4.17}$$

$$\frac{Dw}{Dt} = \frac{\partial w}{\partial t} + u\frac{\partial w}{\partial x} + v\frac{\partial w}{\partial y} + w\frac{\partial w}{\partial z} = -\frac{1}{\rho}\frac{\partial P}{\partial z} + \nu\nabla^2 w + f_z \tag{4.18}$$

The above set of equations is called the Navier-Stokes equations.

Conservation of mass and momentum equations in cylindrical coordinates

Continuity Equation:

$$\frac{\partial \rho}{\partial t} + \frac{1}{r}\frac{\partial(\rho r u_r)}{\partial r} + \frac{1}{r}\frac{\partial(\rho u_\theta)}{\partial \theta} + \frac{\partial(\rho u_z)}{\partial z} = 0$$

Navier-Stokes Equations:

r-direction:

$$\rho\left(\frac{\partial u_r}{\partial t} + u_r\frac{\partial u_r}{\partial r} + \frac{u_\theta}{r}\frac{\partial u_r}{\partial \theta} - \frac{u_\theta^2}{r} + u_z\frac{\partial u_r}{\partial z}\right)$$

$$= -\frac{\partial P}{\partial r} + \rho f_r + \mu\left[\frac{1}{r}\frac{\partial}{\partial r}\left(r\frac{\partial u_r}{\partial r}\right) - \frac{u_r}{r^2} + \frac{1}{r^2}\frac{\partial^2 u_r}{\partial \theta^2} - \frac{2}{r^2}\frac{\partial u_\theta}{\partial \theta} + \frac{\partial^2 u_r}{\partial z^2}\right]$$

θ-direction:

$$\rho\left(\frac{\partial u_\theta}{\partial t}+u_r\frac{\partial u_\theta}{\partial r}+\frac{u_\theta}{r}\frac{\partial u_\theta}{\partial \theta}+\frac{u_r u_\theta}{r}+u_z\frac{\partial u_\theta}{\partial z}\right)$$

$$=-\frac{1}{r}\frac{\partial P}{\partial \theta}+\rho f_\theta+\mu\left[\frac{1}{r}\frac{\partial}{\partial r}\left(r\frac{\partial u_\theta}{\partial r}\right)-\frac{u_\theta}{r^2}+\frac{1}{r^2}\frac{\partial^2 u_\theta}{\partial \theta^2}-\frac{2}{r^2}\frac{\partial u_r}{\partial \theta}+\frac{\partial^2 u_\theta}{\partial z^2}\right]$$

z-direction:

$$\rho\left(\frac{\partial u_z}{\partial t}+u_r\frac{\partial u_z}{\partial r}+\frac{u_\theta}{r}\frac{\partial u_z}{\partial \theta}+u_z\frac{\partial u_z}{\partial z}\right)$$

$$=-\frac{\partial P}{\partial z}+\rho f_z+\mu\left[\frac{1}{r}\frac{\partial}{\partial r}\left(r\frac{\partial u_z}{\partial r}\right)+\frac{1}{r^2}\frac{\partial^2 u_z}{\partial \theta^2}+\frac{\partial^2 u_z}{\partial z^2}\right]$$

4.3 VECTORS, TENSORS, AND CONSERVATION LAWS

The physical quantities can be divided into the scalar, vector, and tensor quantities. Scalar quantities only need the magnitude, whereas vector quantities require magnitude and direction to describe themselves. Tensor quantities on the other hand require both magnitude, direction, and orientation for a complete description. Pressure is defined as a force unit area. Both force and area are the vector quantities, and the ratio of these two vector quantities gives us the pressure. The stress is also defined as force per unit area. When external force is applied on the body of a fluid the stresses are generated due to internal forces. The stress description with the magnitude and direction is incomplete, as one needs to understand the differences in the stresses because of orientation. This makes it compulsory that we use two indexes for stresses, for example, σ_{pd} where the first index p represents the plane in which the stresses act, and the second index d represents the direction of the stress.

Table 4.1
Physical Quantities in Conservation Equations

Scalar Tensor of rank 0	Vector Tensor of rank 1	Tensor Tensor of rank 2 and above
Distance	Displacement	Shear stress
Speed	Velocity	Normal stress
Pressure	Acceleration	
Mass		

Table 4.2
Tensor Multiplication

Tensor Product	Sign	Rank
Dyadic Product (Inner Product)	None	0
Cross Product	$\times$	-1
Dot (or Scalar) Product (Outer Product)	.	-2
Scalar or Double Dot Product (Inner Product)	:	-4

Table 4.3
Rank of Tensor Products

	Rank
Gradient	
∇(Vector)	Tensor
∇(Scalar)	Vector
Divergence	
$\nabla\cdot$ (Vector)	Scalar
$\nabla\cdot$ (Tensor rank 1)	Tensor rank 0
$\nabla\cdot$ (Tensor rank 2)	Tensor rank 1

The word *inner product* is used in vector spaces while the *dot/scalar product* is used in vector algebra and calculus. Dyads are the tensors of rank 2.

Note that ∇ is a vector defined as:

$$\nabla = \frac{\partial}{\partial x}\hat{i} + \frac{\partial}{\partial y}\hat{j} + \frac{\partial}{\partial z}\hat{k}$$

We now analyze the conservation equations term by term to understand more the vector and the denser products appearing in these equations. Starting from the continuity equation we have

$$\frac{\partial \rho}{\partial t} + \nabla\cdot(\rho\vec{V}) = 0$$

If we consider incompressible flow then density will be removed from this equation, and the equation will reduced to form:

$$\nabla(\vec{V}) = 0$$

Table 4.4
Vector and Tensor Form of Continuity Equation

Equation	Vector Form	Tensor Form
Continuity	$\frac{\partial \rho}{\partial t}+\nabla \cdot (\rho \vec{V})=0$ $\frac{\partial \rho}{\partial t}+div(\rho \vec{V})=0$	$\frac{\partial \rho}{\partial t}+\frac{\rho \partial U_k}{\partial x_k}=0$

The divergence of velocity field is zero.

If we consider incompressible flow, then the term

$$\frac{\partial U_k}{\partial x_k}=0$$

has k as dummy index and whenever the index appears twice, the summation rule is applied as

$$\frac{\partial U_1}{\partial x_1}+\frac{\partial U_2}{\partial x_2}+\frac{\partial U_3}{\partial x_3}=0$$

where x_1, x_2, and x_3 are the three space x, y, and z coordinates. U_1, U_2, and U_3 are the u, v, and w velocity components.

Tensorial products in Conservation Equations

The term $\nabla \cdot (V)$ is the divergence of the velocity vector, and is sometimes also written as **div**(V). The divergence operator will reduce the order of the tensor. When we add the ranks we get:

$$\underbrace{1}_{\nabla\ operator\ rank} + \underbrace{(-2)}_{Dot\ rank} + \underbrace{1}_{Velocity\ rank} = 0$$

We now consider the tensor products appearing in the Navier-Stokes equation.

If we take the inner product of ∇ with velocity V or gradient of the velocity vector, we have the summation of ranks as

$$\underbrace{1}_{\nabla\ operator\ rank} + \underbrace{(0)}_{No\ sign} + \underbrace{1}_{Velocity\ rank} = 2$$

indicating that ∇V represents a second order tensor.

Now consider the term $V \cdot \nabla V$, whose summation of ranks would be

$$1+(-2)+2=1$$

This gives us a first order tensor. Note that $V \cdot \nabla V$ actually represents the convective acceleration.

Table 4.5
Vector Form of Navier-Stokes Equation

Navier-Stokes Eq.	Vector Form
x-momentum	$\frac{\partial(\rho u)}{\partial t}+div(\rho u\vec{V})=-\frac{\partial p}{\partial x}+div(\mu\ grad\ u)$
y-momentum	$\frac{\partial(\rho v)}{\partial t}+div(\rho v\vec{V})=-\frac{\partial p}{\partial x}+div(\mu\ grad\ v)$
z-momentum	$\frac{\partial(\rho w)}{\partial t}+div(\rho w\vec{V})=-\frac{\partial p}{\partial x}+div(\mu\ grad\ w)$

We now consider the inner product of ∇ with scalar pressure $\nabla\ p$, which is also indicated as **grad** p, then

$$\underbrace{1}_{\nabla\ operator\ rank}+\underbrace{(0)}_{No\ sign}+\underbrace{0}_{scalar\ rank}=1$$

the rank of this operation is one, indicating that the **grad** p is representing a vector quantity.

$$\nabla p=\mathbf{grad}p=\frac{\partial p}{\partial x}\hat{\imath}+\frac{\partial p}{\partial y}\hat{\jmath}+\frac{\partial p}{\partial z}\hat{k}$$

PROBLEMS

4P-1 Prove that the following equations satisfy continuity equations:

(i)

$$u(x,y)=\left(\frac{2}{x^2+y^2}-\frac{4x^2}{(x^2+y^2)^2}\right)$$

(ii)

$$v(x,y)=\left(\frac{-4xy}{(x^2+y^2)^2}\right)$$

Table 4.6
Tensor Form of Navier-Stokes Equation

$$\rho\left[\frac{\partial U_j}{\partial t}+U_i\frac{\partial U_j}{\partial x_i}\right]=-\frac{\partial P}{\partial x_j}-\frac{\partial \tau_{ij}}{\partial x_i}+\rho g_i$$

$$\tau_{ij}=-\mu\left[\frac{\partial U_j}{\partial x_i}+\frac{\partial U_i}{\partial x_j}\right]+\frac{2}{3}\mu\delta_{ij}\frac{\partial U_k}{\partial x_k}$$

4P-2 Prove that the following equations satisfy continuity equations:

(i)

$$v_r=U_\infty\left[1-\left(\frac{r_o}{r}\right)^2\right]\cos(\theta)$$

(ii)

$$v_\theta=-U_\infty\left[1+\left(\frac{r_o}{r}\right)^2\right]\sin(\theta)$$

4P-3 The x component of velocity of gas deceases gradually as gas moves through the duct as follows:

$$u(x)=220\ (1-\ \tanh(x))$$

where x is the duct length. If flow is steady and two-dimensional, find the component of velocity in y direction.

4P-4 The normal stresses in MPa on fluid particle are:

$$\sigma_{xx}=14x+15$$

$$\sigma_{yy}=12y^2+5xy$$

$$\sigma_{zz}=0$$

How much is bulk stress at (1,10,2)?

4P-5 In a viscous flow, stress tensor at a point is:

$$\tau_{ij}=\begin{bmatrix}-2000 & 3000 & 1500\\ 4568 & 2345 & 1234\\ 2000 & -3000 & 3452\end{bmatrix}\times 10^3\ Pa$$

How much is normal stress at a point on an interface whose normal unit vector is $a = 0.5\,\hat{i} + 0.7\,\hat{j} + 0\,\hat{k}$?

4P-6 In a viscous flow, stress tensor at a point is:

$$\tau_{ij} = \begin{bmatrix} -3000 & 3000 & 1500 \\ 4568 & 2345 & 1234 \\ 2000 & -3000 & 3452 \end{bmatrix} \times 10^3\ Pa$$

What is the thermodynamic pressure at this point?

4P-7 The body force per unit mass distribution is expressed as:

$$f_B = 32x\,\hat{i} + 20\,\hat{j}\ N/kg$$

If the material density is expressed as:

$$\rho = x^2 + 3z\ kg/m^3$$

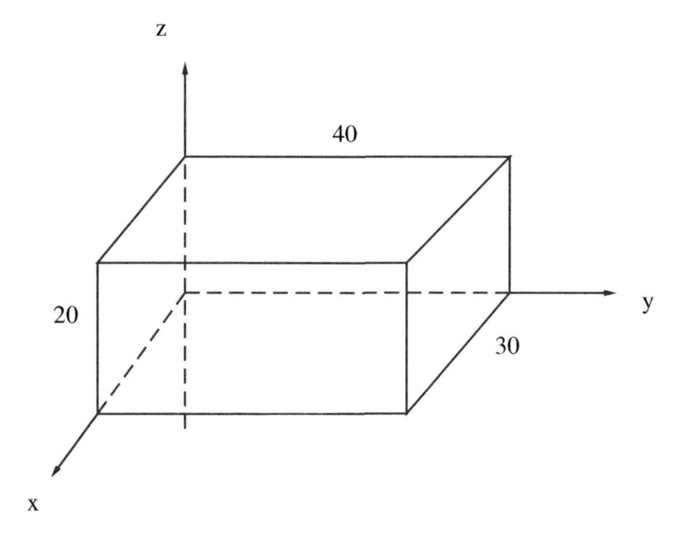

Figure 4.4 Figure of problem 4P-7.

What is the resultant body force for the region shown in Figure 4.4 whose extents are 30 m in x, 20 m in y, and 40 m in z directions.

REFERENCES

R. H. Sabersky, E. G. Hauptman and A. J. Acosta, Fluid Flow: A First Course in Fluid Mechanics, Prentice Hall PTR, USA, 1989.

I. H. Shames, Mechanics of Fluids, 3rd ed, McGraw-Hill Publishers, USA, 1992.

5 Dimensional Analysis and Similitude

In the preceding chapters, we discussed the fluid properties and rudimentary equations for analysis. We will now present a method called similitude in the physics literature and dimensional analysis in the fluid dynamics literature. Over the years, this technique has been applied to many problems, and it has been found to be a very fruitful technique. The similitude or dimensional analysis is not specifically a fluid mechanics technique but can be applied to many physical and biological problems. The basic premise is that the physical quantities can be cast into a dimensionless parameter that can be related together. Thus, instead of comparing quantities individually, we can investigate the problems by making the dimensionless parameters. This chapter will show that by comparing dimensionless parameters together, we can draw a meaningful conclusion.

Learning outcomes: After finishing this chapter, you should be able to:

- Develop a better understanding of the dimensional homogeneity of the equations.
- Know how to form the dimensionless parameters using the Buckingham pi theorem.
- Understand the concept of the geometric and the dynamic similarity in the development of models for experiments.

It happens not infrequently that the results in the form of laws are put forward as novelties on the basis of elaborate experiments, which might have been predicted a priori after a few minutes consideration.

—*Lord Rayleigh*

The procedure of sorting out the combination of various variables in such a manner as to have the dimensionless parameter is called dimensional analysis. Similitude, or dimensional analysis, is a remarkably productive tool that can help investigators to arrive quickly at meaningful results even when a complete model is not known. The analysis must apply to a list of the relevant variables the researcher might consider the most pertinent in the problem. The utility of the techniques when applied to a model or a list of variables can be enhanced by the collateral use of speculative and asymptotic analysis. Dimensional analysis of a list of system variables that define some chemical, physical, biological process, or social or economic problem is an invaluable help in developing the understanding of the experimental or

DOI: 10.1201/9781003315117-5

Table 5.1
Dimensions in Mass, Length, and Time (M-L-T) and Force, Length, and Time (F-L-T)

Quantity	Symbol	Dimensions (M-L-T)	Dimensions (F-L-T)
Length	ℓ	L	L
Time	t	T	T
Mass	m	M	FT^2L^{-1}
Force	F	MLT^{-2}	F
Velocity	V	LT^{-1}	LT^{-1}
Acceleration	a	LT^{-2}	LT^{-2}
Area	A	L^2	L^2
Discharge	Q	L^3T^{-1}	L^3T^{-1}
Pressure	P	$ML^{-1}T^{-2}$	FL^{-2}
Gravity	g	LT^{-2}	LT^{-2}
Density	ρ	$M L^{-3}$	FT^2L^{-4}
Specific weight	γ	$M L^{-2}T^{-2}$	FL^{-3}
Dynamic viscosity	μ	$ML^{-1}T^{-1}$	$FL^{-2}T$
Kinematic viscosity	ν	L^2T^{-1}	L^2T^{-1}
Surface tension	σ_e	$M T^{-2}$	FL^{-1}
Bulk modulus of elasticity	κ	$M L^{-1} T^{-2}$	FL^{-2}
Power	Pw	ML^2T^{-3}	FLT^{-1}
Shear stress	τ	$ML^{-1}T^{-2}$	FL^{-2}

numerically computed data. Table 5.1 lists some important quantities in fluid mechanics with dimensions.

5.1 VASCHY-BUCKINGHAM PI THEOREM

Selection of right variable: A. Vaschy, in 1892, introduced the Pi theorem. John William Strutt, the third Baron Rayleigh (1842–1919) was a well-known British physicist. He refined the method of dimensions, which is now known as Buckinghams Π theorem. According to the procedure, the dimensional analysis starts with the preparation of a list of the individual dimensional variables that can be dependent, independent, and parametric that are presumed to define the behavior of the problem under investigation. The selection of a necessary and sufficient set of variables sometimes requires knowledge gained from experience. It hence is both challenging and uncertain when dealing with a new or unfamiliar aspect of phenomenon or trend.

Minimum number of dimensionless groups: The minimum number of independent dimensionless groups, i, that are required to describe the fundamental and parametric behavior was stated by Buckingham

$$i = n - m$$

where n is the number of variables and m is the number of fundamental dimensions such as mass M, length L, time θ, and temperature T, that are introduced by the variables. Van Driest has shown that in some rare cases, i is actually greater than $n - m$ and it is better to redefine the equation as:

$$i = n - r$$

where r is the maximum number of the chosen variables that cannot be combined to form a dimensionless group. This is equal to the rank of the matrix.

Manipulation of groups: In most cases, the dimensionless numbers need to be adjusted by multiplying or dividing two numbers together. Let's say the result of a dimensional analysis is

$$\zeta(X,Y,Z,W) = 0$$

where X, Y, Z, and W are independent dimensionless groups, an equally valid expression is

$$\zeta\left(XY^{1/2}, Z/Y^2, Z/W, W\right)$$

Dimensional analysis itself does not provide any insight as to the best choice of equivalent dimensionless groupings. Three primary methods of determining a minimal set of dimensionless variables will be described:

i Inspection

ii Combination of the residual variables, one at a time, with a set of chosen variables that cannot be combined to obtain a dimensionless group

iii Algebraic procedure

Example 5.1 - Vaschy-Buckingham Pi Theorem

Example The viscous drag can be, depending on the density, viscosity, velocity, shown as:

$$Drag = f(u, \rho, \mu, L)$$

We would like to find the dimensionless groups relevant for this problem.

Solution

Quantity	Symbol	Dimensions
Length	l	L
Velocity	V	L/T
Force	F	ML/T^2
Density	ρ	M/L^3
Viscosity	μ	M/LT

	F	u	ρ	μ	L
M	1	0	1	1	0
L	1	1	−3	−1	1
T	−2	−1	0	−1	0

$$\mathbf{B} = \begin{bmatrix} 1 & 0 & 1 & 1 & 0 \\ 1 & 1 & -3 & -1 & 1 \\ -2 & -1 & 0 & -1 & 0 \end{bmatrix}$$

The rank of this matrix is 3 and the number of variables is 5, so the dimensionless groups that can be formed are two. Now, here we can intuitively experiment with different combinations. There is no set rule for the selection of possible dimensionless groups, so we just do trial and error to come up with a certain combination.

$$\Pi_1 = u^a \, \rho^b \, L^c \, F$$

$$M^0 \, L^0 \, T^0 = \left(LT^{-1}\right)^a \left(ML^{-3}\right)^b (L)^c \left(MLT^{-2}\right)$$

$$\begin{aligned} M: &\quad 0 = b + 1 \\ L: &\quad 0 = a - 3b + c + 1 \\ T: &\quad 0 = -a - 2 \end{aligned}$$

Solution gives:

$$a = -2, b = -1, c = -2$$

$$\Pi_1 = u^{-2} \, \rho^{-1} \, L^{-2} \, F = \frac{(F/L^2)}{\rho u^2}$$

This pi group is known as drag coefficient c_D.

$$M^0 \, L^0 \, T^0 = \left(LT^{-1}\right)^d \left(ML^{-3}\right)^f (L)^e \left(ML^{-1}T^{-1}\right)$$

$$\begin{aligned} M: &\quad 0 = f + 1 \\ L: &\quad 0 = -d - 3f + e - 1 \\ T: &\quad 0 = -d - 1 \end{aligned}$$

Solution gives:

$$f = -1, d = -1, e = -1$$

$$\Pi_2 = u^{-1} L^{-1} \rho^{-1} \mu = \frac{\mu}{\rho u L}$$

This is reciprocal of a dimensionless number called Reynolds number (Re) in fluid mechanics literature. So the drag coefficient is a function of Reynolds number of the flow.

$$c_D = \mathrm{f(Re)}$$

Example 5.2 - Vaschy-Buckingham Pi Theorem

Example We now consider the modeling of pressure drop in a circular pipe of diameter D and length ℓ. The flow velocity is u and the fluid flowing has density ρ and viscosity μ.

Solution We assume that the pressure drop across the pipe might have the following important variables.

$$\Delta p = p_1 - p_2 = f(\ell, D, \varepsilon, \rho, \mu, u)$$

These are the seven variables.

Quantity	Symbol	Dimensions
Length	ℓ	L
Diameter	D	L
Velocity	u	L/T
Pressure	Δp	M/LT^2
Density	ρ	M/L^3
Viscosity	μ	M/LT
Surface roughness	ε	L

Dimensions	Δp	ℓ	D	ε	ρ	μ	u
M	1	0	0	0	1	1	0
L	−1	0	0	0	−3	−1	1
T	−2	1	1	1	0	−1	−1

The matrix will be:

$$\mathbf{B} = \begin{bmatrix} 1 & 0 & 0 & 0 & 1 & 1 & 0 \\ -1 & 0 & 0 & 0 & -3 & -1 & 1 \\ -2 & 1 & 1 & 1 & 0 & -1 & -1 \end{bmatrix}$$

The rank of this matrix is 3. As we have seven variables so we might have $7 - 3 = 4$ dimensionless groups.

$$\Pi_1 = \frac{\Delta p}{\frac{\rho}{2}u^2},\ \Pi_2 = \frac{\rho u D}{\mu},\ \Pi_3 = \frac{\ell}{D},\ \Pi_4 = \frac{\varepsilon}{D}$$

This shows that pressure drop is related with these dimensionless quantities

$$\frac{\Delta p}{\frac{\rho}{2}u^2} = f\left(\frac{\rho u D}{\mu}, \frac{\ell}{D}, \frac{\varepsilon}{D}\right)$$

One of the pi group is known as the Reynolds number; it is defined as:

$$\mathrm{Re}_D = \frac{\rho u D}{\mu}$$

so we may write the dependence between pi groups as:

$$\frac{\Delta p}{\frac{\rho}{2}u^2} = f\left(\mathrm{Re}_D, \frac{\ell}{D}, \frac{\varepsilon}{D}\right)$$

5.1.1 LIMITATIONS

We must be well aware of the fact that the dimensional analysis is not a panacea for all problems we have in fluid mechanics. The reason is that the dimensional analysis does not give an insight into the physics of the problem. The exact nature of the undetermined function $f(\Pi_1, \Pi_2, \Pi_3, ...)$ in an expression obtained from dimensional analysis has to be established either by experiments or by CFD simulations.

5.2 OTHER APPROACHES FOR DIMENSIONLESS NUMBERS

5.2.1 BALANCE OF FORCES

Many dimensionless numbers can be formed by doing the balance of force analysis. In fluid mechanics, we can write a number of agencies in terms of forces. Some examples follow.

Inertia Force: From Newton's second law of motion we have

$$F = m\,a = \underbrace{\rho L^3}_{mass} \underbrace{\left(\frac{u}{t}\right)}_{acceleration} = \rho L^2 u \left(\frac{L}{t}\right) = \rho L^2 u^2$$

$$F_{inertia} \simeq \rho L^2 u^2$$

Viscous Force: Viscous force is shear stress in fluid times area:

$$F = \tau A = \underbrace{\mu \left(\frac{u}{\not{L}}\right)}_{shear\ stress} \underbrace{L^{\not{2}}}_{area} = \mu L u$$

$$F_{viscous} \simeq \mu L u$$

The ratio of forces will give us different dimensionless numbers. For example, the Reynolds number is defined as ratio of inertia forces to viscous forces as below:

$$\mathrm{Re} = \frac{F_{inertia}}{F_{viscous}} = \frac{\rho(L.\,u)^2}{\mu L u} = \frac{\rho u L}{\mu}$$

Osborne Reynolds (1842–1912) was an Irish engineer who investigated the flow types at the University of Manchester, UK.

In 1908, Arnold Sommerfeld presented a paper on hydrodynamic stability at the 4th International Congress of Mathematicians in Rome, where Sommerfeld proposed that the equation known today as the Orr-Sommerfeld equation, we can introduce a number R, which should be called as "Reynolds'sche Zahl". Later this number is called the Reynolds number in fluid mechanics equation. In Fluid Mechanics, Reynolds number is the most often used parameter and it can help us in declaring flow either as laminar or turbulent. Following are the ranges of Reynolds numbers for the transition from laminar to turbulent flow:

$$1000 \leq \mathrm{Re}_D \leq 3000 \quad (Free\ \ jet)$$
$$3 \times 10^5 \leq \mathrm{Re}_L \leq 3 \times 10^6 \quad (Flow\ \ over\ flat\ plate)$$
$$2 \times 10^5 \leq \mathrm{Re}_D \leq 3.8 \times 10^5 \quad (Flow\ \ over\ cylinder)$$
$$3 \times 10^5 \leq \mathrm{Re}_D \leq 3.5 \times 10^5 \quad (Flow\ \ over\ sphere)$$
$$2.3 \times 10^3 \leq \mathrm{Re}_D \leq 4 \times 10^3 \quad (Flow\ \ inside\ pipe)$$
$$500 \leq \mathrm{Re}_{Rh} \leq 750 \quad (Open\ \ Channel\ Flow)$$

where subscripts L and D refer to characteristic length, which can be the length of the flat plate or diameter of the cylinder/sphere, respectively. Also, for open channel flow, the Reynolds number must be computed based on water depth.

American Petroleum Institute recommends using the following formula for determining the pipe flow critical Reynolds number

$$\mathrm{Re}_{critical} = 3470 - 1370n$$

where n is an index of power law fluid.

Cohesion Force: depends on the surface tension symbolized as σ_e (N/m), defined as:

$$F_{cohesion} = \sigma_e \cdot L$$

An important number in fluid mechanics relating inertia and cohesion forces is called the Weber number; it is defined as:

$$We = \frac{F_{inertia}}{F_{cohesion}} \simeq \frac{\rho L^2 u^2}{\sigma_e \cdot L} \simeq \frac{\rho L\, u^2}{\sigma_e}$$

Table 5.2
Organisms and Their Reynolds Numbers

Biological Organism	Reynolds Number
Bacteria	10^{-6}
Protozoa with flagella	10^{-3}
Tadpoles	10^{2}
Segmented parasitic or predatory worms	10^{3}
Species of oily freshwater fish	10^{5}
Dolphins or sharks	10^{7}
Blue whales (marine mammal)	10^{8}
Humans	10^{6}

This parameter (Weber number) is named after the German naval architect Moritz G. Weber (1871–1951) who first suggested this number. The Weber number is important at liquid-liquid and gas-liquid interfaces. Also, it is useful when the interfaces are in contact with a boundary or the surface tension causes small (capillary) waves and droplets to form.

The Ohnesorge number (Oh) is a dimensionless number that is the ratio of the viscous forces to inertial and surface tension forces. The number was defined by Wolfgang von Ohnesorge in 1936, and it is defined as:

$$Oh = \frac{\sqrt{We}}{\mathrm{Re}} = \frac{\mu}{\sqrt{\rho \sigma_e R_o}}$$

where σ_e is the surface tension and R_o is the characteristic length. The Ohnesorge number determines whether the viscous forces are important relative to surface tension forces. Ohnesorge number is often used in dispersion of liquids in gases and in spray technology. For ink-jet printers, liquids whose Ohnesorge number is in range

$$0.1 \leq Oh \leq 1$$

are able to form the jet that can reach the surface of the paper.

Galileo number is a dimensionless number named after Italian scientist Galileo Galilei (1564–1642) and defined as ratio of gravity forces to the viscous forces. The Galilei number is used in viscous film flow over walls, flow in condensers, or flow in chemical columns.

$$Ga = \frac{gL^3}{\nu^2} = \frac{\mathrm{Re}^2}{Fr}$$

External field force: In fluid, different external field or body forces can be present, like magnetic or gravitational fields. Often gravity is far more important

Table 5.3
Rossby Number for Different Types of Vortices

Vortex	*Ro*
Bathtub vortex	1,000,000
Dust Devils	30,000
Tornado	30,000
Hurricane	1
Low-pressure system	0.1
Ocean circulations	0.005

to bring into consideration, and it is defined in terms of force as:

$$F_{gravity} = mg = \underbrace{\rho L^3}_{mass} g$$

The Froude number is a dimensionless number defined as the square root of the ratio of the inertia force to the gravity force:

$$Fr = \sqrt{\frac{F_{inertia}}{F_{gravity}}} \simeq \sqrt{\frac{\rho L^2 u^2}{\rho L^3 g}} = \frac{u}{\sqrt{Lg}}$$

The number commemorates the English engineer William Froude (1810–1879) who introduced ship model making to investigate the ship hull resistance. Froude number is useful in calculations of hydraulic jump, a phenomenon observed in liquid flows. Also it is useful in design of hydraulic structures, and in ship designs.

Coriolis force

Rossby number (*Ro*) and Ekman (Ek) numbers are defined as:

$$Rossby\ \ number = \frac{F_{inertia}}{F_{coriolis}} = \frac{V}{L \cdot f}$$

Here, V indicates the characteristic velocity, L is a characteristic length, f is the Coriolis parameter, ν is the kinematic viscosity, and ω is a characteristic vorticity related to the rotating system. This ratio is called Rossby number in recognition of the research of the Swedish meteorologist, and it gives a convenient measure of the importance of Coriolis forces in flow. When $Ro > 1$, Coriolis forces will cause a slight modification of the flow pattern; but when $Ro < 1$, the Coriolis forces tendency to oppose any expansion in a lateral plane is likely to be dominant.

$$Ekman\ \ number = \frac{F_{frictional}}{F_{Coriolis}} = \frac{\nu}{L^2 f}$$

Table 5.3 lists the Rossby number for different flow events. The critical Rossby number, at which the influence of the Earths rotation becomes significant is around

$Ro = 10^4$. In most atmospheric and oceanic flows, the inertial forces exceed the frictional forces, and therefore the Rossby number is of greater importance than the Ekman number.

5.2.2 RATIO OF VELOCITIES

The Froude number can also be defined as ratio of velocities. The velocity of gravity waves (c) can be defined as:

$$c^2 = g \cdot h$$

Relating the flow speed and gravity wave speed together we will arrive at Froude number as:

$$Fr = \frac{u_{fluid}}{u_{wave}} = \frac{u}{\sqrt{c}} = \frac{u}{\sqrt{g \cdot h}}$$

Another important number is Mach number, defined as:

$$M = \frac{u}{a}$$

where u is flow velocity and a is acoustic speed or speed of sound. The Mach number is named after the Austrian/Czech physicist Ernst Mach (1838–1916), who had investigated the flow round rifle bullets. The artillerymen of the 18th century noticed that they hear only one sound from a low-speed projectile or canon ball, while when they fired a bayonet, the high-speed projectile gave two loud sounds. It was later realized by Mach's research that the second loud sound is associated with arrival of the bow shock waves that occur when, depending on conditions, a detached and an attached shock wave forms on the bullet.

Professor Sarrau in 1884 first highlighted the significance of the Mach number, but it is now widely called Mach number, for the man who introduced it in 1887. It was also called Cauchy number.

5.2.3 RATIO OF LENGTHS

The Strouhal number is defined as the ratio of distance over which flow is advected or moved in a period to a characteristic width of the structure on which the fluid moved.

$$St = \frac{f \cdot D}{U}$$

In literature the reciprocal is sometimes also used as a definition of Strouhal number, like:

$$St = \frac{Advection\ Distance}{width} = \frac{(U/f)}{D}$$

Strouhal number is important in oscillatory, excited, or periodic flows and it is named in honor of Czechoslovakian scientist Cenek Vincent Strouhal (1850–1922) who discovered that frequency of periodic vortex shedding from a wire in air is proportional to flow speed.

In capillary flows, an important number is the Bond number, or Eötvös number, defined as:

$$Bo = Eo = \frac{\rho g \ell^2}{\sigma_e}$$

where σ_e is the surface energy per unit area in J/m^2 or surface tension.

This parameter helps to characterize the shape of bubbles or drops moving in a surrounding fluid.

5.2.4 RATIO OF MASSES

Christopher Scruton (1911–1990) was an English physicist who introduced the dimensionless number known as the Scruton number. It is defined as:

$$Sc = \frac{Damping \times \dot{m}_{structure}}{\dot{m}_{fluid}} = \frac{m\delta}{\rho D^2}$$

5.2.5 RATIO OF DIMENSIONLESS NUMBERS

Certain parameters are made after combining the dimensionless numbers, for example, Morton number, which is defined as:

$$Mo = \frac{We^3}{Fr^2 \mathrm{Re}^4} = \frac{g\mu_w^2}{\rho\sigma_e^3}$$

The pulsatile or periodic flow of blood in the arteries can be analyzed by breaking down the periodic movement into a series of harmonic components (using Fourier analysis) with each component as a multiple of the fundamental frequency. For such problems, instead of using the Strouhal number, a dimensionless number formed by the product of St and Re is used, which is called the Womersley number. Womersley number is a dimensionless number often used in biofluid mechanics, defined as a product of Reynolds number and Strouhal number, defined as:

$$Wo = \frac{Transient\ Inertia\ Force}{Viscous\ Force}$$

$$Wo = St \cdot \mathrm{Re} = \frac{\rho\omega D^2}{\mu}$$

or

$$Wo = \sqrt{2\pi \cdot \mathrm{Re} \cdot St}$$

Womersley number is an expression of the pulsatile flow forces relative to viscous forces. A typical value of Womersley number in the aorta is of order of 12.

5.2.6 SCALE ANALYSIS

If the physical or biological phenomenon has already been modeled in terms of differential equation then it means that we have already gained an insight into the nature of a problem and this will give us a big leap in coming up with the relevant dimensionless number, or may be we can arrive at the mathematical equalities without even solving the equation. To do so, we use a procedure known as scale analysis or order-of-magnitude analysis. Many engineering problems were simplified just by inspecting the differential equations. We will use the scale analysis in chapter 11 on external boundary layer flows.

5.3 SIMILITUDE

In experimental fluid mechanics, we use the concept of similitude to make the models for the objects under investigation, whose actual sizes are called prototypes. Let's say you want to investigate the drag experienced by a new car model. One option is to have a massive wind tunnel facility where you park the car in the test section and then record the velocities or other data over the vehicle, which could subsequently be used to estimate the drag on the vehicle. However, such an arrangement is not possible for a new airplane or a boat design. Often scaled-down models are made which have *geometric, kinematic,* and *dynamic* similarities with the prototype.

The geometric similitude means that the shape and size are scaled down or scaled up, depending on the relevant dimensionless number. If the geometric similarity is correctly achieved, it is expected that the velocities and acceleration may also be scaled down/up in magnitude, thus creating the same type and aerodynamic forces as those anticipated in actual prototypes. The realization that geometric similitude is an essential aspect for modeling nature has been well recognized for a long time. This was mentioned at the earliest by Galileo (1642).

We need to understand that in flow there are two modes of movement. One is the bulk movement called advection (or convection) and another movement is going on at the molecular level, which we call the momentum diffusion. The momentum diffusion is represented by viscous effects in the flow and advection is represented by the inertia term in the Navier-Stokes equation. The relation between diffusion and convection is shown quantitatively with the concept of dynamic similarity. In modeling, we want to make sure that if we change the geometric scale of the model then phenomenologically the balance between the forces is maintained.

Example 5.3

Example A prototype is being designed for movement through oil at speed of 1 m/s. It is expected that both viscous and gravity forces will be important. A scaled-down model scale of 1:5 is planned to be tested. What viscosity of model liquid do you think is necessary to make both the Froude number and the Reynolds number the same in the model and the prototype? Take oil kinematic viscosity as 4.65×10^{-5} m^2/s.

Solution We compare the Froude numbers

$$\left(\frac{u}{\sqrt{gL}}\right)_m = \left(\frac{u}{\sqrt{gL}}\right)_p$$

$$\left(\frac{u_m}{\sqrt{g.(L_p/5)}}\right) = \left(\frac{u_p}{\sqrt{g \cdot L_p}}\right)$$

$$\left(\frac{u_m}{\sqrt{(L_p/5)}}\right) = \left(\frac{1}{\sqrt{L_p}}\right)$$

$$u_m = \sqrt{\frac{L_p}{5 \cdot L_p}} = \sqrt{\frac{1}{5}} = 0.447m/s$$

We compare the Reynolds numbers

$$\mathrm{Re}_m = \mathrm{Re}_p$$

$$\left(\frac{u_m \rho_m L_m}{\mu_m}\right) = \left(\frac{u_p \rho_p L_p}{\mu_p}\right)$$

$$\left(\frac{0.447\rho_m(1/5)}{\mu_m}\right) = \left(\frac{\rho_p}{\mu_p}\right)$$

$$\frac{0.0894}{\nu_m} = \frac{1}{\nu_p}$$

$$\nu_m = 0.0894 \times 4.65 \times 10^{-5} = 4.15710 \times 10^{-6} m^2/s$$

Example 5.4

Example Water at 15.6°C flows at 4.32 m/s in a 140.4 mm diameter pipe. At what velocity must the fuel oil at 32.2°C flow inside a 60.2 mm pipe if we want to make the two flows to be dynamically similar?

Solution For the dynamic similarity, Reynolds number for water should be the same as that of Reynolds number for oil:

$$\left(\frac{u \cdot d}{\nu}\right)_{water} = \left(\frac{u \cdot d}{\nu}\right)_{oil}$$

$$\frac{4.32 \times 140.4}{1.130 \times 10^{-6}} = \frac{u_{oil} \times 60.2}{2.96 \times 10^{-6}}$$

$$u_{oil} = 26.4m/s$$

is the velocity of oil.

Example 5.5

Example Air at 20°C is to flow through a 590 mm pipe at an average velocity of 1.97 m/s. For dynamic similarity, what pipe diameter carrying water at 15.6°C at 2.40 m/s should be used?

Solution

$$\left(\frac{u \cdot d}{\nu}\right)_{air} = \left(\frac{u \cdot d}{\nu}\right)_{water}$$

$$\frac{1.97 \times 590}{14.86 \times 10^{-6}} = \frac{2.40 \times d_{water}}{1.130 \times 10^{-6}}$$

$$d_{water} = 36.8\ mm$$

Example 5.6

Example A scaled-down one-fifth model of a submarine is planned to be tested in a towing tank filled with sea water. If the prototype submarine moves at 4.20 m/s, at what velocity should the model be towed in sea water for dynamic similarity?

Solution We compare the Reynolds numbers for the model and prototype:

$$\left(\frac{4.20 \times L}{\nu}\right)_p = \left(\frac{u \times L/5}{\nu}\right)_m$$

$$u_m = 21 m/s$$

Example 5.7

Example A mechanical engineering student is investigating the new surfboard design in a water channel. As the surfboard will experience both laminar and turbulent flow, the Reynolds number must be used in scaling. Also, the boat prototype will experience different types of sea waves, so Froude number is important in scaling. Further to that, as the test is conducted in a small water channel, the surface tension will play an important role and Weber number should also be used for the scaling. The prototype of surfboard will have the width of 0.6 m and the depth of the bay is 5 m deep. The depth of the water tunnel is 0.5 m. The water temperature is 15°C.

Solution

Properties: The kinematic viscosity of water at 15°C is 1.2×10^{-4} m^2/s. The surface tension of water is 0.074 N/m. The density of water is 980 kg/m^3.

The water waves speed (celerity) in the bay will be shallow water waves which can be calculated by formula

$$U_p = \sqrt{g \cdot h_{bay}}$$

where h_{bay} is the depth of the water. Based on this speed and using prototype surfboard length $L_p = 0.6$ m, the Reynolds number, Froude number, and Weber number for the prototype can be calculated:

$$\mathrm{Re}_p = \frac{U_p L_p}{\nu} = 3.5 \times 10^6$$

$$Fr_p = \frac{U_p}{\sqrt{gL}} = 2.88$$

$$We_p = \frac{U_p}{\sqrt{\sigma/\rho L_p}} = 624.3$$

Now, for the scaling, we need to compare the $\mathrm{Re}_p = \mathrm{Re}_m$, $Fr_p = Fr_m$, and $We_p = We_m$. The depth of the water tunnel is 0.5 m so the wave speed on the model shall be:

$$U_m = \sqrt{g \cdot h_{tunnel}} = 2.21 m/s$$

By comparing the Reynolds numbers for both model and prototype we estimate the model length (L_m) required as:

$$\mathrm{Re} = \frac{U_p L_p}{\nu} = \frac{U_m L_m}{\nu}$$

$$L_{m Reynolds} = 1.897m$$

We compare the Froude numbers for both model and prototype and estimate the model length required as:

$$\frac{U_p}{\sqrt{gL_p}} = \frac{U_m}{\sqrt{gL_m}}$$

$$L_{m Froude} = 5.774$$

We compare the Weber numbers for both model and prototype and estimate the model length required as:

$$\frac{U_p}{\sqrt{\sigma/\rho L_p}} = \frac{U_m}{\sqrt{\sigma/\rho L_m}}$$

$$L_{m Weber} = 6$$

This shows that in the presence of many forces in the fluid the "correct" choice of characteristic length for geometric similarity becomes a complex problem. In this problem, the length of model according to Froude and Weber numbers is very close, but as per Reynolds number, matching this length will be short. Also it is not easy to dispose Reynolds number owing to the fact that a surfboard will experience different viscous flow regimes.

In Figure 5.1 the bird's wing size is related to its weight. This relationship has been investigated in detail since the 19th century. The famous German physicist Hermann von Helmholtz (1821–1894) considered the similarity law of flying animals in a paper published in 1873. He suggested that the wing loading in birds increases proportionally to the cubic root of their weights (see Figure 5.2).

A lot can be learned if we measure the velocity distributions over birds, insects, and other animals. Nature has already offered billions of designs of varying characteristic lengths, from insects to birds, tadpoles to whales, from fast-swimming

Figure 5.1 The birds' wing sizes are related to the weight of the bird.

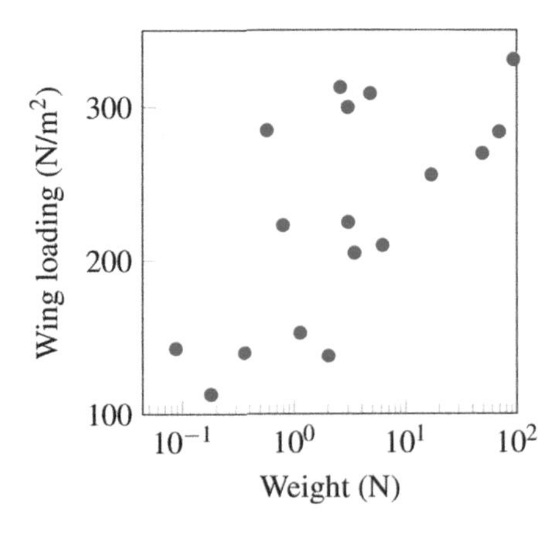

Figure 5.2 The wing loading on various species of birds versus their weight.

Figure 5.3 Bird in a subsonic wind tunnel. *(Author: Professor Jeremy Rayner, University of Leeds; courtesy of Dr Marian Muste, IIHR - Hydroscience & Engineering, University of Iowa., Source: Int. Asso. for Hydro-Environment Engg. and Research.)*

dolphins to fastest running cheetahs, etc. Figure 5.3 shows a bird's flight being investigated in a wind tunnel.

Nature-inspired designs are now a separate field of science called **biomimicry**. Learning from nature and then replicating natural designs, shapes, patterns, processes, and ecosystems can create regenerative designs close to natural human inclinations and natural order. Of course, we also need to use the bio-friendly and biodegradable materials in doing so. **Biomimetic designs** are designs that adapt the nature functions. For example, the use of the shark-skin-inspired surface sheets that can be placed on the airplanes is under investigation.

Table 5.4 lists the order of magnitude for the length of some vortices.

PROBLEMS

5P-1 The capillary rise (h) in a tube of small diameter d is a function of liquid density ρ, the tube diameter d, the gravitational constant g, the contact angle θ, and the surface tension σ of the liquid. Using dimensional analysis, find the dimensionless parameters.

5P-2 A propeller is being tested in a wind-tunnel by a researcher who opined that the thrust (F) depends upon important variables like speed of rotation (ω), speed of advance (U_o), diameter (D), air viscosity (μ), density (ρ), and speed of sound (c). Assuming that he is correct in his intuition, what dimensionless parameter he should use for data analysis to confirm his opinion?

5P-3 The height (h) of the tide that formed after the wind blowing over a lake depends on the average depth (z) of the lake. The wind shear stress is represented by τ. If the length of the lake is L, and γ is the specific weight of the water, how could these variables be related with each other?

Table 5.4
Order of Magnitude for Length of Some Vortices

Quantized vortices in liquified gases	$10^{-8} - 10^{-9}$ cm
Vortices generated by small and winged animals and insects	0.1 cm
Vortices behind leaves	0.1–10 cm
Vortex rings of Cephalopoda	
Dust whirls on the ground	1–10 m
Whirlpools in tidal currents	
Vortex rings in volcanic eruptions	100–1,000 m
Whirlwinds and waterspouts	
Clouds	
Vortices shed from the Gulf Stream	100–2,000 km
Hurricanes	
Ocean internal circulations	2,000–5,000 km
Convection cells inside the Earth	
Jupiter Red spot	5,000–10^5 km
Rings of the planet Saturn	
Sun spots	
Rotation inside of stars	
Galaxies	Light Years

5P-4 The choke-type throttle valve is depicted in Figure 5.4. In this device, the diameter is smaller than the length $d \ll \ell$. The important variables to consider are diameter, length, velocity, density, and absolute viscosity. If diameter (d) and length varies as ($\ell/d = 2 - 16$) with corresponding velocity variation in the range from 0.0003m/s to 0.03 m/s, find the discharge coefficient C_d in terms of relevant variables.

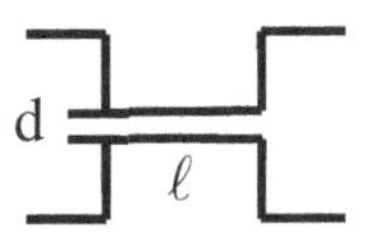

Figure 5.4 Figure of problem 5P-4.

5P-5 The average velocity u at a distance y from the boundary depends on the roughness height ε of the boundary, a length L that specifies the size of the system, the kinematic viscosity ν of the fluid, the mass density ρ of the fluid, and the shearing stress (τ) that the fluid exerts on the boundary. Hence, there is a relationship of the type $f(u,y,\rho,\tau,\varepsilon,\ \nu,\ L) = 0$. Write the dimensions of these quantities in M-L-T.

5P-6 The draining liquid film shown in Figure 5.5, can become wavy and unstable at a critical Reynolds number defined as $\delta U/\nu$. It is expected that the critical Reynolds number will depend on thickness of the falling film (δ), the density (ρ), the viscosity (μ), and the surface tension of the fluid (σ_e). Formulate the dimensionless parameters that are important in this study.

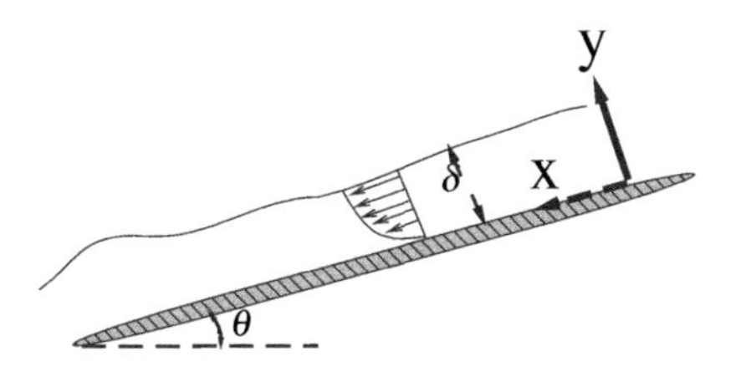

Figure 5.5 Figure of problem 5P-6.

5P-7 Heat transfer by the moving fluid depends on convective heat transfer coefficient h (W/m^{2}K), characteristic length thermal conductivity k (W/m · K), the specific heat c_p (J/kg · K), viscosity μ (Pa · s), velocity (m/s), and density (kg/m^3). The two dimensionless numbers that represent the physics of the problem are $Nu = h \cdot L/k$ and $Pr = c_p \cdot \mu/k$. Can you cast another dimension number using these numbers?

5P-8 Stoke's number is an important dimensionless parameter defined as:

$$St_D = \frac{\rho_p d_p^2 u_{avg}}{18\mu\ell}$$

where ρ is density, d_p is the particle diameter, $u_a vg$ is the average velocity, ℓ is the characteristic length, and μ absolute viscosity. If $St_D<1$, the small particles with $d_p<$ 100 nm follow the flow pathlines, the situation required for inhaler designs when droplets of liquid medicine must reach the lungs instead of depositing on human trachea. Prove that this parameter is dimensionless.

5P-9 Charles Renard (1847–1905), a French aeronautics engineer presented the following formula for the bird flights. According to him, the minimum power required for the bird to fly is

$$\frac{P}{W} \equiv C\sqrt[2]{\frac{W}{\rho S}}$$

where P, W, ρ, and S are the minimum power required by a bird to fly, weight of the bird, density of air, and wing surface area of bird. The C is constant and Renard suggested the value of 0.18. Prove that equation is dimensionally consistent.

5P-10 Dutch mathematician Diederik Korteweg, in 1878, proposed that the material properties of the tube wall are related with the speed of wave propagation in a fluid that is carried by the tube. He proposed the relation:

$$c = \left(\frac{t \cdot E}{2\rho r}\right)^{1/2}$$

where c is the pulse wave velocity, E is the modulus of elasticity of the vessel wall material, t is the wall thickness of the vessel, r is the internal radius of the tube, and ρ is the density of the fluid. This equation is known as Moens-Korteweg equation. Can you prove that the term $(E \cdot t/\rho \cdot r)$ is dimensionally equal to velocity?

Table 5.5
Dimensionless Parameters

Parameter Name	Formula	Definition
Reynolds Number	$\mathrm{Re} = \frac{\rho \cdot U \cdot L}{\mu}$	$\frac{Inertia\ Force}{Viscous\ Force}$
Mach Number	$\frac{U}{a}$	$\frac{Velcoity}{Speed\ of\ Sound}$
Weber Number	$We = \frac{\rho L u^2}{\sigma_e}$	$\frac{Inertia\ Force}{Surface\ Tension\ Force}$
Ohnesorge Number	$Oh = \frac{\mu}{\sqrt{\rho \sigma_e R_o}}$	$\frac{\sqrt{We}}{Re}$
Galileo Number	$Ga = \frac{gL^3}{\nu^2}$	$\frac{Gravity\ Force}{Viscous\ Force}$
Froude Number	$Fr = \frac{u}{\sqrt{Lg}}$	$\frac{\text{Flow Speed}}{\text{Gravity wave speed}}$
Rossby Number	$Ro = \frac{V}{L \cdot f}$	$\frac{Inertia\ Force}{Coriolis\ Force}$
Ekman Number	$Ek = \frac{\nu}{L^2 f}$	$\frac{Frictional\ Force}{Coriolis\ Force}$
Strouhal Number	$St = \frac{f \cdot D}{U}$	$\frac{Vortex\ Shedding\ Frequency}{Frequency\ of\ Bulk\ Flow}$
Bond Number or Eötvös Number	$Bo = Eo = \frac{\rho g \ell^2}{\sigma_e}$	$\frac{Gravitational Force}{Surface\ Tension\ Force}$
Womersley Number	$Wo = St \cdot \mathrm{Re} = \frac{\rho \omega D^2}{\mu}$	$\frac{Transient\ Inertia\ Force}{Viscous\ Force}$
Cavitation Number	$\sigma = \frac{P_v - P}{\frac{\rho}{2} U^2}$	$\frac{P_{fluid} - P_{vapor}}{P_{dynamic}}$
Euler Number	$Eu = \frac{P_2 - P_1}{\frac{\rho}{2} U^2}$	$\frac{P_{s,2} - P_{s,1}}{P_{dynamic}}$
Knudsen Number	$\frac{\lambda_{mfp}}{\ell}$	$\frac{Mean\ Free\ Path}{Characteristic\ Length}$

REFERENCES

E. Buckingham, On physically similar systems: Illustrations of the use of dimensional Equations, Phys. Rev., ser. 2, vol. 4, no. 4, p. 345, 1914.

E. R. Van driest, On Dimensional analysis and the presentation of data in fluid flow problems, J. Appl. Mech., vol. 13, p. A34, 1946.

J. Kline, Similitude and Approximation Theory. New York: McGraw Hill pp. 28, 63, 1965.

S. W. Churchill, The Interpretation and Use of Rate Data: The Rate Process Concept, revised printing. Washington, DC: Hemisphere Publishing Corp., Ch. 10, figs. 18–21, 1979.

H. L. Langhaar, Dimensional Analysis and Theory of Models. New York: John Wiley and Sons, 1951.

R. A. Gore, Clayton T. Crowe, and Adrian Bejan, The geometric similarity of the laminar sections of boundary layer-type flows, vol. 17, no. 4, 465–475, 1990.

H. J. Lugt, Vortex Flow in Nature and Technology. John Wiley and Sons, USA, 1983.

T. V. Karman, Aerodynamics. McGraw Hill, USA, 1968.

6 The Integral Analysis

Many engineering problems can be solved by using the lumped, or integral, analysis. This type of analysis is executed if we are interested in the mean values of velocities instead of the velocity distributions. The lumped analysis is also called the control volume analysis in the fluid mechanics literature. The balance of mass, momentum, and energy can be done through an equation called the Reynolds transport theorem. Also, a special equation called the Bernoulli equation will be discussed, which originated from the first law of thermodynamics.

Learning outcomes: After finishing this chapter, you should be able to:

- Understand differences between differential and integral analysis.
- Develop the lumped analysis equation called Reynolds transport theorem.
- Do the integral analysis of the mass, momentum, and energy balance.
- Learn the basis of Bernoulli equation and its use.

6.1 INTEGRAL FORMULATION OF CONTINUITY EQUATION

Consider a system of fluid is passing through a control volume and flow field and velocity $V(x,y,z,t)$ is generated. The streamlines are shown in Figure 6.1, and system takes a new position in time $t+\Delta t$. An extensive system property N is changing in the system. η is N per unit mass and $\forall$ is the volume of the system. Figure 6.1 shows the balance of conservation quantities.

Three distinct regions are:

I– System at time t

II– System passing

III– System at time $t+\Delta$ t

$$\frac{dN}{dt} = Lim_{\Delta t \to 0}\left\{\frac{(N_{III}+N_{II})_{t+\Delta t}-(N_I+N_{II})_t}{\Delta t}\right\} \tag{6.1}$$

$$\frac{dN}{dt} = Lim_{\Delta t \to 0}\left\{\frac{[(N_{II})_{t+\Delta t}-(N_{II})_t]+(N_{III})_{t+\Delta t}-(N_I)_t}{\Delta t}\right\} \tag{6.2}$$

We isolate the surface net out flow from the transport quantity inside the control volume.

DOI: 10.1201/9781003315117-6

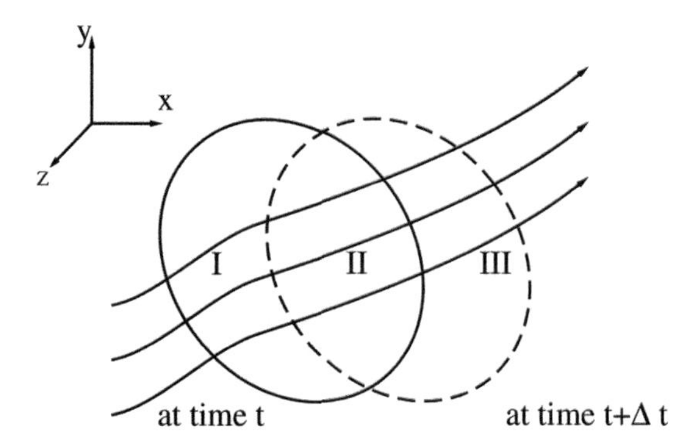

Figure 6.1 The balance of conservation quantities as control volume moves in space.

$$\frac{dN}{dt} = \underbrace{Lim_{\Delta t \to 0}\left\{\frac{[(N_{II})_{t+\Delta t} - (N_{II})_t]}{\Delta t}\right\}}_{control\ volume} + \underbrace{Lim_{\Delta t \to 0}\left\{\frac{[(N_{III})_{t+\Delta t} - (N_I)_t]}{\Delta t}\right\}}_{control\ surface}$$

We consider first the control volume term:

$$Lim_{\Delta t \to 0}\left\{\frac{[(N_{II})_{t+\Delta t} - (N_{II})_t]}{\Delta t}\right\} \tag{6.3}$$

The quantity accumulation inside control volume is related to mass of the control volume:

$$Lim_{\Delta t \to 0}\left\{\frac{[(N_{II})_{t+\Delta t} - (N_{II})_t]}{\Delta t}\right\} \equiv \left(\frac{1}{\Delta t}\right)\left(\frac{N}{m}\right)\rho\forall$$

$$Lim_{\Delta t \to 0}\left\{\frac{[(N_{II})_{t+\Delta t} - (N_{II})_t]}{\Delta t}\right\} = \frac{\partial}{\partial t}\int_{CV} \eta\rho d\forall$$

We now consider the control surface term:

$$Net\ efflux\ rate = Lim_{\Delta t \to 0}\left[\frac{(N_{III})_{t+\Delta t} - (N_I)_t}{\Delta t}\right]$$

$$Lim_{\Delta t \to 0}\left\{\frac{(N_{III})_{t+\Delta t}}{\Delta t}\right\} - Lim_{\Delta t \to 0}\left\{\frac{(N_I)_t}{\Delta t}\right\}$$

Dimensionally, we can formulate the rate of mass flow rate using the density, area, and velocity as:

$$\frac{m}{\Delta t} = \rho \cdot A \cdot V = \left(\frac{kg}{m^3}\right) m^2 \left(\frac{m}{s}\right) = \frac{kg}{s}$$

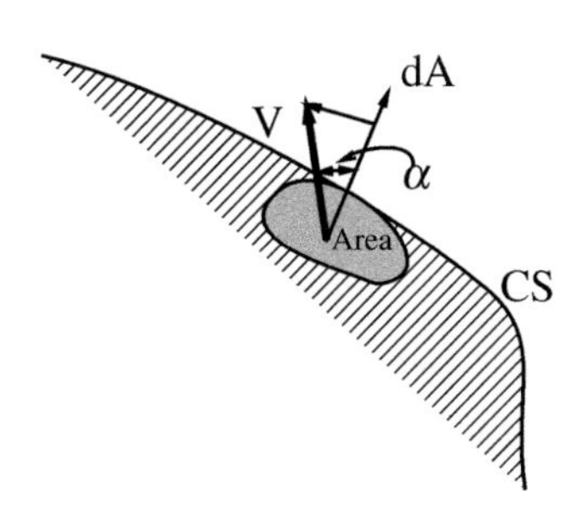

Figure 6.2 The velocity and area vector on the surface of the control volume.

$$Lim_{\Delta t \to 0}\left\{\frac{(N_{III})_{t+\Delta t}}{\Delta t}\right\} = \left(\frac{m}{\Delta t}\right)\left(\frac{N}{m}\right) = \dot{m}\eta = \rho_2 A_2 V_2 \eta_2$$

$$Lim_{\Delta t \to 0}\left\{\frac{(N_I)_t}{\Delta t}\right\} = \rho_1 A_1 V_1 \eta_1$$

$$Lim_{\Delta t \to 0}\left\{\frac{[(N_{III})_{t+\Delta t} - (N_I)_t]}{\Delta t}\right\} = \rho_2 A_2 V_2 \eta_2 - \rho_1 A_1 V_1 \eta_1$$

Figure 6.2 shows the velocity and area vectors. Only the velocity components aligned with area vector are relevant. The volume of fluid that swept out of area dA in time dt is $V\ dA\ cos\ \alpha$, which can be written in vector form as $(\mathbf{V} \cdot dA)$.

Hence we can write:

$$Lim_{\Delta t \to 0}\left\{\frac{[(N_{III})_{t+\Delta t} - (N_I)_t]}{\Delta t}\right\} = \rho_2 A_2 V_2 \eta_2 - \rho_1 A_1 V_1 \eta_1 \equiv \int_{CS} \eta\rho\mathbf{V} \cdot dA$$

$$\frac{dN}{dt} = \int_{CS} \eta\rho\mathbf{V} \cdot dA + \frac{\partial}{\partial t}\int_{CV} \eta\rho d\forall \tag{6.4}$$

This equation is known as **Reynolds transport theorem**. The Reynolds transport theorem, though named after Osborne Reynolds (1842–1912) is, in fact, a three-dimensional generalization of Leibniz integral rule.

We can generate equations for mass, momentum, and energy conservations in integral form by changing the η. For example, since $\eta =$ N/mass, for mass conservation we set $\eta = 1$, leading to an equation:

$$\dot{m} = \int_{CS} \rho\mathbf{V} \cdot dA + \frac{\partial}{\partial t}\int_{CV} \rho d\forall \tag{6.5}$$

for steady flows:

$$\dot{m} = \int_{CS} \rho\mathbf{V} \cdot dA \tag{6.6}$$

These equations are frequently used in fluid mechanics for quick calculations for the mass balance. In a simplified form, for the steady flows, we can write mass balance as:

$$\dot{m}_1 = \dot{m}_2 \tag{6.7}$$

$$\rho_1 A_1 V_1 = \rho_2 A_2 V_2 \tag{6.8}$$

where ρ is the density, A_1 is the cross-sectional area, and V_1 is the velocity. This equation arises from Reynolds transport theorem and is also called the continuity equation. In the case of incompressible flow, the density is constant and we are interested in the balance of volume flow rate instead of mass flow rate. The volume flow rate is indicated by letter Q in this book. The balance of volume flow rate is:

$$Q_1 = Q_2$$

$$A_1 V_1 = A_2 V_2 \tag{6.9}$$

We will now show through examples how we can solve fluid flow problems using Bernoulli's equation and Reynolds transport theorem for mass conservation.

Example 6.1

Example: The vaccine injection problem. A vaccine is being injected into a human arm. The syringe piston is being pushed at the rate of 0.6cm/s. The piston allows the air to move through the peripheral region at a rate of 0.0160 cm^3/s. Find the rate at which the vaccine is injected into the blood. The syringe diameter is 0.5 cm and needle diameter is 0.5 mm.

Solution Consider the syringe as a control volume with the needle included in it. The volume of the control volume is reducing as syringe's piston is being pushed. The pushing of the piston can be taken as mass going into the control volume. From Reynolds transport theorem, we have

$$\int_{CS} \rho \mathbf{V} \cdot dA + \frac{\partial}{\partial t} \int_{CV} \rho d\forall = 0$$

$$+\left(\rho u_{needle} \frac{\pi}{4} d^2\right)_{needle} - \left(\rho u_{piston} \frac{\pi}{4} D^2\right)_{piston} + \rho \frac{d\forall}{dt} = 0$$

$$\underbrace{+\left(\rho u_{needle} \frac{\pi}{4} (0.5 \times 10^{-3})^2\right)_{needle}}_{mass\ out}$$

$$\underbrace{-\left(\rho (0.6 \times 10^{-2}) \frac{\pi}{4} (0.5 \times 10^{-2})^2\right)_{piston}}_{mass\ in}$$

$$+\rho(-0.016 \times 10^{-6}) = 0$$

$$-1.96\times10^{-7}u_{needle}=-1.177\times10^{-7}-0.016\times10^{-6}$$

$$u_{needle}=0.682\text{m/s}$$

Example 6.2

Example Oil is pouring out from a triangular trough at a speed of $V=7\sqrt{(z)}$, where z is the local coordinate as shown in Figure 6.3. The half angle of triangular trough is 35°C. Find the time needed to fully drain the contents of the trough. The trough has a width of $w=3$ ft. The area A at the outlet is 10 in. The density of oil is 870 kg/m^3 and it's depth is 5 ft.

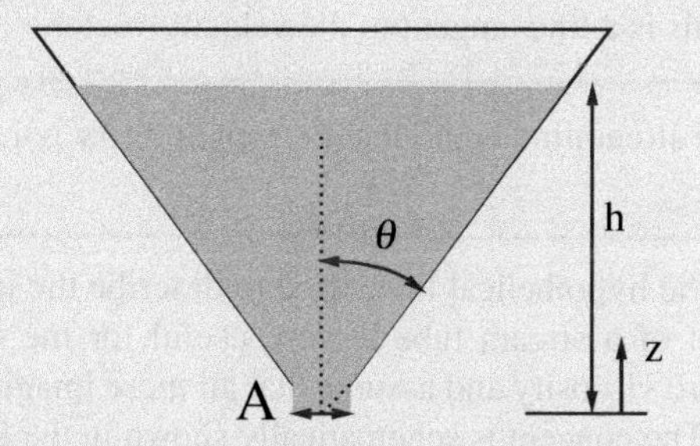

Figure 6.3 Trough drainage problem.

Solution The volume of the trough is

$$\forall=2A_{triangle}\cdot w=\left(\frac{2}{2}z\tan(\theta)\cdot z\right)w=z^2w\tan(\theta)$$

The mass balance in integral formulation is

$$\frac{\partial}{\partial t}\int_{CV}\rho d\forall+\int_{CS}\rho V\cdot dA=0$$

$$\rho\frac{\partial\forall}{\partial t}=-\rho V\cdot A$$

$$\rho\frac{d(z^2\tan(\theta)w)}{dt}=-7\rho\sqrt{z}\left(\frac{10}{144}\right)$$

$$2w\tan(\theta)z\frac{dz}{dt}=-\frac{70}{144}\sqrt{z}$$

$$\left(\frac{-2\times144}{70}w\tan(\theta)\right)\frac{z}{\sqrt{z}}dz=dt$$

$$t = \left(\frac{-2 \times 144}{70} w \tan(\theta)\right) \int_h^0 \sqrt{z} dz$$

$$= \left(\frac{-2 \times 144}{70} w \tan(\theta)\right) \left. \frac{z^{\frac{1}{2}+1}}{\frac{1}{2}+1} \right|_{h=5}^{0}$$

$$t = \frac{8.63 \times 2}{3} 5^{3/2} = 96.6s$$

6.2 STREAM TUBE THEORY

In fluid mechanics, we often describe the flow by an imaginary line known as streamline. Theoretically this is a line tangent to the velocity vector.

Streamline: The streamline is the line tangent at every point to the velocity vector.

Streamlines are the hypothetical lines used to describe the inviscid or ideal fluid motion. The concept of a stream tube is very useful for the solution of the fluid problems, if we ignore viscosity and assume that all these imaginary streamlines can form a stream tube. The concept is schematically shown in the Figure 6.4.

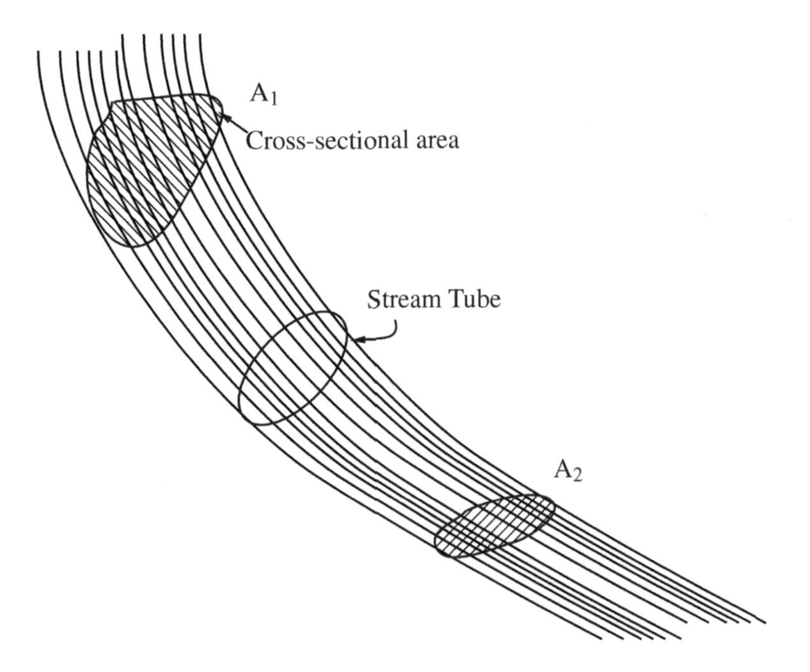

Figure 6.4 The concept of stream tube.

6.3 ENERGY EQUATION

The **law of conservation of energy** is also known as **first law of thermodynamics**, which says that

> *Energy cannot be created or destroyed. It can transfer from one object to another and may change forms.*

Therefore there are several forms of energies, like sound energy, wave energy, solar energy, wind energy, nuclear energy, etc. The energy E is a conserved quantity and will remain the same. Energy can exist in several forms, such as chemical, mechanical, kinetic, potential, electric, nuclear, and magnetic, etc. The total energy of a system (E) on a unit mass basis is denoted by e.

6.4 LUMPED ENERGY ANALYSIS

Consider an open system through which energy and mass are moving. Let's say there is a thermodynamic quantity of interest Φ, which got transported, convected, or diffused due to sources or sinks. Mathematically, one can write the balance of quantity Φ as:

$$\begin{aligned}\frac{D\Phi_{system}}{Dt} &= \underbrace{\sum_i (Convection)_i}_{at\ System\ boundaries} + \underbrace{\sum_j (Diffusion)_j}_{at\ System\ boundaries} \\ &+ \underbrace{\sum_k (Source)_k + \sum_\ell (Sink)_\ell}_{in\ system} \\ &+ \underbrace{\sum_m (Field)_m}_{acting\ on\ System\ volume}\end{aligned} \tag{6.10}$$

Figure 6.5 shows the schematic representation of energy transport processes.

Convection represents the macroscopic transport of the thermodynamic quantities with flow or mass. In a close system, there would be no convection.

Diffusion is a phenomenon that happens due to gradient of the potential. It happens with the molecular movements. Any aspect of diffusion inside a system will be an aspect of internal energy and covered in it. In case of mass transfer, the potential is the concentration difference. In case of heat transfer, the potential is temperature difference.

Souce and sink terms represent the transport due to effects internal to a system. The internal heat generations in the nuclear fuel rods is also an example of the same category.

Field effect represents the transport that arises due to fields affecting directly the volume of the system. Transports that are induced due to gravity or magnetic fields come under this category.

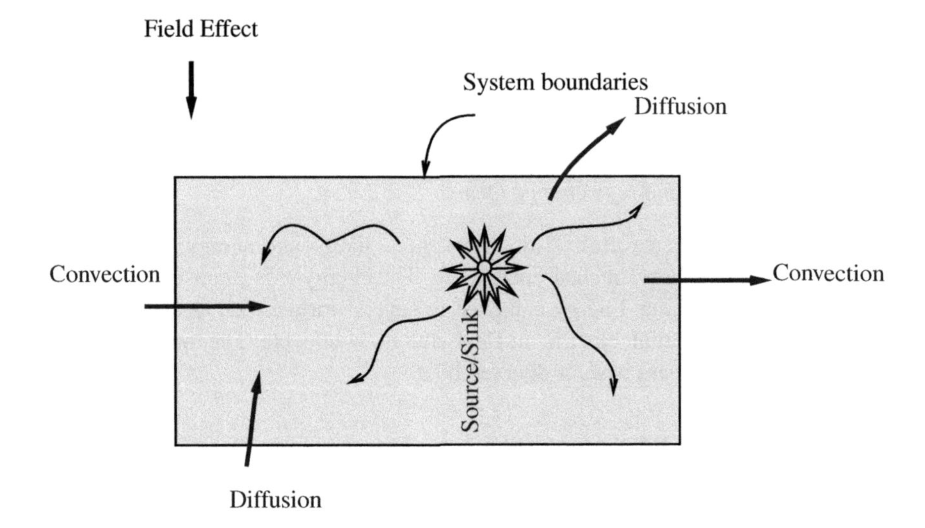

Figure 6.5 The energy transfer through control volume.

$$\left(\frac{DE}{Dt}\right)_{System} - \underbrace{\sum(e\dot{m})_{in} + \sum(e\dot{m})_{out}}_{at\ System\ boundaries} = \sum \dot{Q} + \sum \dot{W} - \underbrace{\sum p\frac{dV}{dt}}_{at\ System\ boundaries} \tag{6.11}$$

In a process the system volume may increase or decrease in time ($dV < 0$ or $dV > 0$). Since $\dot{V} = \dot{m}\upsilon$. The term $P(dV/dt)$ is called the flow work and can be written as $\dot{W}_{flow} = P\dot{m}\upsilon$ and the Eq. can be written as:

$$\left(\frac{DE}{Dt}\right)_{System} - \underbrace{\sum(e\dot{m})_{in} + \sum(e\dot{m})_{out}}_{at\ System\ boundaries} = \sum \dot{Q} + \sum \dot{W} + \underbrace{\sum_{in}(P\upsilon\dot{m}) - \sum_{out}(P\upsilon\dot{m})}_{at\ System\ boundaries} \tag{6.12}$$

The Energy E per unit mass is defined as:

$$e = \left(\mathbf{u} + gz + \frac{\widetilde{u}^2}{2}\right) \tag{6.13}$$

We can define a property enthalpy(h) as $u+P\upsilon$.

$$\left(\frac{DE}{Dt}\right)_{System} \underbrace{-\sum((e+P\upsilon)\dot{m})_{in}+\sum((e+P\upsilon)\dot{m})_{out}}_{at\ System\ boundaries} = \sum\dot{Q}+\sum\dot{W} \tag{6.14}$$

$$\left(\frac{DE}{Dt}\right)_{open} -\sum_{in}\dot{m}\left(h+gz+\frac{\widetilde{u}^2}{2}\right)+\sum_{out}\dot{m}\left(h+gz+\frac{\widetilde{u}^2}{2}\right) = \sum\dot{Q}+\sum\dot{W} \tag{6.15}$$

This is the first law of thermodynamics for open systems where energy and mass can flow through system. However, we are not interested in distribution of energy inside the system. We are interested in overall bulk energy balance. We write $\vec{V}$ in the equation:

$$\left(\frac{DE}{Dt}\right)_{open} -\sum_{in}\dot{m}\left(h+gz+\frac{\vec{V}^2}{2}\right)+\sum_{out}\dot{m}\left(h+gz+\frac{\vec{V}^2}{2}\right) = \sum\dot{Q}+\sum\dot{W}$$

6.5 BERNOULLI EQUATION

The equation of motion, which we developed in the last chapter, known as Navier-Stokes equation is very complicated; therefore, a simplified version of this equation is often used in the engineering analysis. For the case of incompressible flows, the equation of motion and the first law of thermodynamics are the same, which will now be explained in this section. Consider Eq. 6.16:

$$\frac{\partial\vec{V}}{\partial t}+\vec{V}\cdot\nabla\vec{V} = -\frac{1}{\rho}\nabla P+\nu\,\nabla^2\vec{V}+f_y \tag{6.16}$$

Special case: Ignoring the viscous effects, but retaining the inertia force term, we have the reduced form of equation:

$$\frac{\partial\vec{V}}{\partial t}+\vec{V}\cdot\nabla\vec{V} = -\frac{1}{\rho}\nabla P+f_b \tag{6.17}$$

The above equation is called *Euler's equation* in fluid mechanics literature.

Convective acceleration and vorticity: We now expand the convective acceleration term as:

$$(\vec{V}\cdot\nabla)\vec{V} = \nabla\left(\frac{\vec{V}\cdot\vec{V}}{2}\right)-\vec{V}\times(\nabla\times\vec{V}) = \nabla\left(\frac{\vec{V}^2}{2}\right)-\omega\times\vec{V}$$

where ω is vorticity. Interestingly, we can write the acceleration in flow as:

$$\frac{D\vec{V}}{Dt} = \frac{\partial \vec{V}}{\partial t} + \nabla\left(\frac{\vec{V}\cdot\vec{V}}{2}\right) - \omega \times \vec{V}$$

This shows that viscous forces do contribute in the flow acceleration.

Special case: Assuming that there is no rotation of fluid particles present, so neither vorticity term is present nor the viscous diffusion term can be considered. We now write the the equation as:

$$\frac{\partial \vec{V}}{\partial t} + \nabla\left(\frac{\vec{V}^2}{2}\right) - \underbrace{\omega \times \vec{V}}_{zero} = -\frac{1}{\rho}\nabla P + \underbrace{\nu\, \nabla^2 \vec{V}}_{zero} + f_x$$

Substituting the body force into this equation:

$$f_y = \frac{F_b}{m} = \frac{-W}{m} = \frac{-mg}{m} = -g$$

We now integrate the vector form of this equation for only one space coordinate.

$$\frac{\partial \vec{V}}{\partial t} + \frac{1}{2}\frac{d}{ds}(\vec{V}^2) = -\frac{1}{\rho}\frac{\partial P}{\partial s} + f_b$$

$$\int \frac{\partial \vec{V}}{\partial t} ds + \int \frac{1}{2}\frac{d}{ds}(\vec{V}^2) ds = \int \left(-\frac{1}{\rho}\frac{\partial P}{\partial s}\right) dx - \int (g) ds$$

$$\int \frac{\partial \vec{V}}{\partial t} ds + \frac{1}{2}(\vec{V}^2) = -\frac{P}{\rho} - (g)\ s$$

It is customary to write this equation in terms of user-specified datum, which is often represented as z.

$$\int \frac{\partial \vec{V}}{\partial t} ds + \frac{1}{2}(\vec{V}^2) + \frac{P}{\rho} + (g)\ z = 0 \tag{6.18}$$

Eq. 6.18 is called the *unsteady Bernoulli equation* in fluid mechanics literature.

Example 6.3

Example A water reservoir is connected to a long pipe that is exiting in open atmosphere as shown in Figure 6.6. The water in the reservoir has almost zero velocity and pressure; above free surface is maintained at 288 lb_f/ft^2. Derive an equation for the velocity in the long pipe as a function of time. The water level inside the reservoir is maintained at 5 ft and the length of the pipe ℓ is 40 ft.

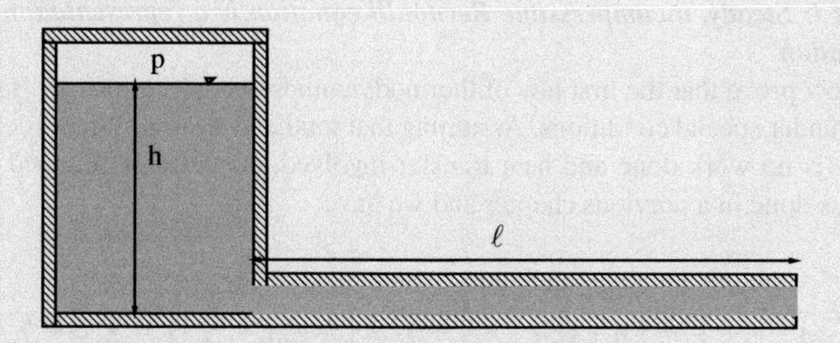

Figure 6.6 Schematic diagram of Example 6.3.

Solution We use the unsteady Bernoulli's equation:

$$\int \frac{\partial \vec{V}}{\partial t} ds + \frac{1}{2}(\vec{V}^2) + \frac{P}{\rho} + g\,z = 0$$

$$\left(\frac{1}{2}(\vec{V}^2) + \frac{P}{\rho} + g\,z\right)_1 = \left(\frac{1}{2}(\vec{V}^2) + \frac{P}{\rho} + g\,z\right)_2 + \int_1^2 \frac{\partial \vec{V}}{\partial t} ds$$

$$\cancel{\frac{1}{2}(\vec{V}_1^2)} + \frac{P_1}{\rho} + g\,z_1 = \frac{1}{2}(\vec{V}_2^2) + \cancel{\frac{P_2}{\rho}} + \int_1^2 \frac{\partial \vec{V}}{\partial t} ds$$

$$\frac{P_1}{\rho} + g\,z_1 = \frac{1}{2}(\vec{V}_2^2) + \frac{\partial \vec{V}}{\partial t} \int_1^2 ds$$

We write $V_2 = V$ and equation is

$$\frac{P_1}{\rho} + g\,h = \frac{1}{2}(\vec{V}^2) + \frac{\partial \vec{V}}{\partial t} \ell$$

$$\frac{1}{\ell}\left[\frac{P_1}{\rho} + g\,h\right] = \frac{\vec{V}^2}{2\ell} + \frac{\partial \vec{V}}{\partial t}$$

The solution of this differential equation is

$$\vec{V}(t) = \sqrt{2\left[\frac{P_1}{\rho} + g\,h\right]} \tanh\left[t\left(\sqrt{\frac{\left[\frac{P_1}{\rho} + g\,h\right]}{2\ell^2}}\right)\right]$$

To cover 50 ft will require large time, and at very large time, $\tanh(\infty) \approx 1$, so we can have an approximate solution

$$\vec{V}(t) = \sqrt{2\left[\frac{P_1}{\rho} + g\,h\right]}$$

Substituting the data we have

$$\vec{V}(t) = 18.22 ft/s$$

Proof I: Steady, incompressible Bernoulli equation is a representation of energy equation

We now prove that the first law of thermodynamics can take a form of Bernoulli equation under special conditions. Assuming that total energy of a system is constant and there is no work done and heat transfer involved, consider the lumped energy analysis as done in a previous chapter and we have:

$$\cancel{\left(\frac{DE}{Dt}\right)_{open}} - \sum_{in} \dot{m}\left(h+gz+\frac{\vec{V}^2}{2}\right) + \sum_{out} \dot{m}\left(h+gz+\frac{\vec{V}^2}{2}\right) = \cancel{\sum \dot{Q} + \sum \dot{W}}$$

$$\sum_{out-in} \dot{m}\left(\mathbf{u}+pv+gz+\frac{\vec{V}^2}{2}\right) = 0$$

$$\sum_{out-in} \left(\mathbf{u}+\frac{p}{\rho}+gz+\frac{\vec{V}^2}{2}\right) = 0$$

Assuming internal energy (**u**) is constant, so net internal energy change is zero.

$$\sum_{out-in} \left(P+\gamma z+\rho\frac{\vec{V}^2}{2}\right) = 0$$

or simply

$$P+\gamma z+\rho\frac{\vec{V}^2}{2} = C \tag{6.19}$$

Note that in case of steady flow, with no work, heat transfer and constant internal energy case the lumped momentum (Eq. 6.18) and lumped energy equations (Eq. 6.19) are the same.

Bernoulli equation in head form is

$$\frac{p}{\gamma}+\frac{\vec{V}^2}{2g}+z = const$$

This shows that pressure, kinetic and potential energies in head form represent the total energy in the head form. This line is sometimes referred as Energy Gradient Line (EGL). Figure 6.7 schematically shows the energy in the fluid.

Example 6.4

Example The gas of density ρ_1 is heated up, and its density is reduced to ρ_2. The gas rises into the chimney and is released into the open atmosphere for dispersion. Consider this flow as inviscid and find the speed of gas coming out of the chimney in terms of inlet velocity. The distance from the heater to the chimney's outlet is h.

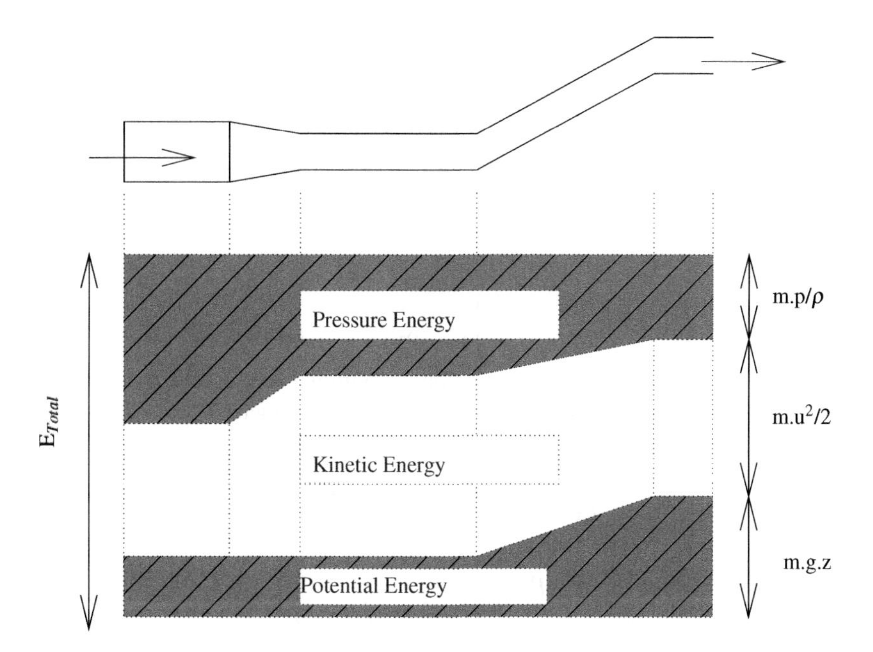

Figure 6.7 Energy Gradient Line (EGL).

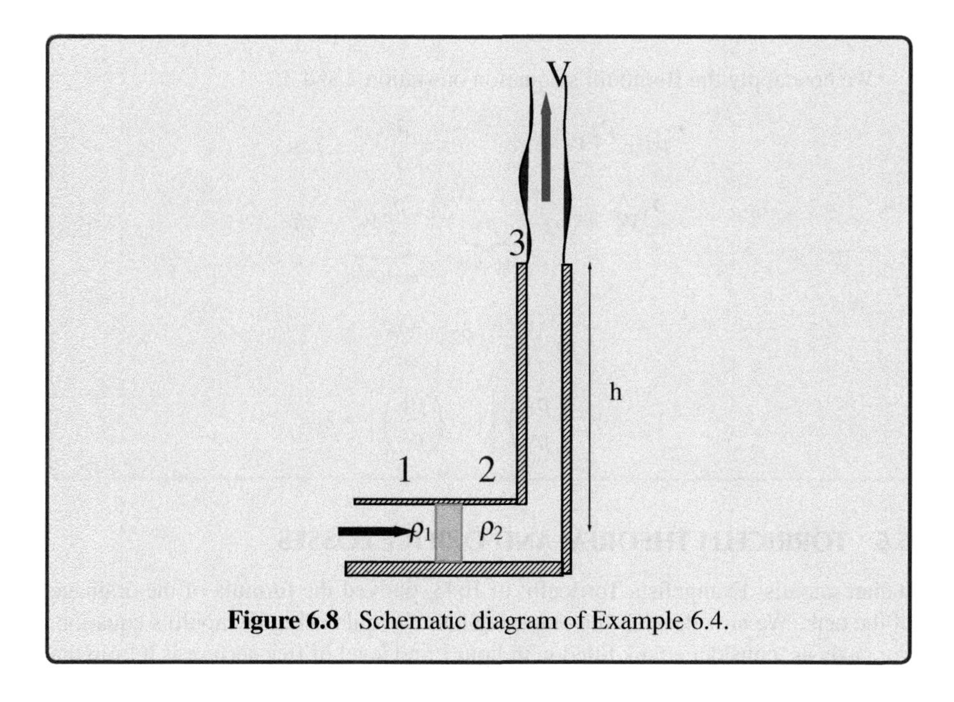

Figure 6.8 Schematic diagram of Example 6.4.

Solution We first apply the continuity equation on station 1 and 2:

$$\rho_1 A_1 V_1 = \rho_2 A_2 V_2$$

Since the area of the heater is the same:

$$\rho_1 V_1 = \rho_2 V_2$$

$$\frac{V_2}{V_1} = \frac{\rho_1}{\rho_2}$$

We now apply the Bernoulli's equation on station 1 and 2. The pressure at station 1 is assume to be the same as atmospheric zero gauge pressure. Also, the height is the same and is ignored in analysis.

$$\cancel{p_1}^{0\ gauge} + \frac{\rho_1}{2} V_1^2 = p_2 + \frac{\rho_2}{2} V_2^2$$

$$p_2 = \frac{\rho_1}{2} V_1^2 \left[1 - \left(\frac{V_2^2}{V_1^2} \right) \left(\frac{\rho_2}{\rho_1} \right) \right]$$

Substituting velocity ratio from the continuity equation we have:

$$p_2 = \frac{\rho_1}{2} V_1^2 \left[1 - \left(\frac{\rho_1}{\rho_2} \right) \right]$$

We now apply the Bernoulli's equation on station 2 and 3:

$$p_2 + \frac{\rho_2}{2} V_2^2 + \gamma z_2 = p_3 + \frac{\rho_3}{2} V_3^2 + \gamma z_3$$

$$\frac{\rho_3}{2} V_3^2 = (p_2 - \underbrace{p_3}_{0\ gauge}) + \underbrace{\frac{\rho_2}{2} V_2^2}_{negligible} - \gamma h$$

$$\frac{\rho_3}{2} V_3^2 = (p_2) - \gamma h$$

$$V_3^2 = V_1^2 \left(\frac{\rho_1}{\rho_2} \right) \left[1 - \left(\frac{\rho_1}{\rho_2} \right) \right] - 2gh$$

6.6 TORRICELLI THEOREM AND ORIFICE LOSSES

Italian scientist Evangelista Torricelli, in 1643, derived the formula of the drainage of the tank. We now develop the emptying tank formula using Bernoulli's equation. For analysis, consider a tank filled with liquid, and level of free surface is h from the base of the outlet.

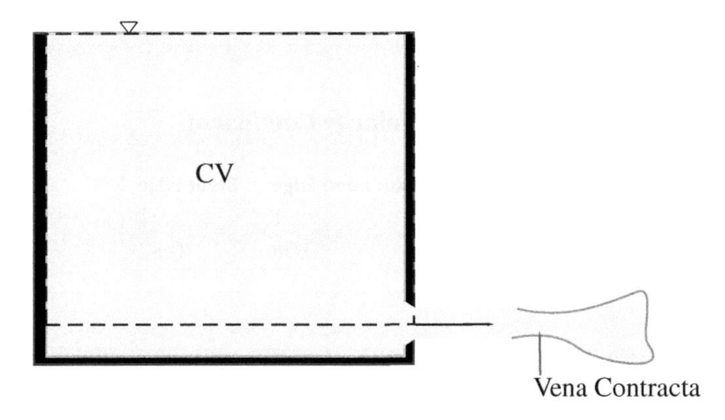

Figure 6.9 Torricelli theorem and the vena contracta.

We apply the Bernoulli's equation at free surface, called station 1, and at the outlet, called station 2:

$$\left(p+\frac{1}{2}\rho v^2+\gamma z\right)_1=\left(p+\frac{1}{2}\rho v^2+\gamma z\right)_2$$

$$\left(\underbrace{p}_{zero\ gauge}+\underbrace{\frac{1}{2}\rho v^2}_{zero}+\gamma z\right)_{free\ surface}=\left(\underbrace{p}_{zero\ gauge}+\frac{1}{2}\rho v^2+\underbrace{\gamma z}_{z=0}\right)_{tank\ outlet}$$

$$v=\sqrt{2gz}$$

If the jet is coming out of an orifice then the flow is slowed down due to friction, and the jet area would be reduced, and the minimum area is called the *vena contracta*. A contraction coefficient is used to describe the loss associated with jet contraction and the velocity is corrected as:

$$v=c_C\sqrt{2gz}$$

The actual flow rate is

$$Q_{actual}=\left(c_v c_C\sqrt{2gz}\right)A_e$$

where c_v is called velocity coefficient and its value is determined experimentally, A_e is the orifice exit area.

The two coefficients are combined into one and the cumulative loss is described by the discharge coefficient.

$$Q_{actual}=\left(c_d\sqrt{2gz}\right)A_e$$

Table 6.1
Typical Values of contraction and Velocity Coefficient

	Sharp Edge	Rounded Edge	Blunt Edge
c_c	0.61	1	1
c_v	0.98	0.98	0.82

Example 6.5

Example A jet is issuing from a tank and the trajectory of the jet is described by x amd y displacements as shown in Figure 6.10. Find the expression for the coefficient of velocity.

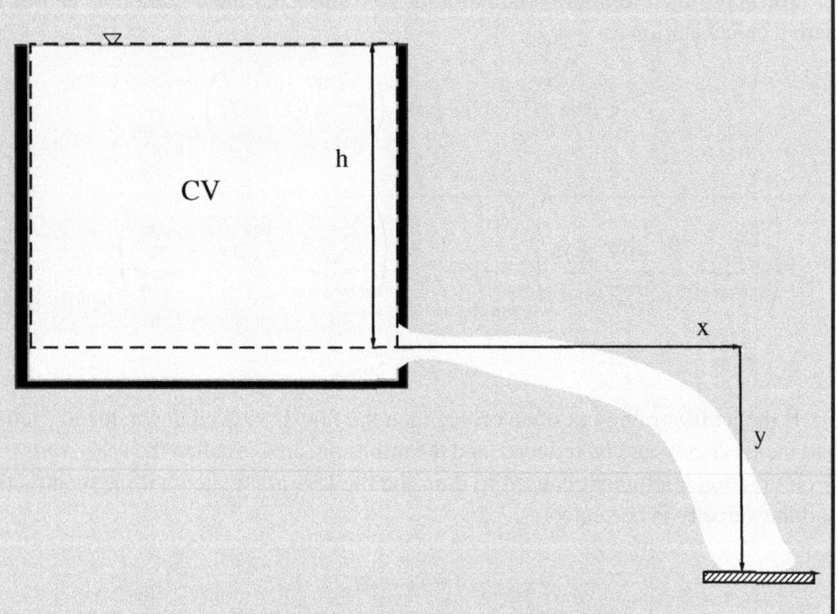

Figure 6.10 Schematic diagram of Example 6.5.

Solution The jet comes out horizontally with x component

$$x = v_{act} \cdot t$$

and in the vertical direction the trajectory is deflected due to the effect of gravity. So,

$$y = \frac{1}{2} g t^2$$

We can form an equaion as:

$$v_{act} = \frac{x}{t} = \frac{x}{\sqrt{(2y/g)}}$$

From the theorem:

$$v_{theo} = \sqrt{2gh}$$

$$c_v = \frac{v_{act}}{v_{theo}} = \frac{x}{\sqrt{(2y/g)}}\left(\frac{1}{\sqrt{2gh}}\right) = \frac{x}{2\sqrt{yh}}$$

Example 6.6

Example A water tank has an orifice located at its base. The water level h is maintained throughout this process. Take D_e and d_j as orifice and jet's diameter. Assume steady incompressible flow if the water is draining through the orifice and a jet of cross-sectional area A(ℓ) is formed at a distance of ℓ from the orifice, if (i) orifice is round edged, (ii) orifice is sharp edged.

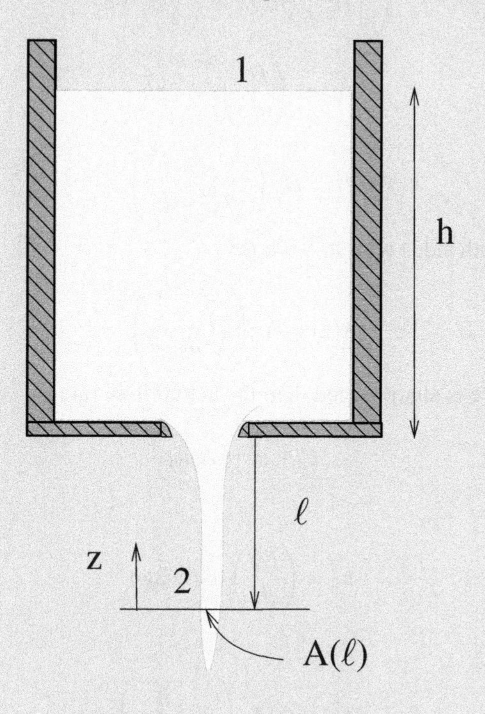

Figure 6.11 Station 1 and 2.

Solution We apply the Bernoulli's equation at station 1 and 2:

$$p_1 + \frac{\rho}{2}u_1^2 + \gamma z_1 = p_2 + \frac{\rho}{2}u_2^2 + \gamma z_2$$

$$\cancel{p_1}^{zero\ gauge} + \cancel{\frac{\rho}{2}u_1^2}^{free\ surface} + \gamma(h+\ell) = \cancel{p_2}^{zero\ gauge} + \frac{\rho}{2}u_2^2 + \gamma(0)$$

$$\gamma(h+\ell) = \frac{\rho}{2}u_j^2$$

$$u_j = \sqrt{2g(h+\ell)}$$

(i) We now apply continuity equation at the orifice and the jet station 2:

$$A_e u_e = A_j u_j$$

$$\left(\frac{\pi}{4}D_e^2\right)u_e = \left(\frac{\pi}{4}d_j^2\right)u_j$$

The exit velocity from the orifice would be following the Torricelli's theorem.

$$\left(\cancel{\frac{\pi}{4}}D_e^2\right)\sqrt{2gh} = \left(\cancel{\frac{\pi}{4}}d_j^2\right)u_j$$

$$u_j = \left(\frac{D_e}{d_j}\right)^2\sqrt{2gh}$$

$$d_j^2 = D_e^2\left(\frac{\sqrt{h}}{\sqrt{(h+\ell)}}\right)$$

Multiplying both sides with $\pi/4$, we get

$$A(\ell) = A_e\left(\sqrt{\frac{h}{h+\ell}}\right)$$

(ii) If the orifice is sharp-edged then the actual flow rate is

$$c_d A_e u_e = A_j u_j$$

This leads to

$$u_j = \left(\frac{D_e}{d_j}\right)^2 c_d\sqrt{2gh}$$

and

$$A(\ell) = c_d A_e\left(\sqrt{\frac{h}{h+\ell}}\right)$$

Example 6.7

Example A tank of cross-sectional area 30 m^2 is being drained with a pipe of 400 mm. The inflow volume flow rate is 1 m^3/s. What shall be the time taken by the tank to reach the level of 3.2 m? The fluid in the tank is water with density of 1,000 kg/m^3.

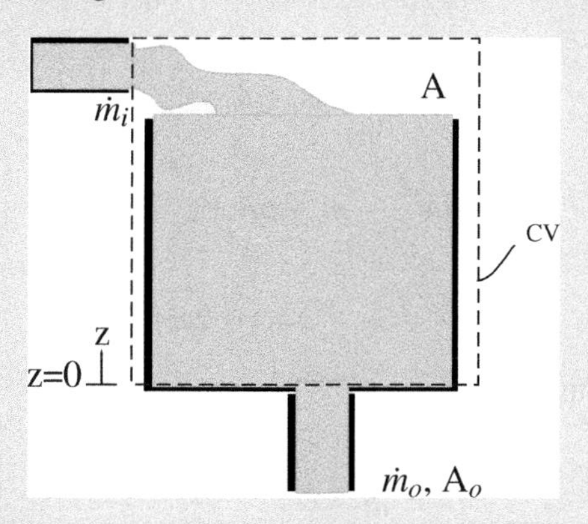

Figure 6.12 Schematic diagram of Example 6.7.

Solution We apply first the Bernoulli's equation at the tank free surface identified as station 1 and outlet, identified as station 2:

$$\left(p+\frac{1}{2}\rho v^2+\gamma z\right)_1=\left(p+\frac{1}{2}\rho v^2+\gamma z\right)_2$$

$$\left(\underbrace{p}_{zero\ gauge}+\underbrace{\frac{1}{2}\rho v^2}_{zero}+\gamma z\right)_{free\ surface}=\left(\underbrace{p}_{zero\ gauge}+\frac{1}{2}\rho v^2+\underbrace{\gamma z}_{z=0}\right)_{tank\ outlet}$$

$$\frac{1}{2}\rho v^2=\gamma z$$

$$v=\sqrt{2gz}$$

This is the outlet velocity, V_o:

$$V_o=\sqrt{2gz}$$

The mass balance as per Reynolds transport theorem is:

$$\int_{CS}\rho\mathbf{V}\cdot dA+\frac{\partial}{\partial t}\int_{CV}\rho d\forall=0$$

Now, since mass is being accumulated in this control volume, the law of conservation of mass dictates that

$$\int_{CS} \rho \mathbf{V} \cdot dA = -\dot{m}_i + \dot{m}_o$$

The time variation of mass inside control volume can be written as

$$\frac{\partial}{\partial t} \int_{CV} \rho d\forall = \frac{dm}{dt} = \rho A \frac{dz}{dt}$$

$$\rho A \frac{dz}{dt} - \dot{m}_i + \dot{m}_o = 0$$

$$\rho A \frac{dz}{dt} - \dot{m}_i + \rho A_o V_o = 0$$

$$\rho A \frac{dz}{dt} - \dot{m}_i + \rho A_o(\sqrt{2gz}) = 0$$

This can be cast into form

$$\frac{dz}{dt} = a\sqrt{z} + b$$

where a and b are:

$$a = \frac{-A_o}{A}(\sqrt{2g})$$

$$b = \frac{\dot{m}_i}{\rho A}$$

We introduce variable $y = dz/dt$:

$$y = a\sqrt{z} + b$$

$$\sqrt{z} = \frac{y}{a} - \frac{b}{a}$$

We differentiate the above equation

$$\frac{dy}{dz} = \frac{1}{2} \frac{a}{\sqrt{z}}$$

We isolate dz from the above equation

$$dz = \left(\frac{2\sqrt{z}}{a}\right) dy$$

Substituting z definition and rearrangement gives

$$dz = \left(\frac{2}{a^2}(1 - \frac{b}{y})\right) ydy$$

Now since

$$y = \frac{dz}{dt}$$

we adjust LHS and write

$$ydt = \left(\frac{2}{a^2}(1 - \frac{b}{y})\right) ydy$$

$$dt = \left(\frac{2}{a^2}(1 - \frac{b}{y})\right) dy$$

Integration of this equation leads to

$$t = \frac{2y}{a^2} - \frac{2b\ln(y)}{a^2}$$

Now insert z back into this equation

$$t = \frac{2(a\sqrt{z}+b)}{a^2} - \frac{2b\ln(a\sqrt{z}+b)}{a^2}$$

Substitution of a and b back into this equation gives

$$t = \left(\frac{A}{A_o}\right)^2 \frac{1}{g}\left(\frac{m_i}{\rho A} - \frac{A_o\sqrt{2gz}}{A}\right) - \frac{m_i}{\gamma}\left(\frac{A}{A_o}\right)^2 \ln\left(\frac{m_i}{\rho A} - \frac{A_o\sqrt{2gz}}{A}\right)$$

Substituting: $z = 3.2$ m, $A_o = 0.12560\,\text{m}^2$, $A = 30\,\text{m}^2$, $\rho = 1{,}000$, $\dot{m}_i = 1{,}000$ kg/s:

$$t = 1695.6s.$$

Note that beyond $z = 3.2$ m, the solution would become imaginary, indicating that further increase in z is impossible under the data set stated in the problem statement.

6.7 PITOT-STATIC TUBE

Consider the pitot-static tube as shown in Figure 6.13. For analysis, we balance the pressure from the streamline to the datum.

$$\rho_m g\Delta h + \rho_f g z_1 + \rho_f g z_2 + P_s = P_o + \rho_f g z + \rho_f g \Delta h$$

where P_s is static pressure, P_o is stagnation pressure, ρ_m is density of the manometer fluid, and ρ_f is density of the flowing fluid.

Rearrangement gives

$$\Delta P = P_o - P_s = (\rho_m - \rho_f) g\Delta h$$

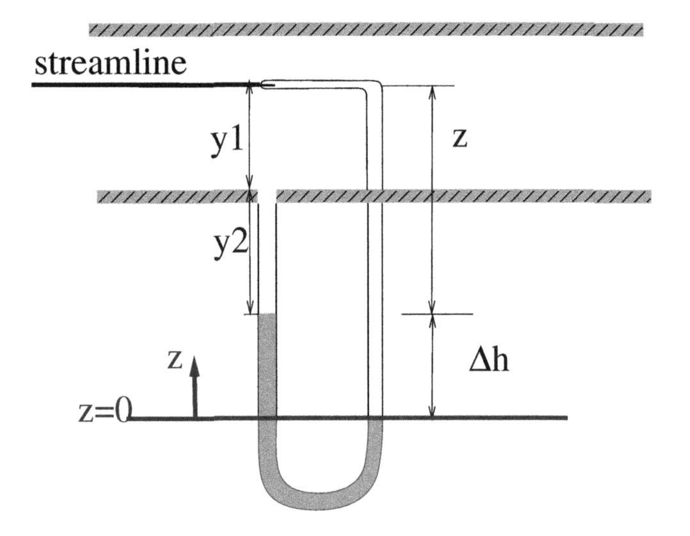

Figure 6.13 Pitot-static tube.

Applying the Bernoulli's equation at the streamline:

$$P_s + \frac{\rho}{2}u^2 + \gamma z = P_o + \underbrace{\frac{\rho}{2}u^2}_{stagnation} + \gamma z$$

This gave

$$u = \sqrt{\frac{2}{\rho}(P_o - P_s)} = \sqrt{\frac{2}{\rho}\Delta P}$$

Example 6.8

Example Water is flowing through a pipe and a pitot-static tube is used to measure the stagnation and static pressure differences. The manometer differential is 60 mm. The manometer fluid is mercury ($\rho = 13,600$ kg/m^3). Find the velocity of the water.

Solution

$$\Delta P = (\rho_m - \rho_f)g\Delta h = (13600 - 1000) \times 9.81 \times 60mm = 7416.36Pa$$

$$u = \sqrt{\frac{2}{\rho}(P_o - P_s)} = \sqrt{\frac{2}{\rho}\Delta P} = 3.85m/s$$

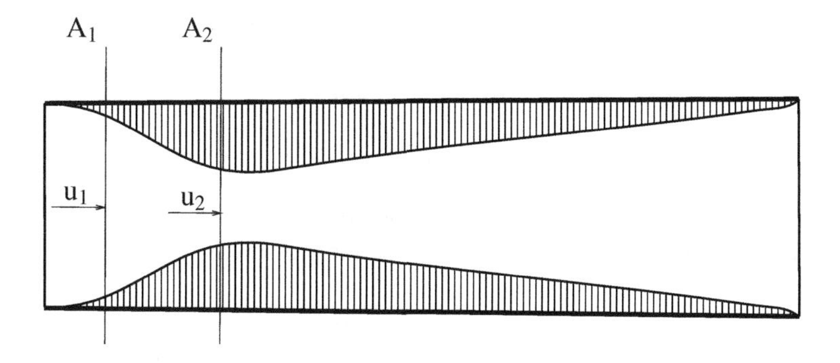

Figure 6.14 The venturimeter.

6.8 DETERMINATION OF FLOW RATES THROUGH VENTURIMETER

Consider the venturimeter shown in Figure 6.14. We apply the law of conservation of mass assuming steady state and incompressible flow:

$$Q_1 = Q_2$$

$$A_1 u_1 = A_2 u_2$$

This gives us velocities as area ratios

$$u_2 = \left(\frac{A_1}{A_2}\right) u_1$$

Now applying the Bernoulli's equation on a single streamline

$$p_1 + \frac{\rho}{2} u_1^2 + \cancel{\gamma z_1} = p_2 + \frac{\rho}{2} u_2^2 + \cancel{\gamma z_2}$$

$$\Delta p = \frac{\rho}{2} u_2^2 - \frac{\rho}{2} u_1^2 = \frac{\rho}{2} \left[u_2^2 - u_1^2\right]$$

Rearranging

$$\Delta p = \frac{\rho}{2} u_1^2 \left[\left(\frac{u_2}{u_1}\right)^2 - 1\right]$$

introducing the area ratio:

$$\Delta p = \frac{\rho}{2} u_1^2 \left[\left(\frac{A_1}{A_2}\right)^2 - 1\right]$$

and velocity can be computed if we measure the pressure at two locations:

$$u = \sqrt{\frac{2\Delta p}{\rho\left[\left(\frac{A_1}{A_2}\right)^2 - 1\right]}}$$

This result is only valid for incompressible flows.

The pressure loss normalized with dynamic pressure is

$$\zeta = \frac{\Delta p}{\frac{\rho}{2}u_1^2} = \left[\left(\frac{A_1}{A_2}\right)^2 - 1\right]$$

The above equation is known as **Carnot's equation.**

There are losses involved and correction is used for this equation and we can rearrange the equation for a throat section as well:

$$Q = c_d\left[\frac{A_2}{\sqrt{1-\left(\frac{A_2}{A_1}\right)^2}}\sqrt{2\left(\frac{p_1 - p_2}{\rho}\right)}\right]$$

where c_d is discharge coefficient. If cross-sections are circular, then $(A_2/A_1)^2 = (D_2/D_1)^4$ and we can define β:

$$Q = c_d\left[\frac{A_2}{\sqrt{1-\beta^4}}\sqrt{2\left(\frac{p_1 - p_2}{\rho}\right)}\right]$$

where $\beta = D_2/D_1$. The factor

$$\frac{1}{\sqrt{1-\beta^4}}$$

is called the velocity-of-approach factor and the actual volume flow rate is

$$Q_{actual} = c_d\left[\frac{A_2}{\sqrt{1-\beta^4}}\sqrt{2\left(\frac{p_1 - p_2}{\rho}\right)}\right]$$

The discharge coefficient and velocity-of-approach factor are usually combined as a flow coefficient, K:

$$K = \frac{c_d}{\sqrt{1-\beta^4}}$$

In engineering practice, the pressure loss is also plot with β.

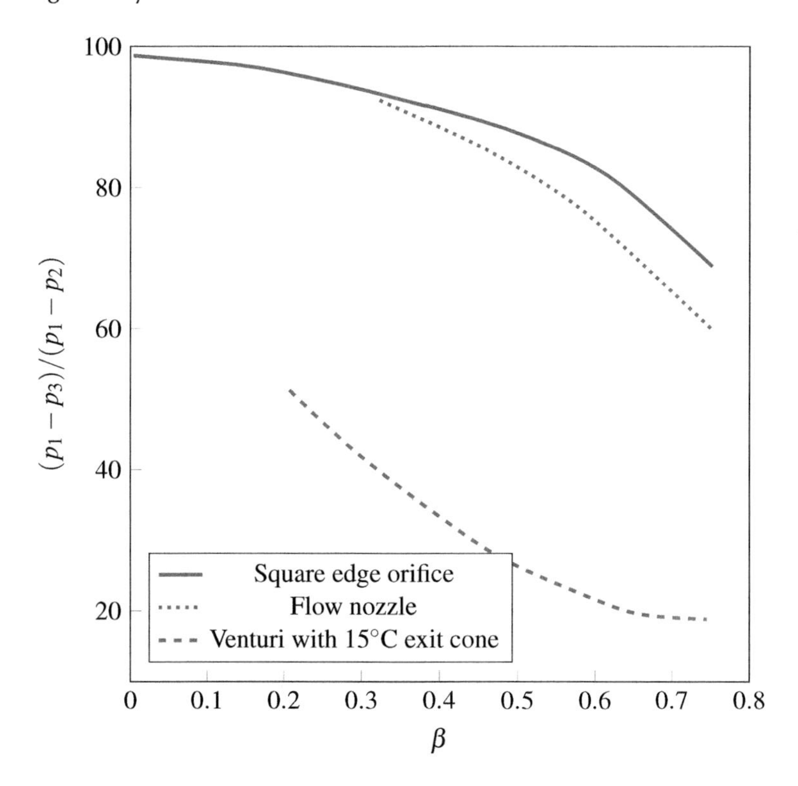

Figure 6.15 Head loss produced by some flow metering devices. The permanent pressure loss as a fraction of the meter differential vs β. Station 1 is upstream of the flow meter device; station 2 is just after the device, and station 3 is downstream of the device.

Table 6.2
Typical Values of c_d for Some Flow Meters

Device	c_d	Accuracy
Venturimeter	0.95-0.98	High
Flow Nozzle	0.7-0.8	Intermediate
Orifice-meter	0.6-0.65	Low

6.9 EXTENDED BERNOULLI EQUATION

The major assumption that there is no viscus force present in the flow limits the use of Bernoulli equation in case of viscous flows. To overcome this issue it was proposed to include an extra head loss term in the equation. The word *head* indicates the level of height above datum. When the fluid moves through the duct and piping system, due to friction, there is a pressure loss and capacity of fluid to reach the higher levels is reduced, thus the pressure loss is referred as head loss. To compensate this loss, a head loss term is added to correct the loss in momentum due to friction.

$$\frac{P_1}{\gamma} + z_1 + \frac{\vec{V}_1^2}{2g} = \frac{P_2}{\gamma} + z_2 + \frac{\vec{V}_2^2}{2g} + \varepsilon$$

where ε is the energy lost and it is also represented as frictional head loss h_f:

$$\frac{P_1}{\gamma} + z_1 + \frac{\vec{V}_1^2}{2g} = \frac{P_2}{\gamma} + z_2 + \frac{\vec{V}_2^2}{2g} + h_f$$

where h_ℓ is the head loss term. The above equation is called the *extended Bernoulli equation* in fluid mechanics literature.

If a pump is present in the fluid pipe network and station 1 is before the pump and station 2 is after the pump, then extended Bernoulli's equation is

$$\frac{p_1}{\gamma} + \frac{\vec{V}_1^2}{2g} + z_1 + h_{pump} = \frac{p_2}{\gamma} + \frac{\vec{V}_2^2}{2g} + z_2 + h_f \qquad (pump\ in\ network)$$

where, h_{pump} is head added by the pump through shaft work.

If a turbine is present in thee fluid pipe network, and station 1 is before the turbine and station 2 is after the turbine, then extended Bernoulli's equation is

$$\frac{p_1}{\gamma} + \frac{\vec{V}_1^2}{2g} + z_1 = \frac{p_2}{\gamma} + \frac{\vec{V}_2^2}{2g} + z_2 + h_{turbine} + h_f \qquad (turbine\ in\ network)$$

where $h_{turbine}$ is head converted by the turbine into shaft work.

Example 6.9

Example A converging duct has an upstream diameter of 80 mm, a downstream diameter of 40 mm, and passes fluid with a density of 1,000 kg/m^3. If the pressure difference across the duct is 80 kN/m^2, determine the downstream velocity and the volume flow rate when flow is ideal and when flow is actual, with $c_v = 0.9$ and $c_c = 0.99$. Find the energy lost if the flow is actual.
Solution We start from ideal Bernoulli's equation:

$$\frac{p_1}{\rho} + gz_1 + \frac{u_1^2}{2} = \frac{p_2}{\rho} + gz_2 + \frac{u_2^2}{2}$$

under ideal conditions:

$$\frac{p_1}{\rho} + \frac{u_1^2}{2} = \frac{p_2}{\rho} + \frac{u_{2i}^2}{2}$$

$$\frac{2(p_1-p_2)}{\rho}=u_{2i}^2-u_1^2$$

From continuity equation we have

$$A_1u_1=A_{2i}u_{2i}$$

$$u_1=\frac{A_{2i}u_{2i}}{A_1}$$

Substituting this into ideal Bernoulli's equation:

$$\frac{2(p_1-p_2)}{\rho}=u_{2i}^2\left(1-\left(\frac{A_2i}{A_1}\right)^2\right)$$

Rearrangement gives

$$u_{2i}=\sqrt{\frac{2(p_1-p_2)}{\rho\left(1-\left(\frac{A_2i}{A_1}\right)^2\right)}}=13.06m/s$$

The ideal flow rate is

$$Q_{ideal}=A_{2i}u_{2i}=\frac{\pi D_2^2}{4}\times u_{2i}$$

$$=\frac{\pi\left(40\times10^{-3}\right)^2}{4}\times13.06=1.6\times10^{-2}\;m^3/s$$

Since

$$Q_{actual}=c_v\cdot c_c\cdot Q_{ideal}$$

$$c_v=\frac{u_{2,actual}}{u_{2i}}=\frac{u_2}{u_{2i}}$$

$$A_{2,actual}=A_2=c_cA_{2i}$$

The energy equation can therefore be recast as

$$2c_v{}^2\frac{(p_1-p_2)}{\rho}=u_2^2\left(1-c_c{}^2\left(\frac{A_{2i}}{A_1}\right)^2\right)$$

$$u_{2,actual}=c_v\sqrt{\frac{2(p_1-p_2)}{\rho\left(1-c_c{}^2\left(\frac{D_{2i}}{D_1}\right)^4\right)}}$$

$$=0.9\sqrt{\frac{2(80\times10^3)}{1000\left(1-0.99^2\left(\frac{40}{80}\right)^4\right)}}=11.75m/s$$

$$Q_{actual}=A_2\cdot u_2=0.01475\,m^3/s$$

$$u_{actual}=\frac{Q_{actual}}{A_1}=2.93\,m/s$$

$$\varepsilon=\frac{\Delta p}{\rho}+\frac{\left(u_{1,actual}^2-u_{2,actual}{}^2\right)}{2}$$

$$\varepsilon=\frac{p_1-p_2}{\rho}+\frac{u_1^2-u_{2i}^2}{2}=\frac{80\times10^3}{1000}+\frac{2.93^2-11.75^2}{2}=15.28\,J/kg$$

6.10 ESTIMATION OF FORCES

6.10.1 INTEGRAL FORMULATION OF MOMENTUM EQUATION

$$\frac{dN}{dt}=\int_{CS}\eta\rho\mathbf{V}\cdot dA+\frac{\partial}{\partial t}\int_{CV}\eta\rho d\forall \tag{6.20}$$

Now as η = N/mass, we have $\eta=\mathbf{V}$ (in case of momentum).

$$\mathbf{F}=\int_{CS}\mathbf{V}\rho\mathbf{V}\cdot dA+\frac{\partial}{\partial t}\int_{CV}\mathbf{V}\rho d\forall \tag{6.21}$$

where **F** is the force as rate of change of momentum is exhibited as force.

We will now analyze some problems using the RTT.

6.10.2 JET'S FORCE ON THE MOVING PLATE

A jet with velocity u is striking a vertical wall as shown in Figure 6.16. The jet has transferred part of its momentum to the plate and due to that the plate started moving with velocity V_p. The impinging jet formed has deflected near the plate. Since the system is unsteady, we impose the $-V_p$ velocity of the system to make the plate stationary. This leads to the velocity of the approaching jet to be defined as $u'=u-V_p$.

$$F_x=\dot{m}(u-V_p)=\rho A(u-V_p)^2$$

and

$$F_y=0$$

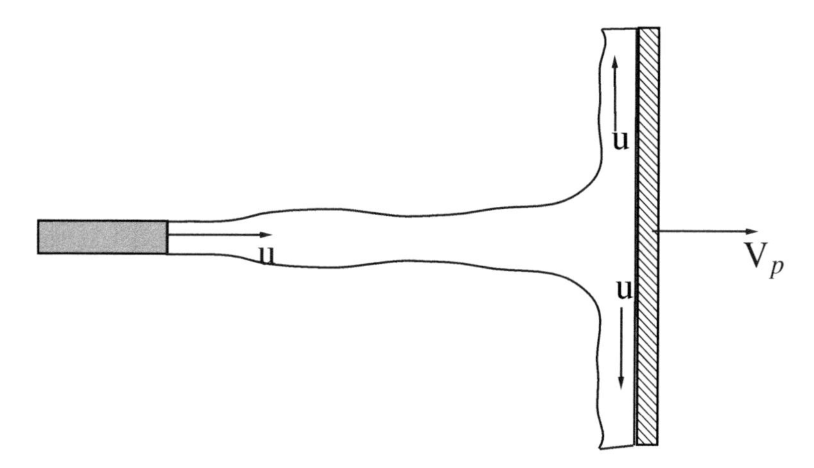

Figure 6.16 Force on a vertical moving plate.

The power needed to push the plate

$$Power = F \cdot V_p = \rho A V_p (u - V_p)^2$$

We now differentiate this power and set the result to zero to obtain the maximum power required:

$$Power = \rho A V_p (u^2 + V_p^2 - 2 \cdot u \cdot V_p)$$

$$Power = \rho A (V_p^3 + u^2 V_p - 2 \cdot u V_p^2)$$

$$\frac{d(Power)}{V_p} = \rho A \left(3{V_p}^2 + u^2 - 4uV_p\right) = 0$$

$$V_p = \frac{u}{3}$$

This shows that plate velocity will be one-third of the jet's velocity when power is maximum.

$$Power_{max} = \frac{4}{27}(\rho A u)u^2 = \frac{4}{27}\dot{m}u^2$$

Also, the kinetic energy of the jet into velocity gives us the power in the jet

$$Power_{in} = \frac{1}{2}\dot{m}u^2$$

The effectiveness of such a device can be measure by taking the ratio of the powers:

$$E = \frac{Power_{max}}{Power_{in}} = \frac{4/27}{1/2} = \frac{8}{27} = 0.2963$$

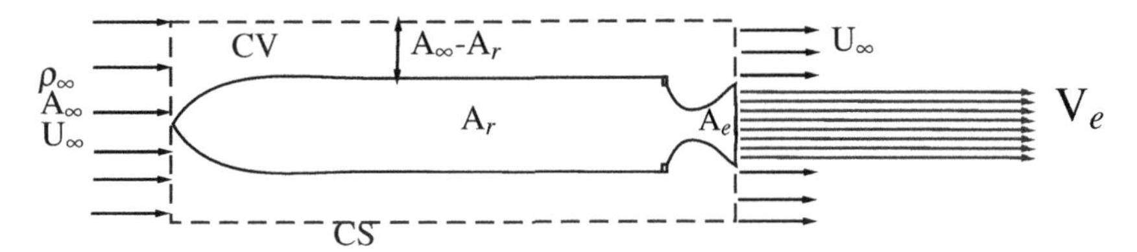

Figure 6.17 Thrust on rocket. A_r is cross-sectional area of rocket and A_e is exit area.

We arrange the of plates that comes in the jet's path, then we can create impulse turbines.

6.10.3 ROCKET THRUST

The rocket nozzle provides the necessary thrust by expanding gases. The control volume has been selected for the analysis of this problem, as shown in Figure 6.17.

$$\dot{m} = \int_{CS} \rho \mathbf{V} \cdot dA + \frac{\partial}{\partial t} \int_{CV} \rho d\forall$$

$$-\rho_\infty A_\infty U_\infty + (A_\infty - A_r)\rho_\infty U_\infty + \rho_e A_e V_e + \dot{m}_{lateral} = 0$$

where $\dot{m}_{lateral}$ is the mass flow rate entering control volume from the lateral sides of the control surface; A_r is the cross-sectional area of the rocket.

$$\mathbf{F} = \int_{CS} \mathbf{V} \rho \mathbf{V} \cdot dA + \frac{\partial}{\partial t} \int_{CV} \mathbf{V} \rho d\forall$$

$$\mathbf{F} = \left[-\rho_\infty A_\infty U_\infty \cdot U_\infty + (A_\infty - A_r)\rho_\infty U_\infty \cdot U_\infty + \rho_e A_e V_e^2\right] + \int_{Lateral} \rho_\infty U_\infty \mathbf{V}_r \cdot dA$$

$$\mathbf{F} = \left[\cancel{-\rho_\infty A_\infty U_\infty \cdot U_\infty} + \cancel{A_\infty \rho_\infty U_\infty \cdot U_\infty} - A_r \rho_\infty U_\infty \cdot U_\infty + \rho_e A_e V_e^2\right] + U_\infty \int_{Lateral} \rho_\infty \mathbf{V}_r \cdot dA$$

$$\mathbf{F} = \left[\rho_e A_e V_e^2 - A_r \rho_\infty U_\infty \cdot U_\infty\right] + U_\infty \int_{Lateral} \rho_\infty \mathbf{V}_r \cdot dA$$

For the case of very large control volume around the rocket, $A_\infty \gg A_r$, and we can safely ignore the lateral momentum associated with mass flow rate:

$$\mathbf{F} \approx \rho_e A_e V_e^2 \underbrace{- A_r \rho_\infty U_\infty \cdot U_\infty}_{ignored} \underbrace{+ U_\infty \int_{Lateral} \rho_\infty \mathbf{V}_r \cdot dA}_{ignored}$$

$$\mathbf{F} \approx \rho_e A_e V_e^2$$

This is the net force/thrust acting on the rocket, if pressure is ignored. If we include the pressure forces, then the net thrust would be

$$\mathbf{F} \approx \rho_e A_e V_e^2 - (P_e A_e - P_\infty A_r)$$

Example 6.10

Example A horizontal circular air jet of diameter 5 in. strikes a conical deflector of 60 degrees as shown in Figure 6.18. A horizontal holding force of 8 lb is needed to keep the cone in its position. Estimate jet's flow rate and the jet's velocity.

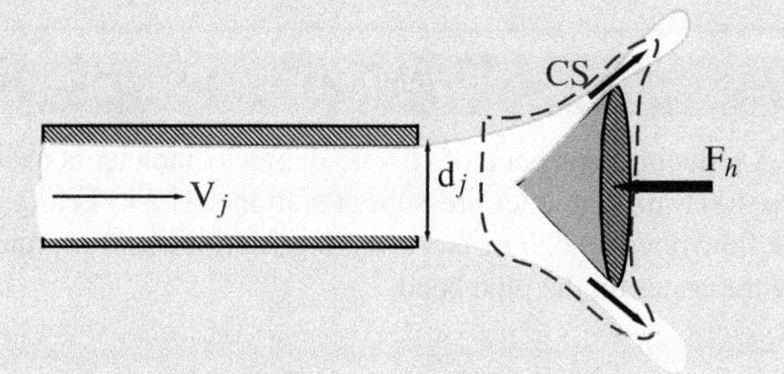

Figure 6.18 Schematic diagram of Example 6.10.

Solution The net force will be equal to the rate of change of momentum.

$$-F_h = -V_j\,(\rho V_j A_1) + -V_2\,(\rho V_2 A_2)$$

$$-F_h = -V_j\,(\rho Q) + -V_2\,(\rho Q) = -F_h$$

From continuity equation

$$Q = V_j A_1 = V_2 A_2$$

$$V_2 = V_j \cos 60^\circ$$

$$V_j = \frac{Q}{A_1}$$

$$-F_h = -V_j \rho Q + V_j \cos 60^\circ \rho Q$$

$$-F_h = -\frac{Q^2}{A_1}\rho + \frac{Q^2}{A_1}\cos 60^\circ \rho$$

$$Q = \sqrt{\frac{F_h A_1}{(1-\cos 60^\circ)\rho}} = \sqrt{\frac{F_h\left(\frac{\pi d_j^2}{4}\right)}{(1-\cos 60^\circ)\rho}}$$

$$Q = \sqrt{\frac{(8lb) \times \pi(5in.)^2}{(1-\cos 60^\circ)(0.00238\frac{slugs}{ft^3})(4)(144\frac{in^2}{ft^2})\left(\frac{1lb}{slug\frac{ft}{s^2}}\right)}}$$

$$Q = 30.3\frac{ft^3}{s}$$

Example 6.11

Example A reducing pipe bend of $\alpha = 45$ degrees diameter is changed from 800 mm to 400 mm. The inlet pressure at station 1 is 250 kPa (gauge). The volumetric flow rate is 0.625 m^3/s. Assuming frictionless flow, find the force exerted by the water on the pipe bend.

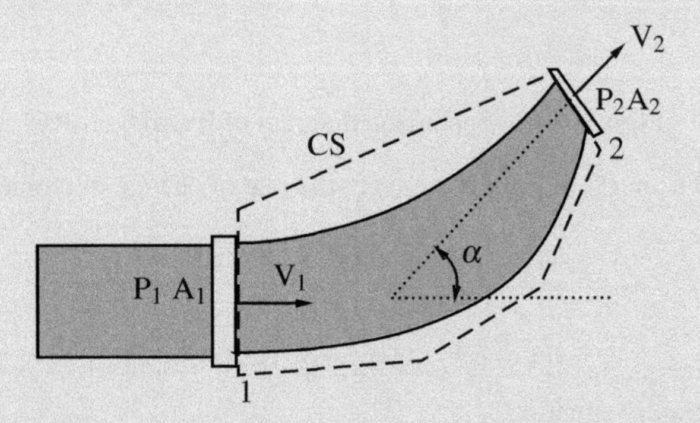

Figure 6.19 Schematic diagram of Example 6.11.

Solution The velocity at pipe inlet is:

$$V_1 = \frac{0.625m^3/s}{\frac{\pi}{4}(0.8)^2} = 1.24m/s$$

The velocity at pipe exit is:

$$V_2 = \frac{0.625m^3/s}{\frac{\pi}{4}(0.4)^2} = 4.97m/s$$

From Bernoulli's equation we have

$$p_2 = p_1 + \frac{1}{2}\rho\left(V_1^2 - V_2^2\right)$$

$$p_2 = 2.5 \times 10^5 + \frac{1000}{2}(1.24^2 - 4.97^2)$$

$$p_2 = 238\ kPa$$

We now apply the momentum balance on the control volume as indicated by control surface (CS) in figure.

The force balance in x-direction is:

$$p_1A_1 - p_2A_2\cos 45^\circ + F_x = \rho Q\left(V_2\cos 45^\circ - V_1\right)$$

$$2.5\times 10^5\left(\frac{\pi}{4}\right)(0.8)^2 - 2.38\times 10^5\left(\frac{\pi}{4}\right)(0.4)^2\cos 45^\circ + F_x$$

$$= 1000\,(0.625)\,(4.97\cos 45^\circ - 1.24)$$

The force balance in y-direction is:

$$-p_2A_2\sin 45^\circ + F_y = \rho Q\left(V_2\sin 45^\circ - 0\right)$$

$$-2.38\times 10^5\left(\frac{\pi}{4}\right)(0.4)^2\sin 45^\circ + F_y$$
$$= 1000\,(0.625)\,(4.97\sin 45^\circ)$$

This gives

$$F_x = -103.09kN$$

$$F_y = 23.34kN$$

$$F_R = \sqrt{\left(-103.09\times 10^3\right)^2 + \left(23.34\times 10^3\right)^2} = 105.7kN$$

The angle between the forces is

$$angle = \tan^{-1}\left(\frac{F_y}{F_x}\right) = -12.76^\circ$$

with negative x direction.

Measured from positive x-coordinate the angle is:

$$\theta = 180 - 12.76 = 167.24^\circ$$

Example 6.12

Example The jet of CO_2 gas strikes the fixed inclined surface and it is deflected as shown in Figure 6.20. If the jet issues from the 200-mm-diameter pipe at 5 m/s, determine the normal force the jet exerts on the inclined surface. The angle of the inclination is 45°C. The density of CO_2 under standard temperature and pressure is 1.87 kg/m^3.

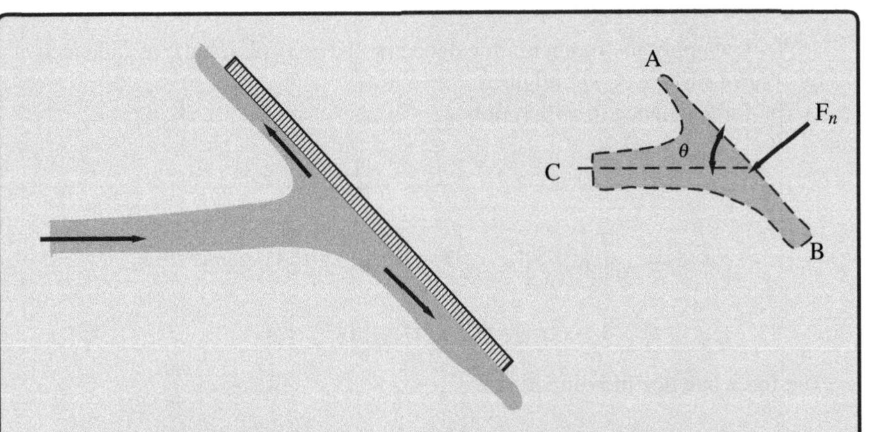

Figure 6.20 Schematic diagram of Example 6.12.

Solution The jet is open to atmosphere, so the $p_A = p_B = p_C = 0$ gauge pressure. From Bernoulli's equation we have

$$\left(\frac{p}{\gamma}+\frac{u^2}{2g}+z\right)_A=\left(\frac{p}{\gamma}+\frac{u^2}{2g}+z\right)_B=\left(\frac{p}{\gamma}+\frac{u^2}{2g}+z\right)_C$$

$$\frac{u_A^2}{2g}=\frac{u_B^2}{2g}=\frac{u_C^2}{2g}$$

$$\frac{(5)^2}{2g}=\frac{u_B^2}{2g}=\frac{u_C^2}{2g}$$

$$u_B=u_c=5\,m/s$$

The control volume is shown in Figure 6.20. We now apply the integral momentum balance:

$$F=\frac{\partial}{\partial t}\int_{\forall} u.d\forall+\int_{CS} u\rho u\cdot dA$$

$$F=\int_{CS} u\rho u\cdot dA=\sum\rho Q$$

The component of rate of change of momentum that will balance the normal force on the wall is:

$$F=-\rho Q_A u_A\sin(\theta)$$

$$F=-1.87\times\frac{\pi}{4}(200\times10^{-3})^2\times5\times5\times\sin(45^o)=1.037\,N$$

PROBLEMS

6P-1 A vessel having varying circular cross section is being drained. Originally the liquid height in vessel was $H_o = 3$ m. If the level of the top surface is sinking at a constant speed of $v_1 = 0.015$ m/s and the area at the outlet is A_2 is 8E-05 m^2, find the volume of liquid drained from the vessel.

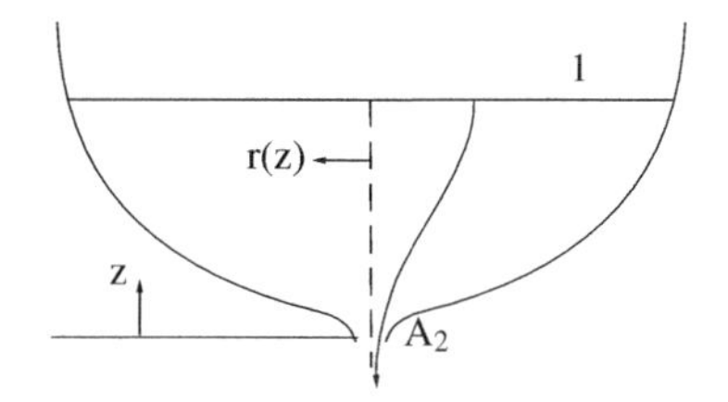

Figure 6.21 Figure of problem 6P-1.

6P-2 A large tank containing water up to a constant level $H = 10$ m is being drained with the help of a diffuser of length ℓ. The diffuser used is shown in Figure 6.22. The inlet diameter of diffuser is d_1 and the outlet diameter is d_2. Find the outlet velocity u_2 at station 2, if d_1 is 10 cm, d_2 is 20 cm, and ℓ is 20 m.

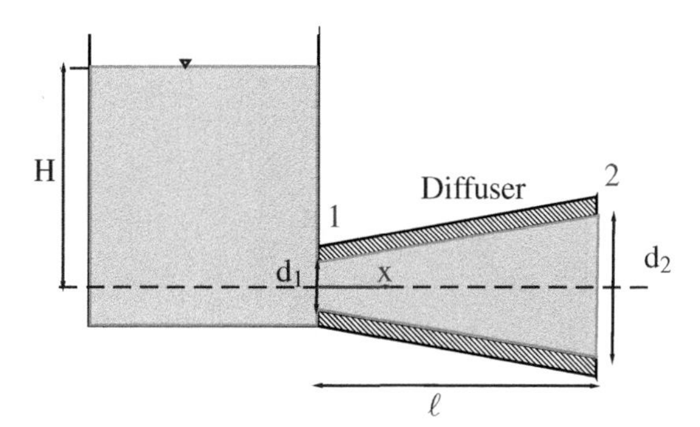

Figure 6.22 Figure of problem 6P-2.

6P-3 A vertical, circular jet of air having 50 mm diameter (d_j) strikes a conical deflector with constant velocity of 40 m/s as shown in Figure 6.23. A vertical holding force of 0.5 N is required to hold the deflector in its position. The conical deflector angle is 35 degrees. Determine the weight of the deflector required to carry out this task. Take density of air as 1.22 kg/m^3.

6P-4 A ladle containing molten steel is discharged into a tundish through a discharge nozzle (mean diameter = 0.04 m) at the bottom. The ladle diameter is $d_1 = 2$ m and the initial metal depth is $z = 3$ m, estimate the time required for emptying the ladle. The length of discharge nozzle is 0.5 m. The density of molten steel can be taken as 7.21×10^3 kg/m^3 and its viscosity is 1.53×10^{-3} kg/m s.

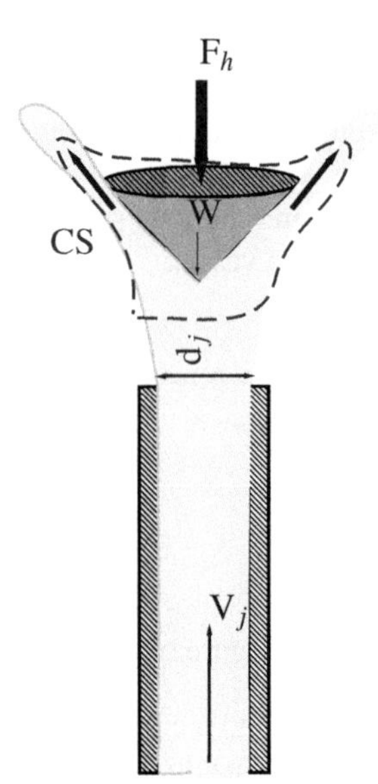

Figure 6.23 Figure of problem 6P-3.

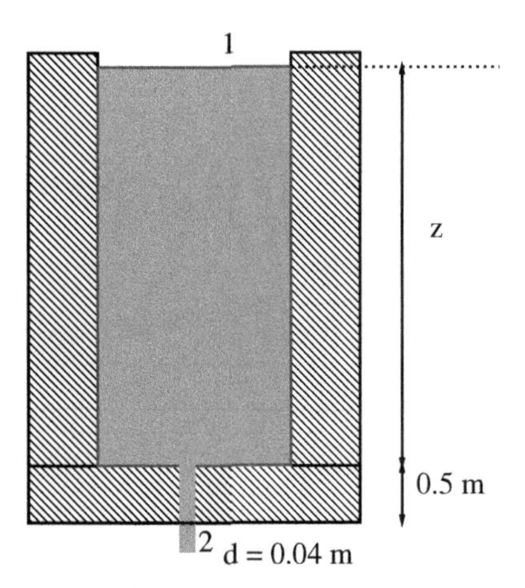

Figure 6.24 Figure of problem 6P-4.

6P-5 Water at constant flow rate is filling up the tank of diameter 2.5 m. Find the time needed to fill up the tank till depth $h = 3$ m, if the velocity of water entering the tank is 0.3 m/s. What will the time be if the diameter is varied from 50 mm to 200 mm?

6P-6 Water flows through the main pipe at mass flow rate of 300 kg/s, and then goes into the double wye with a mean velocity of 3 m/s through branch 1 and an average

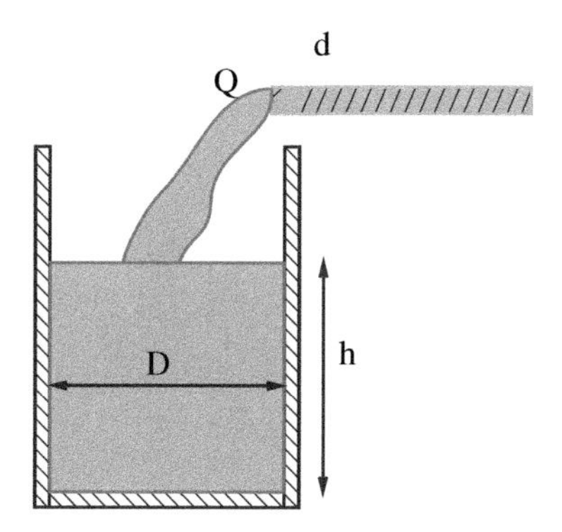

Figure 6.25 Figure of problem 6P-5.

velocity of 2 m/s through branch 2 (Figure 6.26). Determine the average velocity at which it flows through branch 3. The dimensions are as follows:

	d (mm)
Main Pipe	450
Branch 1	250
Branch 2	150
Branch 3	250

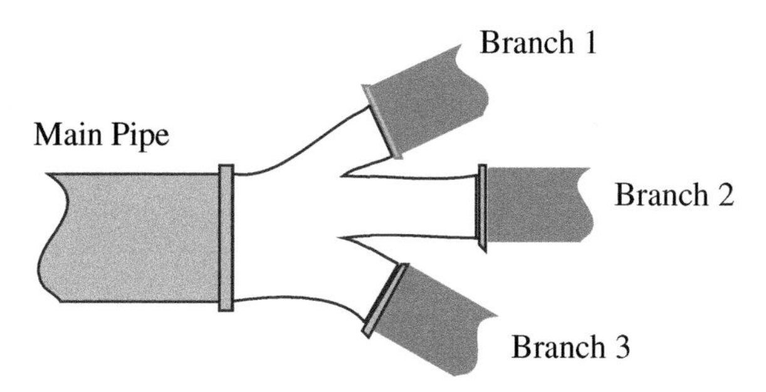

Figure 6.26 Figure of problem 6P-6.

REFERENCE

R. W. Miller, Flow Measurement Engineering Handbook, 3rd ed. New York: McGraw Hill, 1996.

7 Irrotational Flow

Before 1905, the fluid mechanics studies were either experimental or were based on potential flow theory. Although Navier-Stokes equations were available, the analytical solutions for every flow problem were not possible. Researchers have noticed that many flow phenomena and patterns can be analyzed using the stream and velocity potential functions. This chapter focuses on the potential flow theory that arises from the mathematical modeling based on complex variable theory. The main point is that we ignore the viscosity and assume that the flow has circulation instead of vorticity.

Learning outcomes: After finishing this chapter, you should be able to:

- Analyze flows using the hypothetical streamlines.
- Learn the superposition method and develop solutions for complex problems.
- Model complex problems using the stream functions or potential functions.

It has been observed experimentally that viscosity plays an important role for flow near the wall boundary. However, some engineering problems can be solved in a simplified manner by ignoring the viscosity.

7.1 CONCEPT OF STREAM FUNCTION (ψ)

Stream function allows us to wipe out the continuity equation and solve the momentum equation directly for a single variable ψ. The idea of stream function works only if the continuity equation can be reduced to two terms.

In Cartesian coordinates the continuity equation has four terms:

$$\frac{\partial \rho}{\partial t} + \frac{\partial(\rho u)}{\partial x} + \frac{\partial(\rho v)}{\partial y} + \frac{\partial(\rho w)}{\partial z} = 0 \tag{7.1}$$

for steady flows, 2D flows:

$$\frac{\partial(\rho u)}{\partial x} + \frac{\partial(\rho v)}{\partial y} = 0 \tag{7.2}$$

We can define a function ψ that satisfies this equation:

$$\frac{\partial}{\partial x}\left(\frac{\partial \psi}{\partial y}\right) + \frac{\partial}{\partial y}\left(-\frac{\partial \psi}{\partial x}\right) = 0 \tag{7.3}$$

DOI: 10.1201/9781003315117-7

Comparing Eq. 7.3 with this ψ function can be defined as:

$$u = \frac{\partial \psi}{\partial y} \quad (7.4)$$

$$v = -\frac{\partial \psi}{\partial x} \quad (7.5)$$

By this mathematical trick the continuity equation is cast in a function of a single variable.

If we take the curl of velocity vector ($\nabla \times \mathbf{V} = \nabla^2 \psi$) we have:

$$\frac{\partial^2 \psi}{\partial x^2} + \frac{\partial^2 \psi}{\partial y^2} \quad (7.6)$$

We can assume a hypothetical flow for which $\nabla \times \mathbf{V} = \nabla^2 \psi = 0$. This assumption is called ideal flow, and in practice such a region can be assumed for the flows away from the wall.

7.1.1 EQUATION OF STREAMLINES

Lines of constant ψ are streamlines. If elemental arc length **dr** is parallel to **V**, then respective components must be in proportion:

$$\frac{dx}{u} = \frac{dy}{v} = \frac{dz}{w} = \frac{\mathbf{dr}}{\mathbf{V}} \quad (7.7)$$

For 2-D flows: $u\,dy - v\,dx = 0$ (streamline). Introducing stream function as:

$$\frac{\partial \psi}{\partial x}dx + \frac{\partial \psi}{\partial y}dy = 0 = d\psi \quad (7.8)$$

Thus, the change in ψ is zero along a streamline.

At streamline, the value of a stream function is constant.

For a two-dimensional flow, we can satisfy the continuity equation for steady state, incompressible flows:

$$\cancel{\frac{\partial \rho}{\partial t}} + \rho\frac{\partial (u)}{\partial x} + \rho\frac{\partial (v)}{\partial y} = 0 \quad (7.9)$$

$$\frac{\partial (u)}{\partial x} + \frac{\partial (v)}{\partial y} = 0 \quad (7.10)$$

where we can introduced u and v components as:

$$u = \frac{\partial \psi}{\partial y}, v = -\frac{\partial \psi}{\partial x}$$

The function ψ is called the stream function, and for small increments:

$$d\psi = \frac{\partial \psi}{\partial x}dx + \frac{\partial \psi}{\partial y}dy = -vdx + udy$$

If we get the L.H.S. as zero then it is the equation of streamline.

If dy is the gap between two streamlines, then the difference between stream functions can be equated as:

$$\Delta\psi = \int_{y1}^{y2} udy$$

This is equal to volume flow rate per unit gap distance.

Example 7.1

Example Consider the velocity field:

$$u = U$$
$$v = Vsin(2\pi(x - Ut)/\lambda)$$

Find the equation of streamline.

Solution

$$\psi_1 = \int_0^y u\,\mathrm{d}y + f(x,t)$$
$$\psi_1 = U \cdot y + f(x,t)$$

$$\psi_2 = \int_0^x v\,\mathrm{d}x + g(y,t);$$

$$\psi_2 = -\frac{1}{2}\frac{V \cdot \lambda \left(2\cos\left(\frac{\pi \cdot Ut}{\lambda}\right)^2 - 1 - \cos\left(\frac{2 \cdot \pi \cdot (-x+Ut)}{\lambda}\right)\right)}{\pi} + g(y,t)$$

$$\psi = Uy - \frac{V\lambda}{2\pi}\left(2\cos\left(\frac{\pi Ut}{\lambda}\right)^2 - 1 - \cos\left(\frac{2\pi(-x+Ut)}{\lambda}\right)\right)$$

At streamline, ψ will have a specific value, say ψ_o, then

$$y_1 = \frac{\psi_o}{U} + \frac{V\lambda}{2\pi U}\left(2\cos\left(\frac{\psi Ut}{\lambda}\right)^2 - 1 - \cos\left(\frac{2\pi(-x+Ut)}{\lambda}\right)\right)$$

Figure 7.1 shows the streamlines at different values. Note that streamlines never intersect each other.

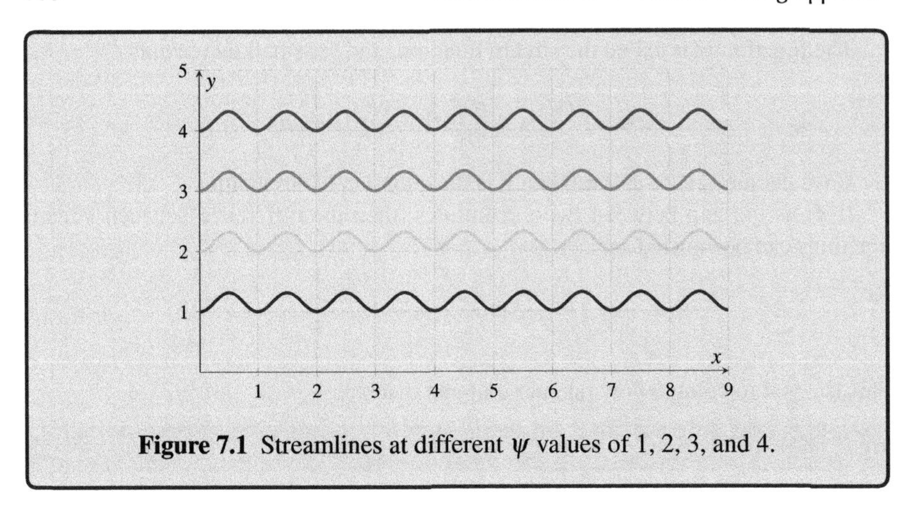

Figure 7.1 Streamlines at different ψ values of 1, 2, 3, and 4.

7.2 POTENTIAL FUNCTION

A vector with zero curl must be a gradient of a scalar function. If $\nabla \times \mathbf{V} = 0$ then $\mathbf{V} = \nabla\phi$ where $\phi = \phi(x,y,z,t)$ is called the potential function. We can define velocities in terms of ϕ as:

$$u = \frac{\partial \phi}{\partial x}, v = \frac{\partial \phi}{\partial y}, w = \frac{\partial \phi}{\partial z} \tag{7.11}$$

Lines of constant ϕ are potential function lines. In an irrotational flow, both ψ and ϕ are perpendicular to each other.

$$d\phi = 0 = \frac{\partial \phi}{\partial x}dx + \frac{\partial \phi}{\partial y}dy = udx + vdy \tag{7.12}$$

Solving:

$$-u/v = -\frac{1}{(\frac{dy}{dx})_{\psi=c}} \tag{7.13}$$

Example 7.2

Example Consider potential function:

$$\phi = A(x^2 - y^2)$$

Is it a harmonic function? What is the stream function?

Solution From incompressible flow

$$\nabla^2\phi = 0$$

$$\frac{\partial}{\partial x}\left(\frac{\partial}{\partial x}\phi(x,y)\right) = 2A$$

$$\frac{\partial}{\partial y}\left(\frac{\partial}{\partial y}\phi(x,y)\right) = -2A$$

$$2A + (-2A) = 0$$

This shows the potential function is harmonic.

$$u = \frac{\partial \phi}{\partial x} = \frac{\partial \psi}{\partial y} = 2Ax$$

$$v = \frac{\partial \phi}{\partial y} = \frac{-\partial \psi}{\partial x} = -2Ay$$

Integration of u and v with y and x respectively gives

$$\psi = 2Ayx$$

7.3 FLOW NET

A flow net consists of two sets of lines that must always be orthogonal (perpendicular to each other): flow lines, which show the direction of flow, and equipotentials (lines of constant head), which show the distribution of potential energy.

(i) In a homogeneous isotropic system, flow lines and equipotentials are always perpendicular and form curvilinear *squares*.

(ii) Equipotentials are always normal to an impermeable boundary. Flow lines are always parallel to an impermeable boundary.

(iii) Equipotentials are always parallel to a constant head boundary. Flow lines are always normal to a constant head boundary.

Let's say a flow field is described by

$$\phi = -2yx$$

and

$$\psi = x^2 - y^2$$

the representative flow net is shown in Figure 7.2.

Stream function in cylindrical and spherical coordinates

Incompressible, planar stream function in cylindrical coordinates:

$$u_r = \frac{1}{r}\frac{\partial \psi}{\partial \theta} = \frac{\partial \phi}{\partial r}$$

$$u_\theta = -\frac{\partial \psi}{\partial r} = \frac{1}{r}\frac{\partial \phi}{\partial \theta}$$

Incompressible, axisymmetric stream function in cylindrical coordinates:

$$u_r = \frac{-1}{r}\frac{\partial \psi}{\partial z}$$

$$u_z = \frac{1}{r}\frac{\partial \psi}{\partial r}$$

Spherical coordinates:

$$u_r = \frac{1}{r^2 \sin\theta}\frac{\partial \psi}{\partial \theta}$$

$$u_\theta = \frac{-1}{r\sin\theta}\frac{\partial \psi}{\partial r}$$

7.4 UNIFORM FLOW

A uniform stream $\mathbf{V} = \mathbf{U}\mathrm{i}$ possesses the stream and potential functions defined as:

$$u = \frac{\partial \phi}{\partial x} = \frac{\partial \psi}{\partial y}$$

$$v = \frac{\partial \phi}{\partial y} = -\frac{\partial \psi}{\partial x} \tag{7.14}$$

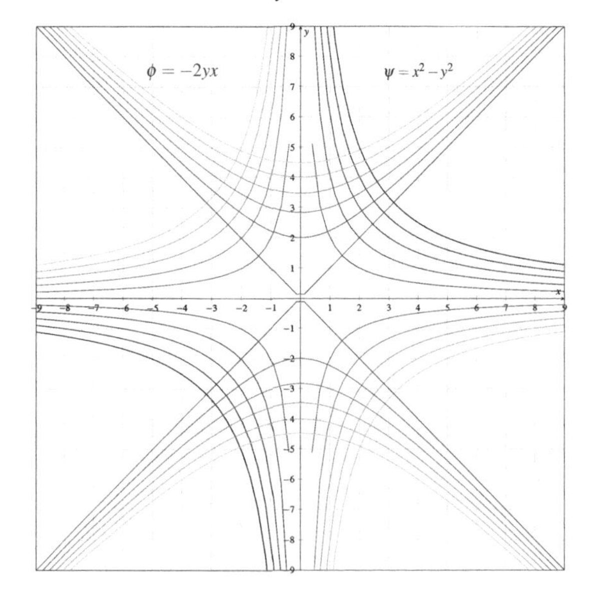

Figure 7.2 The plot of ψ and ϕ.

Figure 7.3 The hurricane is a large vortex with viscous effects. A vortex can be observed in a tub of water.

known as Cauchy-Riemann differential equations. Integration gives:

$$\psi = Uy, \phi = Ux \tag{7.15}$$

Table 7.1
Expression for Uniform Flow

Uniform Flow	**Cartesian Coordinates**	**Polar Coordinates**
Potential Function	$U_\infty x$	$U_\infty rcos(\theta)$
Stream Function	$U_\infty y$	$U_\infty rsin(\theta)$
u-component	U_∞	U_∞
v-component	0	0
u_r-component	$\frac{U_\infty x}{\sqrt{x^2+y^2}}$	$U_\infty cos(\theta)$
u_θ-component	$\frac{-U_\infty y}{\sqrt{x^2+y^2}}$	$-U_\infty sin(\theta)$

7.5 POTENTIAL VORTEX CIRCULATION

A potential vortex approximately describes the motion in a hurricane, tornado, or whirlpool (see Figure 7.3). The viscous forces would cause a particle rotation, and the region of this concentrated vorticity is a real vortex, which can occur both in liquid and gas mediums. The simplified model of this complex flow is accomplished by assuming that viscosity is not of huge importance, and the rotational behavior of the

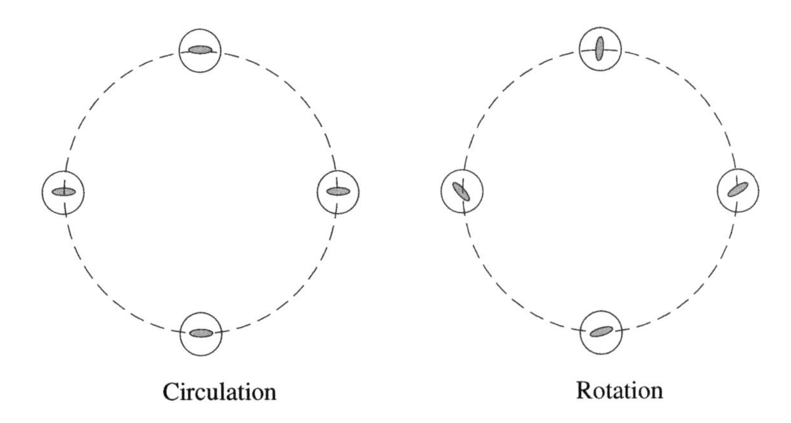

Figure 7.4 The difference between circulation and rotation.

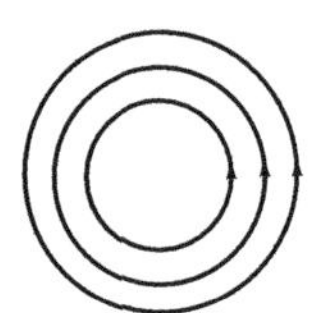

Figure 7.5 The potential vortex circulation.

vortex is approximated as circulation. A schematic representation of the difference between circulation and rotation is depicted in Figure 7.4.

In simplest form, a potential vortex is a two-dimensional motion in which streamlines are the concentric circles, as shown in Figure 7.5.

In real or viscous fluids, the vortex are generated due to fluid particles distortions. However in inviscid or ideal flows, the concept of free vortex gives us a simplified model of the circulatory movement of the fluids. Such a model is desirable, as it still gives us meaningful results for the pressure distribution inside a tornado, whirlpool, or other similar flows.

Real, ideal, and potential flow

Real fluid: The fluid in which viscous effects cannot be ignored.
Potential flow: An idealized flow field in which the vorticity is zero in the entire flow field (except near singular points), and the viscous forces are completely neglected.
Ideal/inviscid flow: An approximated flow field in which there is no diffusion of vorticity.

Table 7.2
Expression for Inviscid Vortex

Ideal Vortex	**Cartesian Coordinates**	**Polar Coordinates**
Potential Function	$\frac{\Gamma}{2\pi}\tan^{-1}\left(\frac{y}{x}\right)$	$\frac{\Gamma}{2\pi}\theta$
Stream Function	$\frac{-\Gamma}{2\pi}\left(\frac{y}{x^2+y^2}\right)$	$-\frac{\Gamma}{2\pi}\ln r$
u component	$\frac{\Gamma}{2\pi}\left(\frac{x}{x^2+y^2}\right)$	$\frac{-\Gamma}{2\pi}\left(\frac{\sin\theta}{r}\right)$
v component	0	$\frac{\Gamma}{2\pi}\left(\frac{\cos\theta}{r}\right)$
u_r component	0	0
u_θ component	$\frac{\Gamma}{2\pi}\left(\frac{1}{\sqrt{x^2+y^2}}\right)$	$\frac{\Gamma}{2\pi}\left(\frac{1}{r}\right)$

Example 7.3

Example A category 4 hurricane is moving in the sea and is about to reach the land. The diameter of the hurricane is 400 miles (643 km) and the eye of the hurricane has a diameter of 30 miles (48.2 km). Determine the tangential velocity distribution in the hurricane if the maximum wind velocity is 216 km/h (60 m/s). Determine also the pressure variation in the hurricane and calculate the minimum pressure the hurricane has at its eye. Assume that the flow field is inside the inner core of the hurricane (the eye of hurricane is behaving like a solid body rotation and the region outside is behaving like a free body rotation). Ignore the translational motion of the hurricane and friction from the sea surface.

Solution

$$\omega = \frac{u_{\theta\max}}{R_i} = \frac{60}{24000} = 0.0025$$

$$K = \omega R_i{}^2 = 0.0025 \times (24 \times 10^3)^2 = 1.44 \times 10^6$$

$$u_\theta\big|_{inner} = r\omega = 0.0025r$$

$$u_\theta\big|_{outer} = \frac{K}{r} = \frac{1.44 \times 10^6}{r}$$

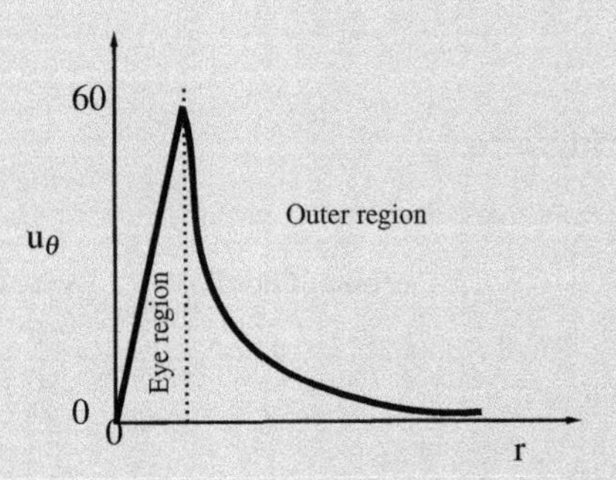

The strength of the vortex is

$$\Gamma = 2.88 \times 10^6 \pi$$

Inner region: We apply the Euler equation in polar cylindrical coordinates in the inner region of the vortex, where $u_\theta = f(r)$ only:

$$\rho \frac{u_\theta^2}{r} = \frac{\partial p}{\partial r}$$

and

$$\frac{1}{r}\frac{\partial p}{\partial \theta} = 0$$

The last equation indicates that pressure is not changing in the periphery of the hurricane, so

$$p \neq p(\theta)$$

$$\frac{\partial p}{\partial r} = \rho \frac{u_\theta^2}{r} = \frac{\rho}{r}(r\omega)^2 = \rho r \omega^2$$

We integrate this equation and obtain pressure as a function of radius

$$p(r) = \frac{\rho r^2 \omega^2}{2} + C_1$$

where C_1 is the constant.

At $r = R_i$, $p = p_o$ so

$$C_1 = p_o - \frac{\rho R_i{}^2 \omega^2}{2}$$

$$p - p_o = \frac{\rho r^2 \omega^2}{2} - \frac{\rho R_i{}^2 \omega^2}{2}$$

$$p - p_o = \frac{\rho u_\theta^2}{2}\left[1 - \left(\frac{R_i}{r}\right)^2\right]$$

Outer region: We now investigate the outer region of the hurricane where flow is behaving like a free vortex. We can apply Bernoulli's equation in the outer region of the hurricane and location far away from the hurricane.

$$\frac{p_{outer\ region}}{\rho} + \frac{u_\theta^2}{2} = \frac{p_{atm}}{\rho}$$

$$p_{outer\ core} = p_{atm} - \rho\frac{u_\theta^2}{2} = p_{atm} - \rho\frac{1}{2}\left(\frac{K}{r}\right)^2$$

$$p_{outer\ core} = p_{atm} - \rho\frac{1}{2}\left(\frac{\omega R_i{}^2}{r}\right)^2$$

$$p_{outer\ core} = p_{atm} - \rho\frac{1}{2}\left(\frac{\omega^2 R_i{}^4}{r^2}\right)$$

$$p_{outer\ core} - p_{atm} = \frac{-\rho u_\theta^2}{2}\left(\frac{R_i}{r}\right)^2$$

$$p_{outer\ core} - p_o = -\frac{\rho u_\theta^2}{2}\left[\left(\frac{R_i}{r}\right)^2 + 1\right]$$

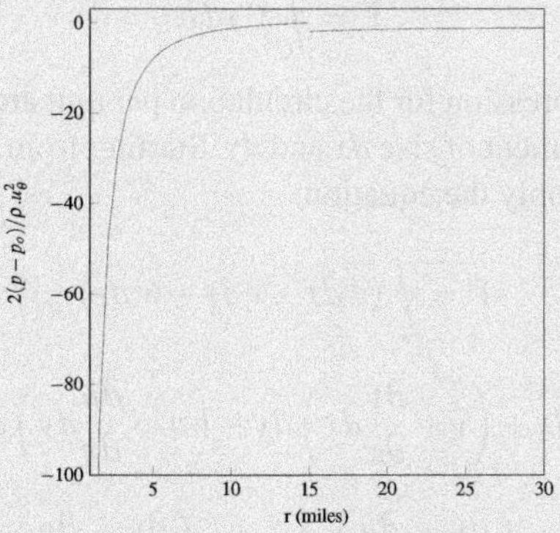

Figure 7.6 Pressure coefficient inside hurricane. The pressure asymptotically reaching 1 atm.

7.6 CIRCULATION AND INVISCID VORTEX

A mathematical concept commonly associated with inviscid vortex motion is that of circulation. The circulation, Γ, is defined as the line integral of the tangential component of the velocity taken around a closed curve in the flow field. In equation form, it can be expressed as

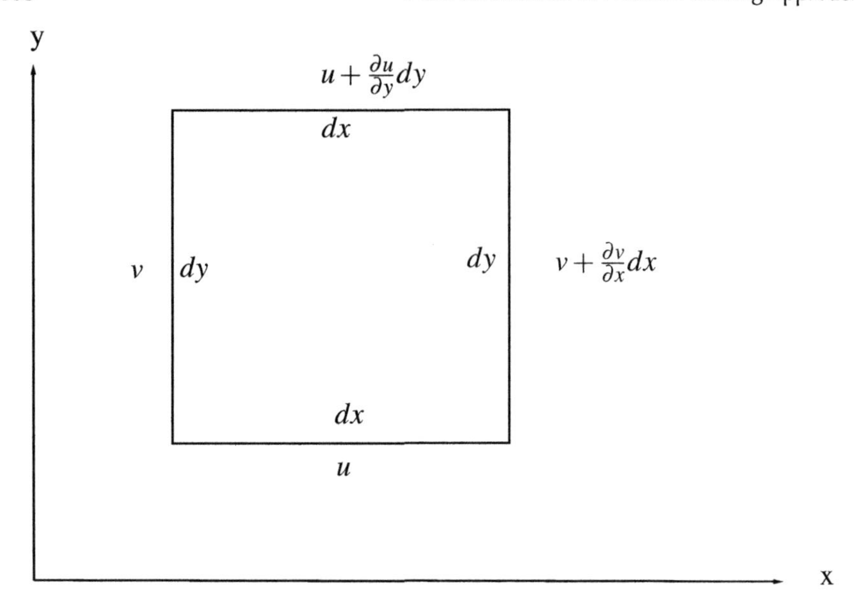

Figure 7.7 Circulation in a closed curve.

$$\Gamma = \oint_C \bar{V} \cdot ds \tag{7.16}$$

We now drive an expression for the circulation per unit area. Consider Figure 7.7, which shows a small element of size dx and dy. Starting from t origin and proceeding counterclockwise, we apply the equation:

$$\Gamma = \oint (u\,dx + v\,dy + w\,dz)$$

$$d\Gamma_z = udx + \left(v + \frac{\partial v}{\partial x}dx\right)dy - \left(u + \frac{\partial u}{\partial y}dy\right)dx - vdy$$

$$d\Gamma_z = \left(\frac{\partial v}{\partial x} - \frac{\partial u}{\partial y}\right)dxdy = \left(\frac{\partial v}{\partial x} - \frac{\partial u}{\partial y}\right)dA_z$$

$$\frac{d\Gamma_z}{dA_z} = \left(\frac{\partial v}{\partial x} - \frac{\partial u}{\partial y}\right)$$

Considering now the circulation around an area of finite size, we see from Figure 7.8 that the line integral around the bounding curve **C** is the algebraic sum of the line integrals around the infinitesimally small fluid elements that comprise the area within **C**. As the interior line of each element is traversed an equal number of times in opposite directions, only the exterior lines of the elements (which comprise curve C)

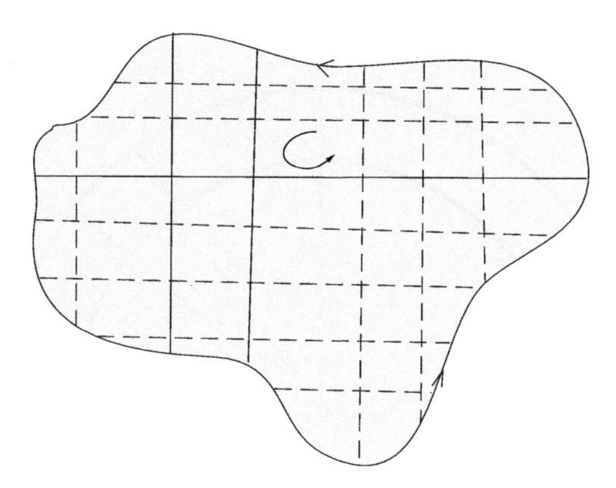

Figure 7.8 According to Green's theorem, the circulation around any closed curve is equal to the sum of the circulations around all the closed curves bounding the elementary areas within the region C.

contribute to the circulation. Therefore, employing the equation,

$$d\Gamma_z = \left(\frac{\partial v}{\partial x} - \frac{\partial u}{\partial y}\right) dxdy = \left(\frac{\partial v}{\partial x} - \frac{\partial u}{\partial y}\right) dA_z$$

we find that the circulation around any finite closed curve C in the $x - y$ plane may in the limit be expressed in terms of a surface integral, namely,

$$d\Gamma_z = \oiint \left(\frac{\partial v}{\partial x} - \frac{\partial u}{\partial y}\right) dxdy$$

We have already derived the expression for vorticity in fluid:

$$\omega_z = \frac{1}{2}\left(\frac{\partial v}{\partial x} - \frac{\partial u}{\partial y}\right)$$

so circulation and vorticity are related to each other as per relation:

$$\frac{d\Gamma_z}{dA_z} = 2\omega_z$$

This relationship was introduced by Stokes, and is called the Stokes theorem.

Note that vorticity is related with rotation of fluid particle, and the concept of circulation is associated with a flow field having billions of fluid particles. The circulation is also called vortex strength. This theorem is used to study the flow inside the pump impellers, blowers, and fans of an aircraft and the flow around an aircraft wing.

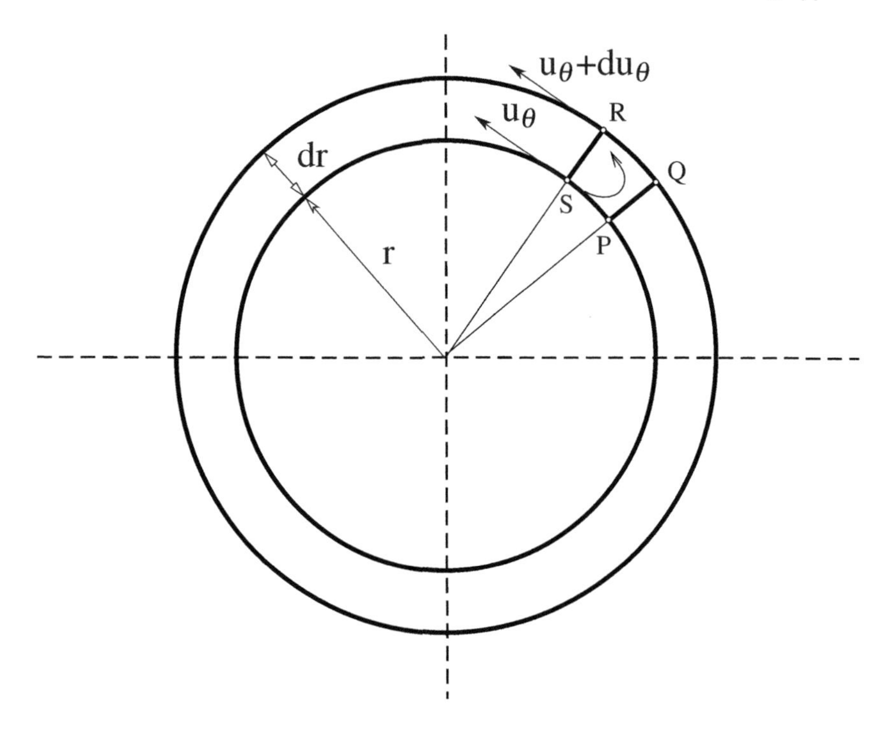

Figure 7.9 Circulation in a close path PQRS.

7.7 CIRCULATION IN FREE VORTEX

Proof: We now prove that the change in circulation in a close connected path of free vortex will be zero. Consider the fluid body in Figure 7.9, indicated by a close path PQRS.

Tangential velocity in a vortex is

$$u_\theta = \frac{\Gamma}{2\pi}\frac{1}{\sqrt{x^2+y^2}} = \frac{\Gamma}{2\pi r}$$

$$u_\theta \cdot r = \frac{\Gamma}{2\pi} = Constant$$

For such ideal fluid we can write:

$$d\Gamma_{PQRS} = (u_\theta + du_\theta)(r+dr)d\theta - u_\theta r\theta$$

$$d\Gamma_{PQRS} = d[u_\theta \cdot r]d\theta$$

Since $u_\theta \cdot r =$ constant, $d\Gamma_{PQRS}$ will be zero

$$d\Gamma_{PQRS} = 0$$

This shows that circulation is constant inside the free vortex.

7.8 SOURCE OR SINK

Table 7.3
Expression for Source (E>0)/Sink(E<0)

Source/Sink	Cartesian Coordinates	Polar Coordinates
Potential Function	$\frac{E}{2\pi}\ln(\sqrt{x^2+y^2})$	$\frac{E}{2\pi}\ln r$
Stream Function	$\frac{E}{2\pi}\tan^{-1}\left(\frac{2ay}{x^2+y^2-a^2}\right)$	$\frac{E}{2\pi}\theta$
u-component	$\frac{E}{2\pi}\left(\frac{x}{x^2+y^2}\right)$	$\frac{E}{2\pi}\frac{\cos(\theta)}{r}$
v-component	$\frac{E}{2\pi}\left(\frac{y}{x^2+y^2}\right)$	$\frac{E}{2\pi}\frac{\sin(\theta)}{r}$
u_r-component	$\frac{E}{2\pi}\left(\frac{1}{\sqrt{x^2+y^2}}\right)$	$\frac{E}{2\pi}\frac{1}{r}$
u_θ-component	0	0

7.9 SUPERPOSITION: RANKINE HALF BODY

If we combine uniform flow and source then we will create a flow pattern that will look like a flow over a curved object, as shown in Figure 7.10.

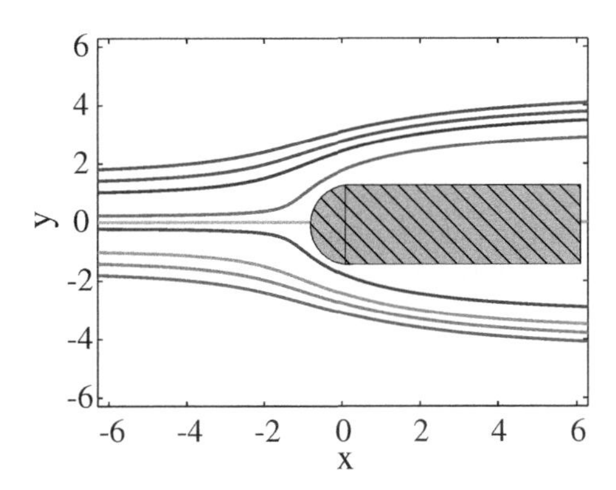

Figure 7.10 Rankine half-body in two dimensions. Plot of equation $\psi = y\,U + \left(\frac{m}{2\pi}\right)\tan^{-1}\left(\frac{y}{x}\right)$.

We use polar coordinates to describe this flow as a combination of uniform flow and a source:

$$\psi = U_\infty r \sin(\theta) + \frac{E}{2\pi}\theta$$

$$\phi = U_\infty r \cos(\theta) + \frac{E}{2\pi}\ln r$$

The source of strength E is positioned at the origin, and a uniform flow is approaching from left to right. There will be a stagnation point where uniform flow and source meet on the axis and the rest of the incoming flow will be deflected sideways, thus giving an appearance of the flow that looks like flow over a two-dimensional body. The velocity at any position on this body can be determined by differentiation of the stream or the potential function. In polar coordinates, we have

$$\psi = U_\infty r \sin(\theta) + \frac{E}{2\pi}\theta$$

$$\phi = U_\infty r \cos(\theta) + \frac{E}{2\pi}\ln r$$

The velocities are:

$$u_r = \frac{1}{r}\frac{\partial \psi}{\partial \theta}$$

$$u_r = \frac{1}{r}U_\infty \, r\cos\theta + \frac{1}{2}\frac{E}{\pi r}$$

$$u_r = U_\infty \, \cos\theta + \frac{1}{2}\frac{E}{\pi r}$$

$$u_\theta = -\frac{\partial \psi}{\partial r}$$

$$u_\theta = -U_\infty \sin\theta$$

The stagnation point is present where the velocity u_θ is zero, which leads to $\sin(\theta) = 0$. We know that this can happen when $\theta = \pi$ or 0. If we take $\theta = 0$ then the u_r equation would have a negative r, which has no meaning, so $\theta = \pi$ is the result. Setting $u_r = 0$, $\theta = \pi$ and solving ψ equation, we get the location of stagnation point as:

$$r = \frac{1}{2}\frac{E}{\pi U_\infty}$$

Substituting this r into ψ relation gives us the value of the stagnation streamline as $E/2$.

$$\psi = U_\infty r \sin(\theta) + \frac{E}{2\pi}\theta = \frac{E}{2}$$

$$\psi = U_\infty y + \frac{E}{2\pi}\tan^{-1}\left(\frac{y}{x}\right) = \frac{E}{2}$$

$$U_\infty y + \frac{E}{2\pi}\theta = \frac{E}{2}$$

Rearranging, we get the equation in terms of y as:

$$y = \frac{\frac{E}{2} - \frac{E}{2\pi}\theta}{U_\infty} = \frac{\frac{E}{2}\left(1 - \frac{\theta}{\pi}\right)}{U_\infty}$$

$$y = \frac{E}{2U_\infty}\left(1 - \frac{\theta}{\pi}\right)$$

We inspect the y at different values of θ:

$$\theta = 0 \qquad y_{\max} = \frac{E}{2U_\infty}$$

The upper ordinate at origin is

$$\theta = \frac{\pi}{2} \qquad y = \frac{E}{4U_\infty}$$

The stagnation point is

$$\theta = \pi \qquad y = 0$$

The lower ordinate at origin is

$$\theta = \frac{3\pi}{2} \qquad y = \frac{-E}{4U_\infty}$$

The Rankine half body will have the maximum thickness at $x \to \infty$

$$\psi_\infty = U_\infty y_\infty + \frac{E}{2\pi}\tan^{-1}\left(\frac{y_\infty}{\infty}\right)$$

$$\psi_\infty = U_\infty y_\infty + \frac{E}{2\pi}\tan^{-1}(0) = \frac{E}{2}$$

$$y_\infty = \frac{E}{2U_\infty}$$

$$thickness = 2y_\infty$$

Example 7.4

Example Consider the flow over a semi-infinite Rankine half body. If the stagnation point is located 1 m upstream of the source, find the body profile and pressure distribution.
Solution The stagnation point is located at 1 m so

$$r = \frac{1}{2}\frac{E}{\pi U_\infty} = 1$$

$$E = 2\pi U_\infty$$

We substitute E into the following equation:

$$\psi = U_\infty r \sin(\theta) + \frac{E}{2\pi}\theta$$

$$\psi = U_\infty r \sin(\theta) + U_\infty \theta$$

For stagnation streamline, the following is valid:

$$\psi = E/2 = \pi U_\infty$$

Rearranging the stream function equation, we get

$$r = \frac{\psi - U_\infty \theta}{U_\infty \sin(\theta)} = \frac{\pi U_\infty - U_\infty \theta}{U_\infty \sin(\theta)} = \frac{\pi - \theta}{\sin(\theta)} \tag{7.17}$$

We now take different θ angles in radians and calculate the r:

Degrees	Radians	r	$r \cdot \cos(\theta)$	$r \cdot \sin(\theta)$
180	π	1	−1	0
170	2.967	1.005	−0.989	0.1745
160	2.792	1.0206	−0.959	0.349
150	2.617	1.0471	−0.9068	0.5235
140	2.443	1.0861	−0.832	0.6981
130	2.268	1.139	−0.7322	0.8726
90	1.57	1.57	0	1.57
80	1.396	1.772	0.3077	1.7453
60	1.047	2.418	1.2091	2.0943
50	0.872	2.961	1.9038	2.2689
32	0.558	4.8744	4.1338	2.583

The first row in the table has the manually inserted data, as we know that $r = 1$ for the stagnation point. For the rest of the angles, the data in the table is computed from Eq. 7.17. Figure 7.11 shows the plot of pressure coefficient.

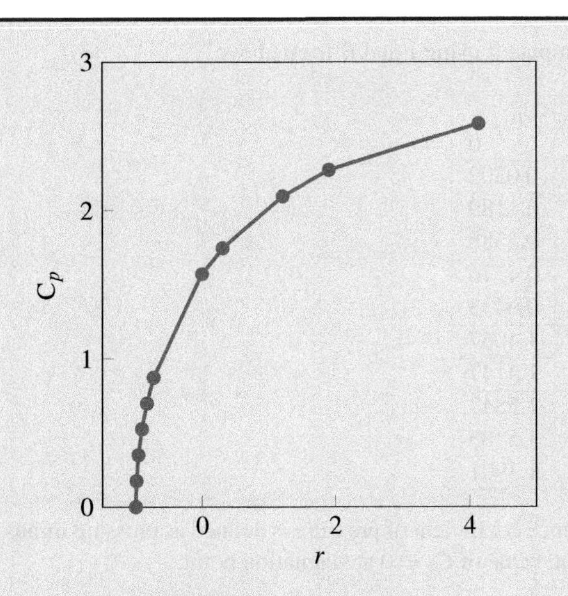

Figure 7.11 Plot of C_p.

We now estimate the pressure distribution over the Rankine half body by using Bernoulli's equation.

$$u_r = U_\infty \cos\theta + \frac{1}{2}\frac{E}{\pi r}$$

with

$$E = 2\pi U_\infty$$

$$u_r = U_\infty \cos\theta + \frac{1}{2}\frac{E}{\pi r}$$

and

$$u_\theta = -U_\infty \sin\theta$$

The total velocity must be used in Bernoulli's equation

$$\vec{V}^2 = {u_r}^2 + {u_\theta}^2$$

$$p_o - p = \frac{1}{2}\rho\vec{V}^2 = \frac{1}{2}\rho({u_r}^2 + {u_\theta}^2)$$

$$C_p = \frac{\Delta p}{\frac{1}{2}\rho U_\infty^2} = \frac{({u_r}^2 + {u_\theta}^2)}{U_\infty^2} = \frac{r^2 + 2r\cos(\theta) + 1}{r^2}$$

$$C_p = 1 + \frac{1}{r^2} + \frac{2\cos(\theta)}{r}$$

We can compute it using r and θ from above.

x	C_p
−1	0
−0.989	0.0302
−0.959	0.1189
−0.9068	0.2588
−0.832	0.4376
−0.7322	0.6433
0	1.4067
0.3077	1.5147
1.2091	1.5847
1.9038	1.5485
4.1338	1.3901

Note that since coefficient of pressure is defined as pressure minus stagnation pressure, the value of $C_p = 0$ at stagnation point.

7.10 SUPERPOSITION: SOURCE AND SINK NEARBY

In some engineering problems the source and sink could be located close to each other, as shown in Figure 7.12. The radial coordinate is defined as:

$$r^2 = (x \mp a)^2 + y^2$$

depending on the position of the source or sink. The strength of the source is $+E$ and the strength of the sink is $-E$.

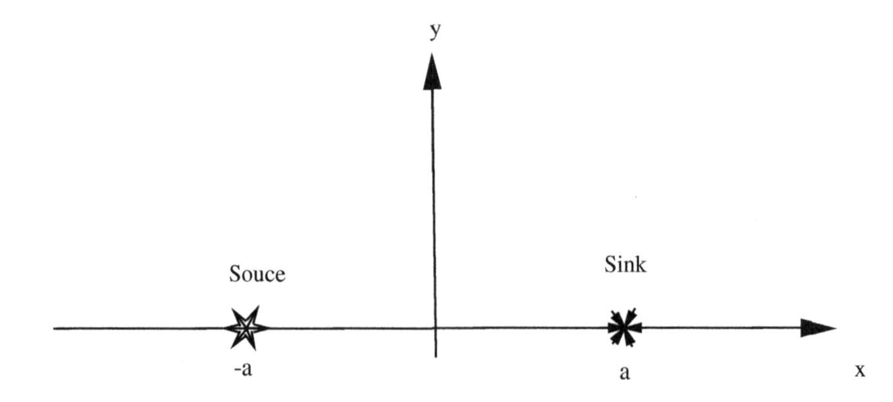

Figure 7.12 The source and sink positioned close to each other.

$$\phi = \frac{E}{2\pi}\ln r_1 - \frac{E}{2\pi}\ln r_2$$

We can construct this flow by writing the combined stream and velocity potential functions in polar coordinates as:

$$\phi = \frac{E}{2\pi}\ln r_1 - \frac{E}{2\pi}\ln r_2$$

and in Cartesian coordinates as:

$$\phi = \frac{E}{2\pi}\ln(\sqrt{(x+a)^2+y^2})) - \frac{E}{2\pi}\ln(\sqrt{(x-a)^2+y^2})$$

where source is located on the x-axis at $(x,y) = (-a,0)$ and sink is positioned at $(x,y) = (a,0)$.

$$\phi = \frac{E}{2\pi}\ln\left(\sqrt{\frac{(x+a)^2+y^2)}{(x-a)^2+y^2)}}\right)$$

Also,

$$\psi = \frac{E}{2\pi}(\theta_1 - \theta_2)$$

$$\psi = \frac{-E}{2\pi}\tan^{-1}\left(\frac{2ay}{x^2+y^2-a^2}\right)$$

7.11 SUPERPOSITION: SOURCE + SINK + UNIFORM FLOW

If we impose the uniform flow over the source and sink combination that we have discussed in a previous section, then the resulting flow pattern is like an oval (see Figure 7.13) and is often called flow over Rankine full body, after W. J. M. Rankine (1820–1872), who first developed the technique of combining flow patterns.

$$\psi = U_\infty y - \frac{E}{2\pi}\tan^{-1}\left(\frac{2ay}{x^2+y^2-a^2}\right)$$

$$\phi = U_\infty x + \frac{E}{2\pi}\ln\left(\sqrt{\frac{(x+a)^2+y^2)}{(x-a)^2+y^2)}}\right)$$

and

$$u = U_\infty - \frac{aE}{2\pi}\left[\frac{x^2-y^2-a^2}{((x+a)^2+y^2)((x-a)^2+y^2)}\right]$$

$$v = -\frac{2aExy}{\pi}\left[\frac{1}{((x+a)^2+y^2)((x-a)^2+y^2)}\right]$$

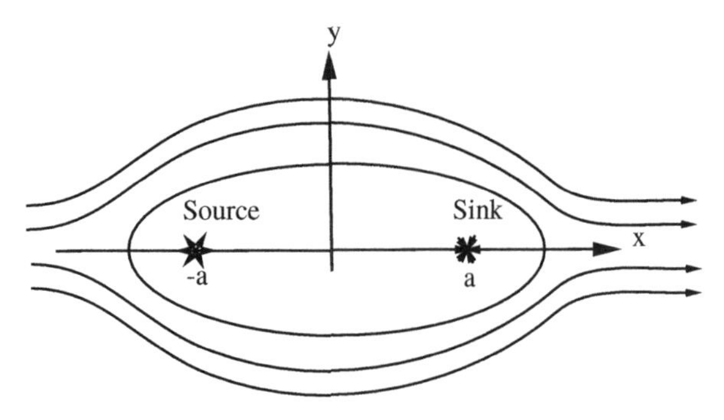

Figure 7.13 Streamlines of flow over Rankine full body.

Example 7.5

Example Consider the uniform flow defined by $U = 15$ m/s with the source and sink having the same strength as $E = 10$. The source and sink are $a = \pm 2$ m apart. What are the values of y if the stream function takes values of 0 and 1?

Solution

$$\psi(x,y) = U \cdot y - \frac{E}{2\,\pi} arctan\left(\frac{2 \cdot a \cdot y}{x^2 + y^2 - a^2}\right)$$

At $x = 0$, $\psi = 1$ and through trial and error, we get

$$\psi(x = 0, y) = 16y - \frac{5}{\pi}\arctan\left(\frac{4y}{x^2 + y^2 - 4}\right)$$

At $x = 1$, $\psi = 1$ and through trial and error, we get $y = 0.0551$ m.

7.12 DOUBLET

Let's say that a source and sink of strength E are approaching each other, as depicted in Figure 7.14.

$$\phi = \frac{E}{2\pi}\ln r_1 - \frac{E}{2\pi}\ln r_2$$

$$\phi = \frac{E}{2\pi}\ln\left(\frac{r_1}{r_2}\right)$$

From trigonometry,

$$r_1 = 2a\cos\theta_1 + r_2\cos(\theta_2 - \theta_1)$$

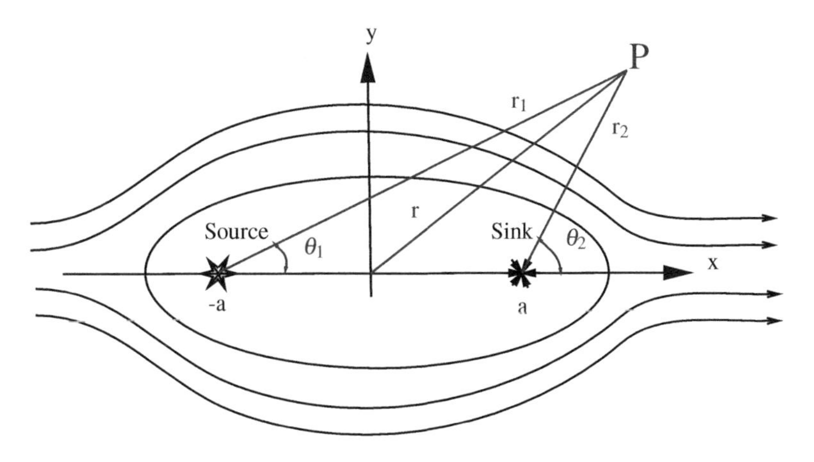

Figure 7.14 The source and sink approaching each other.

$$\phi = \frac{E}{2\pi} \ln\left(\frac{2a\cos\theta_1 + r_2\cos(\theta_2 - \theta_1)}{r_2}\right)$$

As source and sink approach each other:

$$a \to 0, r_2 \to r, \theta_2 \to \theta_1, \cos(\theta_2 - \theta_1) \to 1$$

$$\phi = \frac{E}{2\pi} \ln\left(\frac{2a\cos\theta}{r} + 1\right)$$

This can be approximated as:

$$\phi \approx \frac{2aE}{2\pi}\left(\frac{\cos\theta}{r}\right)$$

$$\phi \approx \frac{M}{2\pi}\left(\frac{\cos\theta}{r}\right)$$

The stream function will be

$$\psi = \frac{E}{2\pi}(\theta_2 - \theta_1)$$

From Figure 7.14

$$r_2 \sin(\theta_2 - \theta_1) = 2a\sin\theta$$

$$\theta_2 - \theta_1 \approx \frac{2a}{r}\sin\theta$$

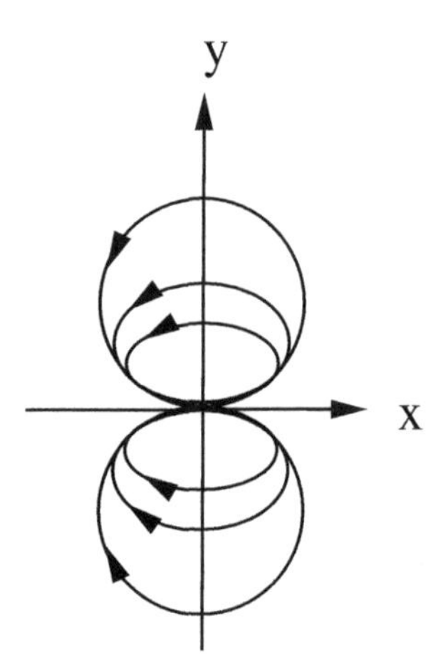

Figure 7.15 A doublet (or dipole) will form when a source and sink approach each other.

$$\psi = \frac{-E}{2\pi}\frac{2a}{r}\sin\theta$$

$$\psi = -\frac{M}{2\pi}\frac{\sin\theta}{r}$$

where $M = 2Ea$.

Figure 7.15 shows the streamlines as source and sink are infinitesimally close to each other.

7.13 FLOW ABOUT A CIRCULAR CYLINDER

Flow about a circular cylinder can be modeled as a combination of uniform flow and doublet. So we may write the stream and potential functions in polar coordinates as:

$$\psi = U_\infty r\sin\theta - \left(\frac{M}{2\pi}\right)\left(\frac{\sin\theta}{r}\right)$$

$$\phi = U_\infty r\cos\theta + \left(\frac{M}{2\pi}\right)\left(\frac{\cos\theta}{r}\right)$$

$$u_r = \frac{1}{r}\frac{\partial\psi}{\partial\theta} = \frac{1}{\cancel{r}}U_\infty\cancel{r}\cos(\theta) - \frac{1}{2}\frac{M\cos(\theta)}{\pi r}$$

$$u_r = \frac{1}{r}\frac{\partial\psi}{\partial\theta} = U_\infty\cos(\theta) - \frac{1}{2}\frac{M\cos(\theta)}{\pi r}$$

For stagnation streamline, we set

Table 7.4
Expression for Doublet

Doublet	Cartesian Coordinates	Polar Coordinates
Potential function	$\left(\frac{M}{2\pi}\right)\left(\frac{x}{x^2+y^2}\right)$	$\left(\frac{M}{2\pi}\right)\left(\frac{\cos\theta}{r}\right)$
Stream function	$\left(\frac{-M}{2\pi}\right)\left(\frac{y}{x^2+y^2}\right)$	$\left(\frac{-M}{2\pi}\right)\left(\frac{\sin\theta}{r}\right)$
u-component	$\left(\frac{-M}{2\pi}\right)\left(\frac{x^2-y^2}{(x^2+y^2)^2}\right)$	$\left(\frac{-M}{2\pi}\right)\left(\frac{\cos 2\theta}{r^2}\right)$
v-component	$\left(\frac{-M}{2\pi}\right)\left(\frac{2xy}{(x^2+y^2)^2}\right)$	$\left(\frac{-M}{2\pi}\right)\left(\frac{\sin 2\theta}{r^2}\right)$
u_r component	$\left(\frac{-M}{2\pi}\right)\left(\frac{x}{(x^2+y^2)^{3/2}}\right)$	$\left(\frac{-M}{2\pi}\right)\left(\frac{\cos\theta}{r^2}\right)$
u_θ component	$\left(\frac{-M}{2\pi}\right)\left(\frac{y}{(x^2+y^2)^{3/2}}\right)$	$\left(\frac{-M}{2\pi}\right)\left(\frac{\sin\theta}{r^2}\right)$

$$\psi = U_\infty r\sin\theta - \left(\frac{M}{2\pi}\right)\left(\frac{\sin\theta}{r}\right) = 0$$

$$r^2 = \frac{M}{2\pi U_\infty}$$

We call this radius r_o, and substitute it into stream function

$$r_o{}^2 = \frac{M}{2\pi U_\infty}$$

$$\psi = U_\infty \sin\theta\left(r - \frac{r_o{}^2}{r}\right)$$

We now find u_θ as:

$$u_\theta = -\frac{\partial\psi}{\partial r}$$

$$u_\theta = -\sin(\theta)U_\infty\left(1 + \frac{r_o{}^2}{r^2}\right)$$

We substitute $r = r_o$, which gives

$$u_\theta = -2U_\infty\sin(\theta)$$

$$u_r = \frac{1}{r}\frac{\partial\psi}{\partial\theta}$$

$$u_r = \frac{1}{r}\cos(\theta)U\left(r - \frac{{r_o}^2}{r}\right) = 0$$

The Bernoulli's equation can now be applied on a streamline with station in a free stream flow and station at an angle $\theta = \pi$ close to the cylinder:

$$\underbrace{p_\infty + \frac{1}{2}\rho U_\infty^2}_{Free\ stream} = \underbrace{p + \frac{1}{2}\rho V^2}_{Near\ cylinder}$$

$$p_\infty + \frac{1}{2}\rho U_\infty^2 = p + \frac{1}{2}\rho[(u_r)^2 + (u_\theta)^2]$$

$$p_\infty + \frac{1}{2}\rho U_\infty^2 = p + \frac{1}{2}\rho[(u_r)^2 + (u_\theta)^2]$$

$$p_\infty - p = \frac{1}{2}\rho[(u_r)^2 + (u_\theta)^2 - U_\infty^2]$$

$$p_\infty - p = \frac{1}{2}\rho[(0)^2 + (-2U_\infty\sin(\theta))^2 - U_\infty^2]$$

$$p_\infty - p = \frac{\rho}{2}U_\infty^2[4\sin^2(\theta) - 1]$$

$$\frac{p - p_\infty}{\frac{\rho}{2}U_\infty^2} = 1 - 4\sin^2(\theta)$$

$$C_p = 1 - 4\sin^2(\theta)$$

Figure 7.16 shows that the pressure distribution across the cylinder is symmetric and thus there will be no shape-associated drag on the cylinder. On the contrary, experience denied this outcome, which was a paradox for the scientific community. We now estimate the drag force by integrating pressure over the area.

$$\frac{F_D}{\ell} = \int_0^{2\pi} p\cos(\theta)r\cdot d\theta$$

where ℓ is the length of the cylinder.

$$\frac{F_D}{\ell} = \int_0^{2\pi}\left\{p_\infty - \frac{\rho}{2}U_\infty^2[4\sin^2(\theta) - 1]\right\}\cos(\theta)r\cdot d\theta$$

Note that

$$\int_0^{2\pi}\cos(\theta)d\theta = 0$$

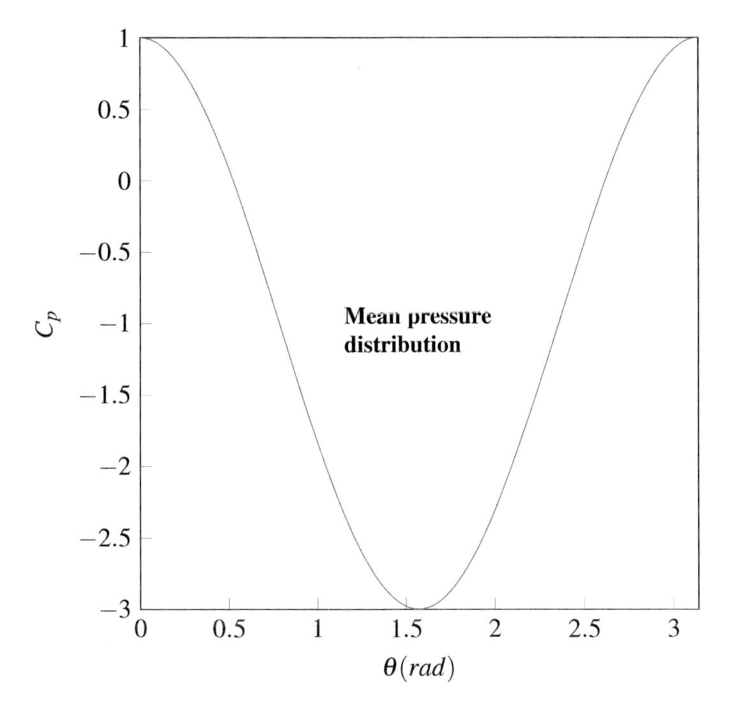

Figure 7.16 Pressure coefficient versus angle over the cylinder surface. Plot of Eq. $C_p = 1 - 4\sin^2(\theta)$.

and

$$\int_0^{2\pi} \cos(\theta)\sin^2(\theta)d\theta = 0$$

so the total drag force is zero across the cylinder.

The estimation of lift force can be done by the equation

$$\frac{F_L}{\ell} = -\int_0^{2\pi} \left\{p_\infty - \frac{\rho}{2}U_\infty^2[4\sin^2(\theta) - 1]\right\} \sin(\theta) r \cdot d\theta$$

but this, too, is zero. So both drag and lift on a stationary cylinder are zero, according to potential flow theory.

7.14 FLOW ALONG A SPINNING CYLINDER

We now investigate the case of a spinning circular cylinder when it is spinning on it's own axis. This case can be modeled as flow over cylinder, which has been discussed in previous sections, plus the superposition of constant circulation on the flow field. The resulting stream function is:

$$\psi = U_\infty \sin\theta \left(r - \frac{{r_o}^2}{r}\right) + \frac{\Gamma}{2\pi}\ln r$$

$$u_r = \frac{1}{r}\frac{\partial \psi}{\partial \theta}$$

$$u_r = \frac{1}{r}\cos(\theta)\, U_\infty \left(r - \frac{{r_o}^2}{r}\right)$$

with substitution of $r = r_o$, $u_r = 0$.

$$u_\theta = -\left.\frac{\partial \psi}{\partial r}\right|_{r=r_o} = -2U_\infty \sin(\theta) - \frac{1}{2}\frac{\Gamma}{\pi r_o}$$

$$C_p = \frac{p - p_\infty}{\frac{\rho}{2}U_\infty^2} = 1 - \left[2\sin(\theta) + \frac{1}{2U_\infty}\frac{\Gamma}{\pi r_o}\right]^2$$

The drag force is zero.

$$\frac{F_D}{\ell} = 0$$

However, the lift force on a revolving cylinder depends on how fast the cylinder is revolving.

$$\frac{F_L}{\ell} = \rho U_\infty \Gamma$$

This is known as the **Kutta-Joukowsky theorem**.

We now set tangential velocity to zero to investigate the stagnation point location.

$$u_\theta = -\left.\frac{\partial \psi}{\partial r}\right|_{r=r_o} = -2U_\infty \sin(\theta) - \frac{1}{2}\frac{\Gamma}{\pi r_o} = 0$$

$$\sin(\theta) = \frac{-\Gamma}{4\pi U_\infty r_o}$$

Also, setting radial velocity to zero:

$$u_r = \frac{1}{r}\cos(\theta)\, U_\infty \left(r - \frac{{r_o}^2}{r}\right)$$

$$u_r = \left(U_\infty - \frac{U_\infty {r_o}^2}{r^2}\right)\cos(\theta)$$

$$u_r = \left(U_\infty - \frac{U_\infty {r_o}^2}{r^2}\right)\cos(\theta) = 0$$

$\cos(\theta) = 0$ and $\theta = \pm\frac{\pi}{2}$ at $r = r_o$.

This shows that stagnation point $\theta = \pm 90^\circ$.

7.15 MAGNUS EFFECT

Newton was the first person to observe, in 1671, that the path of a ball could be influenced by its rotation. Later, in 1742, Robins did systematic experiments with a pendulum to determine the magnitude of the aerodynamic side force acting on a sphere. Tennis, golf, and cricket players know that the flight path can be changed if the ball spins, and that the direction of the rebound will be affected. In 1920, the science of Magnus effects came from sailing technology when the German engineer Anton Flettner introduced his Flettner rotor, and a rotating cylinder was used in the rotor ship. It was found that the Magnus force can have a negative direction in a certain range of Reynolds numbers. Flettner also showed that rotating two tandem cylinders in opposite directions affected the turning of the boat. The boat was built to be used for cheap freighters or fishing boats, but for significant propulsive forces cylinders must be driven fast by the use of ball or roller bearings and this was considered too expensive for fishing boats in the 1920; thus the idea was discarded. Recently, in 2010, the Flettner rotor-based design has been reintroduced by Enercon wind company in their E-Ship 1 design. A similar phenomenon of lift force generation on a spinning sphere is called the Robins effect, which is named after Benjamin Robins (1707–1751).

7.16 CORNER FLOW

The flow in a corner is shown in Figure 7.17.

$$u = \frac{\partial \phi}{\partial x} = \frac{\partial \psi}{\partial y}$$

Table 7.5
Expression for Corner Flow

Corner Flow	Cartesian Coordinates	Polar Coordinates
Potential Function	$\frac{a}{2}(x^2 - y^2)$	$\frac{a}{2}r^2\cos(2\theta)$
Stream Function	axy	$\frac{a}{2}r^2\sin(2\theta)$
u-component	ax	$ar\cos(\theta)$
v-component	$-ay$	$-ar\sin(\theta)$
u_r-component	$\frac{a(x^2-y^2)}{\sqrt{(x^2+y^2)}}$	$ar\cos(2\theta)$
u_θ-component	$\frac{-2axy}{\sqrt{(x^2+y^2)}}$	$-ar\sin(2\theta)$

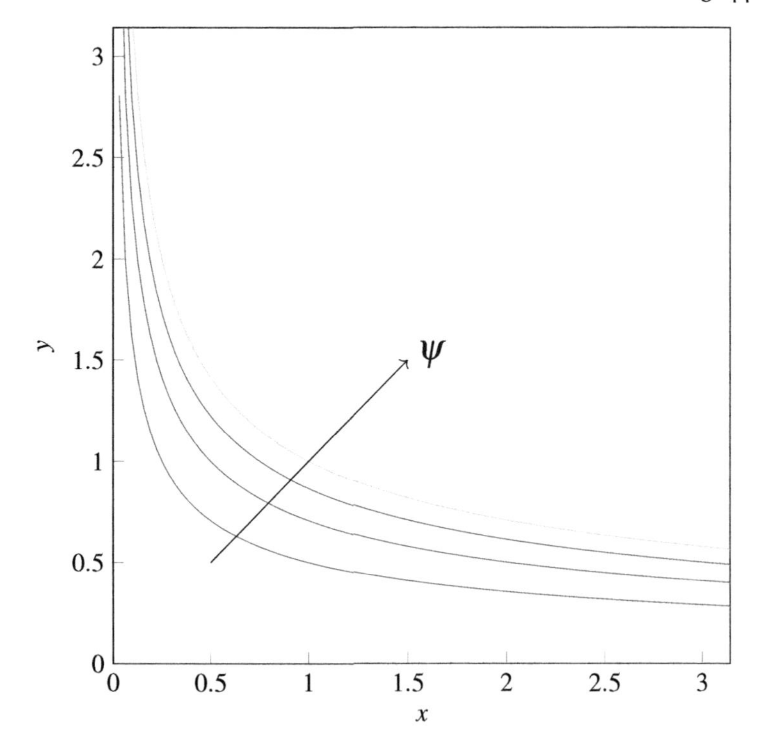

Figure 7.17 Flow in a corner for $\psi = 1, 2, 3$, and 4.

Example 7.6

Example The stream function for flow in a 90° corner is

$$\psi(r,\theta) = 10 \cdot r^2 \cdot \sin(2 \cdot \theta)$$

Is the continuity satisfied? Also, find the velocity at $r = 0.2$ m and $\theta = 30°$ and the equation of streamline.

Solution At $r = 0.2$ m and $\theta = 30°$:

$$u_r = \frac{1}{r} \cdot \frac{d}{d\,\theta}\,\psi(r,\theta) = 20 \cdot r \cdot \cos(2\theta) = 2.0018\, m/s$$

$$u_\theta = -\frac{d}{d\,r}\,\psi(r,\theta) = -20 \cdot r \cdot \sin(2\theta) = -3.463\, m/s$$

The continuity equation is

$$\frac{1}{r}\frac{\partial(r \cdot u_r)}{\partial r} + \frac{1}{r}\frac{\partial u_\theta}{\partial \theta} = 0$$

$$40 \cdot \cos(2\theta) - 40 \cdot \cos(2\theta) = 0$$

Substituting u_r and u_θ proves that continuity is satisfied.

The velocity is

$$V = \sqrt{u_r^2 + u_\theta^2} = 4m/s$$

Now from trigonometry

$$y = r\sin\theta, x = r\cos\theta$$

$$\sin\theta = \frac{y}{r}, \cos\theta = \frac{x}{r}$$

$$\sin(2\theta) = 2 \cdot \sin\theta \cdot \cos\theta = 2\left(\frac{y}{r}\right)\left(\frac{x}{r}\right) = \frac{2xy}{r^2}$$

$$\psi = 10r^2 \sin(2\theta) = 10r^2\left(\frac{2xy}{r^2}\right) = 20 \cdot x \cdot y$$

This gives the equation of streamline.

PROBLEMS

7P-1 A 2D flow field is described by the following velocity components:

$$u = \frac{x}{2t+1}$$

$$v = \frac{y}{t-1}$$

Find the equation of streamline.

7P-2 Consider the following stream function in spherical polar coordinates:

$$\psi(r,\theta) = \frac{2r\,v\,\sin^2\theta}{1 + a - \cos\theta}$$

where a is a constant. Find the streamlines and velocity field.

7P-3 A stream function is defined for streamlines over the sphere as:

$$\psi(r,\theta) = \frac{1}{4}U{r_o}^2\sin^2\theta\left(\frac{a}{r_o} - \frac{3r}{r_o} + \frac{2r^2}{{r_o}^2}\right)$$

where r_o is the radius of the sphere. Find radial and tangential velocity components.

Table 7.6
Catalogue

DESCRIPTION	FLOW	POTENTIAL FUNCTION	STREAM FUNCTION	V COMPONENTS $u(x,y)$	$v(x,y)$	u_r	u_θ
Uniform flow in x direction		$U_\infty x$ $U_\infty r cos(\theta)$	$U_\infty y$ $U_\infty r sin(\theta)$	U_∞ U_∞	0 0	$\frac{U_\infty x}{\sqrt{x^2+y^2}}$ $U_\infty cos(\theta)$	$\frac{-U_\infty y}{\sqrt{x^2+y^2}}$ $-U_\infty sin(\theta)$
Uniform flow in y direction		$U_\infty y$ $U_\infty r cos(\theta)$	$-U_\infty x$ $-U_\infty r cos(\theta)$	0 0	U_∞ U_∞	$\frac{U_\infty y}{\sqrt{x^2+y^2}}$ $U_\infty sin(\theta)$	$\frac{U_\infty x}{\sqrt{x^2+y^2}}$ $U_\infty cos(\theta)$
Stagnation/ Corner Flow		$\frac{a}{2}(x^2-y^2)$ $\frac{a}{2}r^2\cos(2\theta)$	axy $\frac{a}{2}r^2\sin(2\theta)$	ax $ar\cos(\theta)$	$-ay$ $-ar\sin(\theta)$	$\frac{a(x^2-y^2)}{\sqrt{(x^2+y^2)}}$ $ar\cos(2\theta)$	$\frac{-2axy}{\sqrt{(x^2+y^2)}}$ $-ar\sin(2\theta)$
Source($E>0$) Sink($E<0$) $a=$ constant		$\frac{E}{2\pi}\ln(\sqrt{x^2+y^2})$ $\frac{E}{2\pi}\ln r$	$\frac{E}{2\pi}\tan^{-1}\left(\frac{2ay}{x^2+y^2-a^2}\right)$ $\frac{E}{2\pi}\theta$	$\frac{E}{2\pi}\left(\frac{x}{x^2+y^2}\right)$ $\frac{E}{2\pi}\frac{\cos(\theta)}{r}$	$\frac{E}{2\pi}\left(\frac{y}{x^2+y^2}\right)$ $\frac{E}{2\pi}\frac{\sin(\theta)}{r}$	$\frac{E}{2\pi}\left(\frac{1}{\sqrt{x^2+y^2}}\right)$ $\frac{E}{2\pi}\frac{1}{r}$	0 0
Vortex Circulation		$\frac{\Gamma}{2\pi}\tan^{-1}\left(\frac{y}{x}\right)$ $\frac{\Gamma}{2\pi}\theta$	$\frac{-\Gamma}{2\pi}\ln(\sqrt{x^2+y^2})$ $-\frac{\Gamma}{2\pi}\ln r$	$\frac{-\Gamma}{2\pi}\left(\frac{y}{x^2+y^2}\right)$ $\frac{-\Gamma}{2\pi}\left(\frac{\sin\theta}{r}\right)$	$\frac{\Gamma}{2\pi}\left(\frac{x}{x^2+y^2}\right)$ $\frac{\Gamma}{2\pi}\left(\frac{\cos\theta}{r}\right)$	0 0	$\frac{\Gamma}{2\pi}\left(\frac{1}{\sqrt{x^2+y^2}}\right)$ $\frac{\Gamma}{2\pi}\left(\frac{1}{r}\right)$
Doublet		$\left(\frac{M}{2\pi}\right)\left(\frac{x}{x^2+y^2}\right)$ $\left(\frac{M}{2\pi}\right)\left(\frac{\cos\theta}{r}\right)$	$\left(\frac{-M}{2\pi}\right)\left(\frac{y}{x^2+y^2}\right)$ $\left(\frac{-M}{2\pi}\right)\left(\frac{\sin\theta}{r}\right)$	$\left(\frac{-M}{2\pi}\right)\left(\frac{x^2-y^2}{(x^2+y^2)^2}\right)$ $\left(\frac{-M}{2\pi}\right)\left(\frac{\cos2\theta}{r^2}\right)$	$\left(\frac{-M}{2\pi}\right)\left(\frac{2xy}{(x^2+y^2)^2}\right)$ $\left(\frac{-M}{2\pi}\right)\left(\frac{\sin2\theta}{r^2}\right)$	$\left(\frac{-M}{2\pi}\right)\left(\frac{x}{(x^2+y^2)^{3/2}}\right)$ $\left(\frac{-M}{2\pi}\right)\left(\frac{\cos\theta}{r^2}\right)$	$\left(\frac{-M}{2\pi}\right)\left(\frac{y}{(x^2+y^2)^{3/2}}\right)$ $\left(\frac{-M}{2\pi}\right)\left(\frac{\sin\theta}{r^2}\right)$

7P-4 A stream function for a wavy wall is described as:

$$\psi(x,y) = U_o y_o \sin\left(\frac{2\pi x}{\lambda}\right) \exp\left(\frac{-2\pi y}{\lambda}\right) + U_o y$$

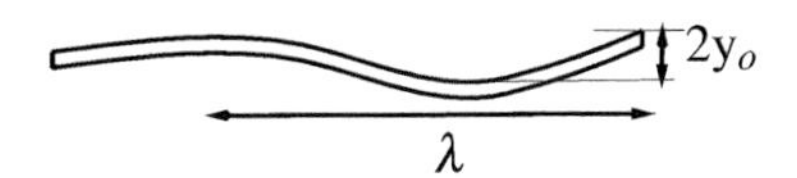

Figure 7.18 Figure of problem 7P-4.

Find the velocity components and equation for the wavy wall.

7P-5 Estimate the lifting force per unit span over a circular cylinder that has a diameter of 0.7 m and is experiencing a free stream velocity of 30 m/s. The maximum velocity close to the surface of the cylinder is 70 m/s. The cylinder is located 5 km above sea level. Take necessary assumptions.

7P-6 The velocity distribution of a liquid flowing in a channel is described as $u = 0.2y^2$ m/s, where y is in meters. Find the potential function for this flow. The channel height h is 1 m. Is the flow rotational or irrotational?

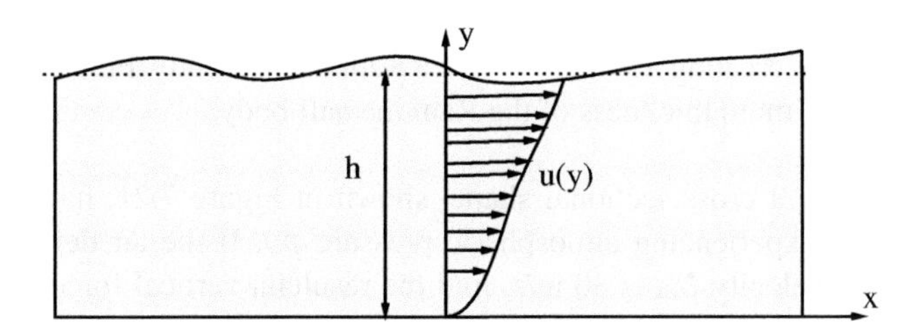

Figure 7.19 Figure of problem 7P-6.

7P-7 A flow is described by the velocity profile:

$$u(y) = 0.2 + 100y^3$$

Is this flow rotational or irrotational?

7P-8 A tornado is sweeping over the houses at a speed of 13 m/s. The eye of the tornado is located at 50 m. If the houses are approximated as a flat surface, located 15 m from the center of the tornado, determine the pressure on top of the houses. Take air density as 1.22 kg/m^3 and model the tornado as free vortex.

7P-9 A source is located at position S shown in Figure 7.20. The flow is moving through two plates that are 2θ apart, where θ is 30 degrees. If potential function is defined as:

$$\phi = \frac{4}{\pi}\ln(r)$$

find the velocity at $r = 10$ m, $\theta = 15°$.

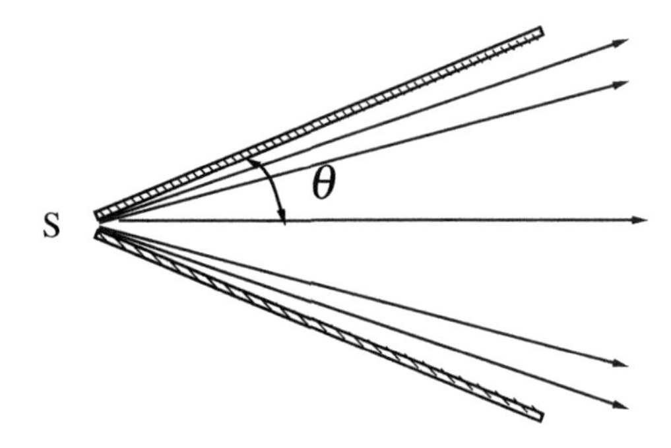

Figure 7.20 Figure of problem 7P-9.

7P-10 A circular cylinder of 500 mm diameter is spinning at its axis in air. The air velocity is 5 m/s. Estimate the lift force per unit length if both the stagnation points coincide. Take air density as 1.22 kg/m^3.

7P-11 A source that is issuing fluid with strength of $E = 3$ m^2/s is swept by a uniform flow of 3 m/s, creating a formation like a Rankine half body. Find the stagnation point and the maximum thickness of the Rankine half body.

7P-12 A hut with a cross-sectional shape, shown in Figure 7.21, having length ℓ and radius r_o is experiencing atmospheric pressure p_∞. If the air density is $\rho_\infty = 1.22$ kg/m^3 and velocity U_∞ is 30 m/s, find the resultant vertical force experienced by the hut. Take $r_o = 2$ m and $\ell = 5$ m.

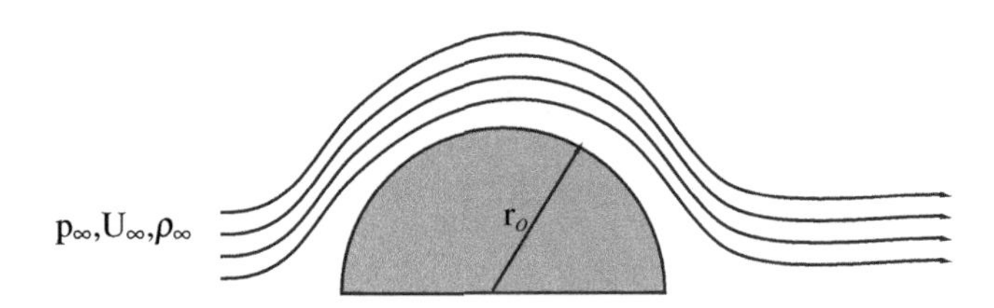

Figure 7.21 Figure of problem 7P-12.

7P-13 A cylinder of 1 m diameter is rotating at angular speed of 60 rad/s. It is experiencing uniformly distributed air at 10 m/s. The pressure within the uniform flow is

100 kPa. Find the maximum and minimum pressure on the surface of this cylinder. Also, estimate the lift force on the cylinder. Take density of air $\rho = 1.22$ kg/m^3.

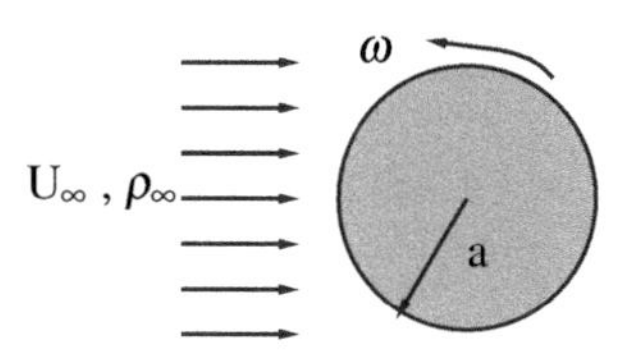

7P-14 A flow field can be represented by a superposition of uniform flow and a source and sink of equal strength ($E = 2\pi$). The uniform velocity is $U = 2$ m/s and $a = 2$. Find the values of ψ at $x = 0$; $y = 1, 2.1, 3, 4$, and at $x = 1$, $y = 1, 2, 3, 4$.

REFERENCES

J. D. Anderson, Fundamentals of aerodynamics, 6th ed., McGrawHill Edition, USA, 2011.

A. H. Shapiro, The Dynamics and Thermodynamics of Compressible Fluid Flow, Vol. I, Ronald Press Company, USA, 1953.

8 Laminar Flows

We will present in this chapter various internal flow problems that can be solved using the Navier-Stokes equations. We will develop the expressions of velocity distributions and then develop the analytical results for the calculation of friction coefficients. We will consider steady, Newtonian, laminar internal flows that completely fill the duct or pipe and are driven by a pressure difference.

Learning outcomes: After finishing this chapter, you should be able to:

- Develop the expression of the velocity profile for some basic internal laminar flow problems.
- Calculate the friction associated with viscous laminar flows.

Fluid flow in circular and noncircular cross-sectional conduits is commonly encountered. Supplying water through ducts is an ancient idea that goes back 5,000 years. In Mohenjo Daro, Pakistan, excavations showed that the city designers had laid underground drains covered by layers of baked bricks, creating a flat floor view to the people standing over the floor. Underground water or liquid transfer systems were also discovered in ancient Egyptian temples at Abusir. Many aqueducts have been found in Europe, which were made primarily to transfer water to the cities during the Roman era. Apart from drainage and water supply, internal flows are encountered in the human body as well. Gotthilf Heinrich Ludwig Hagen and Jean Leonard Marie Poiseuille have investigated the blood flow through human arteries and presented the famous Hagen-Poiseuille velocity profile for laminar flow. The interest in internal laminar flows is no longer an academic inquiry, as newly kindled interest in nanoscale designs has made such an analytical model very important.

In this chapter, we will develop some exact solutions of Navier-Stokes equations for a few practical applications involving flow through ducts, flow through a circular pipe, flow through annular gap, starting flow in pipe, flow due to sudden movement of the plate, etc.

8.1 FLOW BETWEEN PARALLEL PLATES

Often in air-conditioning ducts and industrial applications, the fluid is following in a duct of long width, and such flow can be modeled as flow between parallel plates. The schematic diagram of this situation is depicted in Figure 8.1.

DOI: 10.1201/9781003315117-8

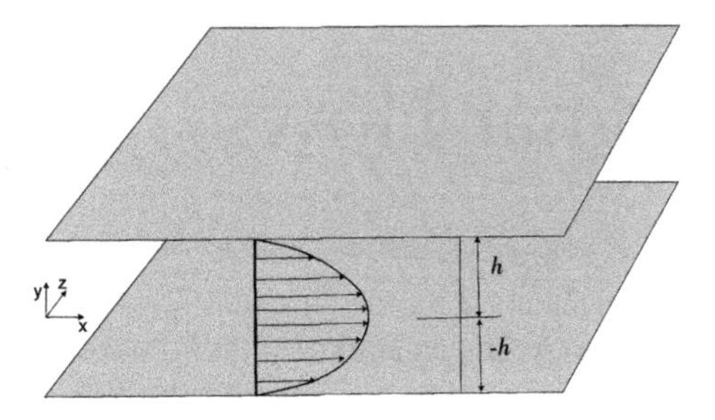

Figure 8.1 Flow between parallel plates or ducts of long width.

Assumptions: The assumptions taken are:

1 Fully developed flow

2 No body forces, $f_b = 0$

3 Nonporous walls, $v = 0$

4 Steady flow

Boundary conditions: Considering the flow presented in Figure 8.1, we can see the boundary conditions are:

1 At $y = 0$, $du/dy = 0$

2 At $y = \pm h$, $u = 0$ (no-slip)

We will now analyze continuity and Navier-Stokes equations with the assumptions of steady, incompressible flow, solid, nonporous walls.

Governing equations: Since there are velocities present in y and z direction, we can reduce the continuity equation for steady, incompressible flows as:

$$\frac{\partial u}{\partial x} + \cancel{\frac{\partial v}{\partial y}}^{v=0} + \cancel{\frac{\partial w}{\partial z}}^{w=0} = 0$$

It shows that velocity is not changing in the flow direction, a condition also called fully developed flow.

Navier-Stokes equation in z direction is

$$\cancel{\frac{\partial w}{\partial t}}^{steady} + \cancel{u\frac{\partial w}{\partial x} + v\frac{\partial w}{\partial y} + w\frac{\partial w}{\partial z}}^{No\ w\ velocity}$$
$$= -\frac{1}{\rho}\frac{\partial P}{\partial z} + \cancel{\nu \nabla^2 w}^{No\ w\ velocity} + \cancel{f_z}^{no\ body\ force}$$

and in y direction is

$$\cancel{\frac{\partial v}{\partial t}}^{steady} + u\cancel{\frac{\partial v}{\partial x}} + \cancel{v\frac{\partial v}{\partial y} + w\frac{\partial v}{\partial z}}^{No\ v\ velocity} = -\frac{1}{\rho}\frac{\partial P}{\partial y} + \cancel{\nu \nabla^2 v}^{No\ v\ velocity} + \cancel{f_y}^{no\ body\ force}$$

This shows that pressure is a function of x direction only, so $p = p(x)$. Now to form x-direction Navier-Stokes equation, we get:

$$\cancel{\frac{\partial u}{\partial t}}^{steady} + u\cancel{\left(\frac{\partial u}{\partial x}\right)}^{fully\ developed} + \underbrace{\cancel{v}\frac{\partial u}{\partial y} + \cancel{w}\frac{\partial u}{\partial z}}_{v=w=0} = -\frac{1}{\rho}\frac{\partial P}{\partial x} + \nu \nabla^2 u + \cancel{f_x}^{no\ body\ force}$$

As flow is steady, incompressible, fully developed, without body forces, x direction equation is reduced to form:

$$\left(\frac{\partial^2 u}{\partial y^2}\right) = \frac{1}{\mu}\frac{\partial P}{\partial x}$$

$$u(y) = \frac{1}{2\mu}\frac{\partial P}{\partial x}\left[y^2 - h^2\right]$$

$$u(y) = \frac{-1}{2}\left(\frac{1}{\mu}\frac{\partial P}{\partial x}\right)\left[y^2 - h^2\right]$$

The substitution β is introduced to make the analysis simple:

$$u(y) = \frac{\beta}{2}\left[h^2 - y^2\right]$$

Mean velocity:

$$u_m = \frac{1}{A}\int_A u(y)dy = \frac{1}{2h(\zeta)}\int_{-h}^{h} u(y)\zeta dy$$

$$u_m = \frac{\beta h^2}{3}$$

where ζ is width and taken as unity.

$$\frac{u(y)}{u_m} = \frac{3}{2}\left[1 - \left(\frac{y}{h}\right)^2\right]$$

Hydraulic diameter:

$$d_H = \frac{4A}{WP} = \frac{4(\zeta \times 2h)}{2(\zeta + 2h)} = \frac{4h}{(1 + \frac{2h}{\zeta})}$$

$$\lim_{\zeta \to \infty} d_H = \lim_{\zeta \to \infty}\frac{4h}{(1 + \frac{2h}{\zeta})} = 4h$$

$$d_H = 4h$$

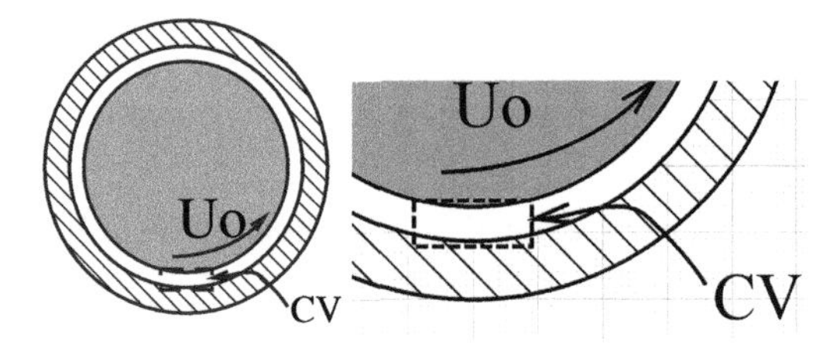

Figure 8.2 Flow between journal and bearing (left) and close up view of the flow between journal and bearing (right).

The skin friction coefficient is:

$$|C_f| = \frac{2\tau}{\rho u_m^2} = \frac{2\beta h}{\rho\left(\frac{\beta h^2}{3}\right)^2} = \frac{18\mu}{h^3\rho\beta}$$

The Reynolds number is:

$$\mathrm{Re} = \frac{4}{3}\frac{\rho\beta h^3}{\mu}$$

$$C_f \cdot \mathrm{Re} = \left(\frac{18\mu}{h^3\rho\beta}\right)\left(\frac{4}{3}\frac{\rho\beta h^3}{\mu}\right) = \frac{18\times 4}{3} = \frac{72}{3} = 24$$

8.2 FLOW BETWEEN PLATES WITH ONE PLATE MOVING

The case of flow between parallel plates is an interesting and practical inquiry. Often such a flow is encountered in flow between journal and a bearing. This branch of fluid mechanics is called lubrication theory, and, along with material science, we call the subject tribology. In this section we discuss a very basic and rudimentary analysis by considering the flow between parallel plates. Figure 8.2 shows the journal and the bearing movement. The flow inside gap region is complex. However we can approximate a small portion of flow as flow between parallel plates, with one moving and one stationary. The flow is historically called the Couette flow (Figure 8.3), named after French physicist Maurice Marie Alfred Couette (d. 1943).

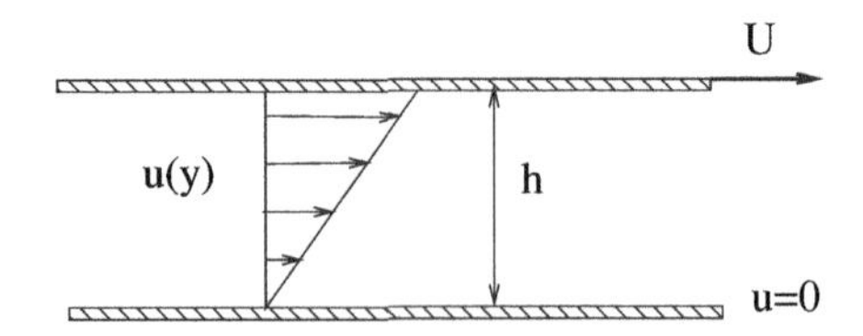

Figure 8.3 The Couette flow with pressure gradient.

The differential equation for flow between parallel plates is also valid for this case.

$$\left(\frac{\partial^2 u}{\partial y^2}\right) = \frac{1}{\mu}\frac{\partial P}{\partial x}$$

We have at $x = 0$, $u = 0$ and at $x =$ h, $u = U$. Applying these boundary conditions and integrating the equation twice would lead to the result:

$$u(y) = \frac{1}{2\mu}\left(\frac{\partial P}{\partial x}\right)y^2 + \frac{C_1}{\mu}y + C_2$$

We have at $x = 0$, $u = 0$:

$$C_2 = 0$$

at $x =$ h, $u = U$:

$$U = \frac{1}{2\mu}\left(\frac{\partial P}{\partial x}\right)h^2 + \frac{C_1}{\mu}h$$

$$C_1 = \frac{U\mu}{h} - \frac{1}{2\mu}\left(\frac{\partial P}{\partial x}\right)h$$

$$u(y) = \frac{1}{2\mu}\left(\frac{\partial P}{\partial x}\right)y^2 + \frac{1}{\mu}\left[\frac{U\mu}{h} - \frac{1}{2\mu}\left(\frac{\partial P}{\partial x}\right)h\right]y$$

$$u(y) = \frac{1}{2\mu}\left(\frac{\partial P}{\partial x}\right)\left[y^2 - hy\right] + \frac{Uy}{h}$$

$$u(y) = \frac{h^2}{2\mu}\left(\frac{\partial P}{\partial x}\right)\left[\left(\frac{y}{h}\right)^2 - \left(\frac{y}{h}\right)\right] + \frac{Uy}{h}$$

The shear stress distribution is

$$\tau = \mu\frac{du}{dy} = \mu\frac{U}{h} + h\left(\frac{\partial P}{\partial x}\right)\left[\left(\frac{y}{h}\right) - \frac{1}{2}\right]$$

If ℓ is the depth of bearing and journal, then flow rate between them is

$$Q = \ell\int_0^h u(y)\cdot dy = \left[\frac{U\cdot h}{2} - \frac{1}{12\mu}\left(\frac{\partial P}{\partial x}\right)h^3\right]\ell$$

$$\frac{Q}{\ell} = \left[\frac{U\cdot h}{2} - \frac{1}{12\mu}\left(\frac{\partial P}{\partial x}\right)h^3\right]$$

Figure 8.4 shows the velocity distribution in the gap of 10 mm. One plate is moving at $U = 10$ m/s. The Reynolds number for the flow can be calculated as

$$\text{Re} = \frac{u_m\cdot h}{\nu} = \frac{h}{\nu}\cdot\frac{Q}{A} = \frac{h}{\nu}\frac{Q}{\ell\cdot h} = \frac{Q}{\ell\nu}$$

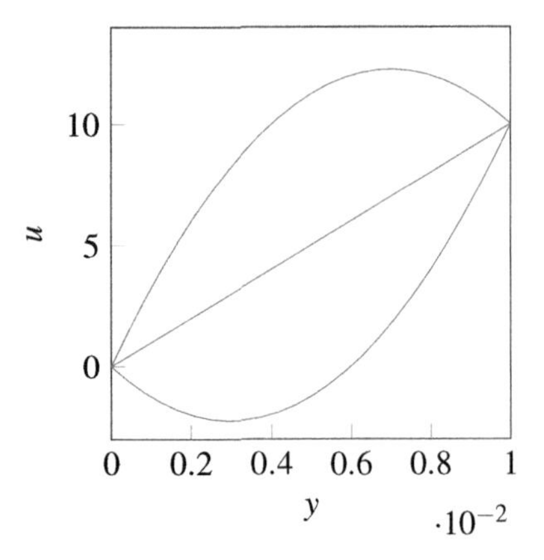

Figure 8.4 Velocity distribution in the gap of 10 mm. One plate is moving at $U = 10$ m/s.

Example 8.1

Example A hydrostatic bearing is supporting the load of 2,000 lb_f/ft across its length. If the flow rate is $Q = 9.46$ L/h per feet, estimate gap height (h), the required width of the bearing (w), and the pressure gradient (dp/dx). The lubrication provided has absolute viscosity $\mu = 2.089 \times 10^{-4} \frac{\text{lbf·s}}{\text{ft}^2}$. Assume fully developed flow with pressure field between the gap is:

$$\text{p(x)} = 35\left(1 - \frac{2x}{w}\right)$$

where x varies from zero to w/2.

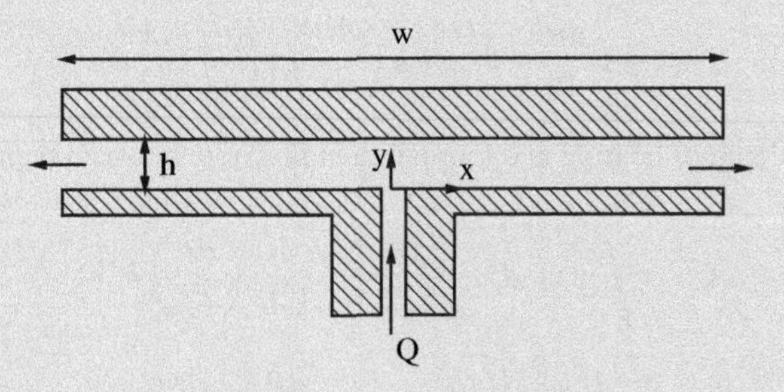

Solution We can apply the equation developed in this section

$$\frac{Q}{\ell} = -\frac{h^3}{12 \cdot \mu} \cdot \left(\frac{dp}{dx}\right)$$

The total force in the vertical direction due to pressure is

$$F = \ell \int p dx$$

$$F = 2 \cdot \ell \cdot \int_0^{\frac{w}{2}} 35 \cdot \left(1 - \frac{2 \cdot x}{w}\right) dx \qquad F = \frac{1}{2} \cdot 35 \cdot \ell \cdot w$$

This must be equal to the applied load. Therefore, with $\ell = 1ft$, $F = 2{,}000$ lb_f

$$w = \frac{2}{35 \times 144} \cdot \frac{F}{\ell} = 0.7936 \cdot \text{ft}$$

The pressure gradient is then

$$\frac{dp}{dx} = -\frac{\Delta p}{w/2} = -\frac{2 \cdot \Delta p}{w} = -2 \times \frac{35 \cdot \text{lb}_\text{f}}{\text{in}^2} \times \frac{1}{0.7936 \cdot \text{ft}} = -88 \cdot 2 \frac{\text{psia}}{\text{ft}}$$

as

$$\frac{Q}{\ell} = -\frac{h^3}{12 \cdot \mu} \cdot \left(\frac{dp}{dx}\right)$$

We rearrange and find h as

$$h = \left(-\frac{12 \cdot \mu \cdot \left(\frac{Q}{\ell}\right)}{\frac{dp}{dx}}\right)^{\frac{1}{3}} = \left(-\frac{12(2.089 \times 10^{-4})\left(\frac{0.334}{3600}\right)}{-88 \cdot 2 \times 144}\right)^{\frac{1}{3}} = 0.000209ft$$

$$h = 0.0025in$$

8.3 HAGEN-POISEUILLE FLOW

Fluid in a long pipe is moving at a steady rate with fully developed conditions under applied constant pressure gradient. The schematic diagram of the flow is shown in Figure 8.5. The velocity profile for such flow is parabolic, as depicted in the figure, which we will derive in this section.

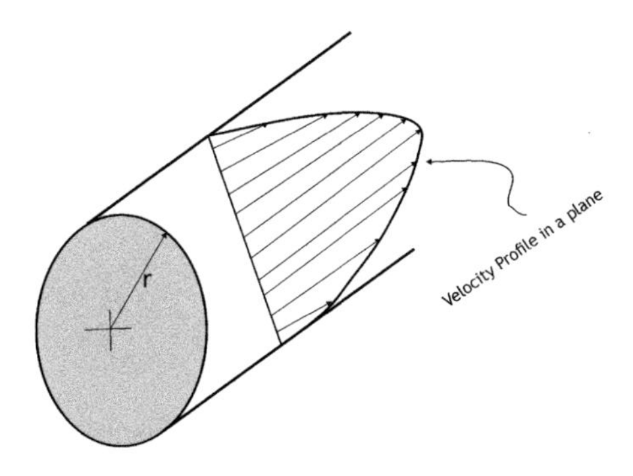

Figure 8.5 Flow through circular pipe or tube.

Assumptions:

1 Fully developed flow

2 No body forces, $f_b = 0$

3 Nonporous walls

4 Steady flow

Governing equations: We analyze the problem in (r, θ, x) coordinates with z as stream-wise coordinate.

Continuity Equation: We begin analysis with continuity equation:

$$\frac{1}{r}\frac{\partial (ru_r)}{\partial r} + \frac{1}{r}\frac{\partial u_\theta}{\partial \theta} + \frac{\partial u_z}{\partial z} = 0$$

since we have no radial and swirl velocities:

$$\cancel{\frac{1}{r}\frac{\partial (ru_r)}{\partial r}}^{\text{no radial velocity}} + \cancel{\frac{1}{r}\frac{\partial u_\theta}{\partial \theta}}^{\text{no swirl}} + \frac{\partial u_z}{\partial z} = 0$$

leaving a result

$$\frac{\partial u_z}{\partial z} = 0$$

hence flow velocity is independent of z-direction.

Navier-Stokes Equations: We now analyze the Navier-Stokes equations with our assumptions: r-direction:

$$\rho\left(\cancel{\frac{\partial u_r}{\partial t}}^{\text{steady}} + \cancel{u_r\frac{\partial u_r}{\partial r}}^{\text{no radial velocity}} + \cancel{\frac{u_\theta}{r}\frac{\partial u_r}{\partial \theta}}^{\text{no swirl}} - \cancel{\frac{u_\theta^2}{r}}^{\text{no swirl}} + u_z\cancel{\frac{\partial u_r}{\partial z}}^{\text{fully developed}}\right)$$

$$= -\frac{\partial P}{\partial r} + \cancel{\rho f_r}^{\text{no body force}} + \cancel{\mu\left[\frac{1}{r}\frac{\partial}{\partial r}\left(r\frac{\partial u_r}{\partial r}\right) - \frac{u_r}{r^2} + \frac{1}{r^2}\frac{\partial^2 u_r}{\partial \theta^2} - \frac{2}{r^2}\frac{\partial u_\theta}{\partial \theta} + \frac{\partial^2 u_r}{\partial z^2}\right]}$$

θ-direction equation with same assumptions as above we have:

$$\rho\left(\cancel{\frac{\partial u_\theta}{\partial t}}^{\text{steady}} + \cancel{u_r\frac{\partial u_\theta}{\partial r} + \frac{u_\theta}{r}\frac{\partial u_\theta}{\partial \theta} + \frac{u_r u_\theta}{r} + u_z\frac{\partial u_\theta}{\partial z}}\right)$$

$$= -\frac{1}{r}\frac{\partial P}{\partial \theta} + \cancel{\rho f_\theta} + \cancel{\mu\left[\frac{1}{r}\frac{\partial}{\partial r}\left(r\frac{\partial u_\theta}{\partial r}\right) - \frac{u_\theta}{r^2} + \frac{1}{r^2}\frac{\partial^2 u_\theta}{\partial \theta^2} - \frac{2}{r^2}\frac{\partial u_r}{\partial \theta} + \frac{\partial^2 u_\theta}{\partial z^2}\right]}$$

This shows that $P \neq P(r, \theta)$, i.e., pressure is not changing in the cross-section of the pipe. As the flow is driven by pressure in absence of the body forces, pressure is a function of z direction only. We now analyze the z-direction Navier-Stokes equation:

$$\rho\left(\cancelto{steady}{\frac{\partial u_z}{\partial t}}+\cancelto{no\ radial}{u_r}\frac{\partial u_z}{\partial r}+\frac{u_\theta}{\cancel{r}}\cancelto{no\ swirl}{\frac{\partial u_z}{\partial \theta}}+u_z\cancelto{fully\ developed}{\frac{\partial u_z}{\partial z}}\right)$$

$$=-\frac{\partial P}{\partial z}+\cancelto{no\ body\ forces}{\rho f_z}+\mu\left[\frac{1}{r}\frac{\partial}{\partial r}\left(r\frac{\partial u_z}{\partial r}\right)+\cancelto{no\ dependence}{\frac{1}{r^2}\frac{\partial^2 u_z}{\partial \theta^2}}+\cancelto{fully\ developed}{\frac{\partial^2 u_z}{\partial z^2}}\right]$$

The above analysis of the equations concludes that $u_z = f(r)$ only. For the sake of simplification, u_z is called u only. Note that inertia force on the L.H.S. in the Navier-Stokes equation is considered to be absent, so only pressure forces are balanced by viscous forces.

$$\frac{\partial^2 u}{\partial r^2}+\frac{1}{r}\frac{\partial u}{\partial r}=\frac{1}{\mu}\frac{\partial P}{\partial z} \tag{8.1}$$

$$\frac{\partial u}{\partial r}=\frac{1}{r\mu}\frac{dP}{dz}\frac{r^2}{2}+\frac{C_1}{r} \tag{8.2}$$

$$u=\frac{dP}{dz}\frac{r^2}{4\mu}+C_1 ln(r)+C_2 \tag{8.3}$$

Boundary conditions: The applicable boundary conditions are:

1. No slip condition at the pipe wall: $u(r_w) = 0$
2. Centerline condition: $(\partial u/\partial r)_{r=0} = 0$

Centerline gradient of velocity profile is zero, so from Eq. 8.3 $C_1 = 0$. At walls velocity is zero, so from Eq. 8.4

$$C_2=-\frac{dP}{dz}\frac{r_w^2}{4\mu} \tag{8.4}$$

$$u=\frac{dP}{dz}\frac{r^2}{4\mu}-\frac{dP}{dx}\frac{r_w^2}{4\mu} \tag{8.5}$$

$$u_z(r)=-\frac{dP}{dz}\frac{1}{4\mu}r_w^2\left[1-(r/r_w)^2\right] \tag{8.6}$$

Figure 8.6 shows the velocity distribution inside a circular pipe.

Mean Velocity

$$\bar{u}=\frac{1}{A}\int_A u_z(r)dA \tag{8.7}$$

$$\overline{u_z}=\frac{1}{\pi r_w^2}\int_0^{r_w} 2u_z(r)\pi r dr \tag{8.8}$$

$$\overline{u_z}=-\frac{dP}{dz}\frac{1}{4\mu}r_w^2\frac{1}{\pi r_w^2}\int_0^{r_w} 2\left[1-(r/r_w)^2\right]\pi r dr \tag{8.9}$$

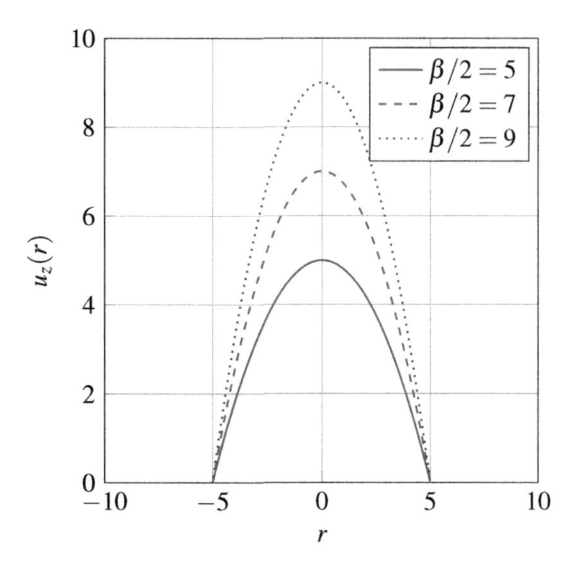

Figure 8.6 Parabolic velocity distribution inside a pipe of radius 5 units. The profile will remain parabolic with increase in pressure gradient.

$$\overline{u_z} = \frac{\beta}{\pi r_w^2}\int_0^{r_w}\left[1-(r/r_w)^2\right]\pi r dr = \frac{\beta}{4} \tag{8.10}$$

where

$$\beta = -\frac{dP}{dz}\frac{1}{2\mu}r_w^2 \tag{8.11}$$

Centerline Velocity

$$u_{max} = u_c = \frac{\beta}{2}\left[1-(r/r_w)^2\right] = \frac{\beta}{2} \tag{8.12}$$

Ratio of centerline velocity to mean velocity is

$$\frac{u_c}{\overline{u}} = \frac{\beta/2}{\beta/4} = 2 \tag{8.13}$$

Shear Stress and Reynolds Number

Shear stress is calculated from Newton's law of viscosity: $\tau_w = \mu(du/dr)$

$$|\tau_w| = \mu(\frac{dP}{dz}\frac{1}{2\mu}r_w^2)(1/r_w) = \mu\beta/r_w \tag{8.14}$$

Reynolds number is defined as $\overline{u}D/\nu$, where D is the diameter of pipe and ν is kinematic viscosity

$$\mathrm{Re}_D = \frac{\overline{u}D\rho}{\mu} = \frac{\beta D\rho}{4\mu} \tag{8.15}$$

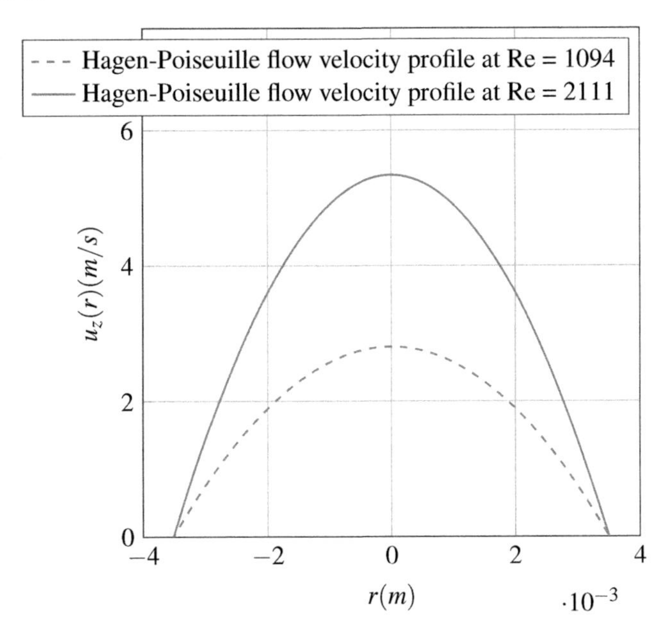

Figure 8.7 Parabolic velocity distribution inside a pipe of radius 0.5 inch for flow of glycerol.

Skin Friction Coefficient

Fanning friction coefficient is defined as:

$$C_f = \frac{|\tau_w|}{\frac{1}{2}\rho \bar{u}^2} = \frac{32\mu}{\rho \beta r_w} \tag{8.16}$$

$$C_f \text{Re}_D = \left(\frac{32\mu}{\rho \beta r_w}\right)\left(\frac{\rho D \beta}{4\mu}\right) = \frac{8D}{r_w} = 16 \tag{8.17}$$

Dary friction coefficient is defined as:

$$f = 4C_f = \frac{64}{\text{Re}_D} \tag{8.18}$$

Stanton and Pannell (1914) have experimentally confirmed the validity of the Hagen-Poiseuille velocity profile for flows with Reynolds number less than 500. A slight departure from standard profile is reported for Reynolds number 500 to 2,000.

Figure 8.7 shows the parabolic velocity distribution inside a pipe of radius 0.5 in. for flow of glycerol at different Reynolds numbers.

Example 8.2

Example Two tanks are connected by two parallel circular pipes. One of the pipes has a diameter 2 cm and another one has a diameter of 5 cm. The pipes are 5 m long. If one of the tanks has higher pressure than the other tank, causing the combined flow rate of 0.1 kg/s, calculate the individual flow rates in two pipes, if tanks contain water with viscosity of 1.0016×10^{-3} Pa · s and density of 998 kg/m^3.

Solution The mean velocity for flow through a pipe or tube is:

$$\overline{u_z} = \frac{\beta}{4} = -\frac{dP}{dz}\frac{1}{8\mu}r_w^2$$

Based on this, the mass flow rate is:

$$\dot{m} = \rho A\overline{u_z} = \rho\left(\pi r_w^2\right)\left(-\frac{dP}{dz}\frac{1}{8\mu}r_w^2\right) = \left(-\frac{dP}{dz}\right)\frac{\pi\rho r_w^4}{8\mu}$$

As the same pressure gradient is driving the flow in the tubes, we can form the equation:

$$\dot{m}_1 + \dot{m}_2 = \left(-\frac{dP}{dz}\right)\left[\frac{\pi\rho r_1^4}{8\mu} + \frac{\pi\rho r_2^4}{8\mu}\right]$$

where r_1 and r_2 refer to two different pipe radii.

$$\dot{m}_1 + \dot{m}_2 = \left(-\frac{dP}{dz}\right)\left[\frac{\pi\rho r_1^4}{8\mu} + \frac{\pi\rho r_2^4}{8\mu}\right] = 0.1$$

$$\Delta p = 1.631 Pa$$

The mass flow rates are:

$$m_1 = 0.0000249$$

$$m_2 = 0.000975$$

Example 8.3

Example An oil with specific gravity 0.85 and dynamic viscosity 0.1 Pa · s flows through a circular pipe of 10 mm diameter with a mean velocity of 2.5 m/s. Determine:

(a) the Reynolds number

(b) the maximum velocity

(c) the volumetric flow rate

(d) the pressure gradient along the pipe

Solution The specific gravity is 0.85, so the oil density is:

$$\rho_{oil} = sg_{oil}\rho_w = 0.85 \times 1000$$

(a) The Reynolds number can be calculated as:

$$\text{Re} = \left(\frac{\rho u d}{\mu}\right)_{oil} = \frac{0.85 \times 1000 \times 2.5 \times 10 \times 10^{-3}}{0.1} = 212.5$$

(b) The maximum velocity is twice that of the mean velocity, so

$$u_{\max} = 2\bar{u} = 5\ m/s$$

(c) The volumetric flow rate is:

$$Q = A\bar{u} = 0.1962 \times 10^{-3} m^3/s$$

(d) The pressure gradient is:

$$\frac{dp}{dx} = \frac{-8\mu\bar{u}}{r_w^2} = -80kPa/m$$

The negative pressure gradient indicates that the pressure decreases with distance.

8.4 STARTING FLOW IN A PIPE

Fluid in a long pipe is at rest at $t = 0$; at this time, a sudden uniform and constant pressure gradient is applied.

At $t \to \infty$ the velocity profile approaches the fully developed flow velocity profile

$$u = u_{max}(1 - r^{*2})$$

where $r^* = r/r_w$ and at the center of the pipe

$$u_{max} = \left(\frac{-dP}{dz}\right)\frac{r_w^2}{4\mu}$$

Assumptions

1 After initial transients, the flow would become fully developed and steady flow will start following Hagen-Poiseuille equation.

2 No body forces, $f_b = 0$

3 Nonporous walls

4 Initially, the fluid is stationary

We analyze the problem in (r,θ,z) coordinates with z as streamwise coordinate x. The streamwise momentum equation:

$$\rho\frac{\partial u}{\partial t} = \underbrace{\frac{-\partial P}{\partial z}}_{inhomogenity} + \mu\left(\frac{\partial^2 u}{\partial r^2} + \frac{1}{r}\frac{\partial u}{\partial r}\right) \tag{8.19}$$

Eq. 8.19 is an inhomogeneous partial differential equation. Inhomogeneity can be removed if we subtract out the steady Poiseuille flow and work with the deviation of u from the Poiseuille flow paraboloid. We define a new variable

$$V = u - u_{steady} = u - u_s = u - u_{max}(1 - r^{*2})$$

where u_s is velocity in steady state.

$$V_t = u_t - \underbrace{(u_s)_t}_{zero}$$

Here subscript t indicates the derivative with t, so V_t is $\frac{\partial V}{\partial t}$. This nomenclature is adopted here in this section.

$$u_t = (-1/\rho)(P_z) + \nu[u_{rr} + u_r/r]$$

Again, u_t is $\partial u/\partial t$, P_z is $\partial P/\partial z$, and u_{rr} is $\partial^2 u/\partial r^2$.

Introducing the dimensionless coordinates

$$t^* = \nu t/r_w^2$$

and

$$r^* = r/r_w$$

$$\frac{\partial u}{\partial t^*} = \frac{r_w^2}{\mu}\left(\frac{-\partial P}{\partial z}\right) + \left[\frac{\partial^2 u}{\partial r^{*2}} + \frac{1}{r^*}\frac{\partial u}{\partial r^*}\right]$$

The subscripts here indicate the derivatives taken.

$$\frac{\partial u}{\partial t^*} = \underbrace{\frac{r_w^2}{\mu}\left(\frac{-\partial P}{\partial z}\right)}_{4u_m} + \left[\frac{\partial^2 u}{\partial r^{*2}} + \frac{1}{r^*}\frac{\partial u}{\partial r^*}\right]$$

$$\frac{\partial u}{\partial t^*} = 4u_m + \left[\frac{\partial^2 u}{\partial r^{*2}} + \frac{1}{r^*}\frac{\partial u}{\partial r^*}\right]$$

Changing the dependent variable to V gives

$$\frac{\partial V}{\partial t^*} + \underbrace{\frac{\partial u_s}{\partial t^*}}_{steady\ so\ zero} = 4u_m + \left[\frac{\partial^2 V}{\partial r^{*2}} + \frac{1}{r^*}\frac{\partial V}{\partial r^*}\right] + \underbrace{\left[\frac{\partial^2 u_s}{\partial r^{*2}} + \frac{1}{r^*}\frac{\partial u_s}{\partial r^*}\right]}_{-4u_m}$$

$$\boxed{\frac{\partial V}{\partial t^*} = \left[\frac{\partial^2 V}{\partial r^{*2}} + \frac{1}{r^*}\frac{\partial V}{\partial r^*}\right]} \tag{8.20}$$

For this equation, we have a new modified set of the boundary conditions:

1 **No Slip Condition** $u(r_w, t) = 0$ or $V(r^*, 0) \neq 0$

2 **Initial Condition** $u(r, 0) = 0$ or $V(r^*, 0) \neq 0 = -u_m(1 - r^{*2})$

3 **Extended time condition** $\partial u/\partial r = 0$ at $r = 0$

The solution of this PDE equation is possible through Separation of Variables technique as:

$$V = X(r^*) \cdot T(t^*)$$

$$V_{t^*} = XT'$$

and

$$V_{r^*} = X'T$$

Now the solution of the following equation is the solution of PDE in Eq. 8.20

$$\frac{T'}{T} = \frac{X''}{X} + \frac{1}{r^*}\frac{X'}{X} = -a^2(Separation\ Constant) \tag{8.21}$$

Looking at the time variable:

$$\frac{1}{T}T_{t^*} = -a^2$$

$$\int \frac{dT}{T} = -\int a^2 dt^*$$

we get

$$\boxed{T = C_1 e^{-a^2 t^*}}$$

and space variable gives

$$\frac{X''}{X} + \frac{1}{r^*}\frac{X'}{X} = -a^2$$

$$X'' + \frac{1}{r^*}X' + a^2 X = 0$$

To solve the last equation we need to introduce another variable into space equation. Let

$$r^* = e^u$$

then

$$du/dr^* = e^{-u}$$

$$\frac{dX}{dr^*} = e^{-u}\frac{dX}{du}$$

$$\frac{d^2X}{dr^{*2}} = \frac{du}{dr^*}\left[e^{-u}\frac{d^2X}{du^2} - e^{-u}\frac{dX}{du}\right]$$

$$\frac{d^2X}{dr^{*2}} = e^{-2u}\left[\frac{d^2X}{du^2} - \frac{dX}{du}\right]$$

Substitution of $\frac{d^2X}{dr^{*2}}$ and $\frac{dX}{dr^*}$ into the space equation gives

$$X'' + \frac{1}{r^*}X' + a^2X = 0$$

$$\frac{d^2X}{du^2} + \frac{a^2}{e^{-2u}}X = 0 \; as \; \left(e^{-u} - \frac{1}{r}\right) = 0$$

or

$$\boxed{X'' + (a^2e^{2u} - 0)X = 0}$$

The solution of this equation is already known

$$Y'' + (k^2e^{2x} - n^2)Y = 0$$

$$Y = \widetilde{C}_1J_n(ke^x) + \widetilde{C}_2Y_n(ke^x)$$

Comparing the two gives

$$k^2 = a^2 \; and \; n^2 = 0$$

Solution of X equation is then:

$$X = \widetilde{C}_1J_o(ae^u) + \widetilde{C}_2Y_o(ae^u)$$

and

$$\boxed{X = \widetilde{C}_1J_o(ar^*) + \widetilde{C}_2Y_o(ar^*)}$$

Figure 8.8 shows the Bessel function of first and second kind. The solution of Eq. 8.20 is then

$$\boxed{V = \underbrace{C_1e^{-a^2t^*}}_{temporal} \cdot \underbrace{[\widetilde{C}_1J_o(ar^*) + \widetilde{C}_2Y_o(ar^*)]}_{spatial}}$$

Applying the boundary conditions $r^* = 0, \; V \; is \; finite$

$$J_o(0) = 1 \; and \; Y_o(0) = \infty \; hence \; X = \widetilde{C}_1J_o(ar^*)$$

$$V = \widehat{C}e^{-a^2t^*}J_o(ar^*)$$

and

$$V = \sum_{n=1}^{\infty} \widehat{C}_ne^{-a_n^2t^*}J_o(a_nr^*) \tag{8.22}$$

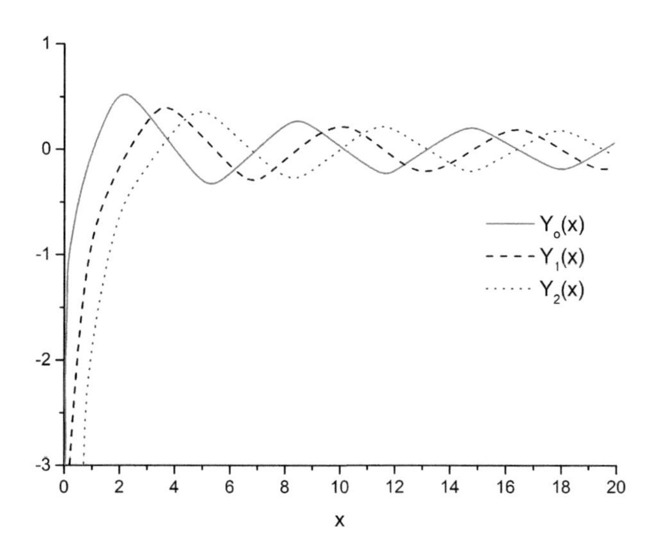

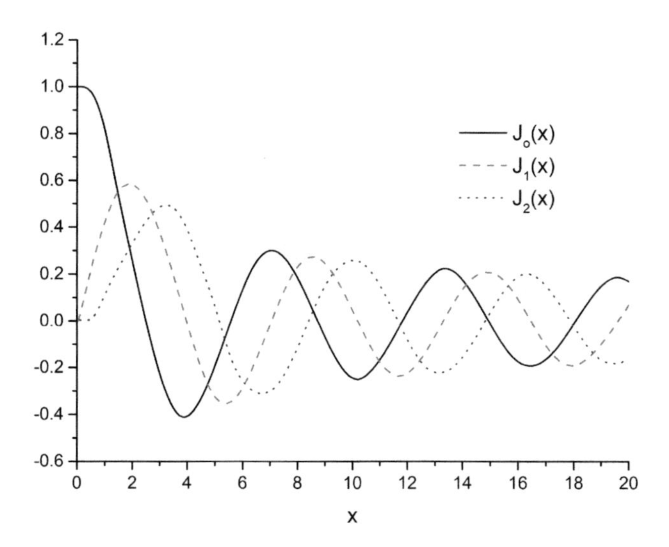

Figure 8.8 Bessel function of first and second kind.

Orthogonality property:

$$\int_0^1 xJ_o(a_i x)J_o(a_j x)dx = \frac{1}{2}J_1(a_i)\ if\ [i = j]\ else\ 0$$

Multiply Eq. 8.22 by $r^* J_o(a_n r^*)$ and integrate the equation from 0 to 1. From orthogonality condition we have:

$$\widehat{C}_n = \frac{V \cdot r^* J_o(a_n r^*)}{e^{-a_n^2 t^*} \int_0^1 r^* J_o(a_n r^*) J_o(a_n r^*) dr^*}$$

Using **initial condition** at $t^* = 0, e^{t^*} = 1$, and $V(r^*, 0) = \not{u} - u_m(1 - r^*)$

$$\widehat{C}_n = \frac{-u_m(1 - r^*) \cdot r^* J_o(a_n \cdot r^*)}{\not{e^{-a_n^2 t^*}} \underbrace{\int_0^1 r^* J_o(a_n r^*) J_o(a_n r^*) dr^*}_{J_1(a_n)/2}}$$

$$\widehat{C}_n = 2 \frac{-u_m(1 - r^*) \cdot r^* J_o(a_n \cdot r^*)}{J_1(a_n)} = \frac{-8u_m}{a_n^3 J_1(a_n)}$$

For more on this, see references.

Hence solution of V is

$$V = \sum_{n=1}^{\infty} \left(\frac{-8u_m}{a_n^3 J_1(a_n)} \right) e^{-a_n^2 t^*} J_o(a_n r^*)$$

As $V = u - u_{steady}$

$$u = \sum_{n=1}^{\infty} \left[\left(\frac{-8u_m}{a_n^3 J_1(a_n)} \right) e^{-a_n^2 t^*} J_o(a_n r^*) \right] + u_m(1 - r^{*2})$$

$$\boxed{\frac{u}{u_m} = \sum_{n=1}^{\infty} \left[\left(\frac{-8}{a_n^3 J_1(a_n)} \right) e^{-a_n^2 t^*} J_o(a_n r^*) \right] + (1 - r^{*2})} \tag{8.23}$$

We defined $X = \widetilde{C}_1 J_o(ar^*)$, applying the wall boundary condition $r^* = 1$ gives $X = 0$ hence $J_o(a_n) = 0$. Hence a_n are the roots of Bessel Function J_o.

Table 8.1
Roots of Bessel Function

n	1	2	3	4	5	6	7
a_n	2.4048	5.5201	8.653	11.7915	14.9309	18.071	21.211

8.5 STOKE'S FIRST PROBLEM

Consider an ultra-thin infinite flat plate (see Figure 8.9). The wall is stationary initially and suddenly accelerates to a constant velocity U_o.

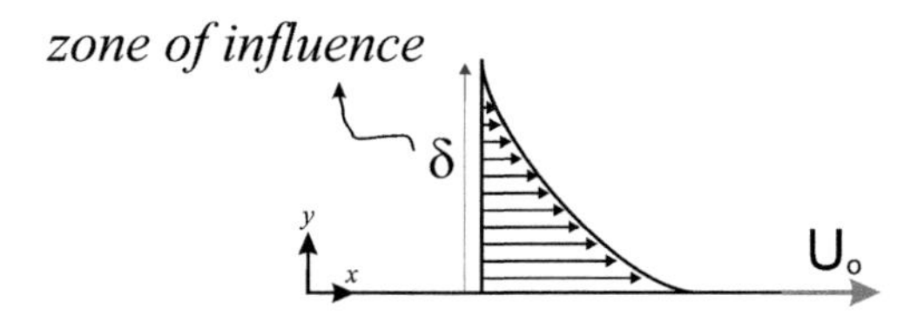

Figure 8.9 Schematic diagram of suddenly started flow.

Assumption

1 Profile is invariant in x direction

$$du/dx = u_x = 0$$

2 No body forces, $f_b = 0$

3 Nonporous walls, $v = 0$

3 Depth has no influence on problem definition; z and w play no role

y-momentum equation:
Analyze y momentum equation:

$$u \underbrace{\cancel{v_x}}_{nonporous\ walls} + \underbrace{\cancel{vv_y}}_{nonporous\ walls} + \underbrace{\cancel{w}}_{2D\ only} v_z$$

$$= -P_y/\rho + \nu \underbrace{\cancel{\nabla^2 v}}_{nonporous\ walls}$$

and

$$P_y = 0$$

Also outside the zone of influence δ, there is no flow ($dP = 0$). That shows that the flow is not pressure driven, instead it is due to applied force only.

x-momentum equation

$$u_t + u \underbrace{\cancel{u_x}}_{invariant\ profile} + \underbrace{\cancel{vu_y}}_{nonporous\ wall} + \underbrace{\cancel{wu_z}}_{2D\ only}$$

$$= \underbrace{\cancel{-P_x/\rho}}_{no\ pressure} + \nu \left(\underbrace{\cancel{u_{xx}}}_{invariant\ profile} + u_{yy} + \underbrace{\cancel{u_{zz}}}_{2D\ only} \right)$$

$$\boxed{u_t - \nu u_{yy} = 0} \tag{8.24}$$

with Bounday Condition
(i) at $y = 0,\ u = U_o$

(ii) at $y \to \infty$, at $u \to 0$
and Initial Condition
at $t = 0,\ u = 0$

Similarity variable: A similarity variable can be defined as

$$\eta = y^a v^b t^c$$

Using the dimensional analysis one may get

$$\boxed{\eta = \frac{1}{2}\frac{y}{\sqrt{vt}}} \tag{8.25}$$

Further we assume that velocity is a function of η only as:

$$u = U_o f(\eta) \tag{8.26}$$

Substitution of this similarity variable in Eq. 8.24 would cast the PDE into an ODE.

Important derivatives:

$$\frac{\partial \eta}{\partial t} = \frac{-1}{2}\frac{y}{2\sqrt{v}}t^{-\frac{1}{2}-1}$$

$$= \frac{-y}{4\sqrt{v}}t^{\frac{-3}{2}} = \frac{-y}{4t\sqrt{vt}}$$

$$\boxed{\frac{\partial \eta}{\partial t} = \frac{-\eta}{2t}}$$

$$\boxed{\frac{\partial \eta}{\partial y} = \frac{1}{2\sqrt{vt}}}$$

$$\frac{\partial u}{\partial t} = \frac{\partial u}{\partial \eta}\frac{\partial \eta}{\partial t}$$

$$\frac{\partial u}{\partial t} = U_o\frac{\partial f}{\partial \eta}\cdot\frac{\partial \eta}{\partial t} = \boxed{-f' U_o \frac{\eta}{2t}}$$

$$\frac{\partial u}{\partial y} = \boxed{\frac{U_o}{2\sqrt{vt}}f'}$$

$$\frac{\partial^2 u}{\partial y^2} = \frac{U_o}{2\sqrt{vt}}f'\left(\frac{1}{2\sqrt{vt}}\right) = \boxed{\frac{1}{4vt}U_o f''}$$

Similarity equation:

$$\boxed{f'' + 2\eta f' = 0} \tag{8.27}$$

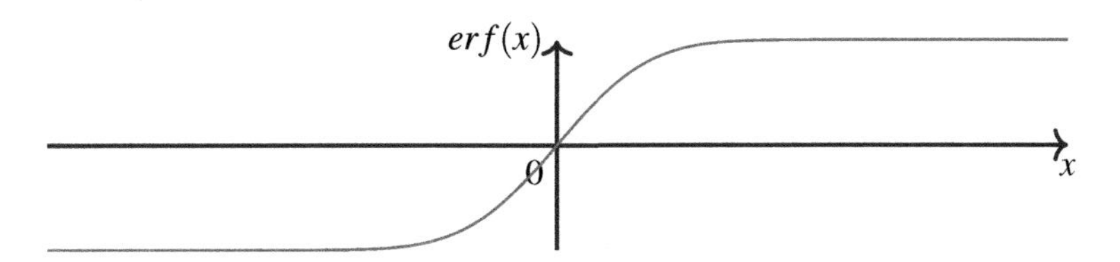

Figure 8.10 The error function.

With boundary conditions
(i) $\eta = 0,\ f = 1$
(ii) $\eta \to \infty,\ f \to 0$

$$-2\eta = \frac{f''}{f'}$$

$$C_1 - \eta^2 = ln(f')$$

$$f' = C_2 e^{-\eta^2}$$

$$f(\eta) = C_2 \int_0^{\eta} e^{-\eta^2} d\eta + C_3$$

from boundary condition (ii)

$$C_3 = 1$$

from boundary condition (i)

$$\underbrace{f}_{0} = C_2 \left[\int_0^{\infty} e^{-\eta^2} d\eta \right] + 1$$

$$C_2 = \frac{-1}{\left[\int_0^{\infty} e^{-\eta^2} d\eta\right]}$$

In mathematics, the complementary error function is a special function that occurs in probability, statistics, and partial differential equations. It is defined as:

$$erfc(x) = \frac{2}{\sqrt{\pi}} \int_x^{\infty} e^{-t^2} dt$$

Figure 8.10 shows the plot of an error function. From *erf* function plot, at $\eta = 0, erf(\eta) = 0$ and $erfc(\eta) = 1 - erf(\eta) = 1 - 0 = 1$ so

$$C_2 = \frac{-1}{\underbrace{erfc(\eta)}_{1} * \sqrt{\pi}/2} = \frac{-2}{\sqrt{\pi}}$$

so

$$f(\eta) = \frac{u}{U_o} = 1 - \frac{2}{\sqrt{\pi}} \int_0^{\eta} e^{-\eta^2} d\eta = 1 - erf(\eta) = erfc(\eta)$$

so finally

$$\boxed{u = U_o erfc\left(\frac{y}{\sqrt{4\nu t}}\right)}$$

From error function plot: at $x = 2$, $erf\ (x)\ \rightarrow 1$, hence $erfc\ (x)\ \rightarrow 0$. As for $\eta \rightarrow \infty\ u\ \rightarrow 0$, we may take $\eta = 2$ as depth of zone of influence and

$$\delta \approx 2\sqrt{4\nu t}$$

$$\delta \approx 4\sqrt{\nu t}$$

Example 8.4

Example What is the depth of momentum diffusion penetration if honey and water are placed between parallel plates 1 m apart and one plate is suddenly moved for 10 s? The absolute viscosity of honey is 19,000 × 10^{-3} Pa·s and density is 1,420 kg/m^3. The absolute viscosity of water is 1.002 × 10^{-3} Pa·s and density is 998 kg/m^3. Assume that the fluid sudden movement is laminar.
Solution From the Stoke's first law we find the penetration depth as:

$$\delta = 4\sqrt{\nu t}$$

For honey, we have

$$\delta_{honey} = 4\sqrt{\left(\frac{19000 \times 10^{-3}}{1420}\right)10} = 0.365m$$

For water, we have

$$\delta_{water} = 4\sqrt{\left(\frac{1.002 \times 10^{-3}}{998}\right)10} = 3.1mm$$

8.6 FLOW IN AN ANNULUS

Flow through an annulus is schematically depicted in Figure 8.11. The flow is laminar, and friction is experienced from the inner and outer walls of the pipes.

We will now develop the expression for fully developed laminar flow in an annular pipe:

$$\frac{\partial^2 u}{\partial r^2} + \frac{1}{r}\frac{\partial u}{\partial r} = \frac{1}{\mu}\left(\frac{dP}{dx}\right)$$

$$\frac{1}{r}\frac{\partial}{\partial r}\left(r\frac{\partial u}{\partial r}\right) = \frac{1}{\mu}\left(\frac{dP}{dx}\right)$$

$$\frac{\mu}{r}\frac{\partial}{\partial r}\left(r\frac{\partial u}{\partial r}\right) = \left(\frac{dP}{dx}\right)$$

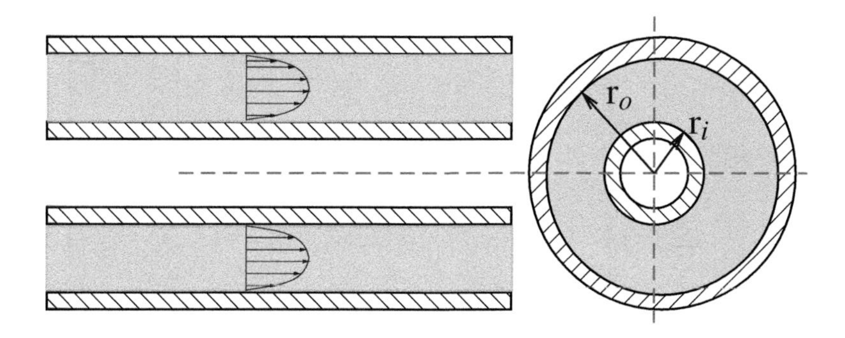

Figure 8.11 The flow through an annulus with inner and outer radii r_i and r_o respectively.

Integrating this will give:

$$u = \left(\frac{dP}{dx}\right)\left(\frac{r^2}{4\mu}\right) + C_1 \ln r + C_2 \tag{8.28}$$

Integrating with the boundary conditions:

(i) at $r = r_i$, $u = 0$

(ii) at $r = r_o$, $u = 0$

we get:

$$u(r) = \left(\frac{dP}{dx}\right)\left(\frac{-r_o^2}{4\mu}\right)\left[1 - \left(\frac{r}{r_o}\right)^2 + B \ln\left(\frac{r}{r_o}\right)\right] \tag{8.29}$$

and B is defined as

$$B = \frac{m^2 - 1}{\ln m}$$

with $m = r_i/r_o$.

We now need to define the hydraulic diameter and Reynolds number for this flow as:

$$d_H = \frac{4A}{WP} = \frac{4\pi(r_o^2 - r_i^2)}{2\pi(r_o + r_i)} = 2(r_o - r_i)$$

$$\text{Re} = \frac{\rho u_m d_H}{\mu} = \frac{2\rho u_m r_o(1 - m)}{\mu}$$

Mean velocity:

$$u_m = \frac{2\pi}{\pi(r_o^2 - r_i^2)} \int_{r_i}^{r_o} r \cdot u(r) \cdot dr$$

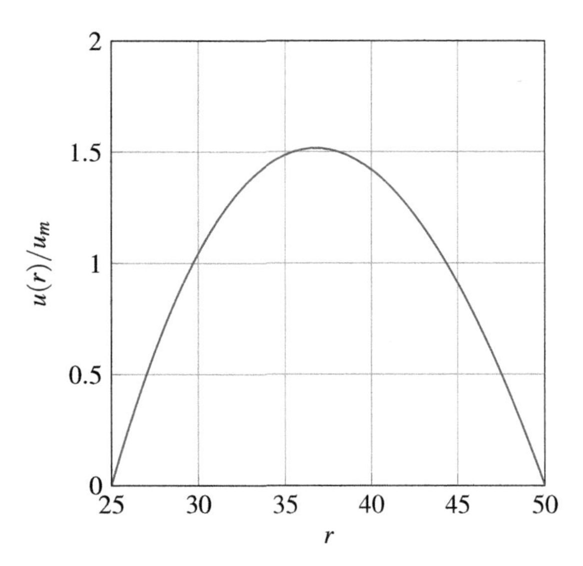

Figure 8.12 Velocity profile of annular flow, $r_i = 25$, $r_o = 50$.

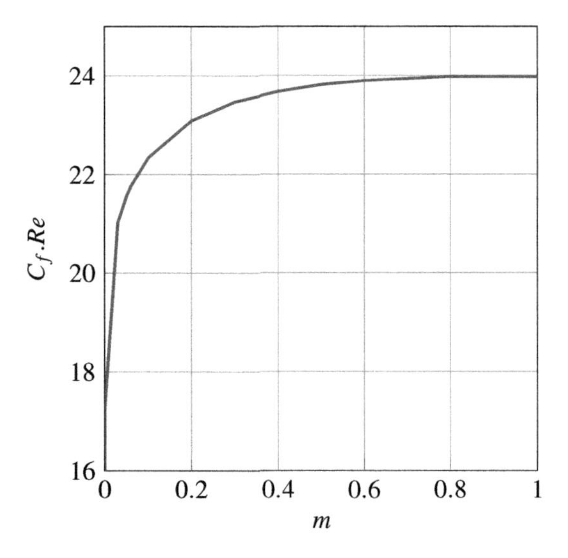

Figure 8.13 $C_f \cdot Re$ distribution for an annular flow (plot of Eq. 8.30).

$$u_m = \frac{-r_o^2}{8\mu}\left(\frac{dP}{dx}\right)\left[1 + m^2 + \frac{(1-m^2)}{\ln(m)}\right]$$

As both the inner and outer walls contribute in friction, we define the mean skin friction coefficient as:

$$\bar{C}_f = \frac{r_i Cf_i + r_o Cf_o}{r_i + r_o}$$

$$\bar{C}_f \mathrm{Re} = \frac{16(1-m)^2}{M} \tag{8.30}$$

where $M = 1 + m^2 - B$,

$$Cf_i = \left.\frac{2\tau_w}{\rho u_m^2}\right|_{r=r_i} = \frac{4\mu}{\rho u_m r_o}\left[\frac{(B/m) - 2m}{1 + m^2 - B}\right]$$

and

$$Cf_o = \left.\frac{2\tau_w}{\rho u_m^2}\right|_{r=r_o} = \frac{-4\mu}{\rho u_m r_o}\left[\frac{B-2}{1 + m^2 - B}\right]$$

When inner radius is approaching zero, the skin-friction factor will be equal to Hagen-Poisuille flow:

$$\lim_{m\to 0} \bar{C}_f \mathrm{Re} = \lim_{m\to 0} \frac{16(1-m)^2}{M} \simeq 16$$

Example 8.5

Example Olive oil is flowing at 70°C through an annular pipe. The inner and outers diameters of the pipe are 35 mm and 50 mm, respectively. The absolute viscosity of olive oil is 0.0181 Pa · s and density is 895 kg/m^3. Find the mean velocity if pressure gradient applied is −40 Pa/m. Also calculate the Reynolds number of flow and skin-friction coefficient.

Solution From the data, $m = 35/50 = 0.7$. We may calculate the mean velocity as:

$$u_m = \frac{-r_o^2}{8\mu}\left(\frac{dP}{dx}\right)\left[1 + m^2 + \frac{(1-m^2)}{\ln(m)}\right]$$

$$u_m = \frac{-(50\times 10^{-3})^2}{8\times 0.0181}(-40)\left[1 + 0.7^2 + \frac{(1-0.7^2)}{\ln(0.7)}\right] = 0.04152\, m/s$$

The Reynolds number is calculated:

$$\mathrm{Re} = \frac{2\rho u_m r_o (1-m)}{\mu} = 1483.42$$

At $m = 0.7$, $C_f \cdot \mathrm{Re} = 23.9$, so $C_f = 0.01611$.

Example 8.6

Example The velocity through an annulus is described by the following relation:

$$u(r) = \left(\frac{dP}{dx}\right)\left(\frac{-r_o^2}{4\mu}\right)\left[1 - \left(\frac{r}{r_o}\right)^2 + B\ln\left(\frac{r}{r_o}\right)\right]$$

and B is defined as

$$B = \frac{m^2 - 1}{\ln m}$$

with $m = r_i/r_o$. Find the location of maximum velocity.

Solution We substitute β for the $\left(\frac{dP}{dx}\right)\left(\frac{-r_o^2}{4\mu}\right)$.

$$u(r) = \beta\left(1 - \frac{r^2}{r_o^2} + B\ln\left(\frac{r}{r_o}\right)\right)$$

Differentiating the equation du/dr and setting it to zero gives:

$$\left(-\frac{2r}{r_o^2} + \frac{B}{r}\right) = 0$$

Substituting B into this equation gives the location of maximum velocity, denoted as R_{max}.

$$R_{\max} = \frac{1}{2}\sqrt{2}\sqrt{\frac{(m^2 - 1)\,r_o^2}{\ln(m)}}$$

If $r_o = 37$ mm and $r_i = 20$ mm, the maximum velocity will occur at 28.063 mm. Note that this value is different from the geometric mean of the radii, i.e., 28.55 mm.

PROBLEMS

8P-1 A liquid film is flowing down over an inclined plane as shown in Figure 8.14, under the action of gravity. Find the velocity distribution inside the falling liquid film as a function of angle of inclination (θ).

8P-2 An oil film of thickness δ is flowing along the wall under the action of gravity as shown in Figure 8.15. The width of the film is 1 m. The density ρ and absolute viscosity μ are 850 kg/m^3 and 9.5E-03 Pa · s, respectively. Find the velocity distribution in the oil film and the volume flow rate. Estimate the shear stress at the wall.

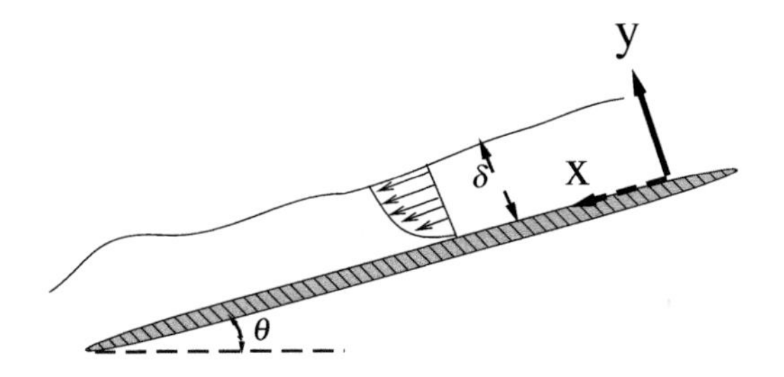

Figure 8.14 Figure of problem 8P.1.

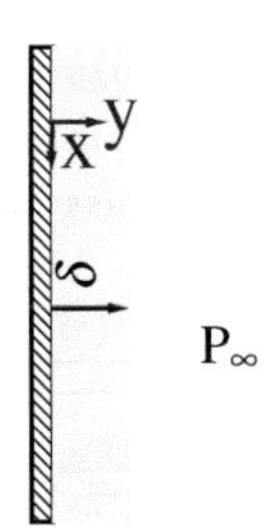

Figure 8.15 Figure of problem 8P-2.

8P-3 Consider the simple power-law model for a non-Newtonian fluid expressed as:

$$\tau_{yx} = \chi \left(\frac{du}{dy} \right)^m$$

where the exponent m is called the flow behavior index and the coefficient χ is called the consistency index. Show that for fully developed laminar flow of a power-law fluid between stationary parallel plates, the velocity profile can be expressed as

$$u(y) = \left(\frac{h}{\chi} \frac{\Delta p}{L} \right)^{1/m} \frac{mh}{m+1} \left[1 - \left(\frac{y}{h} \right)^{(m+1)/m} \right]$$

where y is the coordinate measured from the channel mid-plane.

8P-4 A sealed journal bearing is made from two concentric cylinders. The inner and outer radii are 35 and 36 mm. The gap is filled with oil in laminar motion. The journal length is 100 mm, and it turns at 3,000 rpm. The velocity profile is linear across the gap. The torque needed to turn the journal is 0.2 N $\cdot$ m. Calculate the viscosity of the oil.

8P-5 A fluid flowing between two parallel plates has a velocity profile that can be approximated as shown as Figure 8.16. Find the mean flow velocity if the gap between the plates is h. U_{max} is the maximum velocity and $u(y)$ is the velocity distribution. The plates have width w and have no-slip condition.

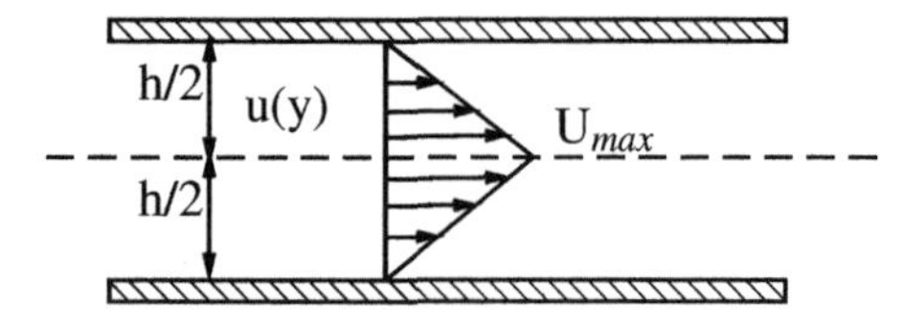

Figure 8.16 Figure of problem 8P-5.

8P-6 Two immiscible fluids are contained between infinite parallel stationary plates that are separated by a gap of $2h$. The two fluid layers are of the same thickness $h = 5$ mm. The dynamic viscosity of the lower fluid is one-fourth of that of the upper fluid. The upper fluid viscosity is $\mu_{upper} = 0.4$ Pa · s. If the applied pressure gradient is -50 kPa/m, find the velocity at the interface. What is the maximum velocity of the flow?

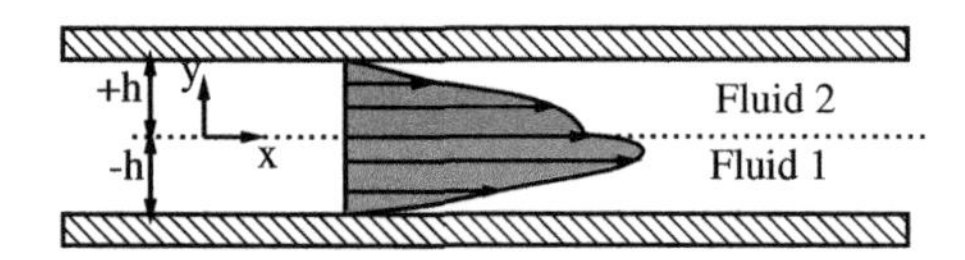

Figure 8.17 Figure of problem 8P-6.

8P-7 A thin rectangular object with a mass of 0.295 g is floating over a thin air film. The narrow gap between the object and the large flat surface is $\delta = 0.135$ mm. The object is $\ell = 9.3$ mm long and $w = 7.32$ mm wide. The object had moved with an initial speed of $U = 1.63$ mm/s. The micro rectangular object is slowed down due to friction. Calculate the time required for the object to lose 3% of its initial speed. Assume the air as Newtonian fluid with linear velocity profile in the narrow gap. Take air absolute viscosity as $\mu = 1.75 \times 10^{-5}$ Pa · s.

8P-8 A stationary housing has a close-fitting rotating drum inside it that formed a shear pump. The assembly created has a small gap δ and the flow in the annular space can be considered as flow between parallel plates. Assume that the depth normal to the diagram is δ. Evaluate the performance characteristics of the shear pump as functions of volume flow rate. The depth of the pump is b.

8P-9 Estimate the kinetic energy coefficient for a fully developed laminar flow in a circular tube, defined as:

$$\alpha = \frac{\int_A \rho u^3 dA}{\dot{m}\bar{u}^2}$$

where $\dot{m} = \rho \cdot \bar{u} \cdot A$

8P-10 Light oil with viscosity $\mu = 0.02$ Pa · s and density 700 kg/m^3 is flowing through an annular pipe at a volume flow rate of 0.3 m^3/h. The inner and outer radii

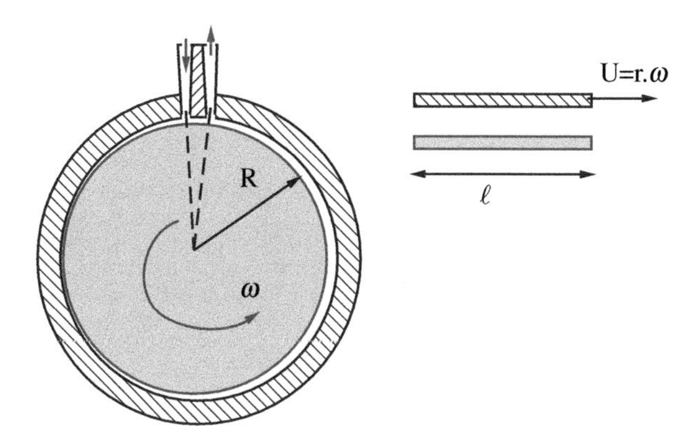

Figure 8.18 Figure of problem 8P-8.

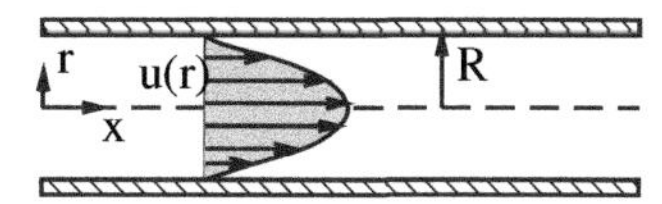

Figure 8.19 Figure of problem 8P-9.

are $r_i = 6$ mm and $r_o = 7$ mm. Find the Reynolds number and the pressure drop through the annular space if the length of the pipe is 20 cm.

REFERENCES

T. E. Stanton and J. R. Pannell, Similarity of motion in relation to the surface friction of fluids, Philosophical Transactions of the Royal Society of London, se. A, Containing Papers of a Mathematical or Physical Character, vol. 214, The Royal Society, 1914, pp. 199–224, http://www.jstor.org/stable/91017.

P. Szymanski, Quelques Solutions Exactes des Equations de I'Hydrodynamique de Fluide Visqueux dans le Cas d'un Tube Cylindrique, J. des Mathe'matiques Pures et Applique's vol. 11, se. 9, pp. 67–107, 1932.

J. G. Knudsen and Donald L. Katz, Fluid Dynamics and Heat Transfer. New York: McGraw-Hill Book Company, Inc., 1958.

9 Introduction to Turbulent Flows

Turbulent flows are desirable in engineering applications as they help in increased momentum and thermal diffusion. Understanding and control of turbulence is important as we can estimate turbulence related drag.

Learning outcomes: After finishing this chapter, you should be able to:
- Define the turbulent flow.
- Understand differences between laminar and turbulent velocity profiles and plot the turbulent velocity distributions in dimensionless coordinates.
- Understand the turbulent eddy scales and secondary flows.
- Understand the difference between three major turbulence simulation approaches.

In many industrial and engineering applications, turbulent boundary layers are more frequent than the laminar flows. The significant contribution in drag on boats, planes, and cars comes from turbulence, and thus the understanding turbulence is crucial for energy-efficient transportation involving less fuel consumption. The atmosphere and oceans also have turbulent flows; thus, it has an enormous contribution in shaping the planet's climate. A slight increase in air turbulence may cause a significant rise in the cost of a flight over the Atlantic Ocean.

9.1 WHAT IS TURBULENCE?

Turbulence is a three-dimensional time-dependent motion in which vortex stretching causes velocity fluctuations. Turbulence is one of the most complex phenomena in nature. Some of the reasons for the complexity are:

1. **Intrinsic spatio-temporal randomness:** Turbulent flows are extremely sensitive to small perturbations/disturbances due to which turbulent flows inherently have no repeatability. However, the statistical properties (not only some means, but almost all statistical properties) of turbulent flows are insensitive to disturbances.

2. **Intermittency:** Turbulence can interact with nonturbulent fluid flow and can appear intermittently in time at certain locations.

3. **Wide range of nonlocally interacting scales:** Turbulent flows are composed of a wide range of scales, from hundreds of kilometers to parts of a millimeter.

DOI: 10.1201/9781003315117-9

4. **Continuous self-production of vorticity and three-dimensionality:** In turbulence, an important role is played by vortex stretching, elongation, spinning, breakup, coalescing, and pairing. Turbulence is a three-dimensional phenomenon, which creates and maintains the turbulence vorticity. Note that a two-dimensional vortex cannot be stretched, hence two-dimensional turbulence does not exist in nature.

5. **Strong diffusion:** A source of energy is required to maintain turbulence (gradients of mean velocity, buoyancy, or other external forces). The energy supply mostly occurs at large scales and its dissipation occurs at small scales. Turbulent fluctuations cause the stretching and distortion of the fluid elements (containing a lump of properties like hot spots of heat, species, momentum, vorticity) until the increase in surface area and the property gradient enables molecular effects to act efficiently. Thus the transport of momentum, heat, species, and mixing is enhanced by several orders of magnitude.

Although turbulence is a very complex flow, fortunately, it can be computed using the purely deterministic equations (Navier-Stokes equations). However, the solution is very sensitive to initial or boundary conditions. The perturbations or disturbances that come from initial conditions, boundary conditions, and external noise can significantly alter the outcome. To study the turbulent flow field, the field variables are treated as random variables and useful information about the behavior can be obtained by standard statistical analysis.

9.2 THE LAW OF THE WALL

Many researchers have investigated the velocity profile in fully developed turbulent flows for a wide range of flow rates, pipe diameters, and fluid properties. A turbulent velocity profile is much different from a laminar velocity profile (see Figure 9.1).

It has been found that plotting the velocity distribution in alternate velocity and length coordinates provides meaningful information. The mean velocity profile in a pipe flow was first obtained by Nikuradse in 1933. The name "law of the wall" was coined by Coles in 1956.

The relevant quantities are the distance from the boundary, the shear stress, and fluid properties like density and viscosity. We now introduce the concept of the

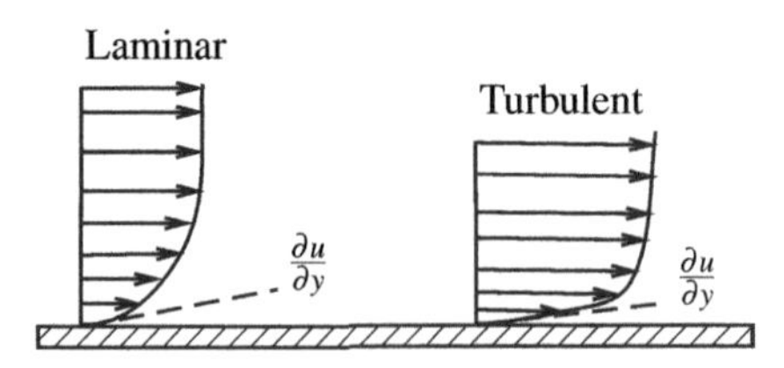

Figure 9.1 Laminar and turbulent velocity profiles. Note that the turbulent profile is steeper near the wall than the laminar velocity profile is.

frictional velocity; it is defined as:

$$u_\tau = \sqrt{\tau_w/\rho}$$

Using this frictional velocity, we can now construct the dimensionless velocity and the wall-normal coordinate, defined as:

$$u^+ = \frac{u}{u_\tau}$$

$$y^+ = \frac{u_\tau \Delta y}{\nu}$$

The experimental data plotted with these dimensionless coordinates prove that different turbulent flows at different Reynolds numbers can exhibit the same velocity distribution. Therefore, it is called the law of the wall, i.e., the experimentally proven velocity distribution near the boundary or surface. This law is valid whether the flow is near the flat plate or in a curved pipe, and therefore it is also called the universal velocity profile. Even though the velocity boundary thickness is very small, it is still necessary to demarcate it into distinct zones to facilitate the description of the turbulent shear layer. The universal velocity profile is not essentially a velocity profile that can traverse the complete turbulent boundary layer, but should be viewed as empirical statements to describe the velocity distribution in different zones.

The turbulent boundary layer can be divided into three regions:

(i) A viscous region close to the surface extending beyond the sublayer to a value of $y^+ = 30$,

(ii) An intermediate region from $y^+ = 30$ to approximately $y/\delta = 0.2$ where the influence of the wall still exists, and

(iii) An outer region characterized by the phenomenon of intermittency.

9.2.1 THE VISCOUS SUBLAYER

The layer immediately adjacent to the surface is called the viscous sublayer. Fluid viscosity plays a dominant role inside this layer. The velocity variation is nearly linear in this zone. For the Reynolds number 10^5–10^6, the thickness of the sublayer can be from 0.000005δ to 0.02δ, where δ is the boundary layer thickness. Turbulent fluctuations are present everywhere in the flow field, whether flow is close to the boundary surface or away from it. The stream-wise fluctuations are the largest, and they are unimpeded by the boundary. There are significant fluctuations present even at 0.0001δ. Although the viscous sublayer is a very thin layer, it plays an important role in the heat and mass transfer applications, especially for high Prandtl number fluids. The velocity profile is almost always linearly described as:

$$u^+ = y^+$$

This linear approximate fits the experimental curve below $y+ = 5$; therefore, this is taken as a width of the sublayer.

9.2.2 THE BUFFER ZONE OR TRANSITION ZONE

The region between $5 \leq y^+ \leq 30$ is called the buffer zone. The experimental data fitting shows that the velocity profile can be cast into the following equation:

$$u^+ = 5\ln y^+ - 3.05 \qquad 5 \leq y^+ \leq 30$$

Often, in the calculations, the following relationship is used:

$$u^+ = 8.7\left(y^+\right)^{1/7}$$

Though the above relation is not a perfect fit to the experimental data, it provides a close match and a mathematical ease in calculations.

Figure 9.2 shows the velocity distributions as suggested by von Karman. The equation

$$u^+ = 8.7\left(y^+\right)^{1/7}$$

matches well with the fully turbulent zone.

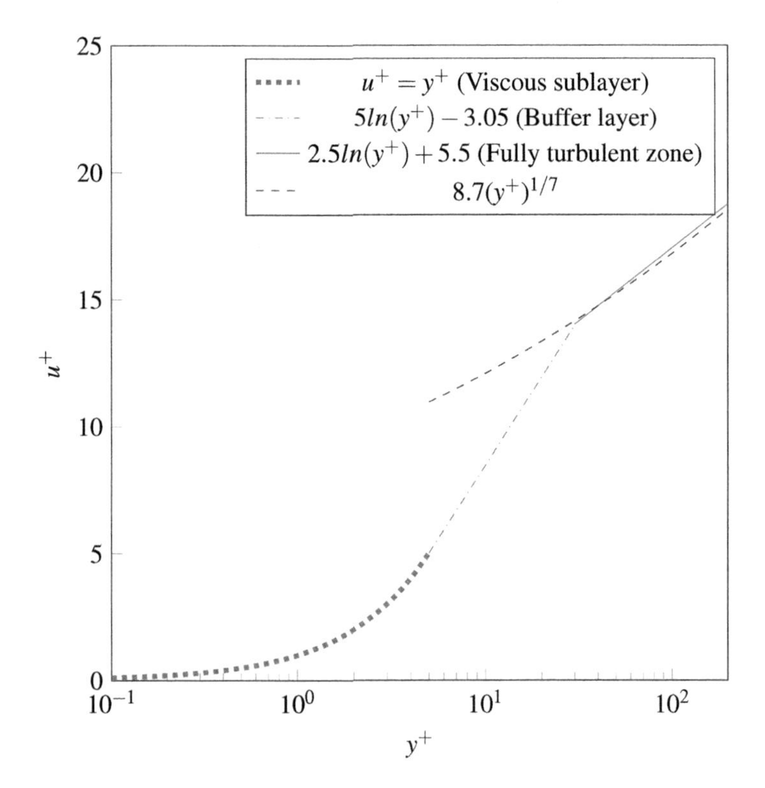

Figure 9.2 Turbulent law of wall according to von Karman.

9.2.3 THE OUTER LAYER

In study of turbulent flow, it would be convenient to define some similarities that would lead to a flow classification. Although the inner region of the turbulent boundary layer is self-similar, the outer region is devoid of self-similar behavior. At the edge of the turbulent boundary layer and at the edge of the turbulent jet/the entertainment region, there are pockets of high fluctuations. In boundary layer flows, the fluctuations are present also in flows at the edge of the boundary layer. Klebanoff (1955) demonstrated that a sharp interface exists between turbulent and nonturbulent flows. This interface has a ragged shape and undulates while traveling downstream. This zone range is between 0.4δ and 1.2δ. Clauser has identified an equilibrium layer that follows the velocity profile relation:

$$\frac{u-U_\infty}{u_\tau} = f\left(\frac{y}{\Delta}\right)$$

$$\Delta = \delta^* \sqrt{\frac{2}{C_f}}$$

where Δ is called defect layer thickness. According to Clauser, for equilibrium layer the parameter β is constant and defined as:

$$\beta = \left(\frac{\delta^*}{\tau_w}\right)\frac{dP_\infty}{dx} \approx \left(\frac{\delta^*}{\tau_w}\right)\left(-\rho U_\infty \frac{dU_\infty}{dx}\right)$$

The negative β value indicates the favorable pressure gradient.

Coles (1956) noted that the deviation or excess velocity of the outer layer has a wake-like shape when viewed from the free stream:

$$\frac{u^+ - u^+_{\log-law}}{U^+_\infty - u^+_{\log-law}} = f\left(\frac{y}{\delta}\right)$$

where $f\left(\frac{y}{\delta}\right)$ is called the wake function, which is normalized to be zero at the wall and unit at the edge of the boundary layer. The velocity profile defined for $y^+ > 30$ is:

$$u^+ = \frac{1}{\kappa}\ln(y^+) + B + \frac{2\Pi}{\kappa} f\left(\frac{y}{\delta}\right)$$

$$u^+ \approx \frac{1}{\kappa}\ln(y^+) + B + \frac{2\Pi}{\kappa}\sin^2\left(\frac{\pi}{2}\frac{y}{\delta}\right)$$

where Π is Cole wake parameter. The law of the wall does not provide the single explicit representation of the average velocity profile in terms of y and δ.

9.3 IS THERE A SINGLE EQUATION AVAILABLE?

Over the years, many researchers have proposed different formulations of the law of the wall. However, there seems to be no physical ground for discriminating between various proposals, and the choice can be made on the grounds of convenience. A law of the wall valid for all y^+ have also been proposed, for example Spalding (1960) proposed the following law:

$$y^+ = u^+ + \exp(-\kappa B)\left[\exp(\kappa u^+) - 1 - \kappa u^+ - \frac{(\kappa u^+)^2}{2} - \frac{(\kappa u^+)^3}{6}\right] \tag{9.1}$$

where $B = 5.5, \kappa = 0.4$.

Van Driest (1956) proposed the law of the wall valid for all y^+:

$$\frac{du^+}{dy^+} = \frac{2}{1+\left\{1+4(\kappa y^+)^2\left(1-\exp\left(\frac{-y^+}{26}\right)\right)\right\}^{1/2}}$$

where $\kappa = 0.4$.

Example 9.1

Example Clauser measured the velocity distribution inside an air turbulent boundary layer at 24°C and 1 atm. Velocity data at position $x = 2.1$ m is given in the table. If boundary layer thickness $\delta = 88.9$ mm and $dU_\infty/dx = 1.06s^{-1}$ at this position, find inner law of wall and Clauser's parameter and Coles wake parameter.

y (mm)	u (m/s)	y (mm)	u (m/s)
2.54	4.92	20.32	6.97
3.81	5.19	22.86	7.22
5.08	5.35	25.4	7.43
6.35	5.54	31.75	8.08
7.62	5.70	38.1	8.60
10.16	5.97	50.8	9.52
12.7	6.25	63.5	9.84
15.24	6.47	76.2	9.89
17.78	6.71	88.9	9.91

Solution First, we try the log-law relation:

$$\frac{u}{u_\tau} = \frac{1}{0.41}\ln\left(\frac{yu_\tau}{\nu}\right) + 5.0$$

For the first data point we have:

$$\frac{4.92}{u_\tau} = \frac{1}{0.41}\ln\left(\frac{u_\tau \times 2.54mm}{1.5\times 10^{-5}}\right) + 5.0$$

We can solve this equation by trial and error, as this is a transcendental equation.

The frictional velocity for the first seven data points is $u_{\tau 1} = 0.331$, $u_{\tau 2} = 0.328$, $u_{\tau 3} = 0.324$, $u_{\tau 4} = 0.373$, $u_{\tau 5} = 0.436$, $u_{\tau 6} = 0.558$, $u_{\tau 7} = 0.676$.

The last point gives $y^+ = 288$, so all these points lay within logarithmic overlap range. The first data point gives $y^+ = 56.19$, which is far from the inner layer. Wall shear stress, based on frequently appearing frictional velocity, is:

$$\tau_w = \rho u_\tau^2 = 1.22 \times (0.328)^2 = 0.131 Pa$$

Displacement thickness: After data fitting, we can numerically integrate the tabulated data to obtain the displacement thickness, which is 14.98 mm.

$$\beta = \left(\frac{\delta^*}{\tau_w}\right)\frac{dP}{dx} \approx \left(\frac{\delta^*}{\tau_w}\right)\left(-\rho U_\infty \frac{dU_\infty}{dx}\right)$$

$$\beta = \left(\frac{\delta^*}{\tau_w}\right)\left(-\rho U_\infty \frac{dU_\infty}{dx}\right)$$

$$\beta \approx \left(\frac{14.98 \times 10^{-3}}{0.131}\right)(-1.22 \times 9.91 \times 1.06)$$

$$\beta = 1.463$$

Coles wake parameter:

$$u^+ \approx \frac{1}{\kappa}\ln(y^+) + B + \frac{2\Pi}{\kappa} f\left(\frac{y}{\delta}\right)$$

The quantity Π is called the Coles parameter. This equation provide the complete velocity distribution for the two-dimensional turbulent boundary layers. Coles law of the wall has a wall function that is prescribed as:

$$u^+ \approx \frac{1}{\kappa}\ln(y^+) + B + \frac{2\Pi}{\kappa}\sin^2\left(\frac{\pi}{2}\frac{y}{\delta}\right)$$

Converting the y in the last term will bring the whole equation in y^+. After substitution, we get:

$$u^+ \approx \frac{1}{0.41}\ln\left(y^+\right) + 5 + Cole_{para}$$

$$Cole_{para} = \frac{2\Pi}{0.41} \cdot \left(\sin\left(\frac{3.14}{2 \cdot (88.9E-03)}\left(\frac{y^+ \cdot 1.5E-05}{0.324}\right)\right)\right)^2$$

Suggesting different Π values and seeing the goodness of the fit, it is found that Π lies between 2 and 3, and a value of 2.2 is a reasonable value for this function. However, we should keep in mind that Π value depends on the flow conditions, and there is no single universal value for Coles wake parameter (see Figure 9.3).

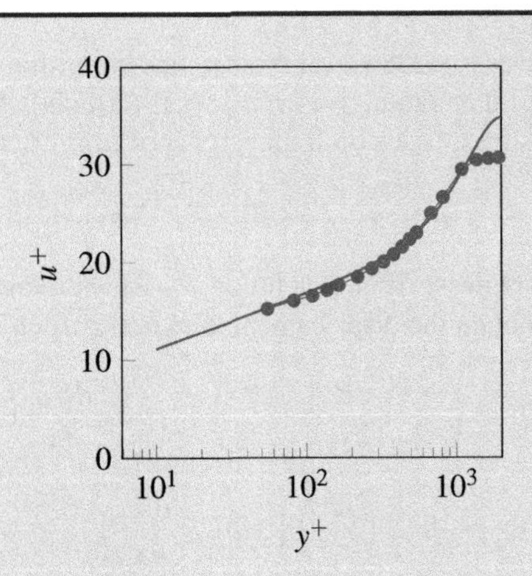

Figure 9.3 Turbulent velocity profile computed from Cole wake law. The $\Pi = 2.2$ is selected in this plot.

9.4 REYNOLDS AVERAGING

Ensemble averaging is the most general type of Reynolds averaging that is suitable for flows that decay in time. It is defined as:

$$< \Psi(x_i,t) > = \lim_{N\to\infty} \frac{1}{N} \sum_{n=1}^{N} \Psi(x_i,t) \tag{9.2}$$

Once the flow becomes statistically stationary (statistics become independent of time) it can be time averaged, which is defined as:

$$\overline{\Psi(x_i)} = \lim_{t\to\infty} \int_{-t/2}^{t/2} \Psi(x_i,t)dt \tag{9.3}$$

According to the Reynolds decomposition of turbulence quantities, we have:

$$\psi = \overline{\Psi} + \psi' \tag{9.4}$$

Example 9.2

Example In Reynolds averaging, the time-average of the mean quantity is considered the same as mean quantity. To understand this assumption, we consider an arbitrary temperature field represented by the following function:

$$T(t) = To \cdot ((1 + 0.5t + 0.01t^2)) + 2 \cdot \cos(85 \cdot t)$$

where To is 293 K. Find the arithmetic mean temperature and the time averaged mean temperature in one cycle $2\pi/\omega$.

Solution The temperature field is oscillating, but the temperature is gradually increasing:

$$T(t) = To \cdot \left(\left(1 + 0.5t + 0.01t^2\right)\right) + 2 \cdot \cos(85 \cdot t)$$

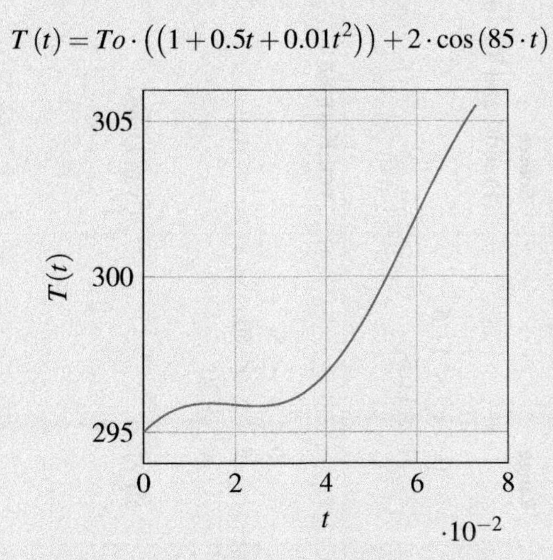

Figure 9.4 Turbulent temperature signal.

Figure 9.4 shows the turbulent temperature signal.
For one cycle, we have

$$\text{cycle} = \frac{(2\pi)}{\omega} = 0.073\, s$$

$$Time\ Average\ of\ T = \frac{1}{0.073} \int_0^{0.073} T(t)\ dt$$

For one cycle, the arithmetic mean temperature is 299.41 K and the time averaged mean temperature is 298.32 K. This change is only 0.364% variation; therefore the assumption that mean of time averaged quantity is the same as mean is justified.

9.4.1 REYNOLDS AVERAGED NAVIER-STOKES EQUATIONS

Figure 9.4 shows a real turbulent velocity data probed at a single point in flow field. According to Reynolds averaging, all the flow quantities, like all velocity components and pressure etc., are fluctuating, as $u = \bar{u} + u'$, $v = \bar{v} + v'$, $w = \bar{w} + w'$, and $p = \bar{p} + p'$. In Figure 9.5 we can bifurcate the velocity $u(t)$ signal into fluctuating and a mean quantity. If the mean is not changing with time then such flow can be locally steady. However, if the mean is changing with time, then such flow is considered unsteady. In turbulent flow, the velocity of the fluid is a continuous random function of time and space. It is a random function because the instantaneous value

Table 9.1
Law of the wall

$u^+ = f(y^+)$	ε^+	Range	Source
$u^+ = y^+$ $u^+ = 2.5\ln y^+ + 5.5$	$\varepsilon^+ = 1$ $\varepsilon^+ = 0.4y^+$	$0 \le y^+ \le 11.5$ $y^+ > 11.5$	Prandtl and Taylor
$u^+ = y^+$ $u^+ = 5\ln y^+ - 3.05$ $u^+ = 2.5\ln y^+ + 5.5$	$\varepsilon^+ = 1$ $\varepsilon^+ = 0.2y^+$ $\varepsilon^+ = 0.4y^+$	$0 \le y^+ \le 5$ $5 \le y^+ \le 30$ $y^+ \ge 30$	von Karman
$\frac{du^+}{dy^+} = \frac{2}{1+\left[1+\frac{4\kappa^2}{y^{+2}}(\zeta)\right]^{1/2}}$ $\zeta = 1 - \exp\left(\frac{-y^+}{A^+}\right)$	$\varepsilon^+ = \frac{-1+\left[1+4\kappa^2 y^{+2}(\zeta)^2\right]^{1/2}}{2}$	all y^+	van Driest

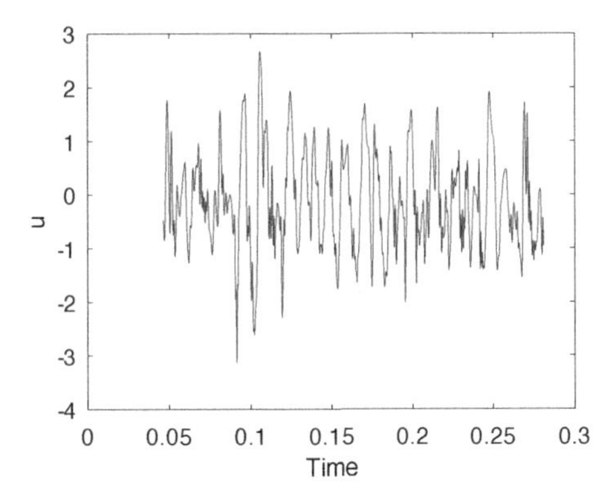

Figure 9.5 A real turbulent velocity data probed at a single point in flow field.

of the fluctuating quantity at any point cannot be predicted from the mean value of the flow quantities. The value of the velocity at any point is distributed according to the laws of probability. If the instantaneous velocity is considered the sum of a mean velocity and fluctuating velocity, the distribution of fluctuating velocity follows the Gaussian or normal distribution.

Following are the averaging rules that have been classified as Reynolds averaging procedure.

Averaging rules:

Rule i: The time-average of all mean quantities is equal to the mean value $\overline{\bar{u}} = \bar{u}$.

Rule ii: All fluctuating quantities are zero when time-averaged, so $\overline{u'} = 0, \overline{v'} = 0, \overline{w'} = 0$, $\overline{\rho'} = 0$, $\overline{p'} = 0$.

Rule iii: The product of mean and fluctuating quantities will be zero as well.

Continuity equation:

$$\frac{\partial(u)}{\partial x} + \frac{\partial(v)}{\partial y} + \frac{\partial(w)}{\partial z} = 0$$

$$\frac{\partial\left[(\bar{u}+u')\right]}{\partial x} + \frac{\partial\left[(\bar{v}+v')\right]}{\partial y} + \frac{\partial\left[(\bar{w}+w')\right]}{\partial z} = 0$$

Applying time-averaging, we get:

$$\frac{\partial \bar{u}}{\partial x} + \frac{\partial \bar{v}}{\partial y} + \frac{\partial \bar{w}}{\partial z} = 0$$

Navier-Stokes equation: We consider the conservation form of the inertia term in Navier-Stokes equation:

$$(\vec{V}\cdot\nabla)u = \frac{\partial(uu)}{\partial x} + \frac{\partial(uv)}{\partial y} + \frac{\partial(uw)}{\partial z}$$

$$\begin{aligned} &\frac{\partial\left[(\bar{u}+u')(\bar{u}+u')\right]}{\partial x} + \frac{\partial\left[(\bar{u}+u')(\bar{v}+v')\right]}{\partial y} \\ &+\frac{\partial\left[(\bar{u}+u')(\bar{w}+w')\right]}{\partial z} = -\frac{1}{\rho}\frac{\partial(\bar{p}+p')}{\partial x} + g_x + \nu\nabla^2(\bar{u}+u') \end{aligned} \tag{9.5}$$

We now average in time every term and apply the rules:

$$\begin{aligned} &\frac{\partial\left[\bar{u}^2+2\bar{u}u'+u'^2\right]}{\partial x} + \frac{\partial\left[\bar{u}\,\bar{v}+\bar{u}\,v'+u'\bar{v}+u'\,v'\right]}{\partial y} \\ &+\frac{\partial\left[\bar{u}\,\bar{w}+\bar{u}\,w'+u'\bar{w}+u'\,w'\right]}{\partial z} = -\frac{1}{\rho}\frac{\partial(\bar{p}+p')}{\partial x} + g_x + \nu\nabla^2(\bar{u}+u') \end{aligned} \tag{9.6}$$

$$\begin{aligned} &\frac{\partial\left[\bar{u}^2+2\bar{u}u'+u'^2\right]}{\partial x} + \frac{\partial\left[\bar{u}\,\bar{v}+\bar{u}\,v'+u'\bar{v}+u'\,v'\right]}{\partial y} + \frac{\partial\left[\bar{u}\,\bar{w}+\bar{u}\,w'+u'\bar{w}+u'\,w'\right]}{\partial z} \\ &= -\frac{1}{\rho}\frac{\partial(\bar{p}+p')}{\partial x} + g_x + \nu\nabla^2(\bar{u}+u') \end{aligned} \tag{9.7}$$

$$\begin{aligned} &\overline{\frac{\partial\left[\bar{u}^2+2\bar{u}u'+u'^2\right]}{\partial x}} + \overline{\frac{\partial\left[\bar{u}\,\bar{v}+\bar{u}\,v'+u'\bar{v}+u'\,v'\right]}{\partial y}} \\ &+\overline{\frac{\partial\left[\bar{u}\,\bar{w}+\bar{u}\,w'+u'\bar{w}+u'\,w'\right]}{\partial z}} = -\frac{1}{\rho}\overline{\frac{\partial(\bar{p}+p')}{\partial x}} + \overline{g_x} + \overline{\nu\nabla^2(\bar{u}+u')} \end{aligned} \tag{9.8}$$

$$\begin{aligned} &\frac{\partial\left[\overline{\bar{u}^2}+\overline{u'^2}\right]}{\partial x} + \frac{\partial\left[\overline{\bar{u}\,\bar{v}}+\overline{u'\,v'}\right]}{\partial y} \\ &+\frac{\partial\left[\overline{\bar{u}\,\bar{w}}+\overline{u'\,w'}\right]}{\partial z} = -\frac{1}{\rho}\frac{\partial(\bar{p})}{\partial x} + g_x + \nu\nabla^2(\bar{u}) \end{aligned} \tag{9.9}$$

We consider now the right-hand side only:

$$\frac{\partial\overline{\bar{u}^2}}{\partial x} + \frac{\partial\overline{\bar{u}\,\bar{v}}}{\partial y} + \frac{\partial\overline{\bar{u}\,\bar{w}}}{\partial z} + \frac{\partial\overline{u'^2}}{\partial x} + \frac{\partial\overline{u'\,v'}}{\partial y} + \frac{\partial\overline{u'\,w'}}{\partial z}$$

From continuity equation RHS can be simplified further and we have:

$$\bar{u}\frac{\partial\bar{u}}{\partial x} + \bar{v}\frac{\partial\bar{u}}{\partial y} + \bar{w}\frac{\partial\bar{u}}{\partial z} + \frac{\partial\overline{u'^2}}{\partial x} + \frac{\partial\overline{u'\,v'}}{\partial y} + \frac{\partial\overline{u'\,w'}}{\partial z} = -\frac{1}{\rho}\frac{\partial(\bar{p})}{\partial x} + g_x + \nu\nabla^2(\bar{u})$$

Finaly, in tensor notation the time-averaged. The time-averaged Navier-Stokes equations are given as:

$$\rho U_j \frac{\partial U_i}{\partial x_j} + \rho \overline{u_i^{'} u_j^{'}} = -\frac{\partial P}{\partial x_i} + \mathbf{F}_i + \frac{\partial}{\partial x_j}\left[\mu\left(\frac{\partial U_i}{\partial x_j} + \frac{\partial U_j}{\partial x_i}\right)\right]$$

where $\mathbf{F}_i$ is body force. Note that the term $\overline{u_i^{'} u_j^{'}}$ arises from the inertia term, but the researchers had modeled it as a shear stress, as it is found that drag in turbulent flow is much higher than laminar flow. Hence the terms $\overline{u_i^{'} u_j^{'}}$ are called Reynolds stresses and merged with viscous stresses.

$$\rho U_j \frac{\partial U_i}{\partial x_j} = -\frac{\partial P}{\partial x_i} + \mathbf{F}_i + \frac{\partial}{\partial x_j}\left(\mu\left(\frac{\partial U_i}{\partial x_j} + \frac{\partial U_j}{\partial x_i}\right) - \rho \overline{u_i^{'} u_j^{'}}\right) \tag{9.10}$$

Here, the capitalized letters are the time-averaged mean velocities.

The $\overline{u_i^{'} u_j^{'}}$ term is an unknown quantity, which originates due to averaging a non-linear inertia term. When the Reynolds averaging is executed over the inertia term in the Navier-Stokes equation, these extra terms appear, which are the time-averaged product of fluctuating velocities. In general, turbulence would cause an increase in shear stress, and the early turbulence researchers modeled these additional terms as stresses, and from then onward, these terms were called Reynolds stresses, even though they are not really stresses but extra inertia force that arise due to turbulence.

9.4.2 SINGLE POINT STATISTICS

A single point turbulence data recorded in time can give a valuable description of this complex phenomenon.

9.4.2.1 Correlations

The time history of a fluctuating function $u^{'}$(t) can be studied by finding the correlation between the values at two different times. Taylor introduced correlation coefficient between the quantities (Taylor, 1935). A time correlation function is defined as:

$$R^t(t, t+\Delta t) = \frac{\overline{u_i^{'}(t) u_j^{'}(t+\Delta t)}}{\sqrt{\overline{(u_i^{'}(t))^2}}\sqrt{\overline{(u_j^{'}(t+\Delta t))^2}}} \tag{9.11}$$

The corresponding time scale can be determined from the known time correlation function:

$$t_{ij} = \frac{1}{2}\int_{-\infty}^{\infty} R^t(t, t+\Delta t) d(\Delta t) \tag{9.12}$$

The integral turbulent time scale can be interpreted as an averaged inverse rotational frequency of the typical big eddy appearing in the spatial location x. Also, it gives the measure of the time interval over which $u^{'}(t)$ remembers its history.

For isotropic turbulence:

$$t = \frac{t_{ii}}{3} \tag{9.13}$$

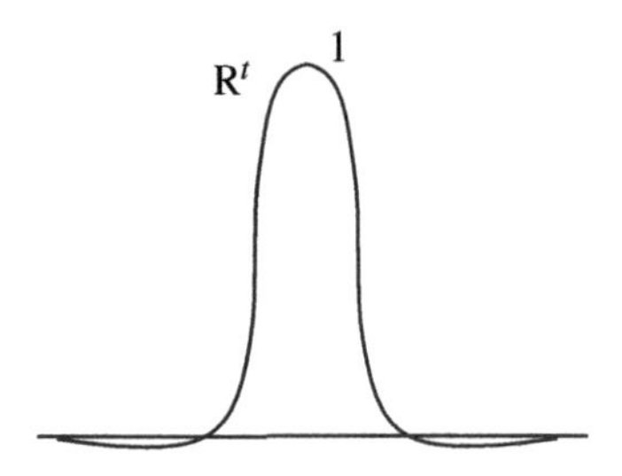

Figure 9.6 The auto-correlation function.

Closely connected to this time is the integral length scale (Λ), which represents the large scales present in the flow field.

The integral length scale is considered to be representative of the largest scales of a turbulent flow, and, as such, it is an important parameter in modern and classical approaches to turbulence theory and numerical simulations. Also, the integral length scale (Λ) can be interpreted as the length scale from which points on the velocity fluctuation are predominantly uncorrelated. The integral length scale (Λ) can be computed with the help of the two-point spatial correlation function for statistically steady (time independent) turbulence as:

$$R^{\Lambda}(x,x+\Delta x) = \frac{\overline{u_i'(x)u_j'(x+\Delta x)}}{\sqrt{\overline{(u_i'(x))^2}}\sqrt{\overline{(u_j'(x+\Delta x))^2}}} \tag{9.14}$$

The size of energy containing eddies depends on the geometry of a spatial domain and on the local intensity of turbulence. This size can be related to the integral turbulent length scale as:

$$\Lambda_{ij} = \frac{1}{2}\int_{-\infty}^{\infty} R^{\Lambda}(x,x+\Delta x)d(\Delta x) \tag{9.15}$$

For isotropic turbulence:

$$\Lambda = \frac{\Lambda_{ii}}{3} \tag{9.16}$$

The concept of local isotropy is inadequate for obtaining the turbulent-energy dissipation, especially in the region near the wall.

Figure 9.6 shows the typical plot of the auto-correlation function.

9.4.2.2 Energy Spectrum

In 1941, Kolmogorov investigated the turbulent eddy spectrum at higher frequencies. He hypothesized that for high Reynolds numbers turbulent flows, the eddies in the universal range transport their energy almost exclusively via inertial forces, and they do not dissipate. This section of the spectrum is called the inertial subrange.

A Fourier transformation of the auto-correlation function is the energy spectrum, defined as:

$$E(f) = \int_{-\infty}^{\infty} e^{-i(2\pi f)\Delta t}[R^t(t,t+\Delta t)]d(\Delta t) \tag{9.17}$$

Compared to laminar flow, the turbulent flow comprises eddies (fluid particles moving together) of different sizes and scales. These eddies significantly affect the overall flow field. The large eddies have high energy and from them this energy is transferred to small eddies. The energy of small scales can be dissipated due to viscous action.

One of the turbulence phenomenon's significant discoveries was identifying energy cascading behavior, which is present in all turbulent flows. The turbulent flow is composed of many flow eddies. We know that an eddy, or vortex, is fluid in rotation. Turbulent flow has a variety of eddy sizes. Some of these eddies are huge in size, like those shed by an airplane at the time of takeoff. Some eddies are of average size, like vortex rings that appear in smoke, and some are so small that we cannot perceive them through naked eye. In turbulent flows, the larger the vortex size, the more energy it has. So the large eddies break down into medium-sized eddies, which further split into small eddies. At the scale of smallest eddies, turbulence energy is dissipated in overcoming fluid viscous effects. Kolmogorov discovered that the slope of energy spectrum at high Reynolds number exhibit the slope of $-5/3$.

Figure 9.7 shows the turbulence spectrum using the turbulence data. Three distant turbulent length scales for eddies are defined to explain this spectrum:

Kolmogorov scales: (η) are the smallest eddies and are usually represented by η in turbulence literature. They are associated with the dissipation phenomenon in turbulence, and are sometimes referred to as dissipation scales. Small eddies behavior is isotropic (nondirectional).

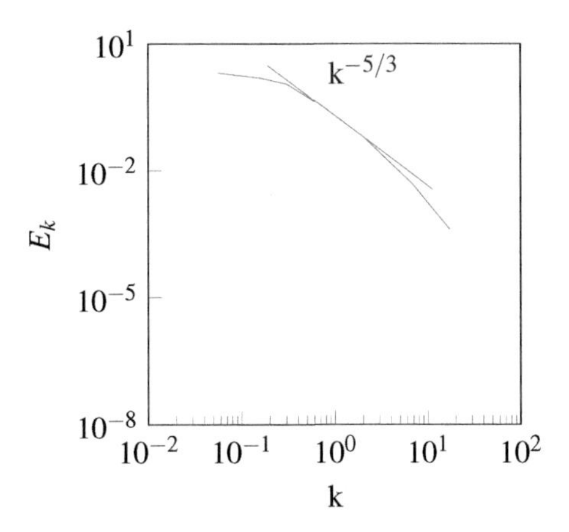

Figure 9.7 The turbulence spectrum using the turbulence data.

Taylor length scales: (λ) are the eddies whose size is larger than Kolmogorov scale but smaller than the large-scale eddies.

Large scales: (Λ) are the eddies that are of the size of the characteristic length/dimension of the object. They are the eddies that are larger than the Taylor microscale and are not strongly affected by viscosity. If flow is moving over a cylinder then these eddies have size in order of magnitude of diameter of cylinder. Large eddies behavior is highly anisotropic, as they cause fluctuations in quantites that differ in different directions.

The ratio of Kolmogorov scale eddies to large-scale eddies is

$$\frac{\eta}{\Lambda} \approx \mathrm{Re}_{\Lambda}^{-3/4}$$

The ratio of Taylor scale eddies to large-scale eddies is

$$\frac{\lambda}{\Lambda} \approx \sqrt{10}\mathrm{Re}_{\Lambda}^{-1/2}$$

The integral length scale (Λ) is a macro-scale that represents the size of the largest eddy at a particular location in a turbulent flow field. The scale can be estimated by recording the time series of turbulent quantities. Roach proposed the relation for estimation of integral length scale as

$$\Lambda = \left[\frac{E(f) \cdot U_{mean}}{4\sigma_{sd}}\right]_{f \to 0}$$

Example 9.3

Example The 900 data points were recorded by probing the turbulent flow field. The mean velocity of signal is 8.105 m/s with standard deviation of $\sigma_{sd} = 0.88$ m/s. The energy spectrum of signal indicates $E(f)$ approaching zero frequency at energy level of 1.5E-2. Find the integral length scale at this probe position.

Solution We can calculate integral length scale as

$$\Lambda = \left[\frac{E(f) \cdot U_{mean}}{4\sigma_{sd}}\right]_{f \to 0}$$

$$\Lambda = \frac{1.5E-2 \times 8.1}{4 \times 0.88^2} = 0.03922\, m$$

This is the scale, or size, of the largest eddy passing at the probe location.

9.5 TURBULENT FLUCTUATIONS

Figure 9.8 shows the turbulent intensities by Klebanoff (1955). Note that the velocity fluctuations are present even in the viscous sublayer (formerly called laminar sublayer); however, due to the no-slip boundary condition at the wall, all velocities and their fluctuations vanish at the wall.

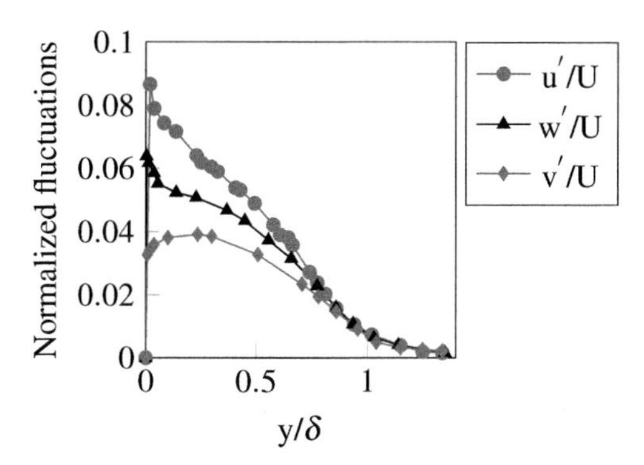

Figure 9.8 Distribution of turbulent intensities by Klebanoff (1955).

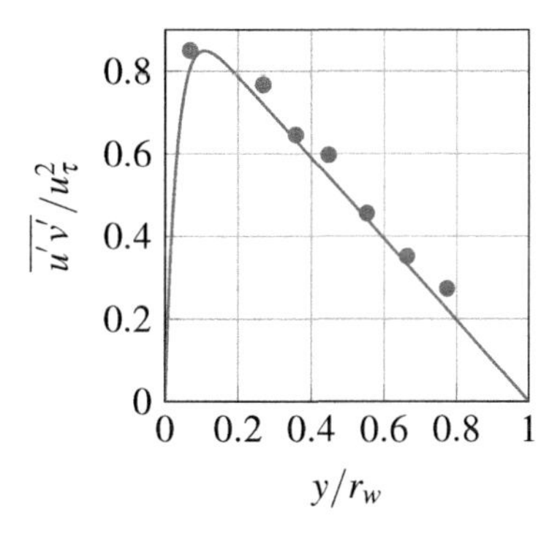

Figure 9.9 Plot of semi-empirical relation for distribution of $\overline{u'v'}/u_\tau^2$ inside a circular pipe at $Re = 3.24 \times 10^6$ as proposed by Pai (1953). Experimental data from Laufer (1954).

The distribution of $\overline{u'v'}/u_\tau^2$ inside a circular pipe at $Re = 3.24 \times 10^6$ is shown in Figure 9.9. Pai proposed the following equation for the prediction of $\overline{u'v'}$ in pipe.

$$\frac{\overline{u'v'}}{u_\tau} = 0.9835\left(1-\frac{y}{r_w}\right)\left[1-\left(1-\frac{y}{r_w}\right)^{30}\right]$$

Turbulent fluctuations are often reported as root-mean-square values and also by turbulent intensity, defined as:

$$Tu = \frac{\sqrt{\frac{1}{3}\left[\overline{u'^2}+\overline{v'^2}+\overline{w'^2}\right]}}{\sqrt{\overline{u}^2+\overline{v}^2+\overline{w}^2}}$$

It is also approximated as follows for isotropic turbulence:

$$Tu = \frac{u_{rms}}{\overline{u}} = \frac{\sqrt{\overline{(u')^2}}}{\overline{u}}$$

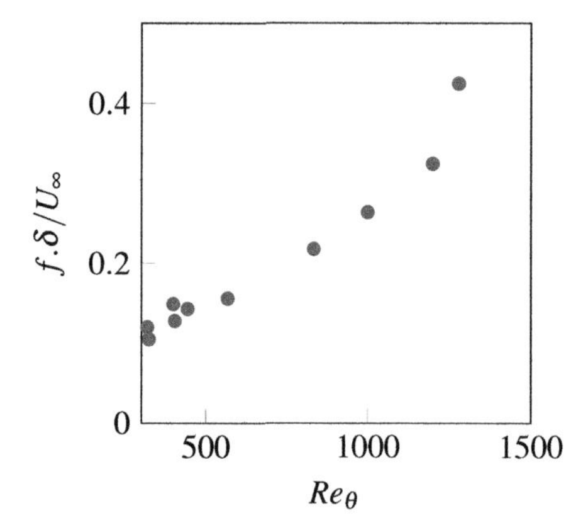

Figure 9.10 The dimensionless bursting frequency scaled with outer flow parameters. Blackwelder et al. (1983), Willmarth et al. (1984), and Kim et al. (1971) data.

Another interesting quantity related with velocity fluctuations is the turbulent kinetic energy:

$$k = \frac{1}{2}\left[\overline{u'^2} + \overline{v'^2} + \overline{w'^2}\right]$$

It has been discovered that turbulent flow has the peculiar bursting phenomenon. Figure 9.10 shows the dimensionless bursting frequency scaled with outer flow parameters vs. Reynolds number.

9.6 TURBULENCE SIMULATIONS

Turbulent flow problems at low Reynolds numbers (Re $<10^3$) can be computed using the Navier-Stokes equations directly without invoking any models. This approach is called the Direct Numerical Simulation (DNS) of turbulent flow. Various DNS data sets are also made available for researchers and students on the internet and can be used to further understand turbulence dynamics. As the Reynolds number increases, the numerical diffusion in codes is too large, and, eventually, it is no longer possible to compute the turbulence scale correctly. For high Reynolds number problems (Re $>10^6$), the available approach at the moment is to use Reynolds Averaged Navier-Stokes (RAN)-based turbulence models. For industrial applications, the Reynolds Averaged Navier-Stokes (RAN)-based turbulence models are still used, and in this approach, the flow is modeled using various assumptions. Since 1970, there has been significant development in the RANS-based modeling approach, and these models are available in Computational Fluid Dynamics (CFD) codes or applications like FLUENT, CFX, and OpenFOAM. A third option is the Large Eddy Simulation (LES) between these two extremes. In this approach, large turbulent scales are computed, and small scales are modeled. LES technique has been proved to be a fruitful option for a wide range of Reynolds numbers. Figure 9.11 shows the difference between the three approaches.

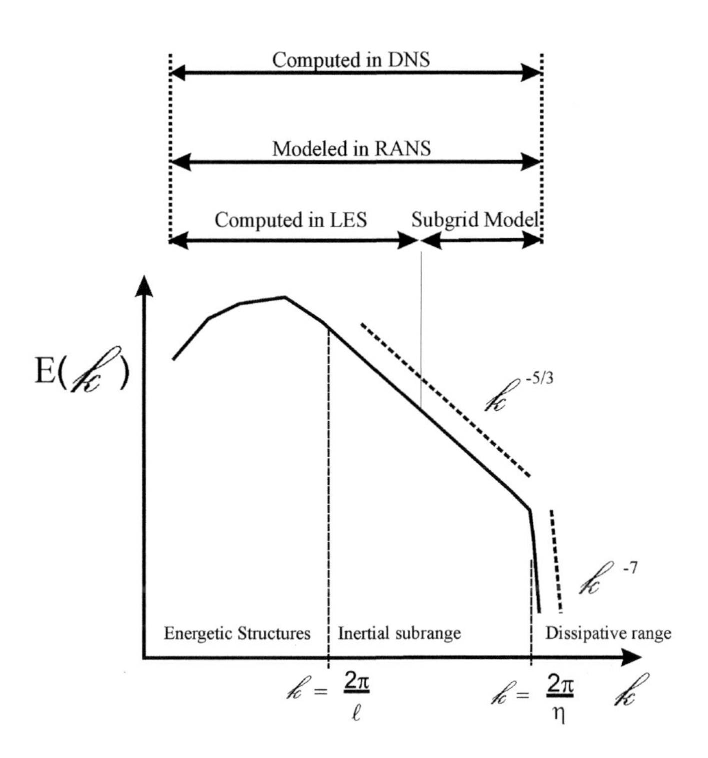

Figure 9.11 Energy spectrum vs. wave number. Various approaches used in CFD for turbulent flow computations.

9.6.1 FLOW THROUGH SQUARE DUCT

In this section we discuss the results for the LES simulation of turbulent flow through square duct using FASTEST code (FASTEST, 2005). Fully developed turbulent flow in square ducts is an intriguing flow because of its transverse (secondary) mean motion. The secondary motion is absent in laminar flow through a square duct; therefore, it must come from the turbulence. The complex flow through a square duct is characterized by the existence of secondary flows of the second kind (as classified by Prandtl) at corners. Although this secondary flow is weak and only about 2–3% of the stream-wise velocity, the effect on wall shear stress and heat transfer is still noticeable. Due to cross-stream gradients, eight stream-wise vortices, two counter-rotating at each corner, with the flow directed toward the corner from the duct center along the corner bisector and toward the duct center along the wall bisector, are created. If one wishes to correctly reproduce the weak secondary flow with a classical one-point-closure statistical modeling approach, then the Reynolds Averaged Navier-Stokes models needs to be employed (Demuren, 1984). It is interesting to note that LES is able to correctly reproduce such statistical quantities (Gavrilakis, 1992).

Table 9.2
Simulation Comparison

Parameters	DNS (Gavrilakis, 1992)	DNS (Huser and Biringen, 1993)	DNS (Vazquez and Metais, 2002)	LES (Madabushi, 1991)	Uddin (2011)
Re_τ	300	600	-	360	600
U_{CL}/U_b	1.33	-	1.29	1.278	1.24
$w_{max}/U_b(\%)$	1.9	-	1.92	2.58	2.12
$(\tau_w/\overline{\tau_w})_{max}$	1.18	1.14	1.11	-	1.14

Figure 9.12(a) shows the mean stream-wise velocity distribution. The DNS study shows that the isotachs bend toward the corners. The occurrence of local $\overline{u}$ maximum at the wall bisector is a low Reynolds number effect (Gavrilakis, 1992). The same is observed via LES simulation. The contours of the mean secondary flow velocity are shown in Figure 9.12(b). In Table 9.2, the simulation results are compared with the previous DNS and LES studies.

DNS simulation shows that $\tau_w/\overline{\tau_w}$ attains maximum at wall bisector. The comparison of the mean velocity profile with law of wall ($u/u_\tau = \frac{1}{\kappa}\ln y^+ + 5.5$) is shown in Figure 9.12(a). The contours of the secondary flow velocity (w/u_τ) in the square duct cross-section are shown in Figure 9.12(b). In a square duct flow the turbulent structures, called streaks, are present. The near wall streaks found through iso-velocity surfaces are plotted in Figure 9.13. The streaks gradually lift up away from the wall. A part of the streak is rapidly ejected from the near-wall region into the outer part of the flow. The duct cross-section shows that at the corners no streaky structure is found, as can be seen in the figure.

Figure 9.14 shows the distribution of law of wall at the wall bisector.

The wall shear stress in the square duct flow is plotted in Figure 9.15. The distribution at the duct corners and at the wall bisector has a waviness in it.

9.7 LAMINAR-TO-TURBULENT TRANSITION

Laminar-to-turbulent transition is a complex flow phenomenon. There is a huge amount of literature on this topic, so here we only discuss a little about it to give the reader a glimpse into its complexities.

Gore et al. (1990) discovered in their experiments on co-axial jets that the length of the laminar co-axial jets (ℓ_{lam}) is related with the wavelength of the first sinusoidal deformation of the jet (λ). It is observed that data were exhibiting a linear relationship between the two quantities, which motivated the research team to investigate the phenomenon in other flow types.

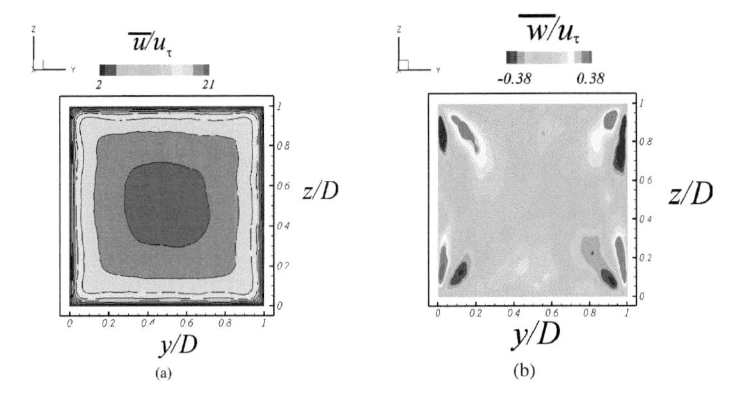

Figure 9.12 (a) Contours of streamwise velocity (u/u_τ) in the square duct cross-section. (notice the distorted isotachs at the corners); (b) contours of the secondary flow velocity (w/u_τ) in the square duct cross-section.

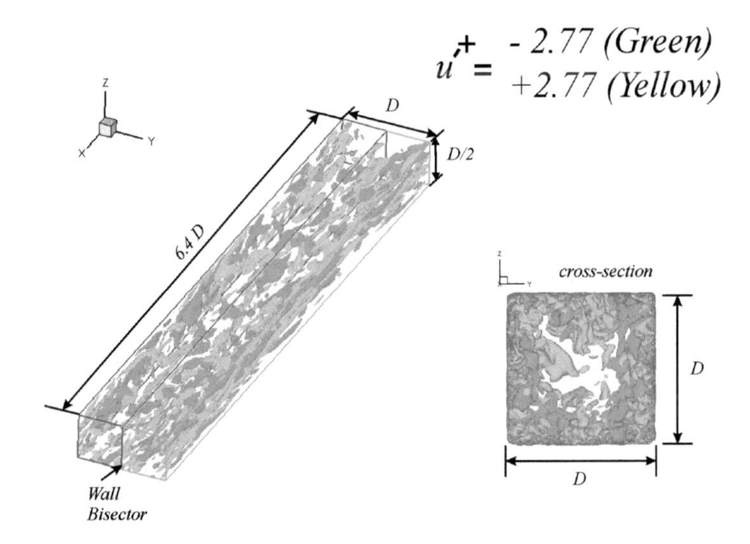

Figure 9.13 Near wall streaks in fully developed square duct.

Further analysis of the data revealed that $\ell_{lam}/\lambda_{lam} \approx \mathcal{O}(10)$ is valid for boundary layer-type flows (Gore et al., 1990).

When laminar flow undergoes transition it all starts with growing disturbances, which become the unstable two-dimensional disturbances called Tollmien-Schlichting waves. These disturbances are amplified in the downstream flow direction; however, it is also possible that these amplified waves are attenuated and the flow remains laminar. So a laminar flow is possible even beyond the traditional upper critical Reynolds number; however, the flow will abruptly change into turbulent

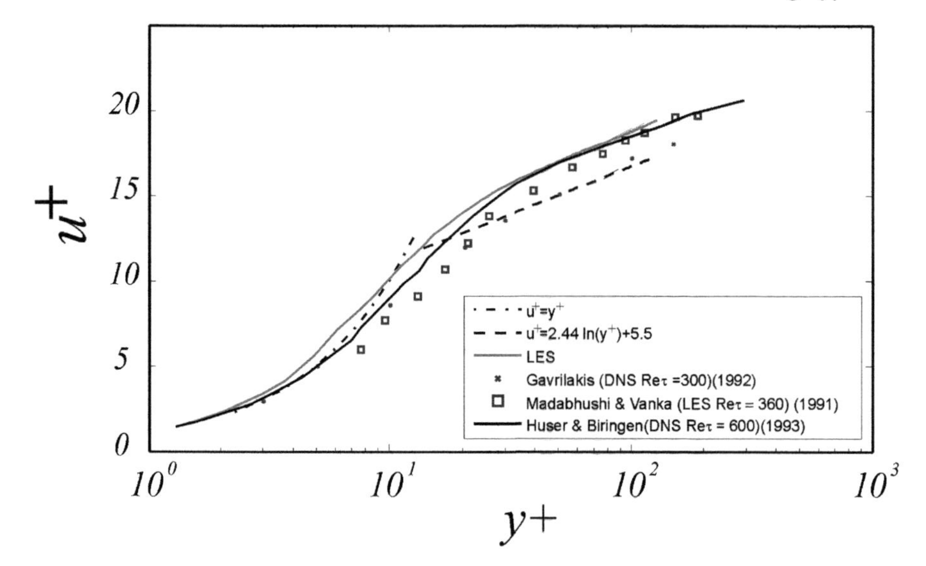

Figure 9.14 Law of the wall at the wall bisector.

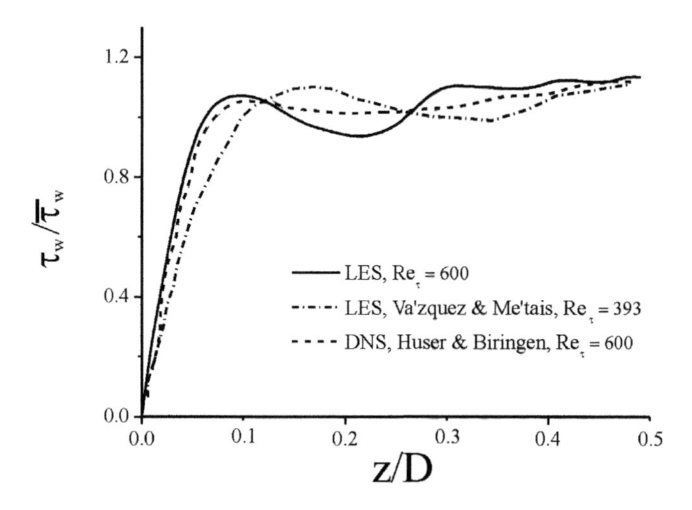

Figure 9.15 Distribution of the shear stress at the wall; $\overline{\tau_w}$ represent the average value of the stress.

flow with the slight disturbance. If the amplitude of the disturbances is large, a non-linear, instability mechanism will transform the Tollmien-Schlichting waves into a three-dimensional form, which leads to the evolution of the **hairpin vortices**. Figure 9.18 shows the transition stages for the flat plate boundary layer. As can be seen, there are spots of localized abrupt changes that occur randomly. These spots are also

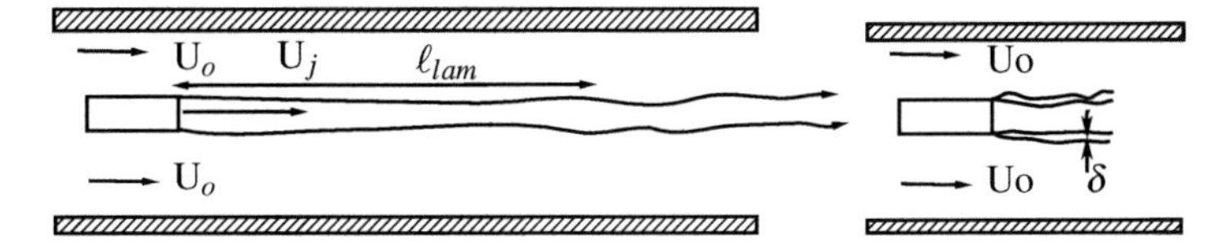

Figure 9.16 Co-axial jet with velocity U_j. The shear layer thickness is δ.

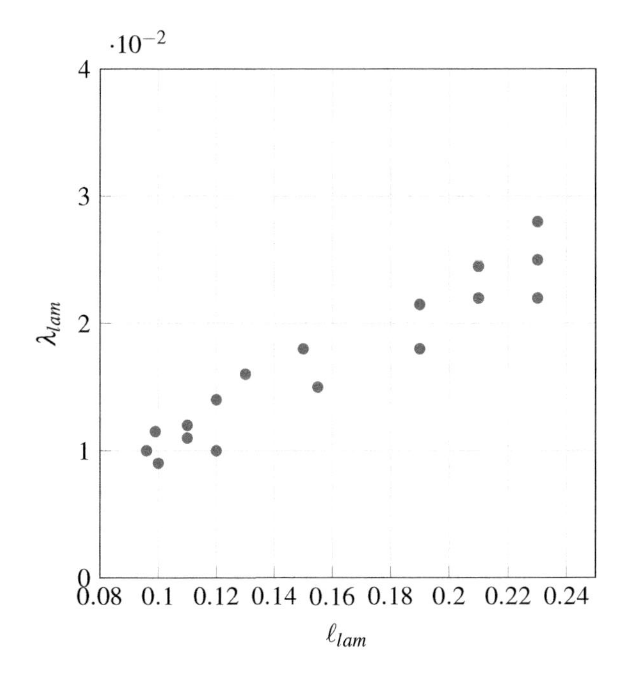

Figure 9.17 Laminar jet length vs size of the wavelength.

advected along the flow as these spots grow in size. These turbulent burst took over the whole flow field of transition boundary layer, and eventually the entire flow was transformed into turbulent flow.

PROBLEMS

9P-1 If $(.)'$ and $\overline{(.)}$ indicate the fluctuating and mean quantities, what is the outcome of an averaging operation using Reynolds-averaging on the following?

(i) $\overline{u'p'}$

(ii) $\overline{u'T'}$

(iii) $\overline{\overline{v}\,\overline{u}}$

(iv) $\overline{\overline{v}p'}$

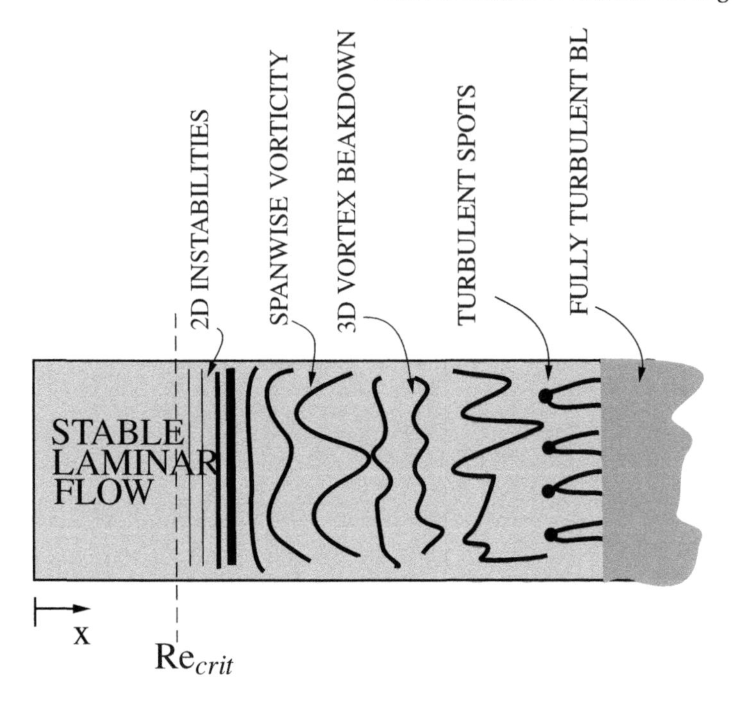

Figure 9.18 Sketch of transition process on flat plate boundary layer (top view).

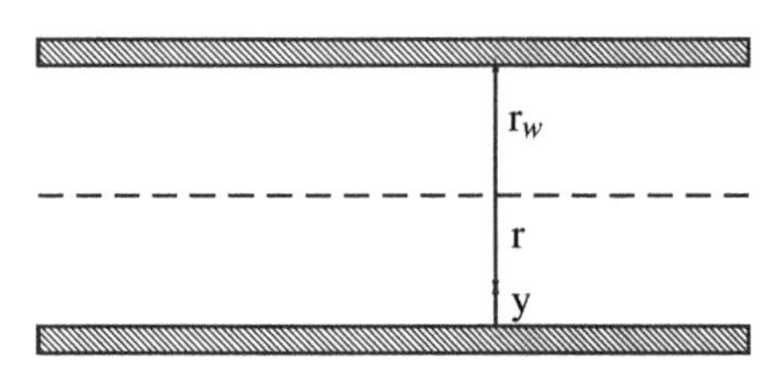

Figure 9.19 Figure of problem 9P-4.

9P-2 Air ($\nu = 1.5 \times 10^{-5}$, $\rho = 1.22$ kg/m^3) is flowing over a flat plate at a speed of $U = 30$ m/s. The flat plate length is 3 m. The shear stress can be calculated from

$$\frac{\tau_w}{\rho U_\infty^2} = \frac{0.0288}{\mathrm{Re}^{1/5}}$$

Find the height of the viscous sublayer and buffer zone from the flat surface in mm.

9P-3 At a given point in a flow field, the instantaneous velocity is given by

$$u_x(t) = 5 + 3 \cdot \sin(\pi t)$$

$$u_y(t) = 3 \cdot \sin(\pi t)$$

Find $\bar{u}_x$, $\bar{u}_y$, $\overline{u_x' u_y'}$, and $\overline{u_x u_y}$.

9P-4 It is found through experiments that turbulent flow velocity profile in a pipe can be expressed as

$$u = U_c \left(\frac{y}{r_w} \right)^{1/n}$$

where n is an index and it can vary from 6 to 9, and U_c is the pipe's centerline velocity. Find the ratio of mean velocity $\bar{u}$ with the centerline velocity in terms of n.

REFERENCES

D. Coles, The law of the wake in the turbulent boundary layer, J. Fluid Mech., vol. 1, pp. 191–226, 1956.

J. Nikuradse, Strömungsgesetze in rauhen rohren, Verein Deutscher Ingenieure, Forschungsheft vol. 361, p. 1–22, 1933.

F. H. Clauser, Turbulent boundary layers in adverse pressure gradients, J. Aero. Sci. 1954.

J. Laufer, The structure of turbulence in fully developed pipe flow, REPORT 1174, 1954.

P. S. Klebanoff, Characteristics of turbulence in a boundary layer with zero pressure gradient, NACA Report 1247, 1955.

G. I. Taylor, Proc. Royal Soc. London, 151A:421, (I-IV), 1935.

A. Demuren and W. Rodi, Calculation of turbulence-driven secondary motion in non-circular ducts, J. Fluid Mech. vol. 140, p. 189, 1984.

A. Huser and S. Biringen, Direct numerical simulation of turbulent flows in a square duct, J. Fluid Mech., vol. 257, 65–95, 1993.

M. S. Vazquez and O. Metais, Large-eddy simulation of the turbulent flow through a heated square duct, J. Fluid Mech., vol. 453, pp. 201–238, 2002.

S. Gavrilakis, Numerical simulation of low-Reynolds-number turbulent flow through a straight square duct. J. Fluid Mech., vol. 244, p.101, 1992.

R. K. Madabushi and S. P. Vanka, Large eddy simulation of turbulence driven secondary flow in a square duct, Phy. Fluids A, vol. 3, no. 11, p. 2734, 1991.

H. T. Kim, S. J. Kline, and W. C. Reynolds, J. Fluid Mech., vol. 50, p. 133, 1971.

R. F. Blackwelder and J. H. Haritonidi, J. Fluid Mech., vol. 152, p. 87, 1983.

W. W. Willmarth and L. K. Sharma, J. Fluid Mech., vol. 142, p. i21, 1984.

Robert A. Gore, Clayton T. Crowe, and Adrian Bejan, The geometric similarity of the laminar sections of boundary layer-type flows, vol. 17, no. 4, pp. 465–475, 1990.

Naseem Uddin, LES of Flow through Square Duct, Institute of Aerospace Thermodynamics, Universitaet Stuttgart, Germany, 2011.

P. E. Roach, The generation of nearly isotropic turbulence by means of grids, Heat and Fluid Flow, vol. 8, no. 2, pp. 82–92, 1986.

FASTEST User Manual, 2005, Department of Numerical Methods in Mechanical Engineering, Technische Universität Darmstadt.

10 Viscous Flow through Conduits

The water we use in commercial and residential housing is pumped through pipes. Extensive piping networks distribute this water with much energy consumed in overcoming the friction through pipes or conduits. Flow-through pipes or conduits are crucial for household and industrial use; oil and natural gas are transported hundreds of miles. Inside the human body, blood is pumped by the heart in arteries and veins. The air from the trachea goes to the lungs. These are all examples of natural and artificial pipe flows.

Learning outcomes: After finishing this chapter, you should be able to:

- Calculate entrance length.
- Calculate the frictional pressure loss for both laminar and turbulent flows.
- Understand the concept of hydraulic diameter and its limitations.
- Compute the pressure loosses associated with entrance, and abrupt expansion and contraction of pipes
- Compute pressure drop in coiled ducts.

We will start this chapter with a discussion of laminar and turbulent flows, the flow in entrance length, and the fully developed flow. We will discuss different empirical relations using entrance length estimation. After that, we review the friction factors proposed for pipe flows, followed by major and minor loss calculations. The industrial piping network is a complex flow network involving pipes, valves, elbows, abrupt expansions, or contractions (see Figure 10.1).

10.1 LAMINAR AND TURBULENT DIFFUSION

The character of flow in a round pipe depends on four variables: fluid density ρ, fluid viscosity μ, pipe diameter D, and average velocity of flow $\bar{u}$. Osborne Reynolds has shown through his famous dye experiment that flow has three distinct characteristics. When he injected dye in the flow, the dye did not diffuse, and he named this flow layered flow. In another type of flow, the dye completely diffused into the flow, and he called this flow a sinuous flow. The latter is now called turbulent flow, and the former is called laminar flow. Between these two types of flows, there is another type of flow, called the transition flow. In this famous Reynolds experiment, Osborne Reynolds employed the streakline visualization, as he injected a colored dye into

DOI: 10.1201/9781003315117-10

Figure 10.1 The piping network.

water that was flowing in a pipe at different speeds. In turbulent flow, the diffusion of dye was strong and the streakline was immediately lost into the diffusing mixture of water and dye. However, in laminar flow, the streakline was maintained and could be visualized by the naked eye.

What is crucial for us is the dimensionless number, called the Reynolds number:

$$\mathrm{Re}_D = \frac{\rho \bar{u} D}{\mu}$$

$$\mathrm{Re}_D = \frac{D\bar{u}}{\nu}$$

where mean velocity can be estimated as:

Mean Velocity

$$\bar{u} = \frac{1}{A}\int_A u_z(r)dA \tag{10.1}$$

It has been found that flow has lower and upper critical Reynolds numbers:

$$2300 < \mathrm{Re}_D < 4000$$

In practical calculations we consider the step transition and treat flow as either laminar or turbulent. So if

$$\mathrm{Re}_D < 2300 \quad \textit{Laminar}$$
$$\mathrm{Re}_D > 2300 \quad \textit{Turbulent}$$

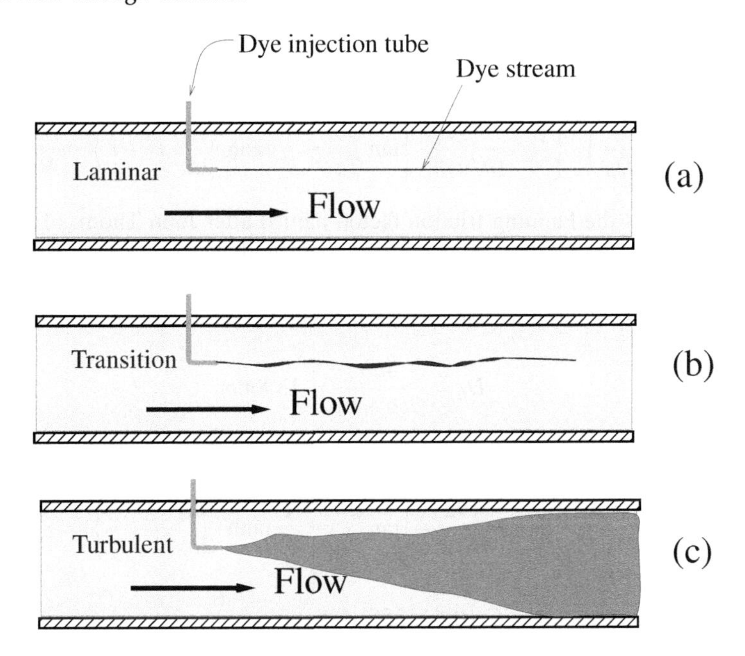

Figure 10.2 Dye injection into (a) Laminar (b) Transition and (c) Turbulent flows.

Example 10.1

Example Consider the flow between two infinitely long parallel plates. In the cross-section, the width of the plate is "b" and the gap between the plates is "a." Find the hydraulic diameter.

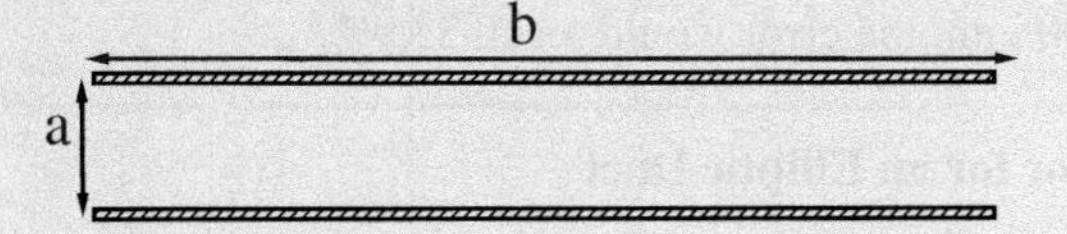

Figure 10.3 Schematic diagram of Example 10.1.

Solution The hydraulic diameter will be:

$$D_h = \frac{4A}{WP} = \frac{4(a.b)}{2b} = 2a$$

Example 10.2

Example Flow is passing through a rectangular conduit with sides $a = 8.9$ mm and $b = 30.8$ mm. Cornish (1928) proposed the relation for friction factor for flow through rectangular ducts of sides a and b:

$$C_f\left(\frac{a}{D_h}\right)^2\left\{1-\frac{192}{(b/a)\pi^5}\left[\tan\frac{\pi b}{2a}-\frac{1}{3^5}\tanh\frac{3\pi b}{2a}+\cdots\right]\right\}=\frac{6}{\text{Re}}$$

where C_f is the Fanning friction factor, named after John Thomas Fanning, and $a < b$. Find the friction factor and estimate the error involved if the friction factor for a circular duct is used along with hydraulic diameter.

Solution The hydraulic diameter for the rectangular duct is:

$$D_h=\frac{4ab}{2(a+b)}=13.8mm$$

The formula gives

$$C_f\left(\frac{a}{D_h}\right)^2\left\{1-\frac{192}{(b/a)\pi^5}\left[\tan\frac{\pi b}{2a}-\frac{1}{3^5}\tanh\frac{3\pi b}{2a}+\cdots\right]\right\}=\frac{6}{\text{Re}}$$

$$C_f(0.415355)(0.8182)=\frac{6}{\text{Re}}$$

$$C_f=\frac{17.65}{\text{Re}}$$

The Darcy friction factor for rectangular duct is $f=4C_f=70.61/\text{Re}$. The Darcy friction factor for a circular duct is $f=64/\text{Re}$.

$$\%\,error=\frac{(70.61-64)}{64}\times 100=10.33\%$$

This error would increase if one of the side lengths increased, for example, for $b=100$ mm, the error would reach 33.9%.

Friction Factor for an Elliptic Duct

$$f=\frac{64}{\text{Re}}\left\{\left(\frac{1+(b/a)^2}{2}\right)\left(\frac{2\pi a}{\zeta}\right)^2\right\}$$

$$\chi=1-\left(\frac{b}{a}\right)^2$$

$$\zeta=2\pi a\left[1-\frac{\chi^2}{4}-\frac{3\chi^4}{64}-\frac{15\chi^6}{2304}+\cdots\right]$$

The hydraulic diameter and Reynolds number can be calculated as

$$D_h=\frac{4\pi ab}{\zeta}$$

$$\text{Re}=\frac{\rho u D_h}{\mu}$$

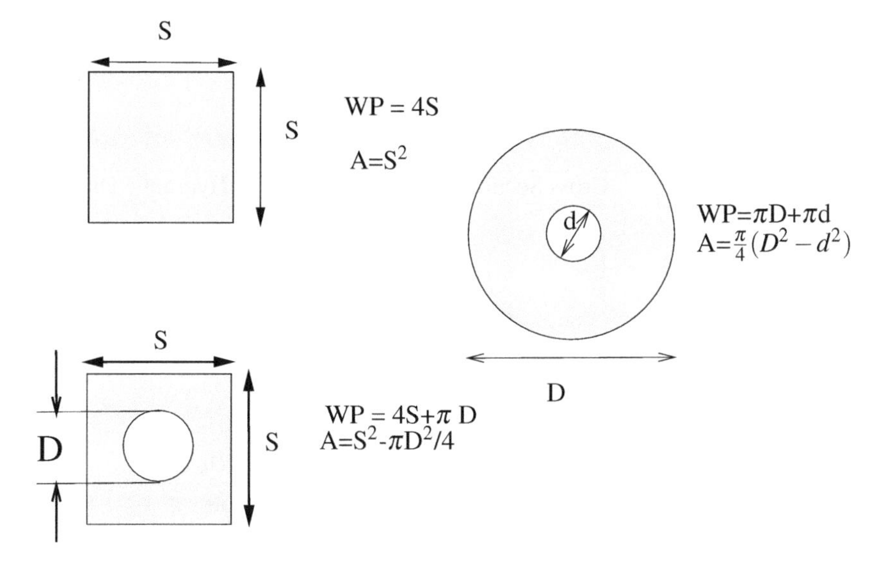

Figure 10.4 Cross-section of some noncircular conduits.

10.2 NONCIRCULAR CONDUITS

In many industrial applications, the pipes are not circular and may have different cross-sections, as shown in Figure 10.4. In such cases, the correlations and charts available in this chapter are still applicable if we calculate the hydraulic diameter as:

$$D_h = \frac{4A}{WP}$$

where D_h is called the hydraulic diameter, A is the cross-section of the noncircular duct, and WP is the wetted perimeter, i.e., the perimeter touched by the fluid. In half-filled liquid ducts, this has to be calculated with care. However, in the case of gases, the wetted perimeter will be the complete perimeter in the cross-section, as gas will spread out and touch all the walls.

Note that the use of hydraulic diameter for the sharp-edged cross-sections like triangular, square, and cusped ducts may lead to significantly unacceptable errors of the order of 35% in turbulent flows friction factors if determined from the circular pipe flow correlations. The error is not substantial for cross-sections that do not have sharp edges. Also, the hydraulic diameter concept gives acceptable results for turbulent flow, but it is not a very reliable approach for laminar flow. Therefore, it is advised to consult the handbooks for sharp-edged cross-sections for improved and accurate results.

10.3 ENTRANCE LENGTH IN LAMINAR FLOWS

We now consider the fluid entering a circular pipe or a duct at a uniform velocity. The fluid in contact with the stationary wall will attain zero velocity. The layer next to

Table 10.1
Hydraulic Diameter of Some Cross-Sections

Name	Cross Section	Area	Hydraulic Diameter
Circle	D	A = $\frac{\pi}{4}D^2$	D
Semicircle	2R	A = $\frac{\pi}{2}R^2$	$D_h = \frac{2\pi R}{\pi+2}$
Sector	R, θ	A = $\frac{\theta}{2}R^2$	$D_h = \frac{2\theta R}{\theta+2}$
Trapezoid	a, b, c, θ	A = $\frac{b}{2}(a+c)$	$D_h = \frac{2b(a+c)}{a+b+c}$
Isosceles Triangle	a, a, θ	A = $\frac{a^2}{2}\sin\theta$	$D_h = \frac{a\sin\theta}{1+\sin\left(\frac{\theta}{2}\right)}$

the surface comes to rest due to wall friction. As the mass flow is always conserved, the flow in the core region will be sped up. This also happens as the boundary layer develops on the wall, and the fluid in the inner core and close to the center starts to be squeezed and sped up. Therefore, there are two regions in the entrance region: the flow velocity and the flow near the wall, where flow is experiencing the boundary layer phenomenon. The zone from the pipe or duct inlet to where the boundary layer combines at the centerline is called the hydrodynamic entrance zone, and the length of this inlet zone is called the hydrodynamic entrance length.

Once the flow becomes fully developed, the velocity profile will become independent of the length of pipe or duct, and mathematically we can write

$$\frac{du}{dx} = 0$$

Laminar Flow Entrance Length Estimation Following are some of the correlations shown in the literature:

$$\left(\frac{L_e}{D}\right)_{lam} \simeq 0.065\text{Re}_D \qquad \textit{Boussinesq} \; \textit{Nikuradse}$$

$$\left(\frac{L_e}{D}\right)_{lam} \simeq 0.06\text{Re}_D \qquad \textit{Asao et al.}$$

Chen et al. (1973) have derived the entrance length for pipe and channel flow valid in a range of $1 < \text{Re} < 2000$.

$$\left.\begin{aligned}\left(\tfrac{L_e}{D}\right)_{lam} &\simeq 0.061\text{Re} + \tfrac{0.72}{0.04\text{Re}+1} \quad (pipe)\\ \left(\tfrac{L_e}{D}\right)_{lam} &\simeq 0.053\text{Re} + \tfrac{0.79}{0.04\text{Re}+1} \quad (channel)\end{aligned}\right\} \text{Chen}(1973)$$

10.4 FRICTION IN THE LAMINAR HYDRODYNAMIC ENTRY LENGTH

In the previous sections we focused attention on the velocity distribution and the friction coefficient at the location that is far from the pipe entrance, where we can consider the flow as fully developed. In case of constant properties flow through a circular pipe the valid differential momentum equation is

$$\rho\left(\frac{\partial u_z}{\partial t} + u_r\frac{\partial u_z}{\partial r} + \frac{u_\theta}{r}\frac{\partial u_z}{\partial \theta} + u_z\frac{\partial u_z}{\partial z}\right)$$

$$= \frac{-dP}{dz} + \rho g_z + \mu\left[\frac{1}{r}\frac{\partial}{\partial r}\left(r\frac{\partial u_z}{\partial r}\right) + \frac{1}{r^2}\left(r\frac{\partial^2 u_\theta}{\partial \theta^2}\right) + \frac{\partial^2 u_z}{\partial z^2}\right]$$

Applying assumptions of steady, fully developed flow with no secondary flow, swirl and body force field effect we have:

$$\rho\left(\cancel{\frac{\partial u_z}{\partial t}} + u_r\frac{\partial u_z}{\partial r} + \cancel{\frac{u_\theta}{r}\frac{\partial u_z}{\partial \theta}} + u_z\frac{\partial u_z}{\partial z}\right)$$

$$= \frac{-dP}{dz} + \cancel{\rho g_z} + \mu\left(\frac{1}{r}\frac{\partial}{\partial r}\left(r\frac{\partial u_z}{\partial r}\right) + \cancel{\frac{1}{r^2}\left(r\frac{\partial^2 u_\theta}{\partial \theta^2}\right)} + \cancel{\frac{\partial^2 u_z}{\partial z^2}}\right)$$

The reduced form of the above equation is:

$$\rho\left(u_r\frac{\partial u_z}{\partial r} + u_z\frac{\partial u_z}{\partial z}\right) = \frac{-dP}{dz} + \mu\left[\frac{1}{r}\frac{\partial}{\partial r}\left(r\frac{\partial u_z}{\partial r}\right)\right]$$

For sake of further simplification, we write u_z as u only in this section, leading to the form:

$$\rho\left(u\frac{\partial u_r}{\partial r} + u\frac{\partial u}{\partial z}\right) + \frac{dP}{dz} = \mu\left[\frac{1}{r}\frac{\partial}{\partial r}\left(r\frac{\partial u_z}{\partial r}\right)\right]$$

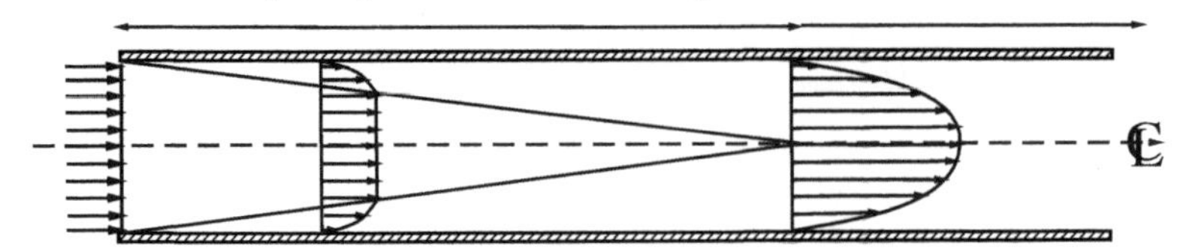

Figure 10.5 Entrance length in pipe flow.

The simplest entry condition would be a uniform floor at the pipe entrance, $x = 0$ as shown in Figure 10.5.

The apparent friction factor will be equal to the pressure drop due to wall friction plus momentum flux change, as the fluids velocity profile changes from entrance flow to the fully developed profile. For the fully developed flow, the Darcy equation is used to calculate pressure drop:

$$\Delta p = f\frac{L}{D}\frac{\rho u_m^2}{2}$$

For this flow the modified Darcy equation will be used:

$$\Delta p_{app} = 4(C_f)_{app}\frac{L}{D}\frac{\rho u_m^2}{2}$$

Boundary condition:

(i) At inlet, $z = 0$, $u = U_o$, $u_r = 0$

(ii) At wall no-slip condition applied, $r = r_o$, $u = 0$, and $u_r = 0$

(iii) At center of pipe, $r = 0$, $du/dr = 0$

Continuity equation in cylindrical coordinates is:

$$\frac{\partial(\rho u)}{\partial z} + \frac{1}{r}\frac{\partial}{\partial r}(\rho r u_r) + \frac{\partial \rho}{\partial t} = 0$$

Considering steady, incompressible flows:

$$\frac{\rho\partial(u)}{\partial z} + \frac{\rho}{r}\frac{\partial}{\partial r}(r u_r) + \cancel{\frac{\partial \rho}{\partial t}} = 0 \tag{10.2}$$

$$\frac{\partial(u)}{\partial z} + \frac{1}{r}\frac{\partial}{\partial r}(r u_r) = 0 \tag{10.3}$$

multiply continuity Eq. 10.3 by $\rho u/r$:

$$(\rho u)\frac{\partial(u)}{\partial z} + \frac{(\rho u)}{r}\frac{\partial}{\partial r}(r u_r) = 0\frac{(\rho u)}{r}\frac{\partial}{\partial r}(r u_r) = -(\rho u)\frac{\partial(u)}{\partial z} \tag{10.4}$$

Consider a special derivative that will be later substituted into momentum equation:

$$\frac{\partial(r\rho uu_r)}{\partial r}$$

which can be expanded like:

$$\frac{\partial(r\rho uu_r)}{\partial r} = (\rho u)\frac{\partial(ru_r)}{\partial r} + (ru_r)\frac{\partial(\rho u)}{\partial r}$$

Divide the whole equation by r:

$$\frac{1}{r}\frac{\partial(r\rho uu_r)}{\partial r} = \frac{(\rho u)}{r}\frac{\partial(ru_r)}{\partial r} + (u_r)\frac{\partial(\rho u)}{\partial r}$$

$$\underbrace{\frac{1}{r}\frac{\partial(r\rho uu_r)}{\partial r}}_{I} = \underbrace{\frac{(\rho u)}{r}\frac{\partial(ru_r)}{\partial r}}_{II} + \underbrace{(u_r)\frac{\partial(\rho u)}{\partial r}}_{III}$$

Term II is replaced by the term in Eq. 10.4:

$$\underbrace{\frac{1}{r}\frac{\partial(r\rho uu_r)}{\partial r}}_{I} = -(\rho u)\frac{\partial(u)}{\partial z} + \underbrace{(u_r)\frac{\partial(\rho u)}{\partial r}}_{III}$$

$$(\rho u)\frac{\partial(u)}{\partial z} = \underbrace{\frac{-1}{r}\frac{\partial(r\rho uu_r)}{\partial r}}_{I} + \underbrace{(u_r)\frac{\partial(\rho u)}{\partial r}}_{III}$$

The above equation will be substituted into the momentum equation.

Momentum equation:

$$\frac{\mu}{r}\frac{\partial}{\partial r}\left(r\frac{\partial u}{\partial r}\right) = \rho u\frac{\partial u}{\partial z} + \rho u_r\frac{\partial u}{\partial r} + \frac{\partial P}{\partial z}$$

$$\frac{\mu}{r}\frac{\partial}{\partial r}\left(r\frac{\partial u}{\partial r}\right) = \underbrace{\frac{-1}{r}\frac{\partial(r\rho uu_r)}{\partial r}}_{I} + \underbrace{(u_r)\frac{\partial(\rho u)}{\partial r}}_{III} + \rho u_r\frac{\partial u}{\partial r} + \frac{\partial P}{\partial z}$$

$$\frac{\mu}{r}\frac{\partial}{\partial r}\left(r\frac{\partial u}{\partial r}\right) = \frac{-1}{r}\frac{\partial(r\rho uu_r)}{\partial r} + (u_r)\frac{\partial(\rho u)}{\partial r} + \rho u_r\frac{\partial u}{\partial r} + \frac{\partial P}{\partial z}$$

$$\frac{\mu}{r}\frac{\partial}{\partial r}\left(r\frac{\partial u}{\partial r}\right) = \frac{-1}{r}\frac{\partial(r\rho uu_r)}{\partial r} + 2\rho u_r\frac{\partial u}{\partial r} + \frac{\partial P}{\partial z} \tag{10.5}$$

Mathematically, we can prove that

$$2\rho u_r\frac{\partial u}{\partial r} = \frac{\partial(\rho uu)}{\partial z} + \frac{2}{r}\frac{\partial(r\rho uu_r)}{\partial r} \tag{10.6}$$

Using the shear stress relationship for two-dimensional flows:

$$\tau_{rz} = \mu \left(\frac{\partial u}{\partial r} + \frac{\partial u_r}{dz} \right)$$

we can form the following equation by multiplying stress stress with r and then differentiating with r:

$$\frac{1}{r}\frac{\partial(r\tau_{rz})}{\partial r} = \mu \frac{\partial^2 u_r}{\partial r \partial z} + \frac{\mu}{r}\frac{\partial(u_r)}{\partial z} + \frac{\mu}{r}\frac{\partial}{\partial r}\left(r\frac{\partial u}{\partial r}\right)$$

or

$$\frac{\mu}{r}\frac{\partial}{\partial r}\left(r\frac{\partial u}{\partial r}\right) = \frac{1}{r}\frac{\partial(r\tau_{rz})}{\partial r} - \mu \frac{\partial^2 u_r}{\partial r \partial z} - \frac{\mu}{r}\frac{\partial(u_r)}{\partial z} \tag{10.7}$$

The terms from Eq. 10.6 and Eq. 10.7 will be substituted into Eq. 10.5:

$$\left[\frac{1}{r}\frac{\partial(r\tau_{rz})}{\partial r} - \mu \frac{\partial^2 u_r}{\partial r \partial z} - \frac{\mu}{r}\frac{\partial(u_r)}{\partial z}\right] = \frac{-1}{r}\frac{\partial(r\rho u u_r)}{\partial r} + \left[\frac{\partial(\rho u u)}{\partial z} + \frac{2}{r}\frac{\partial(r\rho u u_r)}{\partial r}\right] + \frac{\partial P}{\partial z}$$

The term $\frac{\partial(u_r)}{\partial z}$ is ignored, as it tends to zero as flow develops, leading to the form:

$$\left[\frac{1}{r}\frac{\partial(r\tau_{rz})}{\partial r}\right] = \frac{1}{r}\frac{\partial(r\rho u u_r)}{\partial r} + \frac{\partial P}{\partial z} + \frac{\partial(\rho u u)}{\partial z} \tag{10.8}$$

This equation is now double integrated in length and cross-section as:

$$\int_0^z \left\{ \int_0^{r_o} \left\{ \left[\frac{1}{r}\frac{\partial(r\tau_{rz})}{\partial r}\right] = \frac{1}{r}\frac{\partial(r\rho u u_r)}{\partial r} + \frac{\partial P}{\partial z} + \frac{\partial(\rho u u)}{\partial z} \right\} 2\pi r dr \right\} dz$$

Defining the cross sectional area (A_c) as:

$$A_c = \frac{\pi}{4} d_H^2$$

Integration leads to form:

$$(p_o - p_z)A_c = \int_{A_c} \rho u^2 dA_c - \rho U_o^2 A_c + \frac{A_c}{d_H/4}\int_0^z \tau_o dz$$

$$\frac{(p_o - p_z)}{\frac{1}{2}\rho U_o^2} = \frac{\int_{A_c} \rho u^2 dA_c}{\frac{A_c}{2}\rho U_o^2} - \frac{\rho U_o^2 A_c}{\frac{1}{2}\rho U_o^2} + \frac{A_c \int_0^z \tau_o dz}{(d_H/4)\frac{1}{2}\rho U_o^2}$$

The mean skin-friction coefficient is defined as:

$$C_{fm} = \frac{1}{z}\int_0^x \frac{\tau_o}{\frac{1}{2}\rho \bar{u}^2} dz$$

$$(C_f)_{app} = \frac{(p_o - p_z)}{\frac{1}{2}\rho U_o^2}$$

$$(C_f)_{app} = \frac{\Delta p}{\frac{1}{2}\rho U_o^2} = \frac{2}{A_c}\int_{A_c}\left(\frac{u}{U_o}\right)^2 dA_c - 2 + \frac{z}{(d_H/4)}C_{fm}$$

$$(C_f)_{app} = \frac{(d_H/4)}{z}\left[\left(\frac{2}{A_c}\int_{A_c}\left(\frac{u}{U_o}\right)^2 dA_c - 2\right)\right] + C_{fm} \qquad (10.9)$$

The momentum flux change will cause the change in pressure, which is indicated in the first term on the right-hand side. The second term indicates the pressure drop due to friction. This explains the increase in friction in the entrance region of the pipe or duct.

There are two definitions of friction factors used in the literature. One is named after American architect and hydraulic engineer John Thomas Fanning (1837–1911) and another one is named after French engineer Henry Philibert Gaspard Darcy (1803–1858).

Fanning Friction Factor

$$C_f = \frac{\tau_w}{\frac{1}{2}\rho u^2}$$

Darcy Friction Factor

$$f = \frac{\Delta p}{\frac{1}{2}\rho u^2}\left(\frac{D}{\ell}\right)$$

10.5 ENTRANCE LENGTH IN TURBULENT FLOWS

Turbulent boundary layer development in the entrance region of the pipe takes longer than the laminar boundary length development inside the entrance region. The rough estimate is that the turbulent entrance length is from 70 to 100 times the pipe diameter. Following are some of the correlations in the literature:

Turbulent Flow Entrance Length Estimation

$$\left(\frac{L_e}{D}\right)_{turb} \simeq 0.693\mathrm{Re}_D^{1/4} \qquad Latzko\ (1944)$$

$$\left(\frac{L_e}{D}\right)_{turb} \simeq 1.359\mathrm{Re}_D^{1/4} \qquad Zhi-qing(1982)$$

$$\left(\frac{L_e}{D}\right)_{turb} \simeq 4.4\mathrm{Re}_D^{1/6} \qquad \textit{Nikuradse}(1933)$$

$$\left(\frac{L_e}{D}\right)_{turb} \simeq 14.2\log_{10}\mathrm{Re} - 46 \quad \mathrm{Re} > 10,000 \quad \textit{Bowlus \& Brighton}(1968)$$

Figure 10.6 shows different correlations proposed for the estimation of entrance length for turbulent pipe flows. The predictions are compared with experimental data of Barbin and Jones (1963) and reveal that most of the correlations are giving the same result, except the Latzko correlation.

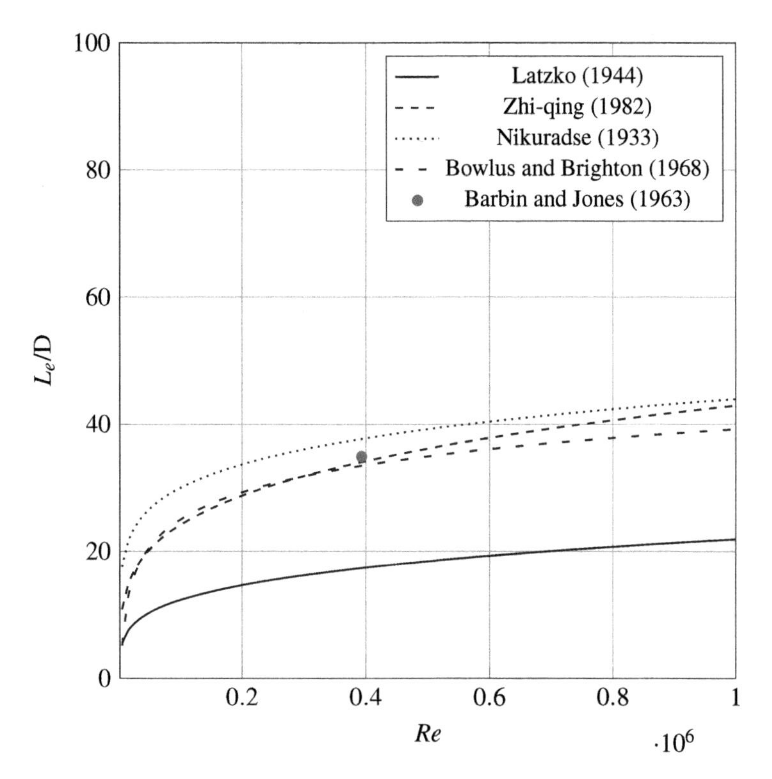

Figure 10.6 Different proposals for entrance length in turbulent pipe flows compared with experimental data of Barbin and Jones (1963).

Example 10.3

Example An organic fluid enters in a circular pipe through a well-rounded entrance whose inside diameter is 3 in. The volume flow rate of organic fluid is 0.05 ft^3/s. Calculate the entrance length by which the flow can be considered as fully developed. The organic fluid has density 1.53 $slug/ft^3$, $\mu = 1.2 \times 10^{-4}$ $lb_f.s/ft^2$.

Solution

$$A = \frac{\pi}{4}D^2 = \frac{\pi}{4}(3/12)^2 = 0.049\ ft^2$$

$$u = \frac{Q}{A} = \frac{0.05}{0.0490} = 1.0191\ ft/s$$

$$Re = \frac{\rho \cdot u \cdot D}{\mu} = \frac{1.53 \times 1.0191 \times (3/12)}{1.2 \times 10^{-4}} = 3248.4$$

The flow is turbulent, as Re > 2300.
Using Nikuradse correlation, we have

$$L_e = 4.4 \times D \times \text{Re}^{1/6} = 4.4\left(\frac{3}{12}\right)(3248.4)^{1/6} = 4.23\ ft$$

Using Latzko correlation, we have

$$L_e = 0.693 \cdot D \cdot \text{Re}^{\frac{1}{4}} = 1.3079\ ft$$

The correlation of Nikuradse is the most widely used.

10.6 THE DARCY-WEISBACH EMPIRICAL EQUATION

The pressure loss (or major loss) in a pipe, tube, or duct can be calculated using an empirical equation:

$$\Delta p = f\left(\frac{\ell}{D}\right)\left(\frac{\rho \bar{u}^2}{2}\right)$$

The equation is named after French engineer Henri Philibert Gaspard Darcy and German engineer Julius Ludwig Weisbach. Here ℓ is the pipe length, D is the diameter of the pipe, and $\bar{u}$ is the mean velocity of the flow. The quantity Δp is also called the major loss in literature. The minor contribution comes from the bends, elbows, fittings, expansions and contractions, etc.

10.7 SMOOTH PIPE'S DARCY FRICTION FACTORS

Blasius correlation:

$$f = \frac{0.3164}{\text{Re}^{1/4}} \qquad (\text{Re} = 4 \times 10^3 - 2 \times 10^5)$$

Nikuradse correlation I:

$$f = 0.0032 + \frac{0.221}{\mathrm{Re}^{0.237}} \qquad (\mathrm{Re} = 10^5 - 3\times10^6)$$

Karman-Nikuradse correlation:

$$f = \frac{1}{\left[2\log_{10}(\mathrm{Re}\sqrt{f}) - 0.81\right]^2} \qquad (\mathrm{Re} = 3\times10^3 - 3\times10^6)$$

Itaya correlation:

$$f = \frac{0.314}{0.7 - 1.65\log_{10}(\mathrm{Re}_D) + (\log_{10}\mathrm{Re})^2}$$

Nikuradse correlation II:

$$\frac{1}{\sqrt{f}} = 4\log(\mathrm{Re}\sqrt{f}) - 0.4$$

McAdams correlation:

$$f \simeq \frac{0.184}{\mathrm{Re}^{1/5}} \qquad (\mathrm{Re} = 3\times10^4 - 10^6)$$

Bhatti and Shah correlations:

$$C_f = A + \frac{B}{\mathrm{Re}_D^{1/m}}$$

Flow	A	B	m
Laminar	0	16	1
Transitional Flow	0.0054	2.30E-08	-0.666
Turbulent Flow	1.28E-03	0.1143	3.2154

Techo, Tickner, and James correlation:

$$\frac{1}{\sqrt{C_f}} = 1.7372\ln\left[\frac{\mathrm{Re}}{1.964\ln(\mathrm{Re}) - 3.8215}\right]$$

This correlation is defined using the Fanning's definition of the friction factor, but it covers the wide range of Reynolds numbers.

Figure 10.7 shows the plot of various friction factor correlations for smooth pipe.

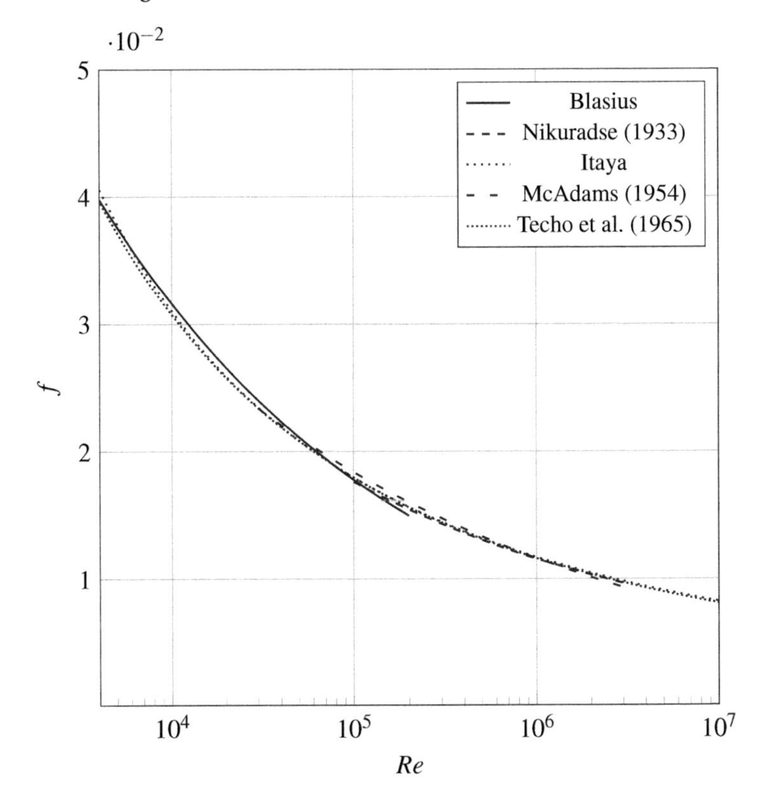

Figure 10.7 Friction factor for turbulent flow through smooth pipe plotted in their validity ranges shows that they give similar results.

Example 10.4

Example Water is flowing through a smooth tube of 10 mm internal diameter and length 5 m. Find the pressure drop in the tube, if:

(i) The mean velocity is 0.10 m/s.

(ii) The mean velocity is 10 m/s.

The viscosity and density of water at 25°C are 8.91×10^4 Pa · s and 997 kg/m^3.

Solution (i) **Water flow with the mean velocity of 0.10 m/s:**
We first calculate the Reynolds number as:

$$\text{Re}_D = \frac{\rho D\bar{u}}{\mu} = \frac{997 \times 10 \times 10^{-3} \times 0.1}{8.91 \times 10^{-4}} = 1118.96$$

As Re < 2300, the flow is laminar. We now estimate the friction factor as:

$$f = 64/Re_D = 0.0571$$

The pressure drop is:

$$\Delta p = f\left(\frac{\ell}{D}\right)\left(\frac{\rho \bar{u}^2}{2}\right) = 142.56Pa$$

(ii) **Water flow with the mean velocity of 10 m/s:**
We calculate the Reynolds number:

$$\text{Re}_D = \frac{\rho D \bar{u}}{\mu} = 1.118 \times 10^5$$

As Re > 2300, the flow is turbulent. We now estimate the friction factor as:

$$f = \frac{0.3164}{\text{Re}_D^{1/4}} = \frac{0.3164}{(1.118 \times 10^5)^{1/4}} = 0.017299$$

The pressure drop is:

$$\Delta p = f\left(\frac{\ell}{D}\right)\left(\frac{\rho \bar{u}^2}{2}\right) = 4.31188 \times 10^5 Pa$$

Figure 10.8 shows the meaning of roughness in the pipe. The pipe is considered smooth as long as the roughness is submerged in the viscous sublayer. If the wall roughness is large and it protrudes out of the viscous sublayer, pipe is considered as rough. The flow in pipe can be divided into three different flow regimes:

(i) **Hydraulically smooth:** In this regime, the roughness is fully submerged inside the viscous sublayer, and thus roughness is not influencing the flow.

(ii) **Hydraulically rough:** In this regime, the roughness is not influencing the flow, and friction factor is solely the function of the Reynolds number.

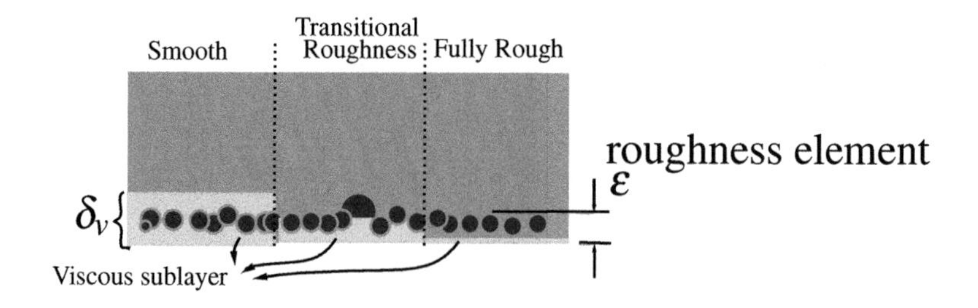

Figure 10.8 The roughness in the pipes.

Table 10.2
Roughness (ε) for Some Pipe Materials

Description	Condition	ε (mm)
Seamless Metal	New drawn copper, brass, or lead	0.0015-0.01
	New drawn aluminum	0.015-0.06
	Commercial steel, light rust	0.02-0.1
	Commercial steel, mildly rusted	0.15-1
	Commercial steel, heavily rusted	>5
Welded Steel	New, smooth, enameled surface	0.006-0.06
	Very poor condition	10-3.5
Iron	New wrought	0.045
	Cast	0.1-1
	Cast, corroded	2-4
Sheet Metal	Ducts with smooth joints	0.02-0.1
Concrete	Smooth	0.025-0.18
	Rough	2.5-9
Rock	Tunnel	600
Glass/Plastic		0.0015-0.01
Rubber	Smooth tube	0.006-0.07
Ceramic		1.4

(iii) **Transitional roughness:** In this regime, the roughness is influencing the flow, and friction factor is a function of both Reynolds number and roughness ratio. This regime is also sometimes referred to as transitional flow, and care should be exercised to avoid mixing it up with laminar-to-turbulent transition.

Table 10.2 lists the wall roughnesses for different materials.

10.8 DARCY FRICTION FACTORS FOR ROUGH PIPES

Following are the correlations that are widely used for rough pipe friction factor.
Transitional flow Colebrook-White correlation:

$$\frac{1}{\sqrt{f}} = 1.74 - 2\log_{10}\left[\frac{2\varepsilon}{D} + \frac{18.7}{\mathrm{Re}\sqrt{f}}\right]$$

Jain correlation (1976):

$$f = \left[1.14 - 2\log_{10}\left(\frac{\varepsilon}{D} + \frac{21.25}{\mathrm{Re}^{0.9}}\right)\right]^{-2}$$

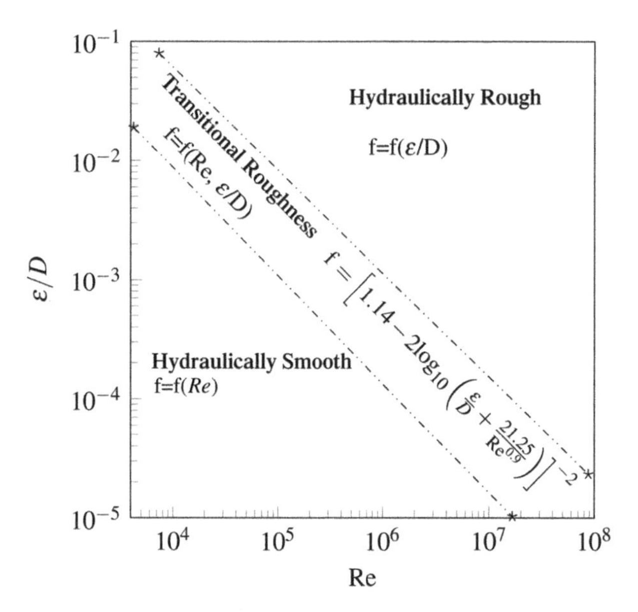

Figure 10.9 Regimes of turbulent pipe flow.

Figure 10.10 shows the Moody diagram for turbulent flow in rough pipes. The smooth pipe line is along the $\varepsilon = 0.00001$ D. For smooth pipe, McAdams and Blasisus correlations are used. For rough pipes, Jain's correlation is used. Note that in the original Moody diagram for the transitional flow, Colebrook-White correlation is used.

Haaland (1983) gave a correlation which can be applied for wide range of flow types:

$$\frac{1}{\sqrt{f}} = \frac{-1.8}{n}\log\left[\left(\frac{6.9}{\text{Re}}\right)^n + \left(\frac{\varepsilon}{3.75D}\right)^{1.11n}\right]$$

where $n = 3$ is suitable for natural gas pipelines and $n = 1$ is good for abrupt transition cases.

Example 10.5 Find Head Loss

Example Determine the head loss for flow of 160 l/s of oil $\nu = 0.00001$ m^2/s, through 500 m of schedule 40 internal diameter 1 in., cast iron pipe having 0.25 mm roughness.

Solution From Table 17.4, schedule 40 pipe with ID 1 in. is 2.664 cm.
$\varepsilon/D = 0.25$ mm$/2.664$ cm $= 0.00938$, or approximately 0.0094.
Reynolds number for this flow is:

$$\text{Re} = \frac{4Q}{\nu\pi D} = \frac{4 \times 0.16m^3/s}{0.00001 \times \pi \times 2.664 \times 10^{-2}} = 7.65 \times 10^5$$

The Re > 2300 and flow is turbulent. We now consult the Moody diagram or any of the correlations for the rough pipe for the estimation of friction factor. Moody Diagram: A rough estimation from the diagram gives Darcy friction factor, $f = 0.038$.

Colebrook-White correlation:

$$\frac{1}{\sqrt{f}} = 1.74 - 2\log_{10}\left[\frac{2\varepsilon}{D} + \frac{18.7}{\mathrm{Re}\sqrt{f}}\right]$$

This correlation requires trial and error to determine the friction factor. As a first approximation, we assume $f = 0.038$.

With further iterations, the value converged to $f = 0.037182$.

Jain correlation (1976)

$$f = \left[1.14 - 2\log_{10}\left(\frac{\varepsilon}{D} + \frac{21.25}{\mathrm{Re}^{0.9}}\right)\right]^{-2}$$

is an explicit correlation in f and gives f = 0.03721.

Haaland correlation (1983) gave a correlation which can wide range of flow types:

$$\frac{1}{\sqrt{f}} = \frac{-1.8}{n}\log\left[\left(\frac{6.9}{\mathrm{Re}}\right)^{n} + \left(\frac{\varepsilon}{3.75D}\right)^{1.11n}\right]$$

With Haaland correlation, $f = 0.0371$ for $n = 1$. It is found via sensitivity analysis that n variation does not bring substantial variation for this case.

This shows that all three correlations are giving the value of 0.0371 and it is very close to our rough estimate of $f = 0.038$ from the Moody diagram. We continue solving the problem using the f value of 0.0371 and estimate the pressure drop from relation:

$$h_f = f\frac{\ell}{D}\left(\frac{\bar{u}^2}{2g}\right)$$

$$h_f = f\frac{\ell}{D}\left(\frac{\bar{u}^2}{2g}\right) = 2.955 \times 10^6 N \cdot m/N$$

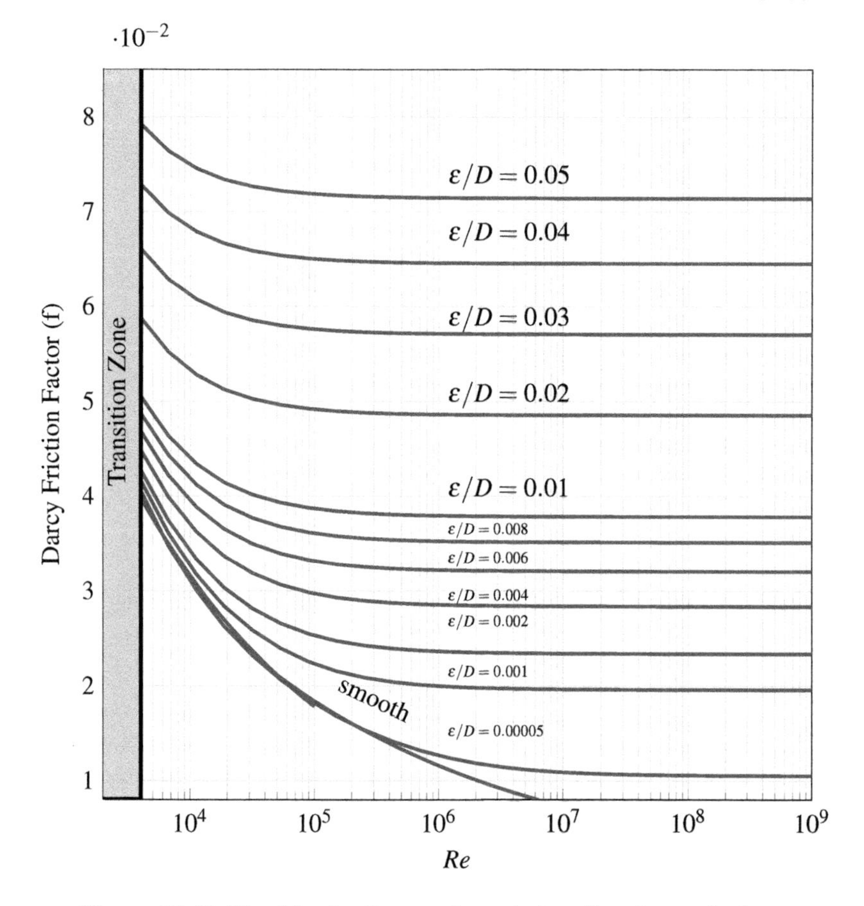

Figure 10.10 The Moody diagram for turbulent flow in rough pipes.

Example 10.6 Find the Flow Rate

Example Water at 20°C ($\nu = 1.13 \times 10^{-6}$ m^2/s) flows through a 500-mm diameter commercial steel pipe, mildly rusted, with roughness 0.15 mm, with a head loss of 7 m in 300 m of pipe length.

Solution

$$h_f = f\frac{\ell}{D}\left(\frac{\bar{u}^2}{2g}\right)$$

Assuming a reasonable value of $f = 0.04$ as a first trial, we can calculate mean velocity as:

$$6 = 0.04\left(\frac{300}{0.5}\right)\left(\frac{\bar{u}^2}{2g}\right)$$

$$\bar{u} = 1.85m/s$$

$$\text{Re} = \frac{\bar{u}D}{\nu} = \frac{1.85 \times 0.5}{1.13 \times 10^{-6}} = 8.19 \times 10^5$$

The correlations give f in the order of magnitude of $f = 0.0151$.

Using this $f = 0.0151$ and repeating this procedure, we get $f = 0.0155$, which gives:

$$\bar{u} = 2.97 \text{ m/s} \quad \text{Re} = 1.3172 \times 10^6.$$

$$Q = A \cdot \bar{u} = 0.584 \text{ m}^3\text{/s}.$$

Example 10.7 Find the Diameter

Example Suggest a pipe diameter that can deliver 466 L/s of a fluid having a viscosity of 1.655E5 m^2/s through a 150-m long pipe with a head loss of 75 m.

Solution For the solution of such problems, we have three unknowns: *D*, mean velocity, and Reynolds number. We insert the flow rate into the head loss equation

$$h_f = f\frac{\ell}{D}\left(\frac{16Q^2}{2g\pi^2 D^4}\right)$$

and rearrange it for diameter:

$$D = \left[\left(\frac{8\ell Q^2}{h_f g\pi^2}\right)f\right]^{1/5}$$

We again use a trial and error procedure, and assuming friction factor for a high Reynolds number flow, we calculate the diameter using the above equation.

$$d = 0.2352 \text{ m}$$

and we calculate the velocity as:

$$u = Q/((3.14(1/4))D^2) = 10.74m/s$$

$$f = \left[1.14 - 2\log_{10}\left(\frac{\varepsilon}{D} + \frac{21.25}{\text{Re}^{0.9}}\right)\right]^{-2}$$

$$\log_{10}\left(\frac{\varepsilon}{D} + \frac{21.25}{\text{Re}^{0.9}}\right) = \frac{\left[1.14 - \left(\frac{1}{f}\right)^{1/2}\right]}{2}$$

We estimate the ratio ε/D using Jain's correlation and find a new value for diameter $d = 0.2355$ m. Using this new diameter we calculate the flow rate, which is 0.467 m^3/s.

This procedure will give erratic values of diameter if the wrong friction factor is assumed.

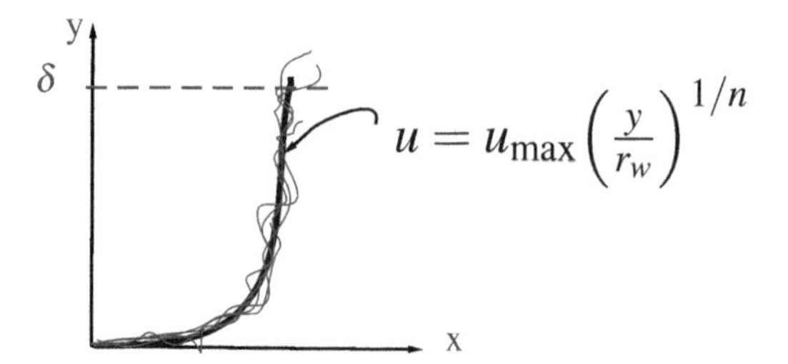

Figure 10.11 The turbulent boundary layers are inherently unsteady. The wavy lines in the sketch are the instantaneous velocity distributions. The time-averaged distribution is shown as a solid line.

10.9 FULLY DEVELOPED TURBULENT VELOCITY PROFILE IN PIPES

The velocity profile for turbulent flows can be

$$\frac{u}{\bar{u}} = \left[1 + 1.43\sqrt{f} + \left(2.15\sqrt{f}\right)\log_{10}\left(\frac{r_w - r}{r_w}\right)\right] \tag{10.10}$$

where friction factor can be calculated from **Jain correlation (1976)** for rough pipes, and r_w is the pipe wall radius.

Prandtl derived the following equation for mean velocity in turbulent pipe flows:

$$u = u_{\text{max}}\left(\frac{y}{r_w}\right)^{1/n}$$

$$\frac{\bar{u}}{u_{\text{max}}} = \frac{2n^2}{(n+1)(2n+1)}$$

$$\frac{u}{\bar{u}} = \frac{(n+1)(2n+1)}{2n^2}\left(\frac{r_w - r}{r_w}\right)^{1/n} \tag{10.11}$$

where y is the distance from the duct wall and $n = f(\text{Re})$. The value of n varies between 6 and 10 for Re = 4000 till 3E6, and mean value of 7 is most widely used.

Figure 10.12 shows laminar and turbulent velocity profiles in the same pipe. We know that if fluid and geometry remain the same, the turbulent velocity is much higher than laminar flows. In Figure 10.12, the velocity normalized with mean velocity is plotted.

10.10 EXTENDED BERNOULLI'S EQUATION IN PIPE NETWORK

The extended Bernoulli's equation was introduced in section 6.9:

$$\frac{P_1}{\gamma} + z_1 + \frac{\vec{V}_1^2}{2g} = \frac{P_2}{\gamma} + z_2 + \frac{\vec{V}_2^2}{2g} + h_f$$

We will now show through example the use of extended Bernoulili equation

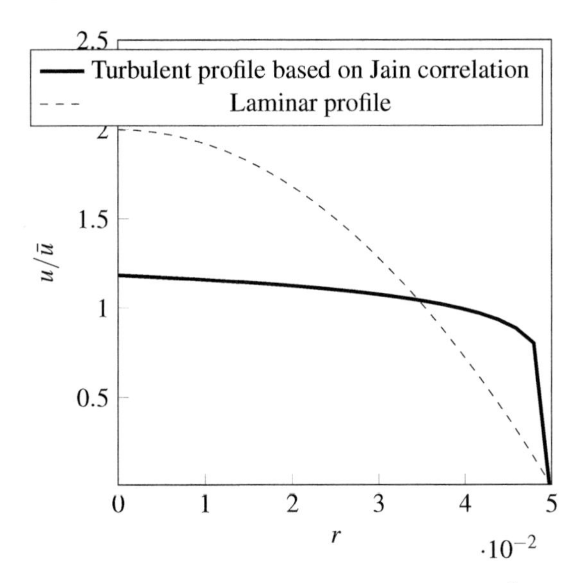

Figure 10.12 Turbulent velocity distribution at Re $= 2 \times 10^5$ (plot of Eq. 10.10). The laminar distribution normalized with mean velocity; $u/\bar{u}$ is independent of the pressure gradient.

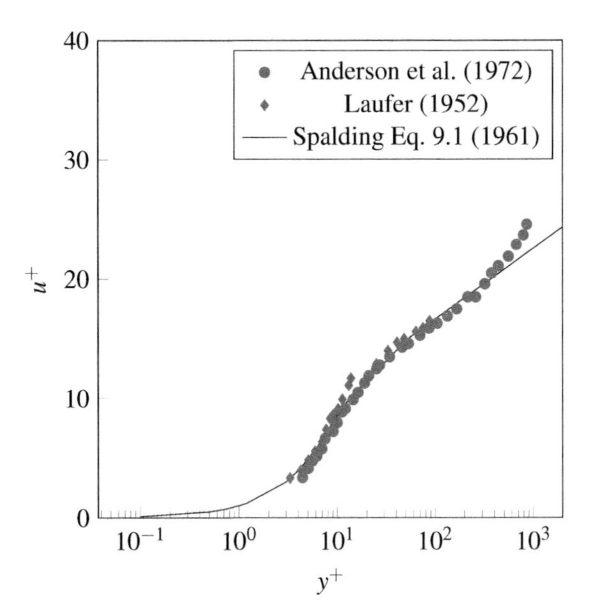

Figure 10.13 Turbulent velocity profile in pipe with dimensionless coordinates.

Example 10.8

Example A pipe of length $\ell = 6000$ ft is connected to a reservoir of oil. The head of oil (ν = 1E-04 ft^2/s) in the reservoir is 8 ft. The pipe diameter is 5 in. Find the flow Reynolds number, assuming steady incompressible viscous flow.

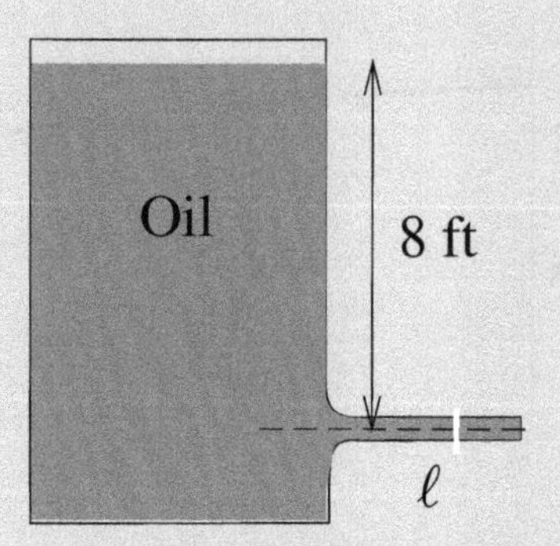

Figure 10.14 Schematic diagram of Example 10.8.

Solution We define:

Station 1: Pipe inlet
Station 2: Pipe outet

Since flow is steady, incompressible flow, according to continuity equation:

$$\rho_1 A_1 V_1 = \rho_2 A_2 V_2$$
$$V_1 = V_2 = V$$

We apply extended Bernoulli's equation at pipe inlet called station 1, and at the pipe's outlet called station 2:

$$\frac{P_1}{\gamma} + z_1 + \frac{\vec{V}_1^2}{2g} = \frac{P_2}{\gamma} + z_2 + \frac{\vec{V}_2^2}{2g} + h_f$$

$$\frac{P_1}{\gamma} + \cancel{z_1} + \cancel{\frac{\vec{V}_1^2}{2g}} = \cancel{\frac{P_2}{\gamma}} + \cancel{z_2} + \cancel{\frac{\vec{V}_2^2}{2g}} + h_f$$

$$\frac{P_1}{\gamma} = h_f \tag{10.12}$$

For laminar flow:

$$f = \frac{64}{\text{Re}} = \frac{64}{\frac{V \cdot d}{\nu}}$$

and head loss is

$$h_f = \frac{64}{\frac{V \cdot d}{\nu}} \left(\frac{\ell}{D} \right) \frac{V^2}{2g} = 3.43453V$$

We now apply the Bernoulli's equation at oil surface and station 1 (approximately just before the flow enters pipe):

$$\frac{(0)}{\gamma} + (8) + \cancel{\frac{\vec{V}^2}{2g}}_{\text{zero velocity}} = \frac{P_1}{\gamma} + \frac{\vec{V}^2}{2g}$$

$$\frac{P_1}{\gamma} = 8 - \frac{\vec{V}^2}{2g} \tag{10.13}$$

We now introduce the pressure in terms of velocity into the above equation

$$8 - \frac{\vec{V}^2}{2g} = 3.43V \tag{10.14}$$

The solution of this equation is $V = 0.897$ m/s

$$f = \frac{0.01536}{V} = 0.01712$$

and the Reynolds number is Re = 3737.5.

This shows that assumption of laminar flow is not valid for this problem. Assuming that the pipe is smooth and using the friction factor:

$$f = \frac{0.3164}{(Re)^{\frac{1}{4}}}$$

$$h_f = f \left(\frac{\ell}{D} \right) \frac{V^2}{2g} = 9.04831V^2$$

$$8 - \frac{\vec{V}^2}{2g} - 9.04831V^2 = 0$$

This gives $V = 0.939$ m/s. Based on this, the Reynolds number is Re = 3912.5.

In practical applications, Nomogram are also used, which have the vertical scales for internal diameter, volume flow rate, and head loss. Nomograms are developed as handy tools for quick estimations in the field. However, for large complex pipe networks, careful analysis is required.

10.11 MINOR LOSSES

Losses that occur because of local disturbances in the flow-through conduits are called minor losses in fluid mechanics literature. Such disturbances can be sudden

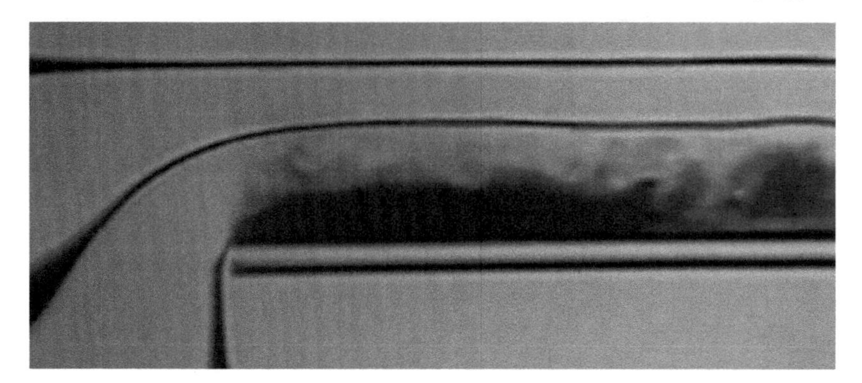

Figure 10.15 Sharp inlet would cause the frictional loss. (*Author: Hunter Rouse; courtesy of Dr. Marian Muste, IIHR - Hydroscience & Engineering, University of Iowa. Source: Intl. Assoc. for Hydro-Environment Engr. and Research.*)

and gradual changes in the conduits cross-section, also the flow through elbows, valves, etc. In turbulent flows, the losses are more than the laminar flows due to eddies. In the case of very long conduits, the contribution of losses is usually small compared with major friction losses. However, if the length of pipes is short, the minor losses are compared with major losses and thus, should be considered. Head loss in a diverging, incompressible flow section is much larger than that in the converging flow section as the flow accelerates. The minor losses were investigated, and the loss coefficients for several cases were tabulated as K factors. The minor loss coefficient is defined as:

$$\Delta p_{loss} = K\left(\frac{\rho \bar{u}^2}{2}\right)$$

10.11.1 ENTRANCE LOSSES

Fluid would lose energy when it would enter a large tank and leave the pipe, or vice versa. In such cases, the shape of the entrance plays an important role. Figure 10.15 shows the eddies that form when fluid enters a pipe. The eddies are formed because of local disturbance of the flow, leading to the formation of small-scale eddies. Because of such a phenomenon, the energy carried by the moving stream is reduced. Table 10.3 lists loss coefficients for some common pipe flow inlet configurations. The loss coefficient for a submerged pipe exit (K) is close to unity. The β is close to unity for fully developed turbulent pipe flow, and for fully developed laminar pipe flow, it is close to 2. In the case of laminar flow, the loss coefficient K can also be approximated as β.

$$K = \beta\left[1 - \frac{A_{small}}{A_{l\,arg\,e}}\right]^2 = \beta\left[1 - \left(\frac{d}{D}\right)^2\right]^2$$

Figure 10.16 shows the K factor variation for a protruding rounded entrance. Here, R is the radius of curvature at the entrance, and D is the diameter of the entrance.

Table 10.3
Loss Coefficient for Some Common Pipe Flow Inlet Configurations

Reentrant ($K = 0.8$) $t < d, \ell = 0.1d$	Sharp Edge ($K = 0.5$)	Rounded Edge ($K = 0.03 - 0.5$)

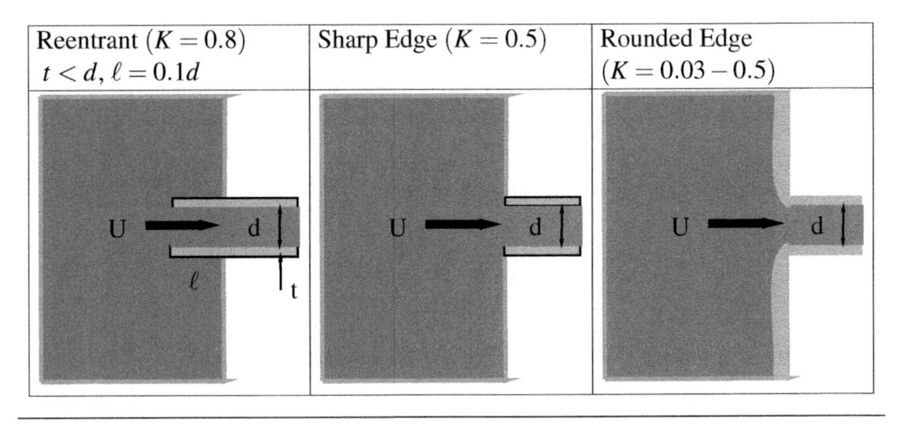

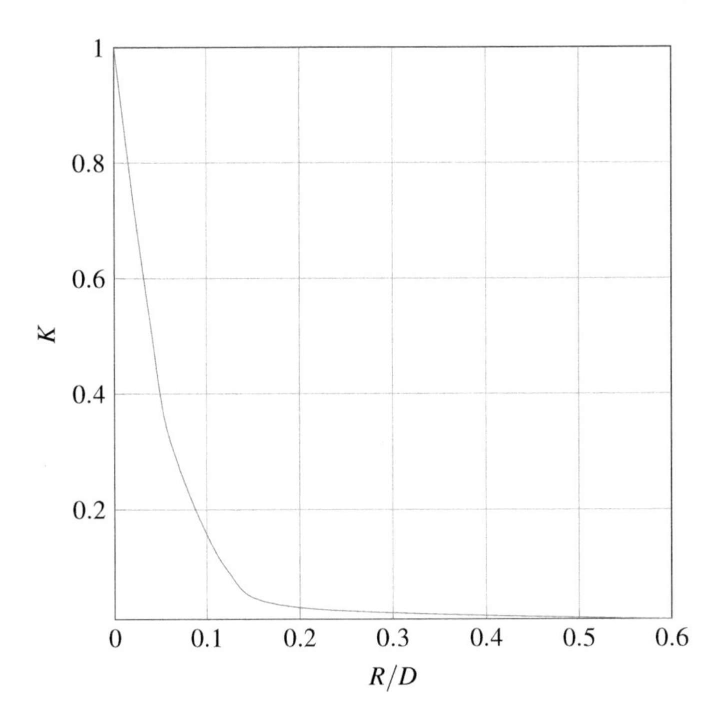

Figure 10.16 K factor for protruding rounded entrance. R is radius of curvature at entrance, and D is diameter of entrance.

Example 10.9

Example Water flows from a larger pipe, diameter $D_1 = 200$ mm, into a smaller one, diameter $D_2 = 50$ m through a reentrant. Assuming steady, incompressible flow, find the head loss. The volume flow rate is 0.03 m^3/s.

Solution

$$A_1 = \frac{\pi}{4}\cdot D_1^2 \quad A_1 = 0.0314 m^2$$
$$A_2 = \frac{\pi}{4}\cdot D_2^2 \quad A_2 = 0.001962 m^2$$
$$Q = 0.03\ m^3/s$$

We use the extended Bernoulli's equation:

$$\frac{p_1}{\rho}+\frac{V_1^2}{2}-\frac{V_2^2}{2}-\frac{p_2}{\rho}=K\cdot\frac{V_2^2}{2}$$

From continuity $Q = V_1 \cdot A_1 = V_2 \cdot A_2$ and also $\dfrac{p_1 - p_2}{\rho} = \dfrac{\rho\cdot g\cdot h}{\rho} = g\cdot h$ where h is the head loss.

Hence

$$g\cdot h+\frac{1}{2}\cdot\left(\frac{Q}{A_1}\right)^2-\frac{1}{2}\cdot\left(\frac{Q}{A_2}\right)^2=K\cdot\frac{1}{2}\cdot\left(\frac{Q}{A_2}\right)^2$$

Solving for h

$$h=\frac{\left(\frac{Q}{A_2}\right)^2}{2\cdot g}\cdot\left[1+K-\left(\frac{A_2}{A_1}\right)^2\right]$$
$$h = 21.39\ \text{m}$$

10.11.2 PRESSURE LOSS IN ABRUPT CONTRACTION

Piping systems also involve abrupt and gradual expansion/contraction sections. Figure 10.17(a) shows the movement of flow pathlines as the abrupt section encountered the flow. The losses in pressure are much larger in the case of abrupt expansion and contraction because of flow separation. The loss coefficient for the case of an abrupt expansion is approximated as

$$K=\left[1-\frac{A_{small}}{A_{large}}\right]^2=\left[1-\left(\frac{d}{D}\right)^2\right]^2$$

Example 10.10

Example Water flows through a 3-in. diameter pipe that suddenly contracts to a 2 in. diameter. The pressure drop across the contraction is 0.7 psi. Determine the volume flow rate.

Solution For steady, incompressible flow, we have the extended Bernoulli's equation

$$\left(\frac{p_1}{\rho}+\frac{V_1^2}{2}+g\cdot z_1\right)=\left(\frac{p_2}{\rho}+\frac{V_2^2}{2}+g\cdot z_2\right)+h_{Loss}$$

$$h_{Loss}=\left(\frac{p_1}{\rho}+\frac{V_1^2}{2}+g\cdot z_1\right)-\left(\frac{p_2}{\rho}+\frac{V_2^2}{2}+g\cdot z_2\right)$$

where

$$h_{Loss}=K\cdot\frac{V_2^2}{2}$$

From continuity

$$V_1=V_2\cdot\frac{A_2}{A_1}=V_2\cdot A_{Ratio}$$

Hence

$$\left(\frac{p_1}{\rho}+\frac{V_2^2\cdot A_{Ratio}{}^2}{2}\right)-\left(\frac{p_2}{\rho}+\frac{V_2^2}{2}\right)=K\cdot\frac{V_2^2}{2}$$

This gives

$$V_2=\sqrt{\frac{2\cdot(p_1-p_2)}{\rho\cdot(1-A_{Ratio}^2+K)}}$$

$$A_{Ratio}=\left(\frac{D_2^2}{D_1^2}\right)=\left(\frac{2}{3}\right)^2=0.444$$

From Figure 10.17 we have $K=0.4$.

$$V_2=\sqrt{\frac{2\cdot(p_1-p_2)}{\rho\cdot(1-A_{Ratio}^2+K)}}=\sqrt{\frac{2\cdot(0.7\times 144)}{1.94\cdot\left(1-(0.444)^2+0.42\right)}}=9.29\ ft/s$$

$$Q=\frac{\pi}{4}\cdot D_2{}^2\cdot V_2=0.2027\ ft^3/s$$

$$Q=448.83\times 0.2010\ ft^3/s=90.98\ gpm$$

10.11.3 PRESSURE LOSS IN SUDDEN ENLARGEMENT

For cases of sudden expansion, the K factor can be calculated as:

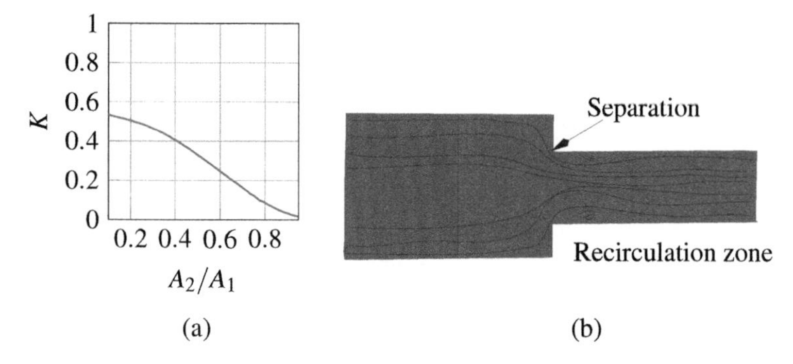

Figure 10.17 (a) K for abrupt contraction $A_2 < A_1$; (b) the schematic diagram.

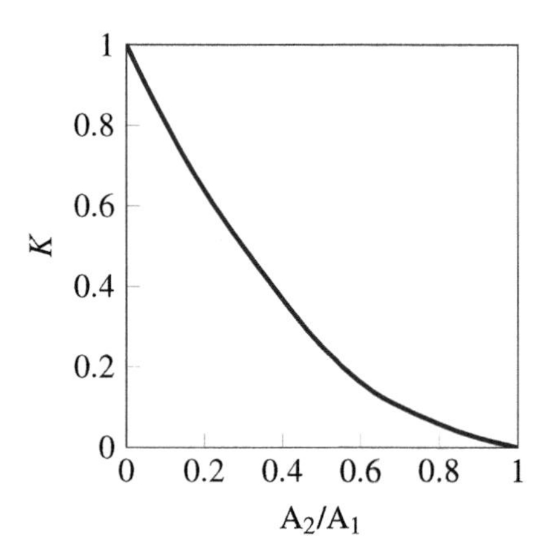

Figure 10.18 K factor for sudden enlargement for $A_2 > A_1$.

$$K = \left[1 - \left(\frac{A_1}{A_2}\right)\right]^2$$

Note that $A_2 > A_1$.

$$K = \left(\frac{1}{C_c} - 1\right)^2$$

Example 10.11

Example Air at standard conditions flows through a sudden expansion in a circular duct causing the pressure reduction of 50 Pa. The upstream and downstream duct diameters are 100 mm and 300 mm, respectively. Find the average speed of the air just before sudden enlargement of the duct, and find the volume flow rate.

Solution We use the extended Bernoulli's equation:

$$\left(\frac{p_1}{\rho}+\frac{V_1^2}{2}+g\cdot z_1\right)=\left(\frac{p_2}{\rho}+\frac{V_2^2}{2}+g\cdot z_2\right)+h_{Loss}$$

$$h_{Loss}=\left(\frac{p_1}{\rho}+\frac{V_1^2}{2}\right)-\left(\frac{p_2}{\rho}+\frac{V_2^2}{2}\right)$$

where minor head loss is defined as

$$h_{Loss}=K\cdot\frac{V_1^2}{2}$$

The inlet velocity is

$$V_1=\sqrt{\frac{2\cdot(p_2-p_1)}{\rho\cdot(1-A_{Ratio}^2-K)}}=\sqrt{\frac{2\times 50}{1.22\cdot\left(1-(0.111)^2-0.8\right)}}=20.89m/s$$

The area ratio is

$$A_{Ratio}=\left(\frac{D_2^2}{D_1^2}\right)=\left(\frac{2}{3}\right)^2=0.111$$

The flow rate can be calculated as

$$Q=\frac{\pi}{4}\cdot D_1{}^2\cdot V_1=0.164\,m^3/s$$
$$Q=448.83\times 0.164m^3/s=73.63\,gpm$$

10.11.4 GRADUAL ENLARGEMENT

An abrupt change in a duct cross-section can be converted into gradual contraction to reduce the losses (see Figure 10.19). Table 10.5 lists the angle of gradual enlargement along with diameter ratios. The angle of enlargement can take a value as small as 2 degrees.

10.11.5 FLOW-THROUGH BENDS AND VALVES

The loss coefficient K for flow-through bends and valves is tabulated in Tables 10.6 and 10.7.

10.11.6 FLOW-THROUGH ORIFICE

The pressure drop due to the orifice plate can be used to measure the flow. The orifice is a thin circular plate with a hole in its center. The plate is clamped between pipe

Table 10.4
Loss Coefficient *K* for Sudden Enlargement

$\bar{u}_1$ (m/s) D_2/D_1	0.6	1.2	3	4.5	6	9	12
1	0	0	0	0	0	0	0
1.2	0.11	0.1	0.09	0.09	0.09	0.09	0.08
1.4	0.26	0.25	0.23	0.22	0.22	0.21	0.2
1.6	0.4	0.38	0.35	0.34	0.33	0.32	0.32
1.8	0.51	0.48	0.45	0.43	0.42	0.11	0.4
2	0.6	0.56	0.52	0.51	0.5	0.48	0.47
2.5	0.74	0.7	0.65	0.63	0.62	0.6	0.58
3	0.83	0.78	0.73	0.7	0.69	0.67	0.65
4	0.92	0.87	0.8	0.78	0.76	0.74	0.72
5	0.96	0.91	0.84	0.82	0.8	0.77	0.75
10	1	0.96	0.89	0.86	0.84	0.82	0.8
∞	1	0.98	0.91	0.88	0.86	0.83	0.81

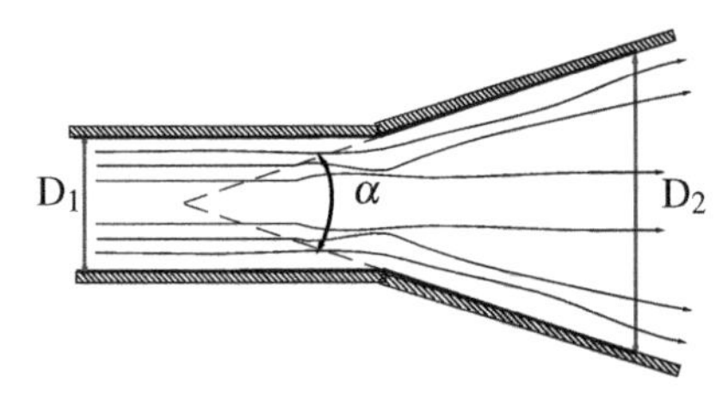

Figure 10.19 *K* factor for gradual enlargement. α is the imaginary cone angle.

flanges. The orifice plate is usually a sharp-edged device. The major disadvantages of the orifice are its limited capacity and the large pressure loss.

$$C_{orifice} = \underbrace{0.5959 + 0.031\beta^{2.1} - 0.184\beta^8}_{C_\infty} + \underbrace{\frac{91.71\beta^{5/2}}{\mathrm{Re}_D{}^{3/4}}}_{\Delta C_{Reynolds}} + \underbrace{\frac{0.09L_1\beta^4}{1-\beta^4} - 0.0337L_2\beta^3}_{\Delta C_{tap}}$$

where
L_1 = dimensionless correction for upstream tap location
L_2 = dimensionless correction for downstream tap location
$\beta = d/D$ = ratio of orifice diameter to pipe diameter

Table 10.5
Loss Factor *K* along with Angle of Enlargement and Diameter Ratio

D_2/D_1	2°	6°	10°	15°	20°	25°	30°	35°	40°	45°	50°	60°
1.1	0.01	0.01	0.03	0.05	0.1	0.13	0.16	0.18	0.19	0.2	0.21	0.23
1.2	0.02	0.02	0.04	0.09	0.16	0.21	0.25	0.29	0.31	0.33	0.35	0.37
1.4	0.02	0.03	0.06	0.12	0.23	0.3	0.36	0.41	0.44	0.47	0.5	0.53
1.6	0.03	0.04	0.07	0.14	0.26	0.35	0.42	0.47	0.51	0.54	0.57	0.61
1.8	0.03	0.04	0.07	0.15	0.28	0.37	0.44	0.5	0.54	0.58	0.61	0.65
2	0.03	0.04	0.07	0.16	0.29	0.38	0.46	0.52	0.56	0.6	0.63	0.68
2.5	0.03	0.04	0.08	0.16	0.3	0.39	0.48	0.54	0.58	0.62	0.65	0.7
3	0.03	0.04	0.08	0.16	0.31	0.4	0.48	0.55	0.59	0.63	0.66	0.71
0	0.03	0.05	0.08	0.16	0.31	0.4	0.49	0.56	0.6	0.64	0.67	0.72

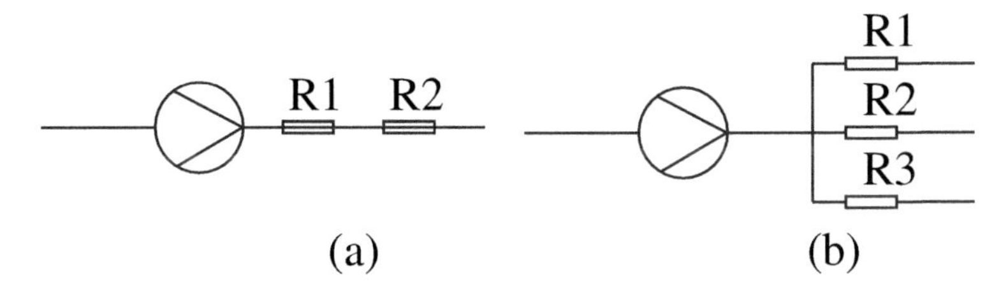

Figure 10.20 Pipes in series and parallel. R_i is the pipe resistance.

In practice, the following correlations are also used (Miller, 1997):

$$C_{orifice} = \underbrace{0.5959 + 0.0312\beta^{2.1} - 0.184\beta^{8}}_{C_\infty} + \underbrace{\frac{91.71\beta^{5/2}}{\mathrm{Re}_D{}^{3/4}}}_{\Delta C_1}$$

$$C_{nozzle} = 0.9975 - \frac{6.53\beta^{0.5}}{\mathrm{Re}_D{}^{0.5}}$$

These relations are valid for $0.25 < \beta < 0.75$ and $10^4 < \mathrm{Re} < 10^7$.

10.12 PIPES IN SERIES AND IN PARALLEL

In the real world, the piping network is often complex, with several pipes connected to each other. Here we focus on pipes connected in series and pipes connected in parallel (see Figure 10.20). In the case of pipes connected in series, the mass flow rate passing through the pipes is the same. However, in the case of pipes connected in parallel, the mass flow rate is divided.

Table 10.6
***K* for Bends and Branches**

Description	K
90 degree flanged	0.3
90 degree threaded	0.9
90 degree miter bend without vanes	1.1
90 degree miter bend with vanes	0.2
45 degree threaded elbow	0.4
180 degree return bend flanged	0.2
180 degree threaded	1.5
Tee flanged branch flow	1
Tee threaded branch flow	2
Tee line flow flanged	0.2
Tee line flow threaded	0.9
Threaded union	0.08

Considering the Darcy-Weisbach equation, we have the pressure drop as:

$$\Delta p = f\left(\frac{\rho \bar{u}^2}{2}\right)\left(\frac{\ell}{d}\right)$$

$$\Delta p = f\left(\frac{\rho \bar{u}^2}{2}\right)\left(\frac{\ell}{d}\right)\left(\frac{A^2}{A^2}\right)$$

$$= f\left(\frac{\rho Q^2}{2}\right)\left(\frac{\ell}{d}\right)\left(\frac{1}{A^2}\right)$$

$$\Delta p = \left(\frac{8 f \ell \rho}{d^5 \pi^2}\right) Q^2$$

This shows that pressure drop in a pipe is related with Q^2. We can draw an analogy between the electric circuits and compare Ohm's law and the above equation.

$$\Delta Voltage = I.R$$

Here, R is resistance. This analogy shows that for pipe flow, the equivalent flow resistance is

$$R = \frac{\Delta p}{Q^2} = \frac{8 f \ell \rho}{d^5 \pi^2}$$

$$R_{T,series} = \sum R_i = R_1 + R_2$$

Table 10.7
***K* Factor for Flow Through Valves**

Description	K
Globe valve (fully open)	10
Gate valve (fully open)	0.2
Angle valve (fully open)	5
Ball valve (fully open)	0.05
Angle valve (1/4 closed)	0.3
Ball valve (1/2 close)	2.1
Swing check valve	2
Swing check valve (3/4 close)	17

In case of parallel flows, we can write for both major and minor losses

$$\Delta p = \sum_{i=1}^{n}\left[f\left(\frac{L}{d}\right)\left(\frac{\rho}{2}u^2\right)\right]_i + \sum_{j=1}^{m}\left[K\left(\frac{\rho}{2}u^2\right)\right]_j$$

where i is the number of pipe branches and m indicates the number of devices contributing toward minor losses.

$$\Delta p = \underbrace{\left\{\sum_{i=1}^{n}\left[f\left(\frac{L}{d}\right)\left(\frac{\rho}{2}\frac{1}{A^2}\right)\right]_i + \sum_{j=1}^{m}\left[K\left(\frac{\rho}{2}\frac{1}{A^2}\right)\right]_j\right\}}_{R} Q^2$$

The term indicted by R is the resistance through the pipes. Wagner (2012) proposed the following equation for total parallel resistance:

$$R_{T,parallel} = \frac{1}{\Sigma\left(\frac{1}{\sqrt{R_i}}\right)^2}$$

For pipe in parallel, there are many unknowns. First, the flow rate through branches is unknown. Also, the head loss through branches is unknown. Therefore, assumptions are often made. If we assume that there are two branches, as shown in Figure 10.21, and head loss through branches is the same, then

$$h_{f,1} = h_{f,2}$$

$$f_1\left(\frac{L_1}{d_1}\right)\left(\frac{u_1^2}{2g}\right) = f_2\left(\frac{L_2}{d_2}\right)\left(\frac{u_2^2}{2g}\right)$$

$$\frac{u_1^2}{u_2^2} = \left(\frac{f_2}{f_1}\right)\left(\frac{L_2}{L_1}\right)\left(\frac{d_1}{d_2}\right)$$

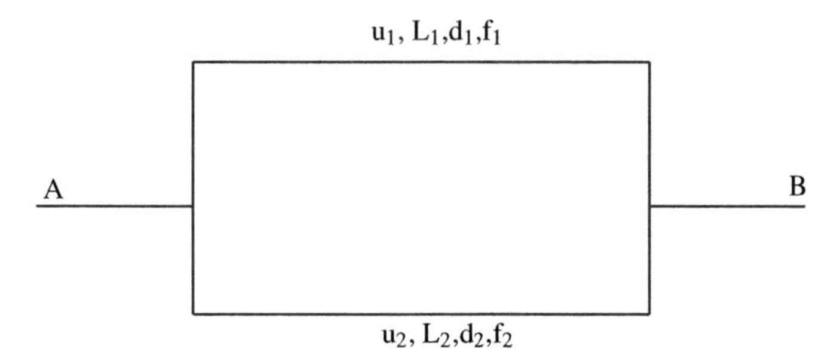

Figure 10.21 Flow through two branches in parallel.

$$\frac{u_1}{u_2} = \sqrt{\left(\frac{f_2}{f_1}\right)\left(\frac{L_2}{L_1}\right)\left(\frac{d_1}{d_2}\right)}$$

The above equation can be used to estimate pressure drop through individual pipe branches with the assumption that head loss for these branches is the same.

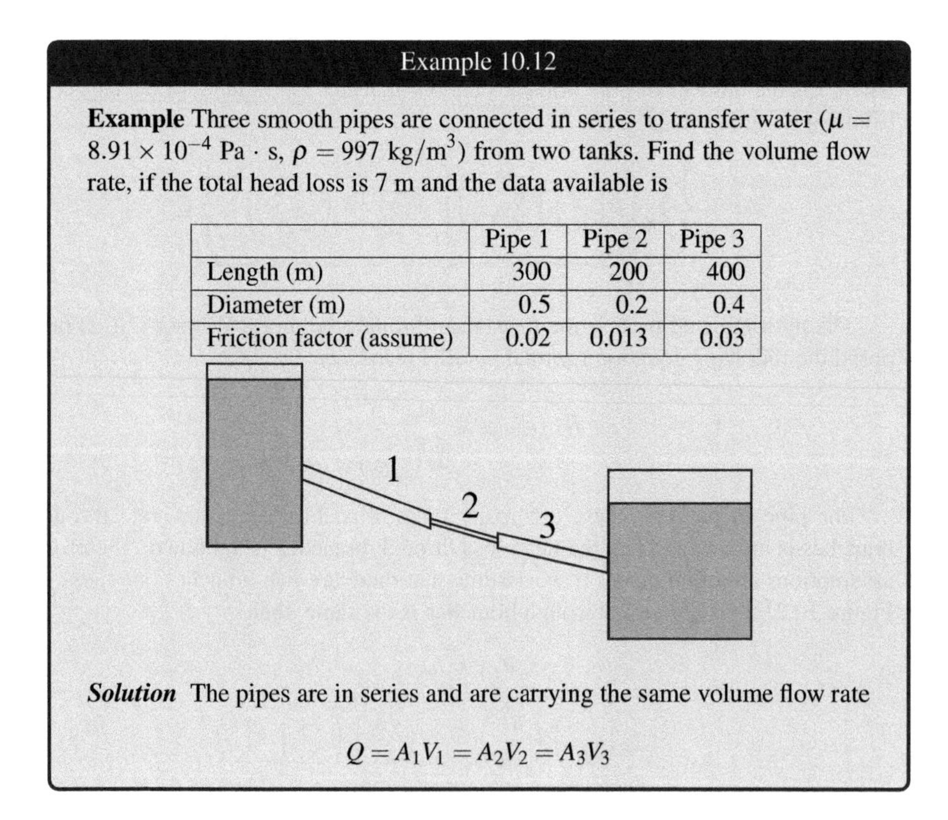

Example 10.12

Example Three smooth pipes are connected in series to transfer water ($\mu = 8.91 \times 10^{-4}$ Pa $\cdot$ s, $\rho = 997$ kg/m^3) from two tanks. Find the volume flow rate, if the total head loss is 7 m and the data available is

	Pipe 1	Pipe 2	Pipe 3
Length (m)	300	200	400
Diameter (m)	0.5	0.2	0.4
Friction factor (assume)	0.02	0.013	0.03

Solution The pipes are in series and are carrying the same volume flow rate

$$Q = A_1V_1 = A_2V_2 = A_3V_3$$

The head loss through individual pipes is

$$h_1 = f_1 \left(\frac{L_1}{d_1}\right) \frac{V_1^2}{2g}$$

$$h_2 = f_2 \left(\frac{L_2}{d_2}\right) \frac{V_2^2}{2g}$$

$$h_3 = f_3 \left(\frac{L_3}{d_3}\right) \frac{V_3^2}{2g}$$

Assuming the friction factor for smooth pipes, the total head loss will be

$$h_T = h_1 + h_2 + h_3$$

$$h_T = f_1 \left(\frac{L_1}{d_1}\right) \frac{V_1^2}{2g} + f_2 \left(\frac{L_2}{d_2}\right) \frac{V_2^2}{2g} + f_3 \left(\frac{L_3}{d_3}\right) \frac{V_3^2}{2g}$$

$$h_T = f_1 \left(\frac{L_1}{d_1}\right) \frac{V_1^2}{2g} + f_2 \left(\frac{L_2}{d_2}\right) \frac{1}{2g} \left(\frac{A_1 V_1}{A_2}\right)^2 + f_3 \left(\frac{L_3}{d_3}\right) \frac{1}{2g} \left(\frac{A_1 V_1}{A_3}\right)^2$$

$$h_T = 0.611 V_1^2 + 25.8 V_1^2 + 3.73 V_1^2$$
$$7 = 0.611 V_1^2 + 25.8 V_1^2 + 3.73 V_1^2$$

$$V_1 = 0.48 m/s$$

Using V_1, we can calculate V_2 and V_3. The Reynolds number is around 10^5. Using the Reynolds number, the friction factor can be estimated, and after some trial and error, V_1 is 0.39 m/s. The volume flow rate is

$$Q = 0.0778 m^3/s$$

We can also solve this equation using flow rate

$$\left(\rho \cdot g \cdot h_L - \left(\frac{8 \cdot f_1 \cdot L_1 \cdot \rho}{d_1{}^5 \cdot 3.14^2} + \frac{8 \cdot f_2 \cdot L_2 \cdot \rho}{d_2{}^5 \cdot 3.14^2} + \frac{8 \cdot f_3 \cdot L_3 \cdot \rho}{d_3{}^5 \cdot 3.14^2}\right) Q^2\right) = 0$$

This gives flow rate as

$$Q = 0.0778 m^3/s$$

Example 10.13

Example The flow through a main pipe carrying water ($\rho = 999.5\ \text{kg/m}^3$) at volume flow rate of 0.075 m^3/s is divided into three parallel branches. Find the pressure drop for this network, if the friction factors, pipe lengths, and minor loss coefficients involved are:

	$d(m)$	f	$L(m)$	$\sum K$
Branch 1	0.1	0.018	13	2.8
Branch 2	0.06	0.017	17	3.5
Branch 3	0.2	0.019	15	4.3

Solution The resistance for the pipe with major and minor losses is

$$R = \left[f\left(\frac{L}{d}\right) + K\right]\left(\frac{\rho}{2}\frac{1}{A^2}\right)$$

$$R_1 = 4.168 \times 10^7$$

$$R_2 = 5.20 \times 10^8$$

$$R_3 = 2.90 \times 10^6$$

The total resistance is

$$R_T = \frac{1}{\left(\frac{1}{\sqrt{R_1}} + \frac{1}{\sqrt{R_2}} + \frac{1}{\sqrt{R_3}}\right)^2} = 1.61E6\frac{Pa}{(m^3/s)^2}$$

$$\Delta p = R_T \cdot Q^2 = 9.11kPa$$

Example 10.14

Example The flow through a main pipe carrying water ($\rho = 999.5\ \text{kg/m}^3$) at a volume flow rate of 0.075 m^3/s is divided into two parallel branches. If the friction factors, pipe lengths, and minor loss coefficients involved are:

	$d(m)$	f	$L(m)$
Branch 1	0.1	0.018	13
Branch 2	0.06	0.017	17

Find the flow rates through branches.

Solution Assuming the pressure drop through the pipes is the same, we have

$$\frac{u_1}{u_2} = \sqrt{\left(\frac{f_2}{f_1}\right)\left(\frac{L_2}{L_1}\right)\left(\frac{d_1}{d_2}\right)}$$

This gives

$$u_1 = 1.434u_2$$

The volume flow rate balance gives

$$Q = u_1A_1 + u_2A_2$$

This gives

$$u_1 = 7.637\text{m/s}$$

$$u_2 = 5.323\ m/s$$

and the flow rates through branches is

$$q_1 = 0.0599\text{m}^3/s$$
$$q_2 = 0.0150\ \text{m}^3/s$$

10.13 SIMILITUDE CONSIDERATIONS

Example 10.15

Example The valve coefficient K for a 1.96-ft diameter prototype is required to be determined from the test data obtained for a similar model of 0.6 m. The fluid used in the test was water at 70°F with velocities of 3.28 ft/s (1 m/s) and 8.20 ft/s (2.5 m/s). What is the range of speeds for the prototype, if the prototype valve is used with air at 80°F?

Solution The valve coefficient is defined as:

$$K = \frac{\Delta p}{\rho u^2/2}$$

Air at 80° F = 26.66°C, $\nu = 1.67 \times 10^5$ ft²/s

Water at 70°F = 21.11°C. Table of water properties in appendix[a] gives: ρ_w = 998 kg/m³, $\mu_w = 1.02 \times 10^3$ Pa · s

The minimum Reynolds number on this model is:

$$(\text{Re})_{m,w} = \frac{998 \times 1m/s \times 0.6}{1.02 \times 10^{-3}} = 587 \times 10^3$$

and the maximum Reynolds number on this model is:

$$(\text{Re})_{m,w} = 1467 \times 10^3$$

Since

$$(\text{Re})_m = (\text{Re})_p$$

$$\left(\frac{uD}{\nu}\right)_m = \left(\frac{uD}{\nu}\right)_p$$

The range of air velocities is:

$$u_{p,\min} = (\mathrm{Re})_{m,w}\left(\frac{\nu}{D}\right)_p = 587 \times 10^3 \left(\frac{1.67 \times 10^{-5} ft^2/s}{1.96 ft}\right) = 5 ft/s$$

$$u_{p,\max} = 12.5 ft/s$$

[a] Appendix can be found online at https://routledge.com/9781032324531

10.14 PRESSURE LOSS IN A COILED TUBE

In order to increase heat transfer, flow is sometimes moved through a coiled tube. This leads to an increase in the secondary velocities, and thus overall heat transfer would be increased. However, there is an associated pressure loss, which can be expressed using the Dean number, defined as

$$De = \mathrm{Re}\sqrt{\left(\frac{a}{R}\right)}$$

where a is the radius of the coiled tube and R is the radius of curvature of the coil itself.

Srinivasan and Nandapukar (1970) proposed the following correlation for friction factor in coiled tubes compare with straight tubes

$$\frac{f_c}{f_s} = \begin{cases} 1 & De < 30 \\ 0.419 De^{0.275} & 30 < De < 300 \\ 0.1125 De^{0.5} & De > 300 \end{cases}$$

This is valid for the range 7 < R/a < 104.

For the coiled tube, we can no longer use $\mathrm{Re}_{crit} = 2,300$; instead the upper critical Reynolds number must also be calculated as

$$\mathrm{Re}_{crit} = 2100\left[1 + 12\left(\frac{R_{min}}{a}\right)^{-0.5}\right]$$

where R_{min} is the minimum radius of curvature of a coil.

Example 10.16

Example An organic fluid (kinematic viscosity $\nu = 3\text{E-}06\ m^2/s$) is flowing in a coiled smooth tube of diameter 10 cm. The constant radius of curvature of the coil is 80 cm, and the total length of the coiled tube is 50 m. If the volume flow rate is 0.03 m^3/s, Find the head loss for flow through the coiled tube.

Solution We calculate the velocity of flow through the tube as

$$u = \frac{Q}{A} = \frac{0.03}{\frac{\pi}{4}\left(\frac{10}{100}\right)^2} = 3.82 m/s$$

$$\mathrm{Re} = \frac{u \cdot d}{\nu} = \frac{3.82 \times 0.1}{3 \times 10^{-6}} = 1.27 \times 10^5$$

$$\mathrm{Re}_{crit} = 2100\left[1 + 12\left(\frac{R}{a}\right)^{-0.5}\right] = 8400$$

As Re>Re_{crit}, flow is turbulent.

We now calculate the Dean number as

$$De = \mathrm{Re}\sqrt{\frac{a}{R}} = 1.27 \times 10^5 \sqrt{\frac{5}{80}} = 31847$$

Note that De >300.

For a smooth tube, the friction factor can be calculated from Techo et al. (1965) relation

$$f_s = \left(\frac{1}{\ln\left(\frac{Rey}{1.96 \cdot \ln(Rey) - 3.8215}\right) \cdot 1.7372}\right)^2 = 0.00427$$

For a coiled tube, the friction factor f_c is

$$f_c = f_s \cdot 0.1125 \cdot De^{0.5} = 0.0859$$

$$h_f = \frac{f_c \cdot L}{d} \cdot \frac{u^2}{2 \cdot 9.81} = 31.98\mathrm{m}$$

PROBLEMS

10P-1 Pai (1953) found that the following empirical relation can describe the turbulent velocity profile in whole pipe:

$$\frac{u}{U_{\max}} = 1 - \frac{51}{2500}\left(\frac{r}{r_w}\right)^2 - \frac{1}{4}\left(\frac{r}{r_w}\right)^{32}$$

for flow of water at Reynolds number of $Re = 3.24 \times 10^6$. Find the expression of mean velocity, if the radius of pipe is $r_w = 25$ mm.

10P-2 A lubricating oil of viscosity 20 cP and SG = 0.8 is passing through an annulus of inner and outer radii of 8 and 12 mm, respectively. The volume flow rate is 1 m^3/hr. Find the pressure drop through this device.

10P-3 Find the frictional velocity if the shear stress at the river bed is estimated to be 0.01 ρ, where ρ is density of river water.

10P-4 Calculate the entrance length using different correlations given in this chapter, if flow $Re = 2,000$.

10P-5 Calculate the pressure drop for laminar flow of $Q = 0.03$ m^3/s, inside a 10-m long elliptic duct at velocity of 0.1 cm/s. The semi-major axes a and b are 5 mm and 10 mm, respectively. The properties are $\mu = 0.91\text{e-}3$ Pa · s; $\rho = 800$ kg/m^3.

10P-6 Water is flowing in a half-filled pipe of diameter d and length ℓ. The vertical height of the inclination is Δh. Find the head loss through the pipe if the length of pipe is 1,000 m, diameter is 2 m, surface roughness is 2 mm, and volume flow rate is 0.5 m^3/s. Take viscosity of liquid as 10^{-6} m^2/s.

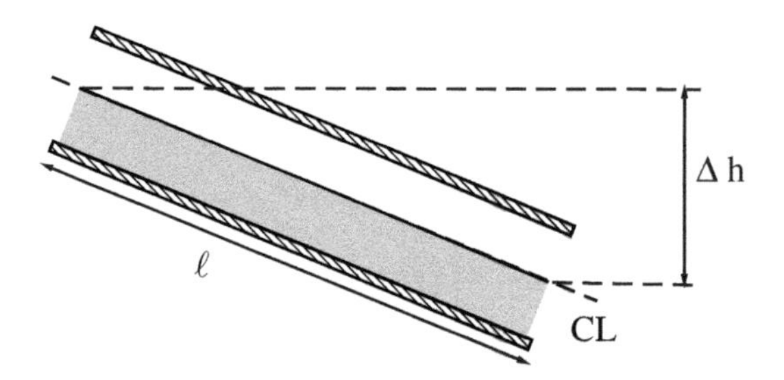

Figure 10.22 Figure of problem 10P-6.

10P-7 Water is being pumped from a tank where the water level is maintained at a height of 0.5 m. The water is then converted into a fountain by passing it through a nozzle of diameter 0.05 m. The maximum height the water reached is $h_3 + h_2$, where h_2 is 1 m and h_3 is 20 m. The piping diameter is 0.15 m. The total pipe length is 10 m and elbow head loss factor is 0.15. Find the velocity of the jet, the head loss, and the pressure rise across the pump. The surrounding air would cause a reduction in the fountain speed by a factor of 0.8. Assume a smooth pipe, sharp-edged entrance, and 90° flanged elbow.

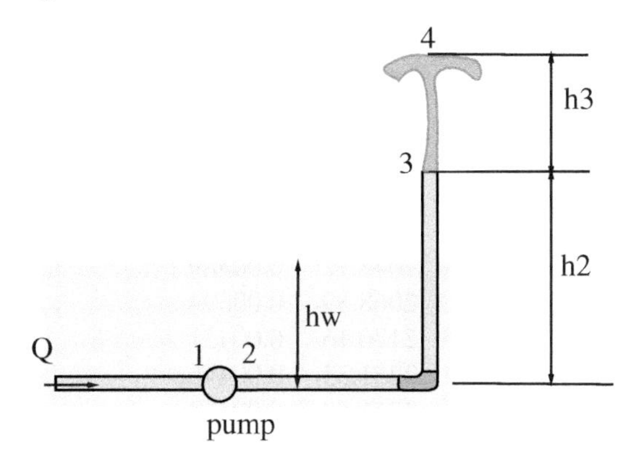

Figure 10.23 Figure of problem 10P-7.

10P-8 The laminar flow data from an experiment in which the pipe velocity is gradually increased is listed below. Assume that the frictional loss of pressure in a pipe is following the relation $\Delta P = k \cdot v^n$. Find k if the value of index n is 1.72.

$Velocity(m/s)$	$\Delta PLoss(Pa)$
2.45	22.913
3.71	42.55
5.18	65.46
10.16	229.13
11.03	255.31
13.29	346.97
17.69	572.83
21.36	831.42
24.43	1083.46
27.69	1338.78
28.16	1414.07
30.64	1662.84

10P-9 The data of water flowing in a pipe of diameter 18.09 mm is given below. Plot the data and give comments.

Re	$\Delta P/\rho u_m^2$
1365.10	0.00611
1406.21	0.00578
1431.88	0.00548
1640.98	0.00486
2008.05	0.00394
2042.72	0.00389
2068.39	0.00394
2170.16	0.00431
2231.33	0.00446
2311.80	0.00462
2526.58	0.00484
2591.72	0.00486
2879.47	0.00495
3278.18	0.00482
4226.15	0.00423
5242.18	0.00391
7723.76	0.00358

10P-10 It takes 20 s for 8.193 mL of water to flow through a 1.17-mm diameter capillary tube viscometer, as shown in Figure 10.24. Calculate the Reynolds number of flow through this tube.

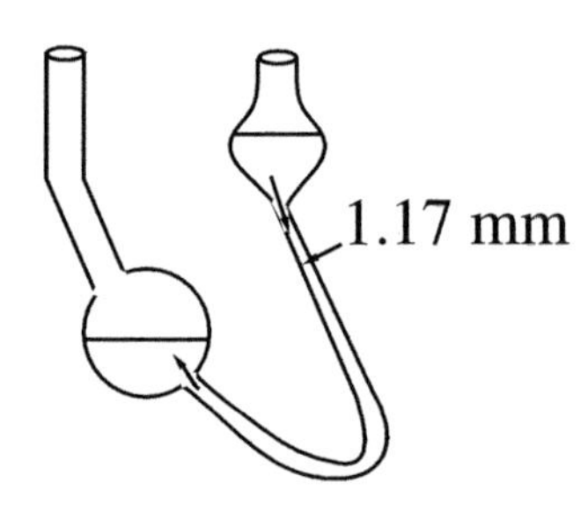

Figure 10.24 Figure of problem 10P-10.

10P-11 Water flows from a tank to atmosphere through a very short pipe. Assuming steady, find the volume flow rate at the exit if the tank area is 5,000 mm^2, the exit pipe area is 400 mm^2, and the water height from free surface till exit is 3 m.

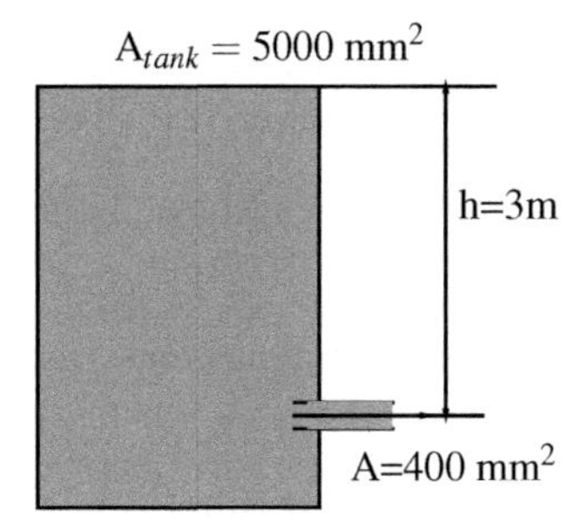

Figure 10.25 Figure of problem 10P-11.

10P-12 Flow straighteners are placed in the section upstream of the test section of the wind tunnel, which consist of an array of hexagonal ducts. The straightener section is 0.5 m long and the side of a hexagon is 1 cm. The inlet bulk velocity to straightener is 5 m/s. Calculate the Reynolds number.

REFERENCES

P. S. Anderson, W. M. Kays, and R. J. Moffat, The Turbulent Boundary Layer on a Porous Plate: An Experimental Study of the Fluid Mechanics for Adverse Free Stream Pressure Gradients, Report No. HMT-15, Department of Mechanical Engineering, Stanford University, CA, 1972.

J. Laufer, The Structure of Turbulence in Fully Developed Pipe Flow. NACA Report 1174, 1952.

S. Itaya, "New experimental formula of the coefficient of pipe friction," J. of JSME, 48[332], 84 (1945), [in Japanese].

H. Schlichting, Boundary Layer Theory. New York: McGraw-Hill, 1960.

H. Latzko, Heat Transfer in a Turbulent Liquid or Gas Stream, NACA TM 1068, Oct. 1944.

A. R. Barbin and J. B. Jones, Turbulent flow in the inlet region of a smooth pipe, J. of Basic Engr., Trans. ASME, vol. 85, ser. D, pp. 29–34, Mar. 1963.

D. A. Bowlus and J. A. Brighton, Incompressible turbulent flow in the inlet region of a pipe, J. of Basic Engr., ASME Tech. Brief, 1968.

R. Y. Chen, Flow in the Entrance Region at Low Reynolds Numbers, J. of Fluids Engr., vol. 95 no. 1, p. 153, 1973.

W. Zhi-qing, Study on correction coefficients of laminar and turbulent entrance region effect in round pipe, Appl. Math. Mech. 3, pp. 433–446, 1982.

R. P. Benedict, N. A. Carlucci, and S. D. Swetz, Flow losses in abrupt enlargements and contractions, J. of Engr. for Power, vol. 88 no. 1, 1966.

J. Nikuradse, Strmungsgesetze in rauhen Rohen, Forschung. Arb. Ing.-Wes., (361), 1933.

R. W. Miller. Flow Measurement Engineering Handbook, 3rd ed. New York: McGraw-Hill, 1997.

McAdams, Heat Transmission, McGraw Hill, USA, 1954.

R. Techo, R. R. Tickner, and R. E. James, An accurate equation for the computation of the friction factor for smooth pipes for the Reynolds number, J. Appl. Mech. vol. 32, p. 443, 1965.

R. T. Szczepura, Flow characteristics of an axisymmetric sudden pipe expansion - results obtained from the turbulence studies rig. Part 1 mean and turbulence velocity results. Cegb Berkeley Tprd/B/0702/N85, 1985.

P. S. Srinivasan, S. S. Nandapukar, and F. A. Holland, Friction factors for coils, Trans. Inst. Chem. Engrs. (London), vol. 48, T156-Tl61, 1970.

J. Weisbach, Die Experimental-Hydraulik, p. 133, Freiberg: J. S. Englehardt, 1855.

V. L. Streeter, ed., Handbook of Fluid Dynamics. New York: McGraw-Hill, 1961.

W. Wagner, Rohrleitungstechnik. 11. Aufl., Wurzburg: Vogel Verlag, 2012.

11 External Boundary Layer Flows

Fluid flow over objects frequently occurs in engineering applications, for example, flow over cars, trucks, buildings, towers, and missiles. This flow differs significantly from the internal flow in conduits. The primary reason for studying external flows is to estimate the drag and lift forces. The reduction in drag is directly connected with efficient fuel consumption.

Learning outcomes: After finishing this chapter, you should be able to:

- Calculate the skin friction drag force associated with flow over a flat surface.
- Understand the effects of laminar and turbulent flow on the drag coefficients.
- Calculate the drag force for combined laminar and turbulent flow over a flat plate.
- Formulate the momentum integral equation, which can be used for different flow types and boundary conditions.

In the Third International Congress of Mathematicians held in 1904 in Heidelberg, **Ludwig Prandtl** presented his paper and explained that for a fluid of small viscosity, such as air or water, the viscosity will still substantially affect the flow in a thin layer near the surface. Viscosity can be neglected outside this layer, and the flow can be treated as nonviscous, or inviscid. Prandtl called this thin layer near the surface the *Grenzschicht*, now translated in English as the boundary layer. Thus, for more than 100 years, thousands of scientists across the globe have dedicated their efforts to developing an understanding of external flow drag, both via experiments and through analytical modeling.

Irrotational flow theory agrees well with the experimental data as far as pressure is concerned. However, it does not give an acceptable drag prediction (zero drag). In 1905, L. Prandtl provided the explanation of how vorticity diffuses from a solid boundary and is swept into the surrounding inviscid flow. Boundary layer theory is actually a thin layer approximation.

Step 1: Compute inviscid flow around an object and obtain the pressure distribution.
Step 2: Compute the flow inside t boundary layer.

Displacement thickness: Displacement thickness is the distance by which a wall would have to be displaced outward, in a hypothetical frictionless flow, to maintain the same mass flux as in an actual flow. Figure 11.1 schematically depicts this

DOI: 10.1201/9781003315117-11

$$(\dot{m}/A)_{Viscous} = (\dot{m}/A)_{inviscid}$$

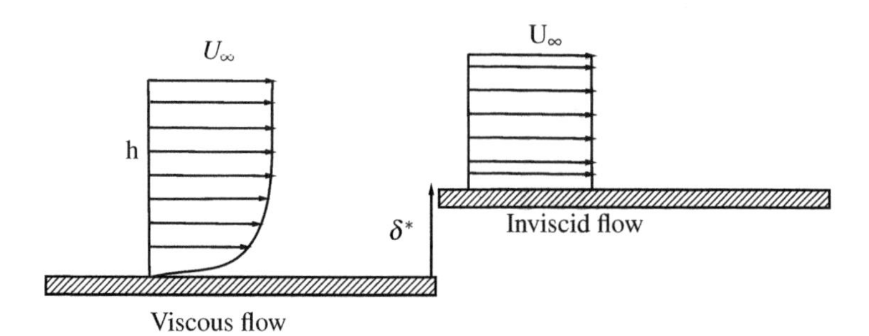

Figure 11.1 The concept of displacement thickness in boundary layers is related with mass flux in viscous and inviscid flows.

definition.

$$\int_0^h u dy = U_\infty(h - \delta^*)$$

$$\delta^* U_\infty = U_\infty h - \int_0^h u dy$$

$$\delta^* U_\infty = U_\infty \int_0^h dy - \int_0^h u dy \tag{11.1}$$

Momentum thickness: Momentum thickness is the distance by which streamlines are displaced due to the presence of a boundary layer. Figure 11.2 shows schematically the deflection of streamlines.

$$\delta^* U_\infty = \int_0^h (U_\infty - u) dy$$

$$\delta^* = \int_0^\delta (1 - \frac{u}{U_\infty}) dy \tag{11.2}$$

Loss of momentum due to boundary layer = $\dot{M}_A$ -$\dot{M}_B$

$$\rho U_\infty^2 \theta = \rho U_\infty^2 h - \int_0^h \rho u^2 dy - \rho U_\infty^2 \delta^*$$

$$\theta = \int_0^\delta \frac{u}{U_\infty}(1 - \frac{u}{U_\infty}) dy \tag{11.3}$$

Drag is equal to reduction in momentum. Drag =

$$\int_0^x \tau_{wall} b dx$$

where b is the width of the plate

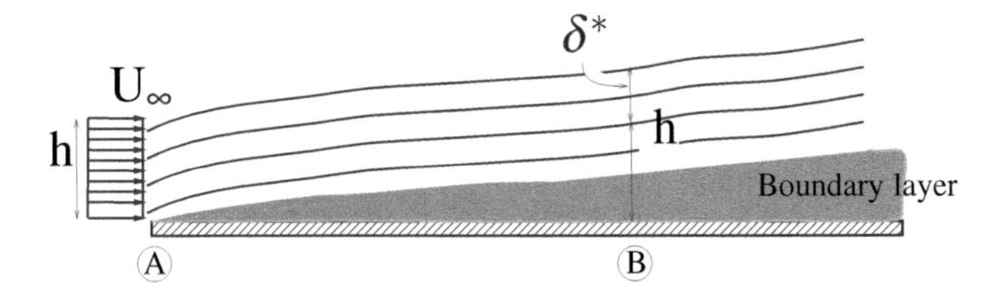

Figure 11.2 Streamlines displacement due to formation of the boundary layer indicated graphically by displacement thickness.

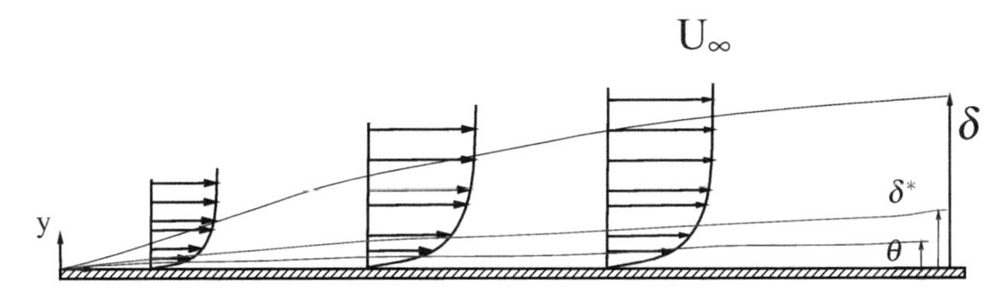

Figure 11.3 Various thicknesses inside boundary layer.

$$\frac{d(Drag)}{dx} = \rho b U_\infty^2 \frac{d\theta}{dx}$$

$$\frac{d(Drag)}{dx \cdot b} = \tau_{wall} = \rho U_\infty^2 \frac{d\theta}{dx}$$

The usefulness of this relationship lies in the ability to easily obtain approximate boundary layer results by using rather crude assumptions. For example, if we knew the detailed velocity profile in the boundary layer, we could evaluate the drag.

Shape factor (H): The ratio of displacement thickness to momentum thickness is known as shape factor and it gives a quick measure to determine the type of boundary layer. For laminar flows, H is less than or equal to 2.6 and for turbulent flows it is 2.25 and beyond.

11.1 ORDER OF MAGNITUDE OR SCALE ANALYSIS

Boundary layer can be considered as a zone where the diffusion of vorticity by viscous effects is significant compared with convective/advective effects. We now estimate different terms in the Navier-Stokes equations and make decsions about their importance for external flat plate boundary layer flows. This approach is also called the boundary layer approximation. The conservation equations for a steady, incompressible, 2D boundary layer flow are:

$$\frac{\partial u}{\partial x} + \frac{\partial v}{\partial y} = 0 \tag{11.4}$$

$$u\frac{\partial u}{\partial x} + v\frac{\partial u}{\partial y} + w\frac{\partial u}{\partial z} = -\frac{1}{\rho}\nabla P + \nu\nabla^2 u + f_x \tag{11.5}$$

$$u\frac{\partial v}{\partial x} + v\frac{\partial v}{\partial y} + w\frac{\partial v}{\partial z} = -\frac{1}{\rho}\nabla P + \nu\nabla^2 v + f_y \tag{11.6}$$

We can estimate the order of magnitude of various terms in the above set of equations. We know that x cannot be larger than L; velocity cannot be larger than the free stream velocity for non accelerating flow conditions. For the boundary layer, the viscous zone cannot be larger than the thickness of the boundary layer. Hence, the characteristic dimensions are:

(i) $x \approx L$

(ii) $y \approx \delta$

(iii) $u \approx U_\infty$

We consider analyzing the viscous term of Navier-Stokes equation $\nabla^2 u$:

$$\nu\frac{\partial^2 u}{\partial x^2} + \nu\frac{\partial^2 u}{\partial y^2}$$

$$\nu\frac{U_\infty}{L^2} + \nu\frac{U_\infty}{\delta^2}$$

The order of diffusion in the vertical direction is much higher than the diffusion in the streamwise direction.

$$\frac{\partial^2 u}{\partial y^2} >> \frac{\partial^2 u}{\partial x^2}$$

From the continuity equation:

$$v \approx \frac{\partial u}{\partial x}dy \equiv \frac{U_\infty \delta}{L}$$

We now analyze the inertia force in the x-momentum equation:

$$u\frac{\partial u}{\partial x} \equiv U_\infty \frac{U_\infty}{L} \equiv \frac{U_\infty^2}{L}$$

The inertia force in y-direction in x-momentum equation gives:

$$v\frac{\partial u}{\partial y} \equiv \frac{U_\infty \delta}{L}\frac{U_\infty}{\delta} \equiv \frac{U_\infty^2}{\delta}$$

This shows that both of the inertia force terms have the same order of magnitude, and both must be retained in the x-momentum equation.

We analyze the first term of inertia force in y-momentum equation:

$$u\frac{\partial v}{\partial x} \equiv U_\infty \frac{U_\infty \delta}{L}\frac{1}{L} \equiv \frac{\delta U_\infty^2}{L^2}$$

We analyze the second term of inertia force in y-momentum equation:

$$v\frac{\partial v}{\partial y} \equiv \left(\frac{U_\infty \delta}{L}\right)\frac{\left(\frac{U_\infty \delta}{L}\right)}{\delta} \equiv \frac{1}{\delta}\left(\frac{U_\infty \delta}{L}\right)^2 \equiv \frac{\delta U_\infty^2}{L^2}$$

This shows that both of the inertia force terms in y-momentum equations have the same order of magnitude, and both must be retained in the y-momentum equation.

Introducing this in the momentum equation, the RHS shows that the order of the terms in the convective term is the same!

Comparing RHS and LHS:

$$Re = L^2/\delta^2 (function\ of\ geometry)$$

This is an important result, and it shows that for a flat plate boundary layer, the thickness of the boundary layer is a function of the Reynolds number. This analysis led us to a reduced form of the conservation equations:

$$\frac{\partial u}{\partial x} + \frac{\partial v}{\partial y} = 0 \tag{11.7}$$

$$u\frac{\partial u}{\partial x} + v\frac{\partial u}{\partial y} = -\frac{1}{\rho}\frac{\partial P}{\partial x} + \nu\left(\frac{\partial^2 u}{\partial y^2} + \cancel{\frac{\partial^2 u}{\partial x^2}}\right) \tag{11.8}$$

$$u\frac{\partial v}{\partial x} + v\frac{\partial v}{\partial y} = -\frac{1}{\rho}\frac{\partial P}{\partial y} + \nu\left(\frac{\partial^2 v}{\partial y^2} + \cancel{\frac{\partial^2 v}{\partial x^2}}\right) \tag{11.9}$$

Note that we will ignore the streamwise viscous diffusion terms, considering them less significant than the wall normal viscous diffusion terms. These equations are valid for steady, incompressible, laminar boundary layer flows over a flat plate.

11.2 BLASIUS SOLUTION FOR LAMINAR EXTERNAL FLOWS

We solve the boundary layer approximate equations:

$$\frac{\partial u}{\partial x} + \frac{\partial v}{\partial y} = 0$$

$$u\frac{\partial u}{\partial x} + v\frac{\partial u}{\partial y} = -\frac{1}{\rho}\frac{\partial P}{\partial x} + \nu\left(\frac{\partial^2 u}{\partial y^2}\right)$$

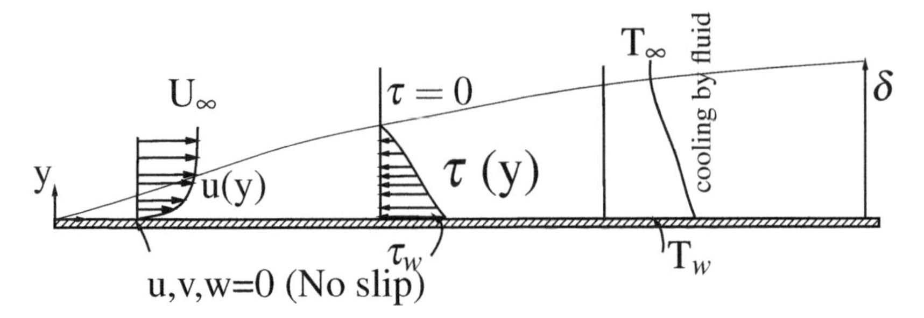

Figure 11.4 Velocity, shear stress, and temperature distributions inside boundary layer flow.

$$u\frac{\partial v}{\partial x}+v\frac{\partial v}{\partial y}=-\frac{1}{\rho}\frac{\partial P}{\partial y}+\nu\left(\frac{\partial^2 v}{\partial y^2}\right)$$

With the boundary conditions:

(i) $u(x,0)=v(x,0)=0$ (no-slip condition)

(ii) $u(x,\infty)=U_o$ (matching condition)

(iii) $u(0,y)=U_y$ (leading edge condition)

Figure 11.4 explains the boundary conditions for the boundary layer flows. Equations are parabolic in character (parabolized Navier-Stokes equations) and have no separation (no elliptic character).

By applying a coordinate transformation, Blasius reduced the partial differential equations to an ordinary differential equation that he was able to solve and provide a similarity solution.

It can be argued that in dimensionless form, the boundary layer velocity profiles on a flat plate should be similar regardless of the location along the plate. That is,

$$\frac{u}{U_o}=g\left(\frac{y}{\delta}\right)$$

We introduce the dimensionless similarity variable

$$\eta=\left(\frac{U_o}{\nu x}\right)^{1/2}y$$

and the stream function

$$\psi=(\nu xU_o)^{1/2}f(\eta)$$

$$u=U_of'(\eta),v=(\nu U_o/4x)^{1/2}(\eta f'-f)$$

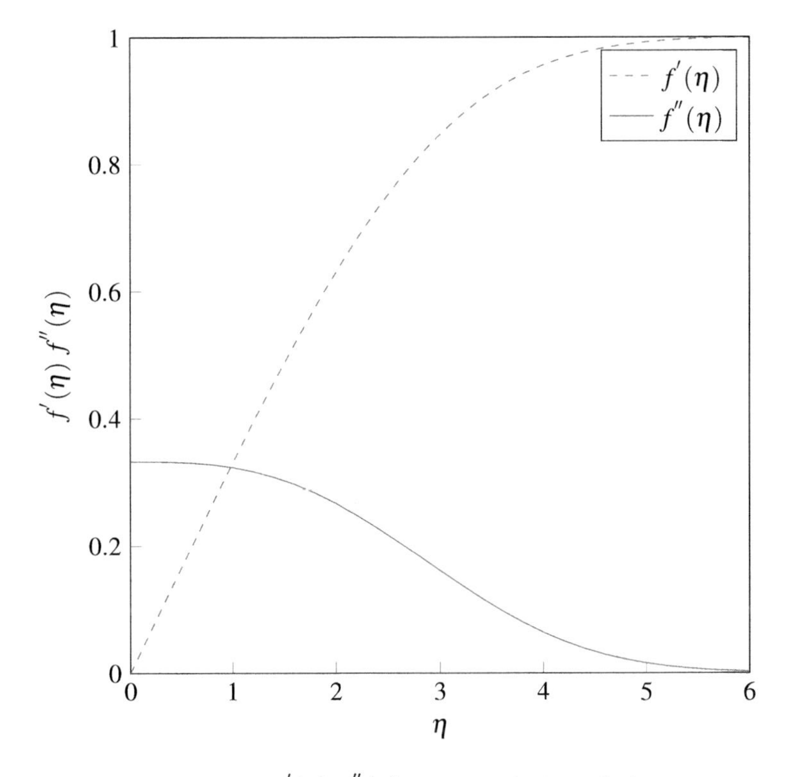

Figure 11.5 $f'(\eta)\, f''(\eta)$ vs. η in Blasius solution.

Substituting this in the boundary layer equations, a third order ordinary differential equation is obtained (boundary value problem):

$$2f''' + ff'' = 0$$

Conditions: $f, f' = 0$ at $y = 0$, $f' \to 1$ as $\eta \to \infty$. The equations are solved numerically, and Figure 11.5 shows the distribution of function f and f' inside the boundary layer. The solution at discrete points is tabulated in Table 11.1. This solution is well-known in the literature as the Blasius solution.

Now that we have the solution, we can substitute the solution into the similarity variable:

$$\eta = \left(\frac{U_o}{\nu x}\right)^{1/2} y$$

At $\eta = 5$ the u/U_o is 0.99

$$\frac{\delta}{x} = \frac{5}{\sqrt{Re}}$$

Table 11.1
Exact Solution of the ODE $2f''' + ff'' = 0$

η	f(η)	f′(η)	f″(η)	η	f(η)	f′(η)	f″(η)
0	0.0000	0.0000	**0.3326**	3.2	1.5713	0.8772	0.1392
0.2	0.0067	0.0665	0.3325	3.4	1.7494	0.9029	0.1179
0.4	0.0266	0.1330	0.3320	3.6	1.9322	0.9245	0.0981
0.6	0.0598	0.1992	0.3306	3.8	2.1189	0.9422	0.0801
0.8	0.1063	0.2651	0.3279	4	2.3089	0.9566	0.0642
1	0.1658	0.3303	0.3235	4.2	2.5014	0.9681	0.0505
1.2	0.2383	0.3944	0.3171	4.4	2.6959	0.9770	0.0389
1.4	0.3235	0.4569	0.3083	4.6	2.8920	0.9838	0.0294
1.6	0.4210	0.5175	0.2971	4.8	3.0893	0.9889	0.0218
1.8	0.5303	0.5756	0.2833	**5**	3.2875	**0.9926**	0.0159
2	0.6510	0.6307	0.2671	5.2	3.4863	0.9953	0.0113
2.2	0.7824	0.6823	0.2486	5.4	3.6855	0.9972	0.0079
2.4	0.9237	0.7300	0.2283	5.6	3.8851	0.9985	0.0054
2.6	1.0741	0.7735	0.2066	5.8	4.0849	0.9994	0.0036
2.8	1.2328	0.8126	0.1841	6	4.2849	1.0000	0.0024
3	1.3988	0.8472	0.1614				

With the velocity profile known, it is an easy matter to determine the wall shear stress, where the velocity gradient is evaluated at the plate. The value of $(\partial u/\partial y)_{wall}$ at $y = 0$ can be obtained from the Blasius solution to give:

$$\tau_{wall} = 0.322 U_o^{3/2} \sqrt{\rho \mu / x} \tag{11.10}$$

This equation gives the shear stress distribution at the wall for laminar flow.

11.3 INTEGRAL ANALYSIS FOR EXTERNAL BOUNDARY LAYER FLOWS

We now consider a body of revolution over which axisymmetric flow is moving. The boundary layer thickness δ is much smaller then the radius of curvature, R, of the object. If R is too large, in case of external flow, the flow can be treated as flow over a flat plate. Figure 11.6 shows the control volume used for the analysis. The free stream velocity is a function of the streamwise distance x along the surface. The height of the control volume is ΔY, and the thickness of the control volume is δx.

The mass flow rates:

1. Blowing:

$$\dot{m}_1 = \upsilon_o \rho_o A = (R\delta\phi)\delta x \upsilon_o \rho_o$$

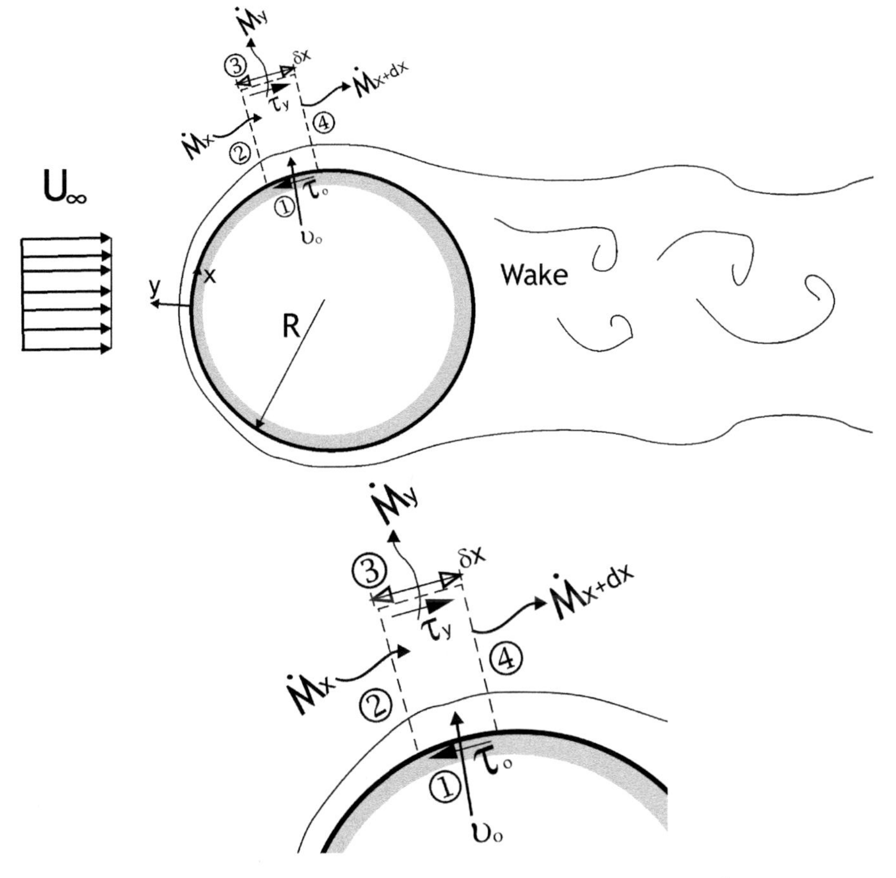

Figure 11.6 Control volume for boundary layer over the cylinder. $\dot{M}$ is momentum flow rate.

2. At inlet of control volume:

$$\dot{m}_2 = (R\delta\phi)\int_0^y \rho u dy$$

3. Using Taylor series, the mass going out of CV is estimated as:

$$\dot{m}_4 = \dot{m}_{inlet-left} + \frac{d(\dot{m}_{inlet-left})}{dx}\delta x$$

$$= (R\delta\phi)\int_0^y \rho u dy + \frac{d}{dx}\left[(R\delta\phi)\int_0^y \rho u dy\right]\delta x$$

4. Mass flow rate at the top outlet of CV:

$$\dot{m}_3 = (R\delta\phi)\delta x \upsilon_y \rho_y$$

We now do the mass balance and obtain the function:

$$\sum_{out} \dot{m} - \sum_{in} \dot{m} = 0$$

The angle $\delta\phi$ is not a function of x, so dividing the equation with $(R\delta\phi)\delta x$ will lead to

$$\begin{aligned}&\cancel{(R\delta\phi)\int_0^y \rho u dy - (R\delta\phi)\int_0^y \rho u dy}\\ &-\frac{d}{dx}\left[(R\delta\phi)\int_0^y \rho u dy\right]\delta x\\ &+(R\delta\phi)\delta x \upsilon_o \rho_o - (R\delta\phi)\delta x \upsilon_y \rho_y = 0\end{aligned}$$

$$\upsilon_y \rho_y = \frac{-1}{R}\frac{d}{dx}\left[\int_0^y R\rho u dy\right] + \upsilon_o \rho_o \tag{11.11}$$

Momentum flux balance associated with stream-wise velocity:

$$\dot{M}_x\big|_2 = (R\delta\phi)\int_0^y \rho u^2 dy$$

$$\dot{M}_{x+\delta x}\big|_4 = (R\delta\phi)\int_0^y \rho u^2 dy + \frac{d}{dx}\left[(R\delta\phi)\int_0^y \rho u^2 dy\right]\delta x$$

$$\dot{M}_y\big|_3 = (R\delta\phi)\delta x \upsilon_y \rho_y u_y$$

Since the rate of change of momentum is euqal to force, we have:

$$\sum_{net} \dot{M} + F_{pressure} + F_{viscous} = 0$$

$$\begin{aligned}&(R\delta\phi)\delta x \upsilon_y \rho_y u_y + \frac{d}{dx}\left[(R\delta\phi)\int_0^y \rho u^2 dy\right]\delta x\\ &+\frac{dP}{dx}(R\delta\phi)\delta x \Delta Y + (\tau_o - \tau_y)(R\delta\phi)\delta x = 0\end{aligned}$$

$$\upsilon_y \rho_y u_y + \frac{1}{R}\frac{d}{dx}\left[(R)\int_0^y \rho u^2 dy\right] + \frac{dP}{dx}\Delta Y + (\tau_o - \tau_y) = 0$$

$$(\tau_y - \tau_o) - \frac{dP}{dx}\Delta Y = \upsilon_y \rho_y u_y + \frac{1}{R}\frac{d}{dx}\left[(R)\int_0^y \rho u^2 dy\right] \tag{11.12}$$

We now substitute Eq. 11.11 into the above equation:

$$(\tau_y - \tau_o) - \frac{dP}{dx}\Delta Y = \left(\frac{-1}{R}\frac{d}{dx}\left[R\int_0^y \rho u dy\right] + \upsilon_o \rho_o\right) u_y + \frac{1}{R}\frac{d}{dx}\left[(R)\int_0^y \rho u^2 dy\right]$$

For ΔY sufficiently large,

$$u_y = U_\infty, \tau_y = 0$$

$$-\tau_o - \frac{dP}{dx}\Delta Y = \left(\frac{-1}{R}\frac{d}{dx}\left[R\int_0^\infty \rho u dy\right] + \upsilon_o \rho_o\right) U_\infty + \frac{1}{R}\frac{d}{dx}\left[(R)\int_0^\infty \rho u^2 dy\right]$$

From Bernoulli's equation:

$$P + \frac{1}{2}\rho_\infty U_\infty^2 = Const$$

$$-\frac{dP}{dx} = \rho_\infty U_\infty \frac{dU_\infty}{dx}$$

$$-\tau_o = \frac{1}{R}\frac{d}{dx}\left[R\int_0^\infty \rho u^2 dy\right] - \frac{U_\infty}{R}\frac{d}{dx}\left[R\int_0^\infty \rho u dy\right] + \upsilon_o \rho_o U_\infty + \int_0^\infty \rho_\infty U_\infty \frac{dU_\infty}{dx} dy$$

$$\tau_o = \frac{U_\infty}{R}\frac{d}{dx}\left[R\int_0^\infty \rho u dy\right] - \frac{1}{R}\frac{d}{dx}\left[R\int_0^\infty \rho u^2 dy\right] + \int_0^\infty \rho_\infty U_\infty \frac{dU_\infty}{dx} dy - \upsilon_o \rho_o U_\infty$$

The preceding equation is called the momentum integral equation, valid for boundary layer flows.

Now if we assume that R is not a function of x and it's large, then the problem will approximate to flow over a flat plate and can be simplified to a form:

$$\tau_o = -\frac{d}{dx}\left[\int_0^\infty \rho u^2 dy\right] + U_\infty \frac{d}{dx}\left[\int_0^\infty \rho u dy\right] + \int_0^\infty \rho_\infty U_\infty \frac{dU_\infty}{dx} dy - \upsilon_o \rho_o U_\infty$$

$$\tau_o = \underbrace{-\frac{1}{R}\frac{d}{dx}\left[R\int_0^\infty \rho u^2 dy\right]}_{I} \underbrace{-\upsilon_o\rho_o U_\infty}_{II} + \underbrace{\frac{U_\infty}{R}\frac{d}{dx}\left[R\int_0^\infty \rho u dy\right]}_{III} + \underbrace{\rho_\infty U_\infty \frac{dU_\infty}{dx}\delta}_{IV}$$

For simplification purposes, we take:

$$\beta(x,y) = R\int_0^\infty \rho u dy$$

$$\frac{d}{dx}\left[\beta(x,y)U_\infty\right] = U_\infty \frac{d\beta(x,y)}{dx} + \beta(x,y)\frac{dU_\infty}{dx}$$

$$U_\infty \frac{d\beta(x,y)}{dx} = \frac{d}{dx}\left[\beta(x,y)U_\infty\right] - \beta(x,y)\frac{dU_\infty}{dx}$$

$$\frac{U_\infty}{R}\frac{d}{dx}\left[R\int_0^\infty \rho u dy\right] = \frac{1}{R}\frac{d}{dx}\left[U_\infty\left(R\int_0^\infty \rho u dy\right)\right] - \frac{1}{R}\frac{dU_\infty}{dx}\left(R\int_0^\infty \rho u dy\right)$$

We now consider only terms I and III after the above substitution:

$$= \underbrace{+\frac{U_\infty}{R}\frac{d}{dx}\left[R\int_0^\infty \rho u dy\right]}_{III} \underbrace{-\frac{U_\infty}{R}\frac{d}{dx}\left[\frac{R}{U_\infty}\int_0^\infty \rho u^2 dy\right]}_{I}$$

$$= -\frac{1}{R}\frac{d}{dx}\left[R\int_0^\infty \rho u^2 dy\right] + \frac{U_\infty}{R}\frac{d}{dx}\left[R\int_0^\infty \rho u dy\right]$$

$$= -\frac{1}{R}\frac{d}{dx}\left[R\int_0^\infty \rho u^2 dy\right] + \frac{1}{R}\frac{d}{dx}\left[U_\infty\left(R\int_0^\infty \rho u dy\right)\right] - \frac{1}{R}\frac{dU_\infty}{dx}\left(R\int_0^\infty \rho u dy\right)$$

$$= -\frac{1}{R}\frac{d}{dx}\left[R\int_0^\infty \rho u^2 dy - U_\infty\left(R\int_0^\infty \rho u dy\right)\right] - \frac{1}{R}\frac{dU_\infty}{dx}\left(R\int_0^\infty \rho u dy\right)$$

$$= -\frac{1}{R}\frac{d}{dx}\left[R\int_0^\infty \rho(u^2 - U_\infty u)dy\right] - \frac{1}{R}\frac{dU_\infty}{dx}\left(R\int_0^\infty \rho u dy\right)$$

$$= -\frac{1}{R}\frac{d}{dx}\left[RU_\infty^2\int_0^\infty \rho\left(\frac{u^2 - U_\infty u}{U_\infty^2}\right)dy\right] - \frac{1}{\not{R}}\frac{dU_\infty}{dx}\left(\not{R}\int_0^\infty \rho u dy\right)$$

$$= -\frac{1}{R}\frac{d}{dx}\left[RU_\infty^2\int_0^\infty \rho\frac{u}{U_\infty}\left(\frac{u - U_\infty}{U_\infty}\right)dy\right] - \frac{1}{\not{R}}\frac{dU_\infty}{dx}\left(\not{R}\int_0^\infty \rho u dy\right)$$

$$= -\frac{\rho}{R}\frac{d}{dx}\left[RU_\infty^2\int_0^\infty \frac{u}{U_\infty}\left(\frac{u - U_\infty}{U_\infty}\right)dy\right] - \frac{1}{\not{R}}\frac{dU_\infty}{dx}\left(\not{R}\int_0^\infty \rho u dy\right)$$

$$= \frac{\rho}{R}\frac{d}{dx}\left[RU_\infty^2\theta\right] - \frac{dU_\infty}{dx}\left(\int_0^\infty \rho u dy\right)$$

Forces balanced will take form:

$$\tau_o = -\underbrace{\upsilon_o\rho_o U_\infty}_{II} + \underbrace{\frac{\rho}{R}\frac{d}{dx}\left[RU_\infty^2\theta\right]}_{V} - \underbrace{\frac{dU_\infty}{dx}\left(\int_0^\infty \rho u dy\right)}_{VI} + \underbrace{\rho_\infty U_\infty\frac{dU_\infty}{dx}\delta}_{IV}$$

Consider terms IV and VI only:

$$\Rightarrow +\underbrace{\rho_\infty U_\infty\frac{dU_\infty}{dx}\delta}_{IV} - \underbrace{\frac{dU_\infty}{dx}\left(\int_0^\infty \rho u dy\right)}_{VI}$$

$$= \int_0^\infty \rho_\infty U_\infty\frac{dU_\infty}{dx}dy - \frac{dU_\infty}{dx}\left(\int_0^\infty \rho u dy\right)$$

$$= \frac{dU_\infty}{dx}\int_0^\infty (\rho_\infty U_\infty - \rho u)dy$$

$$= \rho\frac{dU_\infty}{dx}\int_0^\infty (U_\infty - u)dy$$

$$= \rho\frac{1}{U_\infty}\frac{dU_\infty}{dx}\int_0^\infty \left[1 - \frac{u}{U_\infty}\right]dy$$

$$= \frac{\rho}{U_\infty}\frac{dU_\infty}{dx}\delta^*$$

The force balance takes a new form:

$$\tau_o = -\upsilon_o \rho_o U_\infty + \frac{\rho}{R}\frac{d}{dx}\left[RU_\infty^2\theta\right] + \frac{\rho}{U_\infty}\frac{dU_\infty}{dx}\delta^*$$

$$\tau_o = -\upsilon_o \rho_o U_\infty + \frac{\rho}{R}\frac{d}{dx}\left[RU_\infty^2\theta\right] - \frac{\rho}{U_\infty}\frac{dU_\infty}{dx}\delta^*$$

Expand term 2:

$$\frac{1}{R}\frac{d}{dx}\left[RU_\infty^2\theta\right] = \frac{1}{R}\left[U_\infty^2\frac{d}{dx}[R\theta] + R\theta\frac{d}{dx}\left[U_\infty^2\right]\right]$$

$$= \frac{1}{R}\left[U_\infty^2\left(R\frac{d\theta}{dx} + \theta\frac{dR}{dx}\right) + R\theta\frac{d}{dx}\left[U_\infty^2\right]\right]$$

$$= U_\infty^2\frac{d\theta}{dx} + \frac{\theta U_\infty^2}{R}\frac{dR}{dx} + \theta\frac{d}{dx}\left[U_\infty^2\right]$$

New force balance:

$$\frac{\tau_o}{\rho U_\infty^2} = -\frac{\upsilon_o \rho_o}{\rho U_\infty} + \frac{d\theta}{dx} + \frac{\theta}{R}\frac{dR}{dx} + \frac{1}{U_\infty}\frac{dU_\infty}{dx}(2\theta + \delta^*)$$

$$\frac{\tau_o}{\rho U_\infty^2} = -\frac{\upsilon_o \rho_o}{\rho U_\infty} + \frac{d\theta}{dx} + \theta\left[\frac{1}{R}\frac{dR}{dx} + \frac{1}{U_\infty}\frac{dU_\infty}{dx}(2 + \frac{\delta^*}{\theta})\right]$$

The ratio of displacement thickness to momentum thickness is called shape factor (H):

$$H = \frac{\delta^*}{\theta}$$

Also, for $R \longrightarrow \infty$, we can reduce the equation for the flat plate:

$$\frac{\tau_o}{\rho U_\infty^2} = -\frac{\rho_o \upsilon_o}{\rho U_\infty} + \frac{d\theta}{dx} + \theta\left[\frac{1}{U_\infty}\frac{dU_\infty}{dx}(2 + H)\right]$$

11.4 LAMINAR FLOW WITHOUT PRESSURE GRADIENT

The non accelerating laminar boundary layer flow over a flat plate occurs when zero pressure gradient is applied. In case of no suction and blowing, the momentum integral equation would reduce to the form:

$$\frac{\tau_o}{\rho U_\infty^2} = \frac{d\theta}{dx}$$

Many engineering problems involving laminar boundary layer flows can be solved just by measuring the velocity profile in the near wall region. We will now demonstrate that a judiciously selected "reasonable" profile can predict the skin friction drag comparable to Blasius's exact solution. We may assume a polynomial as:

$$\frac{u}{U_\infty} = \alpha_o + \alpha_1\left(\frac{y}{\delta}\right) + \alpha_2\left(\frac{y}{\delta}\right)^2 + \alpha_3\left(\frac{y}{\delta}\right)^3$$

(i) At $y = 0,\ (u, v) = (0, 0)$ [No slip condition]

(ii) At $y = 0,\ \tau_w = du/dy$ [Newton's law of viscosity]

(iii) At $y = \delta,\ du/dy = 0$ [Edge of boundary layer]

(iv) At $y = 0$, from momentum equation for zero pressure gradient flows, we have:

$$\cancel{u\frac{\partial u}{\partial x}} + \cancel{v\frac{\partial u}{\partial y}} = \cancel{-\frac{1}{\rho}\frac{\partial p}{\partial x}} + \nu\left(\frac{\partial^2 u}{\partial y^2}\right)$$

$$\left(\frac{\partial^2 u}{\partial y^2}\right) = 0$$

Using the above conditions, we arrive at coefficients:

$$\frac{u}{U_\infty} = (0) + \frac{3}{2}\left(\frac{y}{\delta}\right) + (0)\left(\frac{y}{\delta}\right)^2 - \frac{1}{2}\left(\frac{y}{\delta}\right)^3$$

$$\frac{u}{U_\infty} = \frac{3}{2}\left(\frac{y}{\delta}\right) - \frac{1}{2}\left(\frac{y}{\delta}\right)^3$$

Displacement thickness:

$$\delta^* = \int_0^{\delta(x)} \left[1 - \frac{u}{U_\infty}\right] dy = \int_0^{\delta(x)} \left[1 - \left(\frac{3}{2}\left(\frac{y}{\delta}\right) - \frac{1}{2}\left(\frac{y}{\delta}\right)^3\right)\right] dy$$

$$\delta^* = \frac{3\delta}{8}$$

Momentum thickness:

$$\theta = \int_0^{\delta(x)} \frac{u}{U_\infty}\left[1 - \frac{u}{U_\infty}\right] dy = \frac{39}{280}\delta(x)$$

Shear stress:

$$\tau_w = \mu\frac{du}{dy} = \mu U_\infty\left(\frac{3}{2\delta(x)} - \frac{3}{2}\frac{y^2}{\delta(x)^3}\right)_{y=0} = \mu U_\infty\left(\frac{3}{2\delta(x)}\right)$$

From the momentum equation, we have:

$$\frac{\tau_w}{\rho U_\infty^2} = \frac{d\theta}{dx}$$

$$\frac{\mu U_\infty}{\rho U_\infty^2}\left(\frac{3}{2\,\delta(x)}\right) = \left(\frac{39}{280}\frac{d}{dx}\delta(x)\right)$$

Table 11.2
Comparison of Various results Obtained Using Reasonable Velocity Profiles for a Laminar Boundary Layer with the Blasius's Exact Solution ($\eta = y/\delta$)

$\frac{u}{U_\infty}$	$\frac{\delta\sqrt{Re_x}}{x}$	$\frac{\delta^*\sqrt{Re_x}}{x}$	$\frac{\theta\sqrt{Re_x}}{x}$	$\frac{\tau_w\sqrt{Re_x}}{\rho U_\infty^2}$
Blasius solution	5	1.721	0.664	0.332
$2\eta - \eta^2$	5.48	1.826	0.730	0.365
$\frac{3}{2}\eta - \frac{1}{2}\eta^3$	4.64	1.740	0.646	0.323
$2\eta - 2\eta^3 + \eta^4$	5.84	1.751	0.685	0.343
$sin(\pi\eta/2)$	4.79	1.743	0.655	0.328

$$\frac{\mu U_\infty}{\rho U_\infty^2}\left(\frac{3}{2}\right)\left(\frac{280}{39}\right)dx = \delta d\delta$$

$$\frac{\mu U_\infty}{\rho U_\infty^2}\left(\frac{6}{2}\right)\left(\frac{280}{39}\right)\frac{x}{x^2} = \frac{\delta^2}{x^2}$$

$$\frac{\mu}{\rho U_\infty}\left(\frac{6}{2}\right)\left(\frac{280}{39}\right)\frac{x}{x^2} = \frac{\delta^2}{x^2}$$

$$\frac{\delta}{x} = \sqrt{\left(\frac{280}{13}\right)\left(\frac{\mu}{\rho U_\infty x}\right)} = \frac{4.46}{\sqrt{Re_x}}$$

Note that this value of 4.64 is under-predicting the value of 5 (exact solution) just by 7.2%. Another reasonable profile:

$$\frac{u(y)}{U_\infty} = \sin\left(\frac{1}{2}\frac{\pi y}{\delta(x)}\right)$$

can predict a much closer value of 4.79.

Example 11.1

Example 11.1 Water is flowing over a thin splitter plate that is 12 in. long and 4 ft wide, installed in a wind tunnel. The splitter plate can be treated as a small flat plate. The free stream velocity is $U_\infty = 4$ ft/s. Approximate the velocity profile in the boundary layer as parabolic and following relation

$$\frac{u}{U_\infty} = 2\eta - \eta^2$$

where $\eta = y/\delta$. Estimate the total drag experienced by the splitter plate. The water kinematic viscosity is $\nu = 1.08 \times 10^{-5}$ ft^2/s and $\rho = 1.94$ slug/ft^3.

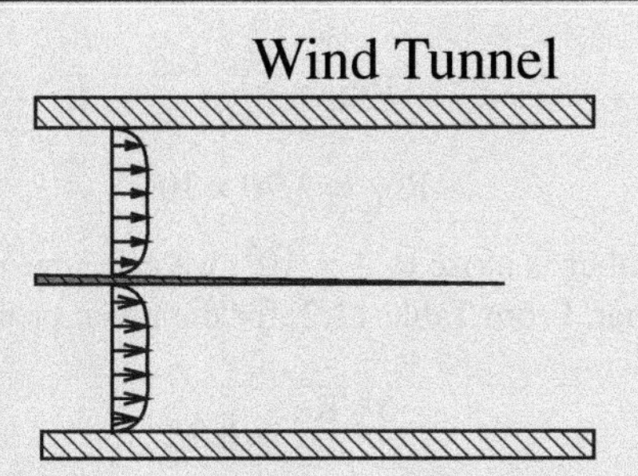

Figure 11.7 Flow over splitter plate.

Solution We make the assumptions that

1. Splitter plate is flat and hence $dp/dx = 0$, and $U_\infty =$ constant
2. Boundary layer thickness δ is a function of x only.
3. Flow is incompressible.
4. Flow is Steady flow.

The momentum integral equation can be simplified for a flat plate

$$\frac{\tau_w}{\rho} = \frac{d}{dx}\left(U_\infty^2 \cdot \theta\right)$$

where

$$\theta = \int_0^\delta \frac{u}{U_\infty} \cdot \left(1 - \frac{u}{U_\infty}\right) dy$$

Form momentum integral equation for flow over a flat plate with $U_\infty =$ const

$$\tau_w = \rho \cdot U_\infty^2 \cdot \frac{d\theta}{dx}$$

The drag force is then

$$F_D = \int \tau_w dA = \int_0^L \tau_w \cdot w dx$$

$$F_D = \int_0^L \rho \cdot U_\infty^2 \cdot \frac{d\theta}{dx} \cdot w\, dx$$

$$F_D = \rho \cdot U_\infty^2 \cdot w \cdot \int_0^{\theta_L} 1\, d\theta$$

$$F_D = \rho \cdot U_\infty^2 \cdot w \cdot \theta_L$$

The Reynolds number over the plate is

$$\mathrm{Re}_L = \frac{U_\infty \cdot L}{\nu}$$

$$\mathrm{Re}_L = 3.70 \times 10^5$$

As Reynolds number is close to 3×10^5, we can approximate the flow over the plate as laminar. From Table 11.2, for the given profile

$$\frac{\theta\sqrt{\mathrm{Re}}}{L} = 0.73$$

$$\theta_L = 0.01439 \cdot in$$

$$F_D = \rho \cdot U_\infty^2 \cdot w \cdot \theta_L$$

$$F_D = 0.14893 \cdot \mathrm{lb}_f$$

11.5 INTEGRAL ANALYSIS FOR TURBULENT BOUNDARY LAYER FLOWS

The momentum integral equation that we derived in the previous section is independent of the type of flow field. Considering the momentum equation again,

$$\frac{\tau_o}{\rho U_\infty^2} = -\frac{\upsilon_o \rho_o}{\rho U_\infty} + \frac{d\theta}{dx} + \theta\left[\frac{1}{R}\frac{dR}{dx} + \frac{1}{U_\infty}\frac{dU_\infty}{dx}(2 + \frac{\delta^*}{\theta})\right]$$

Taking a value of 1.29 for the shape factor for no suction or blowing and variable free stream velocity, the equation valid for the flat plate is

$$\frac{\tau_o}{\rho U_\infty^2} = \frac{d\theta}{dx} + \theta\left[\frac{1}{R}\frac{dR}{dx} + \frac{1}{U_\infty}\frac{dU_\infty}{dx}(2 + \frac{\delta^*}{\theta})\right]$$

Approach 1: Assuming that the turbulent profile for buffer layer

$$u^+ = 8.7\left(y^+\right)^{1/7}$$

is valid inside the whole boundary layer, we can cast the equation into the form:

$$\frac{\tau_o}{\rho U_\infty^2} = 0.0125\left(\frac{\theta U_\infty}{\nu}\right)^{-1/4} \tag{11.13}$$

giving us the form:

$$0.0125\left(\frac{\theta U_\infty}{\nu}\right)^{-1/4} = \frac{d\theta}{dx} + \theta\left[\frac{1}{R}\frac{dR}{dx} + \frac{1}{U_\infty}\frac{dU_\infty}{dx}(2 + 1.29)\right]$$

Integrating this equation would lead to form:

$$\theta = \frac{0.036\nu^{1/5}}{RU_\infty^{3.29}} \left(\int_0^x R^{5/4} U_\infty^{3.86} dx \right)^{4/5}$$

Now for a flat plate with constant free stream velocity, we have:

$$\theta = \frac{0.036\nu^{1/5}}{U_\infty^{3.29}} \left(U_\infty^{3.86} x \right)^{4/5}$$

$$\frac{\theta}{x} = \frac{0.036\nu^{1/5}}{U_\infty^{0.202}} \frac{1}{x^{1/5}} \approx \frac{0.036\nu^{1/5}}{U_\infty^{0.2}} \frac{1}{x^{1/5}}$$

$$\frac{\theta}{x} = \frac{0.036}{\mathrm{Re}_x^{1/5}}$$

Now inserting the momentum thickness expression into Eq. 11.13 we obtain

$$\frac{C_f}{2} = \frac{0.0287}{\mathrm{Re}_x^{1/5}}$$

The relation is valid till $\mathrm{Re} = 10^6$.

Approach 2: A similar type of relation can be obtained for the turbulent boundary layer thickness using empirical data-fitted one-seventh power law

$$u/U_\infty = (y/\delta)^{1/7}$$

and substituting the shear stress distribution from the experimental data:

$$\tau_{wall} = 0.0225\rho U_\infty^2 \left(\frac{U_\infty \delta}{\nu} \right)^{-1/4}$$

Substituting in momentum integral equation, we obtain:

$$\frac{\delta}{x} = \frac{0.37}{Re^{1/5}}$$

Shear stress in terms of Reynolds number is:

$$\frac{\tau_{wall}}{\rho U_\infty^2} = \frac{0.0288}{Re^{1/5}}$$

$$\frac{C_f}{2} = \frac{0.0288}{\mathrm{Re}_x^{1/5}}$$

This formula can be used to find the local skin friction coefficient.

11.6 COMBINED LAMINAR AND TURBULENT FLOW

In previous sections, we discussed the laminar and turbulent boundary layers separately over a flat plate. Often the plate length is long enough, and the flow goes into the transition from laminar to turbulent flow. The transition from laminar to turbulent flow in the boundary layer depends on the local critical Reynolds numbers. There are two critical Reynolds numbers: lower critical Reynolds numbers and upper critical Reynolds numbers. On the flat plate, the lower critical value at which the laminar boundary layer becomes unstable is less well-defined than the critical Reynolds number for the pipe flow. In literature, a different range of Reynolds numbers was suggested. The most widely held view is that the transition occurs between $\mathrm{Re} = 3 \times 10^5$ and $\mathrm{Re} = 5 \times 10^5$.

The transition boundary layer analysis is omitted in this book. We will assume that the laminar flow is instantly converted into turbulent flow, and the whole transition boundary layer is treated like a turbulent boundary layer. Figure 11.8 shows the skin-friction drag coefficient in laminar and turbulent flows. Notice that there is a drop is skin friction coefficient, and the need to form a combined relation for wall distribution of shear stress is evident.

11.6.1 SKIN FRICTION FOR COMPLETE SURFACE

We have already established the skin friction coefficients relations for laminar and turbulent boundary layers. Using Blasius's solution, we obtain the surface averaged skin friction coefficient for laminar boundary layer:

$$C_{F,laminar} = \frac{1.328}{\sqrt{\mathrm{Re}_L}}$$

where C_F is the surface-averaged skin friction coefficient. We obtain the surface averaged skin friction coefficient for turbulent boundary layer:

$$C_{f,x} = \frac{2 \times 0.0288}{\left(\frac{U \cdot x}{\nu}\right)^{\frac{1}{5}}}$$

$$C_F = \frac{1}{L}\int_0^L C_{f,x}dx = \frac{0.072}{\left(\frac{U \cdot L}{\nu}\right)^{1/5}}$$

Some researchers have proposed the relation:

$$C_{F,turbulent} = \frac{0.074}{\mathrm{Re}_L^{1/5}}$$

which we prefer to use here, as it gives a slightly higher value of skin friction coefficient. We can also do the surface averaging for the entire length of the surface:

$$C_F = \frac{1}{L}\left[\int_0^{x_{cr}} C_{f,laminar}dx + \int_{x_{cr}}^L C_{f,turbulent}dx\right]$$

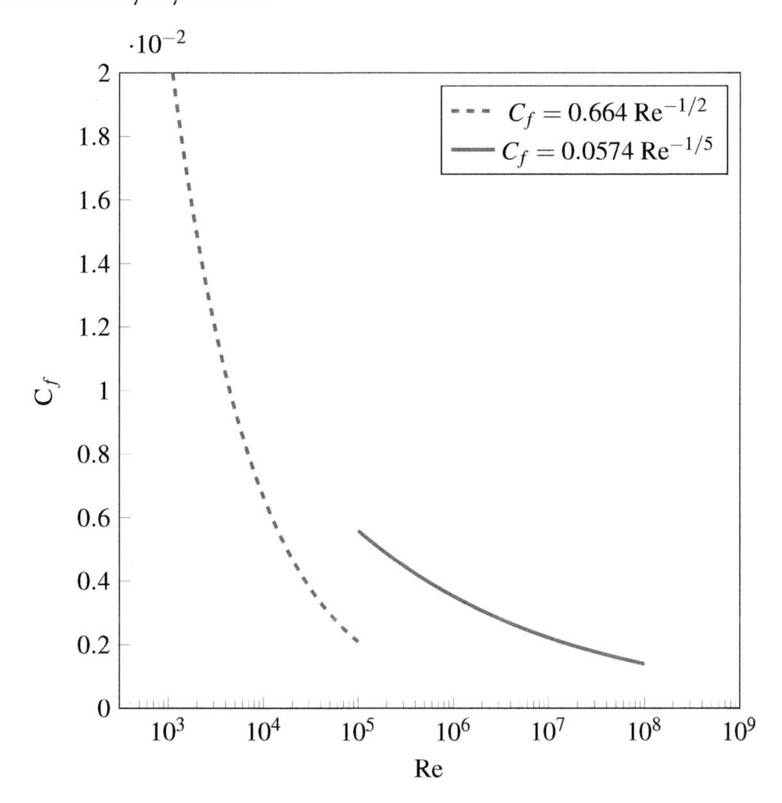

Figure 11.8 Skin-friction drag coefficient in laminar and turbulent flows. A rise in drag is expected as flow goes into transition.

which leads to mean skin friction coefficient for the entire surface

$$C_F = \frac{0.074}{\mathrm{Re}_L^{1/5}} - \frac{1742}{\mathrm{Re}_L} \quad 5 \times 10^5 \leq \mathrm{Re}_L \leq 10^7$$

This relation is valid for smooth surfaces.

11.6.2 MOMENTUM THICKNESS-BASED APPROACH

A better way to identify the transition is to define the Reynolds number based on momentum thickness. Also, if we consider combined laminar and turbulent flows, the estimation of momentum thickness will help us form a relationship between the quantities. For any boundary layer flow without pressure gradient, the local skin friction coefficient is related to momentum thickness

$$\frac{C_f(x)}{2} = \frac{\tau_w}{\rho U_\infty^2} = \frac{d\theta}{dx}$$

We now integrate this equation for the length of the plate and define the C_F for friction coefficient integrated over the whole surface as:

$$Total\ drag = \int \tau_w dx = \int \rho U_\infty^2 \frac{d\theta}{dx} dx = \rho U_\infty^2 \theta$$

which leads to equation:

$$\frac{C_F}{2} \approx \frac{\theta}{x}$$

The experimental measurements shows that

$$\frac{C_F}{2} \approx \frac{\theta}{x} \approx \frac{0.074}{2} \mathrm{Re}_L^{-1/5}$$

Now if x_o is the virtual origin of the turbulent boundary layer then

$$\frac{\theta}{x - x_o} = \frac{0.074}{2} \underbrace{\left[\frac{(x - x_o)U_\infty}{\nu}\right]^{-1/5}}_{\mathrm{Re}}$$

since $x = x_t, \theta = \theta_t$:

$$x_t - x_o = \frac{\theta_t^{5/4} U_\infty^{1/4}}{0.037^{5/4} \nu^{1/4}}$$

$$x_t - x_o = \frac{0.664^{5/4} \nu^{3/8} x_t^{5/8}}{0.037^{5/4} U_\infty^{3/8}} = 36.9 x_t \left(\frac{\nu}{x_t U_\infty}\right)^{3/8}$$

$$\left(\frac{x_o}{x_t}\right) = 1 - \left(\frac{36.9}{\mathrm{Re}_t^{3/8}}\right)$$

Example 11.2

Example Air flows over a 5-m long flat plate at a speed of 20 m/s. The density of air is 1.22 kg/m^3 and kinematic viscosity is 1.5×10^{-5} m^2/s. The flow near the leading edge is laminar till 0.3 m, where a small slot in the plate causes the flow to become turbulent. Ignoring the depth of the slot, investigate the drag experienced by the flow if the abrupt transition caused a 20% increase in the momentum thickness compared to laminar boundary layer momentum thickness. Find the total skin friction drag experienced by the plate.

Solution We have:

$x =$ length of the plate (5 m)

$x_t =$ location of transition (0.3 m)

x_o = virtual origin of turbulent boundary layer

In this example, at the transition location:

$$\theta_{lam} \neq \theta_{turb} \; or \; \theta_t$$

as assumed in the previous section.

We calculate first the Reynolds number till the laminar boundary layer:

$$Re = \frac{x_t \cdot U_\infty}{\nu}$$

This gives Re = 4×10^5, so momentum thickness associated with laminar boundary layer

$$\theta_{lam} = 0.664 x_t \left[\frac{1}{(U_\infty x_t)/\nu} \right]^{1/2} = 0.664 \times 0.3 \left[\frac{1}{(20 \times 0.3)/1.5 \times 10^{-5}} \right]^{1/2}$$

$$\theta_{lam} = 0.315mm$$

This value will be increaed by 20%, so

$$\theta_t = 0.3779mm$$

The virtual origin is estimated as:

$$x_o = x_t - \left(\frac{Re^{\frac{1}{5}} \cdot (1 + \%age) \cdot \theta_{lam}}{0.037 \cdot x_t^{\frac{1}{5}}} \right)^{\frac{5}{4}} = 0.195$$

Now we compute momentum thickness again at the end of the plate:

$$\frac{\theta}{x - x_o} = \frac{0.074}{2} \left[\frac{(x - x_o) U_\infty}{\nu} \right]^{-1/5}$$

$$\theta = 0.784mm$$

$$C_F \approx \frac{2\theta}{x} = 0.00313$$

11.6.3 SURFACE ROUGHNESS EFFECT

It is found that the skin friction coefficient increases several-fold in magnitude on rough surfaces in cases of turbulent flows. Schlichting proposed a curve fit for empirical data of average friction coefficient for turbulent flow over a flat plate as:

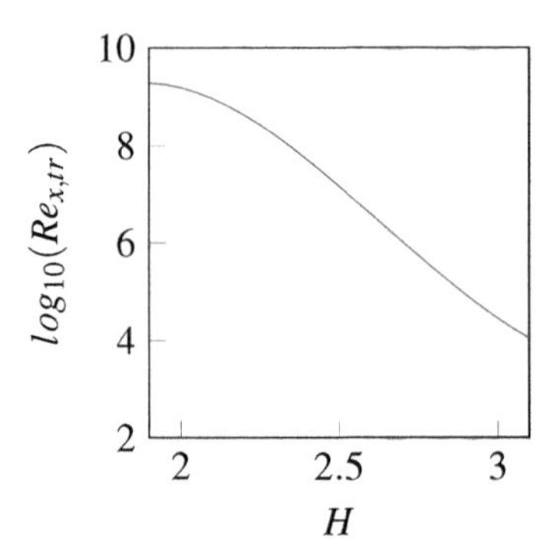

Figure 11.9 Transition Reynolds number vs. H.

$$C_f = \left[1.89 - 1.62\log\left(\frac{\varepsilon}{L}\right)\right]^{-2.5}$$

where ε is the surface roughness and L is the length of the surface.

11.6.4 SHAPE FACTOR (H) AND TRANSITION

The shape factor is an important parameter to study the growth of a boundary layer, and it has been correlated with the transition Reynolds number as well.

Figure 11.9 shows the variation of transition Reynolds number versus the shape factor. Cebeci amd Smith (1974) proposed the correlation in terms of Reynolds number based on stream-wise distance x.

$$\mathrm{Re}_\theta = 1.174\mathrm{Re}_x^{0.46}\left[1 + \frac{22400}{\mathrm{Re}_x}\right] \qquad 10^5 < \mathrm{Re}_x < 4\times10^7$$

Michel also proposed relation in terms of stream-wise Reynolds number as:

$$\mathrm{Re}_\theta = 2.9\left(\mathrm{Re}_{x,tr}^{0.4}\right)$$

For a flat plate boundary layer, the minimum $\mathrm{Re}_{x,tr}$ is 2×10^4, so Re_θ lies between 152 and 236. According to Cebeci and Smith (1974), their correlation is more accurate, which gives the value of $\mathrm{Re}_\theta = 236$. Also, Kays et al. suggested to use $\mathrm{Re}_\theta = 162$ for boundary layer transition cases in general. Bejan has analyzed the data for several type of flows and found that minimum $\mathrm{Re}_\lambda \simeq 94$ for laminar boundary layer flows. This can be used to calculate the size of wavelength in transition modeling.

PROBLEMS

11P-1 A spillway crest discharges 2.83 m^3/s on a flat surface that is inclined at 45°. The depth of water at the top is 1.524 m, and the velocity profile is:

$$\frac{u}{U_o} = \left(\frac{y}{\delta}\right)^{1/7}$$

Find the mean shear stress on the inclined surface, if further down the depth is 1.13 m and the velocity outside is 8.8 m/s.

11P-2 The test section of the wind tunnel is square with 4 ft sides. The length of test section is 10 ft. What is the pressure drop in water gauge at the end of the test section.

11P-3 Find the power wasted if the flow is moving in a diverging conical pipe. At a certain section, the diameter is 10 ft. The boundary layer is 0.5 in. thick and the velocity profile inside boundary layer is

$$\frac{u}{U_o} = \frac{1}{2}\left[1 - \cos\left(\frac{\pi y}{\delta}\right)\right]$$

11P-4 Air flows in the entrance region of a square duct of a wind tunnel, as shown in Figure 11.10. The velocity is uniform, $U_o = 30$ m/s, and the duct is 50 cm × 50 cm. At a section, marked as station 2, which is 1 m downstream from the entrance, the displacement thickness on each wall measures 0.9 mm. Find the pressure difference between stations 1 and 2.

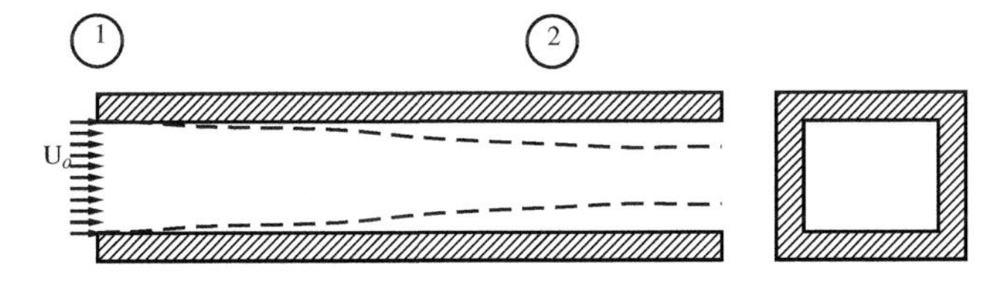

Figure 11.10 Figure of problem 11P-4.

11P-5 A laminar boundary layer velocity profile over a flat plate is expressed as

$$\frac{u}{U} = \frac{3}{2}\frac{y}{\delta} - \frac{1}{2}\left(\frac{y}{\delta}\right)^3$$

Does this distribution mathematically satisfy the boundary conditions applicable to the boundary-layer? Estimate δ^* and θ in terms of δ.

11P-6 Air flows through a cylindrical pipe of diameter $D = 120$ mm. At a section 1, a few meters from the entrance, the turbulent boundary layer is of thickness $\delta_1 = 5.43$mm. Farther downstream the boundary layer is of thickness $\delta_2 = 27$ mm. The velocity profile in the boundary layer is following the 1/7th power law. If the velocity in the inviscid central core is $U_1 = 10.5$ m/s, find the velocity at station 2, U_2, and the pressure drop between the two sections.

11P-7 Water flows with uniform velocity of 200 ft/s in the entrance region of a square duct that is 5 in. per side. At a section 1 ft downstream from the entrance, the displacement thickness, δ^*, on each wall measures 0.035 in. Determine the pressure change between sections 1 and 2.

11P-8 Air flows over a horizontal smooth flat plate of 3 m in length at free stream velocity of $U = 50\,\mathrm{m/s}$. The width of the plate is 0.8 m. The boundary layer is made turbulent by a trip wire at $x_t = 0.4$ m. Estimate the skin friction drag on the portion of the plate between $x_t = 0.4$ m and the trailing edge of the plate. The velocity profile is well represented by the $\frac{1}{7}$th power expression. Take $\rho = 1.22\,\mathrm{kg/m^3}$; $\nu = 0.146 \times 10^{-4}\,\mathrm{m^2}/s$.

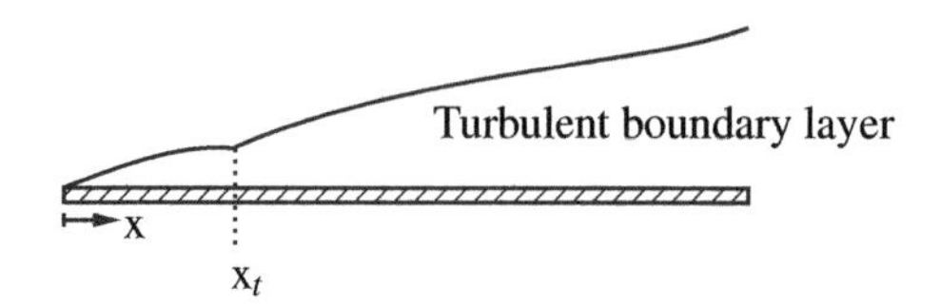

Figure 11.11 Figure of problem 11P-8.

11P-9 The laminar boundary layer over a flat plate is represented as:

$$\frac{u\,(y)}{U} = \sin\left(\frac{\pi}{2} \cdot \frac{y}{\delta}\right)$$

Find the displacement thickness, momentum thickness, and shear stress if the length of the plate is 2 m, and free stream velocity is 3 m/s. The fluid is air at 20°C.

11P-10 Turbulent flow over a flat plate is represented as

$$\frac{u(y)}{U_\infty} = \left(\frac{y}{\delta}\right)^{1/n}$$

where n is an index, which can take values 6, 7, 8, and 9. Find the shape factor (H) for the different values of n.

REFERENCES

Thwaites, Approximate Calculation of the Laminar Boundary Layer, The Aeronautical Quarterly, Volume 1 , Issue 3 , 01 November 1949 , pp. 245–280.

W. M. Kays, M. E. Crawford, and B. Weigand, Convective Heat and Mass Transfer. McGraw-Hill, USA, 2004.

Tuncer Cebeci, and A. M. O. Smith, Analysis of Turbulent Boundary Layers, Academic Press, USA, 1974.

Adrian Bejan, Convective Heat Transfer, J. Wiley, USA, 1995.

12 Free Shear Flows

Flows in fluid are classified as wall-bounded or free shear flows. Wall-bounded flows are the flows through duct or pipe. Free shear flows are the free jets, impinging jet, wall jet, and mixing layers. The behavior of these flows is quite different from wall-bounded flows.

Learning outcomes: After finishing this chapter, you should be able to:
- Identify the jet as laminar, transitional, or turbulent.
- Understand a jet's instabilities and vortical structures.
- Understand the dynamics of an impinging jet.
- Understand the use of a synthetic jet.

12.1 FREE JET

The shear layer that emanates from a nozzle exit contains a high vorticity. Large-scale structures of a free jet will lose coherence as the distance from the outlet increases. These structures will give rise to small-scale turbulence. The free jet structure can be classified on the basis of behavior of these structures. The free jet may consist of a potential core region (contains initial shear layer, till ξ/D is $\approx 4-5$), flow development region ($\xi/D \approx 5-8$), and the fully developed region after ($\xi/D \approx 8$), where ξ is the distance from the jet outlet.

Different flow regimes of a jet are (Viskanta, 1993):

The dissipated laminar jet (Re $<$ 300)

A fully laminar jet (300 $<$ Re $<$ 1000)

A transition or semi-turbulent jet (1000 $<$ Re $<$ 3000)

A fully turbulent jet (Re $>$ 3000)

A free jet (depicted in Figure 12.1) consists of large eddies that are repetitive in structure; remain coherent for downstream distances, which are very much greater than their length scales; and contribute greatly to the properties of the turbulent flows. These structures are generally called coherent structures. The jet's shear layer near the outlet is initially dominated by the Kelvin Helmholtz (KH) instability mechanism. The KH instabilities arise due to the fast moving jet flow and slow moving surrounding fluid. In a circular jet, the ring vortices are generated due to KH instabilities. Such instabilities occur frequently in commonly observable flows and can be seen in clouds (see Figure 12.2).

DOI: 10.1201/9781003315117-12

Figure 12.1 The jet spreading and entrainment phenomenon. (*Author: Hunter Rouse; Courtesy of Dr. Marian Muste, IIHR - Hydroscience & Engineering, University of Iowa. Source: Intl. Assoc. for Hydro-Environment Engr. and Research.*)

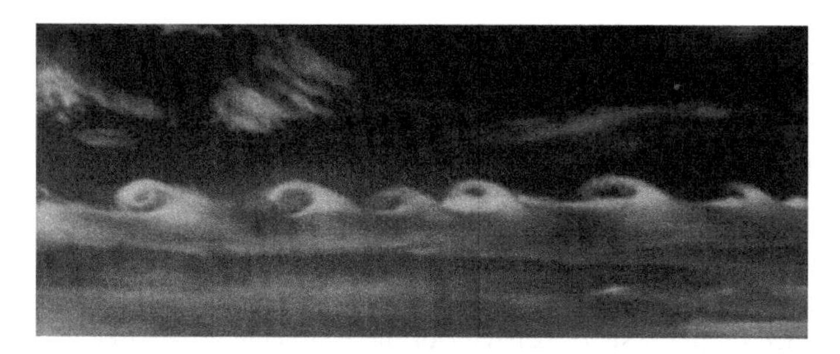

Figure 12.2 Kelvin-Helmholtz instabilities in clouds. (Courtesy of Dr. Marian Muste, IIHR - Hydroscience & Engineering, University of Iowa. Source: Intl. Assoc. for Hydro-Environment Engr. and Research.)

The formation of a ring vortex is attributed to the most amplified frequency of small disturbance which grows exponentially. The large-scale structures originating from the shear layer instabilities of a Kelvin-Helmhotz (KH) type significantly affect the later development of the jet. The KH instabilities that exist between the fast moving jet flow and surrounding slow moving fluid also enhance the entrainment of the surrounding fluid. The instabilities give rise to the formation of the ring vortex.

12.2 2D LAMINAR FREE JET

Figure 12.3 shows the two-dimensional, incompressible, laminar free jet as it issues from a slot. The jet is assumed to be non-heated and there is no co-flow or strong entrainment effects surrounding the jet. We analyze this jet, starting with continuity and Navier-Stokes equations:

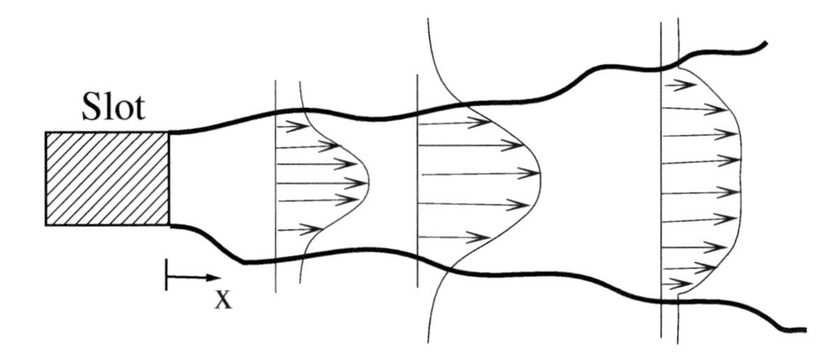

Figure 12.3 The free jet. Note the velocity reduction as the jet progresses into the medium.

The laminar jet flow equations are:
Continuity:

$$\frac{\partial u}{\partial x} + \frac{\partial v}{\partial y} = 0$$

Parabolized Navier-Stokes equation:

$$\left(u\frac{\partial u}{\partial x} + v\frac{\partial u}{\partial y}\right) = \nu\left(\frac{\partial^2 u}{\partial x^2}\right)$$

If the continuity equation is multiplied by ρu and added to the momentum equation, the result is:

$$\left(\frac{\partial uu}{\partial x} + \frac{\partial uv}{\partial y}\right) = \nu\left(\frac{\partial^2 u}{\partial x^2}\right)$$

The boundary conditions are:

$$u(x,\infty) = 0, v(x,0) = 0$$

Note that pressure forces are neglected and only the centerline normal diffusion is considered present. Using these boundary conditions and integration with respect to y gives:

$$\rho\frac{d}{dx}\int_{-\infty}^{\infty} u\,u\,dy = 0$$

$$\rho\int_{-\infty}^{\infty} u\,u\,dy = constant = J$$

J represents the flux of momentum per unit time per unit span.

Schlichting (1933) has provided the exact solution for the laminar free jet spreading as:

$$u = \frac{2}{3x^{1/3}}a^2\text{sech}^2(a\eta)$$

where

$$\eta = \frac{y}{3x^{2/3}\sqrt{\nu}}$$

and

$$a = 0.8255\left(\frac{\sqrt[3]{J}}{(\rho\mu)^{1/6}}\right)$$

J is the momentum flux. Since sech(0) = 1, so maximum jet velocity is in the center of the jet and this velocity is reducing with $x^{1/3}$.

$$u_{\max} = \frac{2}{3x^{1/3}}a^2$$

The velocity distribution in terms of maximum velocity is:

$$u = u_{\max}\text{sech}^2\left[0.27516\left(\frac{\sqrt[3]{J}}{(\rho\mu)^{1/6}}\right)\left(\frac{y}{x^{2/3}\sqrt{\nu}}\right)\right]$$

Mass flow rate of the laminar free jet is:

$$\dot{m} = \sqrt[3]{36J\rho\mu x}$$

Also the location at which the jet's velocity is half of the centerline (or maximum) velocity is called the jet's half width, defined as:

$$b = \frac{21.8\cdot(x\mu)^{\frac{2}{3}}}{(\rho J)^{\frac{1}{3}}}$$

Example 12.1

Example The air jet is issuing from the slot. At $x = 20$m the maximum velocity is 0.11 m/s. Find the

(i) jet's width

(ii) mass flow rate per unit jet's half-width.

Take temperature of air as 15°C.
Solution

$$u_{\max} = \frac{2}{3x^{1/3}}0.8255\left(\frac{\sqrt[3]{J}}{(\rho\mu)^{1/6}}\right)$$

$$J = \left(\left(\frac{u_{max}}{0.454}\right)^3\cdot\rho\cdot\mu\cdot x\right)^{0.5}$$

$J = 0.249\times10^{-3}$

The jet's half-width is

$$b = \frac{21.8 \cdot (x\mu)^{\frac{2}{3}}}{(\rho J)^{\frac{1}{3}}}$$

$b = 0.0760$ m.
The mass flow rate per unit jet's half width is:

$$\dot{m} = \sqrt[3]{36J\rho\mu x} = 0.00340 \frac{kg/s}{m}$$

12.2.1 SELF-PRESERVING JET

Figure 12.4 shows some important features of the free jets. The jet, when it comes out of the pipe, nozzle, or orifice, will have the high shear stress region near the walls. There will be a core region in which shear stress will be small or negligible; it is called the potential core. The length of the potential core extends till six to seven diameters of the jet. Downstream of the potential core, the jet's velocity looks more like a Gaussain distribution. Afterward, around 20 jet's diameter, the jet will exhibit a *self-preserving* behavior.

The velocity of the self-preserving jet can be modeled as:

$$\frac{\bar{u}}{U_{\max}} = f\left(\frac{y}{b}\right) \qquad \textit{Plane jet}$$

$$\frac{\bar{u}}{U_{\max}} = f\left(\frac{r}{b}\right) \qquad \textit{Round axisymmetric jet}$$

12.2.2 JET'S HALF WIDTH

Free jets exhibit some peculiar behaviors, which are helpful in developing the mathematical models. It has been found through experimental studies that the jet's half

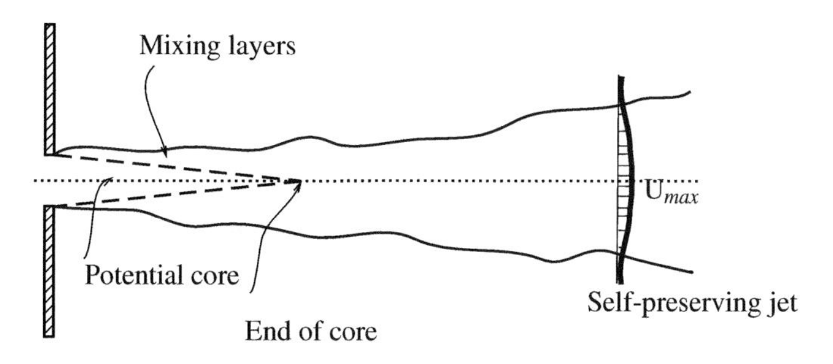

Figure 12.4 The development of the free jet.

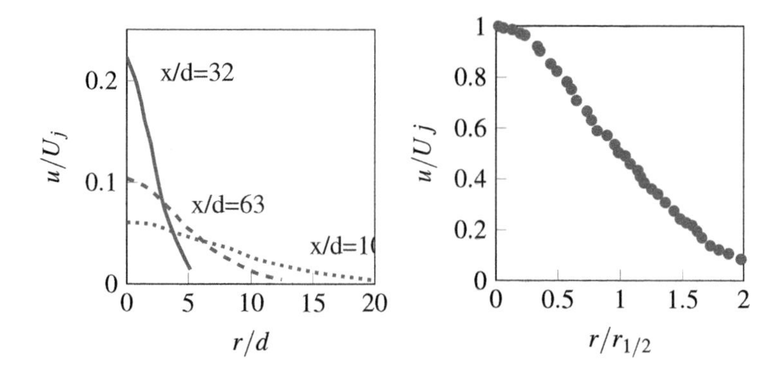

Figure 12.5 (Left) Jet's mean velocity at various locations. U_j is the jet's centerline velocity, d is the diameter of the jet, and x is the distance from the jet's outlet. (Right) Jet's mean velocity at various locations vs. the locations normalized with the jet's half radius ($r_{1/2}$).

width can be used to scale the characteristic length. Figure 12.5 shows the plot of velocity of a circular jet when plotted along with radial coordinates and when the same data is plotted with normalized coordinates by the jet's half width. It is evident from Figure 12.5 that the jet's velocity data at different locations can collapse on a single curve. This brings in the Reynolds number defined based on the jet's half width and the jet's spreading rate, also defined in terms of the jet's half width.

The Reynolds number and spreading rate for the self-similar jet can be defined as:

$$\mathrm{Re}_j = \frac{U_j(x) r_{1/2}(x)}{\nu}$$

$$S = \frac{dr_{1/2}(x)}{dx} = const$$

12.2.3 JET MODES

Experimentally, the passage frequency of these structures can provide information about the sizes and shapes of the jet structures. In literature, distinct jet modes, based on the most amplified frequency, are defined. These modes are based on the jet outlet conditions. One is called the shear-layer mode, and the other one the jet-column mode. The initial region of jet shear layer can be considered as parallel or two dimensional if the ratio of jet diameter to initial momentum thickness is large. In this case, the proper scaling parameter is momentum thickness (see Hussain, 1983). This mode, represented by the Strouhal number (scaled with momentum thickness), is called a shear layer mode. Gutmark and Ho (1983) have indicated that the shear layer mode can occur in the range of the Strouhal numbers.

$$St_\theta = \frac{f_o\theta}{U_o} = 0.01 - 0.018 \tag{12.1}$$

$$0.01 \leq St_\theta \leq 0.018$$

where St_θ is defined as $f_o\delta^{**}/U_o$. Here f_o is the most amplified frequency and U_o is the jet's velocity. This mode is important in case of jet emergence from a converging nozzle. However, if the jet emerges in a fully developed turbulent flow state from a long pipe, the momentum thickness is of order of the jet radius, and the jet does not exhibit the shear-layer mode; instead it will attain the jet column mode. The most amplified frequency in the jet column mode scaled with diameter is called the preferred mode of the jet, defined by the range of Strouhal numbers:

$$St_D = \frac{f_oD}{U_o} = 0.25 - 0.85 \tag{12.2}$$

$$0.25 \leq St_D \leq 0.85$$

where St_D is defined as f_oD/U_o.

According to Hussain (1983), the preferred mode of an axisymmetric jet can occur in the range of the Strouhal numbers $St_D = 0.3 - 0.5$. Gutmark & Ho (1983) have found that the preferred mode depends upon the initial conditions. Liu and Sullivan (1996) have found that the dimensionless natural frequency of a jet decreases as it evolves. The natural frequency of impinging jet is the same for free jet if $x/D > 1.2$. In a fully developed region of free jet, there is no potential core. The flow experiences the decay in centerline velocity.

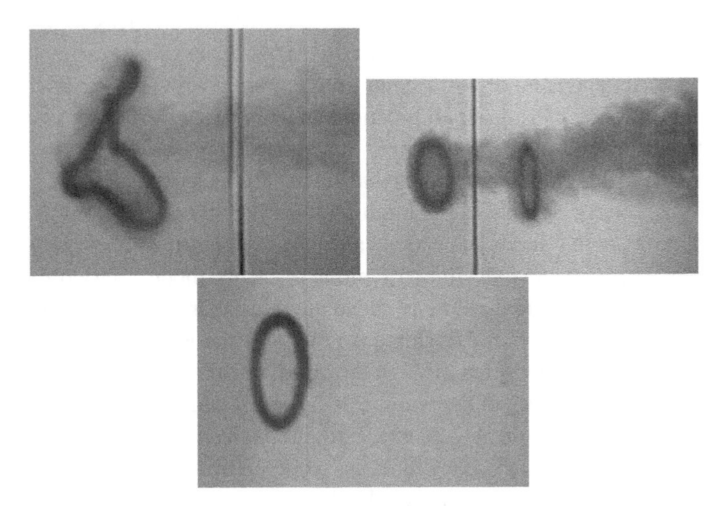

Figure 12.6 Pairing of ring vortex as flow moves from right to left and two rings merge together. (Author Hunter Rouse. Courtesy of Dr. Marian Muste, IIHR - Hydroscience & Engineering, University of Iowa. Source: Intl. Assoc. for Hydro-Environment Engr. and Research.)

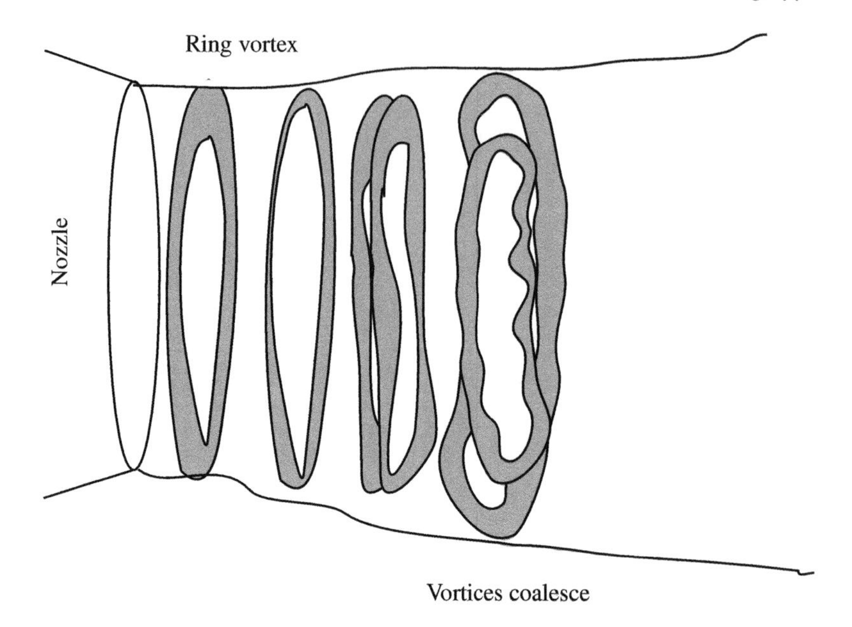

Figure 12.7 The ring vortex in free jet and its coalescing, or pairing, phenomenon.

As the free jet develops, the ring vortices can also merge together, a phenomenon known as vortex pairing. The pairing phenomenon can occur in impinging jets as well. It is an intrinsic feature of the development of the free jet and occurs irregularly in natural jets (non-excited jets) (Popiel, 1991). Stable vortex pairing has been found to occur at two distinct modes. One at $St_\theta \approx 0.012$, which gives thin vortex rings when the boundary layer is laminar, and called shear layer mode pairing, and another at $St_D \approx 0.85$ independent of initial conditions, which gives thick vortex rings, known as jet column mode pairing. In all practical jets, pairing, if it occurs, is known to be complete before $x/D \approx 2$.

There are different flow structures that are present in a jet, usually classifed on the basis of characterstic geometric configuration of the stuctures in physical space. A *structure representing* mode in general can be axisymmetric or non-axisymmetric. A ring vortex represents an axisymmetric mode (varicose mode). Varicose mode is the same as axisymmetric mode but can cause alternate pairing and lead to bifurcating jet. Helical mode is the non-axisymmetric mode of the jet. Both axisymmetric and helical modes can be identified through the cross-correlation between the two halves of the surface, or "planes," of the volume at various locations with respect to jet axis. A peak in cross-correlation at the center will tell us that two halves are in phase with each other and are therefore axisymmetric, while an off-center peak implies that the halves are out of phase and the jet is helical.

12.3 IMPINGING JET

The dynamics of flow and heat in an impinging jet are very complex, involving jet stagnation, ring vortex impact, acceleration of flow, and then deceleration. An

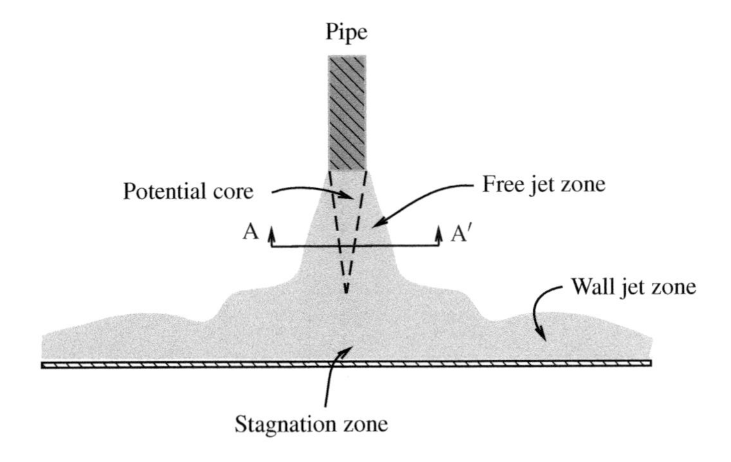

Figure 12.8 The jet impingement phenomenon.

impinging jet is used in a variety of engineering applications, like drying, cooling, vertical take-off, etc. The gas turbines component cooling by impinging jets is not possible at every location, as the construction of an impinging jet system weakens the structural strength. They are used only at the locations where the thermal loads are excessively high, like turbine guide vanes, rotor blades, rotor disks, combuster case and combuster walls, etc. Compact high intensity coolers are the kind of heat-exchange devices, where an array of impinging jets strike the heated surface with orifices. The meandering flow through plates creates further impingement possibilities and this in turn improves the heat transfer. They are used in applications where the primary criterion is the weight of exchanger, for example, air conditioning units in airplanes. They are used in numerous industrial applications for mass transfer, for example, drying, mine ventilation, tunneling operations (mass transfer), paint spraying, cavitation drilling, shielded arc welding, laser cutting, and welding. They are also important for safety engineering studies like accidental release of flammable and toxic materials.

The experimental investigations of impinging jets are numerous. There is a vast body of literature on this topic, covering entrainment, turbulence, cross flow, swirl, nozzle-to-plate distance, inlet configurations, impingement angles, turbulence statistics, etc.

12.3.1 HIEMENZ FLOW

We now present the simplified model of two-dimensional planar laminar flow impingement of a target wall. This flow has importance in many engineering application like the case of two-dimensional bodies with rounded noses or leading edges and drying and cooling applications, etc. The flow field is depicted in Figure 12.9. In 1911, Hiemenz formulated this problem and gave the solution numerically. The

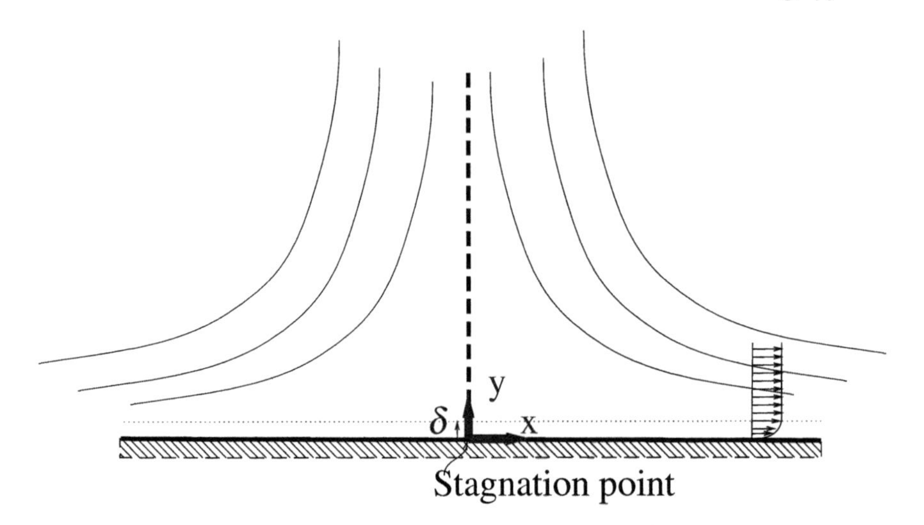

Figure 12.9 Heimenz's two-dimensional stagnation point flow.

assumptions taken are that the flow is steady and the flow is approaching the flat wall of a cylinder of infinite radius compared to the outlet of the jet.

The inviscid flow solution for this case has the stream function:

$$\psi = axy$$

This is the equation of deflected streamline:

$$u = \frac{\partial \psi}{\partial y} = ax$$

$$v = \frac{\partial \psi}{\partial x} = ay$$

and the velocity vector is:

$$\vec{V} = u\hat{i} + v\hat{j} = a(x\hat{i} + y\hat{j})$$

where a is an arbitrary constant.

We can apply Bernoulli's equation:

$$\underbrace{p_o + \frac{\rho}{2}(0)^2}_{stagnation\ point} = p + \frac{\rho}{2}(\vec{V})^2$$

$$p_o - p = \frac{\rho}{2}(\vec{V})^2$$

$$p_o - p = \frac{\rho}{2}(a(x\hat{i} + y\hat{j}))^2$$

We have only one boundary condition for this flow: at $y = o$, $u, v = 0$, i.e., the no-slip wall boundary condition. Note that vorticity generated at the solid surface is not diffusing much due to the opposing acceleration of the deflected flow on the wall. So we introduced another function $f = f' = 0$ for this flow and postulate that there exists a self-similar solution for viscous flow if one defines stream function as:

$$\psi = axy = xf(\eta)\sqrt{a\nu}$$

where

$$\eta = y\sqrt{\frac{a}{\nu}}$$

$$u = axf'$$

$$v = -af\sqrt{\nu}$$

$$p_o - p = \frac{\rho}{2}\left[a^2x^2 + 2a\nu f' + a\nu f^2\right]$$

Only, two-dimensional Navier-Stokes equation can be used

$$f''' + ff'' - \left(f'\right)^2 + 1 = 0$$

$$f(0) = 0, f'(0) = 0, f'(\infty) = 1$$

The numerical solution of this equation gives the result $\eta = 2.4$ at $f'' = 0.99$ so we can write the

$$\eta = y\sqrt{\frac{a}{\nu}}$$

by inserting η value and replacing y with δ:

$$\delta \simeq 2.4\sqrt{\frac{\nu}{a}}$$

Example 12.2

Example Flow at the leading edge of the wing section of an airplane is laminar. Calculate the boundary-layer thickness in the stagnation zone if leading-edge radius of airfoil is 170 mm at a flight speed of 200 m/s. The air kinematic viscosity is 1.5×10^{-5} m^2/s.

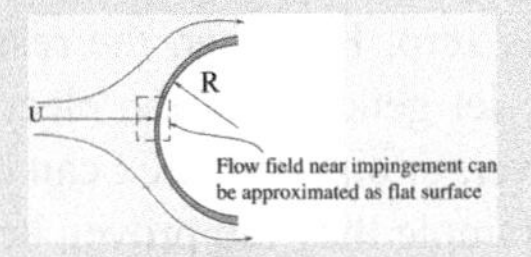

Figure 12.10 Schematic diagram of Example 12.2.

Solution For impinging plane jet the dominant velocity is

$$u = a \cdot x \cdot f'$$

and for case of flow around a cylinder, the dominant velocity is tangential velocity:

$$u_\theta = 2U \sin\theta = 2 \cdot U \cdot \theta$$

which is valid for small angles. Also, we know from trigonometry

$$x = R \cdot \theta$$

so in the cylinder case

$$u_\theta = 2U\frac{x}{R}$$

We now compare the impingement zone velocity with the velocity of flow around the cylinder

$$a \cdot x \cdot f' \approx 2U\frac{x}{R}$$

At the edge of boundary layer f' is 0.99 and we can approximate as

$$a \approx \frac{2U}{R}$$

With $U = 200$ m/s and $R = 170$ mm we have

$$a = \frac{2 \cdot U}{R} = \frac{2 \times 200}{170 \times 10^{-3}} = 2352$$

and

$$\delta = 2.4\sqrt{\frac{\nu}{a}}$$

$$\delta = 2.4\sqrt{\frac{1.5 \times 10^{-5}}{2352}} = 0.00019162m = 190\mu m$$

12.4 SYNTHETIC JETS

As synthetic jet is generated by moving the diaphragm or a boundary in such a manner that the net mass flux is zero. However, the resulting jet has a significant momentum flux. For synthetic jet generation, the electro-mechanical devices, like an actuator, a piezoelectric driver, or a piston, etc., can be used. Synthetic jets have numerous applications. For example they are proven helpful in case of flow control of propulsive jets, control of lift and drag forces in airfoils, reduction of skin-friction coefficient, and modification of aerodynamic characteristics of bluff bodies. The flow control on airfoils is a promising approach to control flows. The control mechanism can be passive control devices or active devices. The examples of passive devices are mechanical flaps or slats, while the active control may be based on pneumatically controlled devices that may introduce blowing or suction in the flow field. The synthetic jet actuator comes in the category of active control device. Mechanical synthetic jet actuators are composed of a moveable diaphragm, a narrow cavity, and an orifice. By generating the forward and backward motion of the diaphragm, the fluid is ejected from the cavity, which will form a vortical structure.

Example 12.3

Example An aircraft has mean wing chord of 7 m and a flap chord of 1.6 m. The wind approach speed is about 800 km/hr with the kinematic viscosity of the air 1.5E-5 m^2/s. It is proposed to locate an array of MEMS actuators near the location of minimum pressure, which corresponds to 150 mm from the leading edge of the flap. Actuators will introduce the puffs of synthetic jets into the boundary layer, which will ultimately trigger the formation of another new boundary layer on the flap underneath the separated boundary layer from the main wing. The phenomenon is depicted in Figure 12.11. This will make sure that the flap boundary layer is disturbed enough so that it can go into early transition to turbulence. What is the minimum spacing for the span-wise placement of the MEMS actuators to generate the synthetic jets.

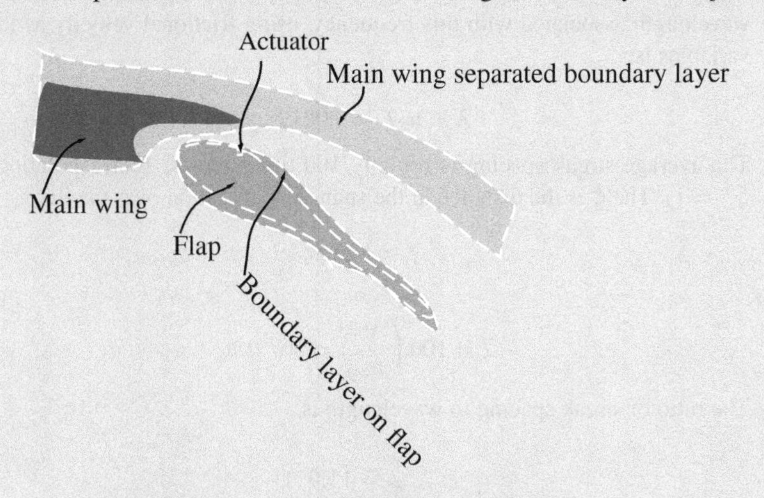

Figure 12.11 Schematic diagram of Example 12.3.

Solution The aircraft is flying at $U_\infty = 800$ km/hr (222.22 m/s). The flow Reynolds number at this location ($x = 150$ mm) is:

$$\mathrm{Re}_x = \frac{222.2 \times 150 \times 10^{-3}}{1.5 \times 10^{-5}} = 2.22 \times 10^6$$

Now, using the following equation we estimate the local skin friction coefficient and shear stress:

$$C_f = \frac{\tau_w}{\frac{1}{2}\rho U_\infty^2} = [2\log_{10}(\text{Re}) - 0.65]^{-2.3} = 0.003267$$

$$\tau_w = \frac{0.003267 \times 1.22}{2}(222.22)^2 = 98.42\, N/m^2$$

The frictional velocity is:

$$u_\tau = \sqrt{\tau_w/\rho} = \sqrt{08.42/1.22} = 8.98 m/s$$

The Reynolds number based on momentum thickness is:

$$\text{Re}_\theta = \frac{\theta U_\infty}{\nu} = 29,478$$

A Kim et al. (1971) formula can be used to estimate the bursting frequency. The mean bursting period is:

$$\frac{T_B u_\tau^2}{\nu} = 0.65\text{Re}_\theta^{0.73}$$

Using the mean bursting period, the frequency, of bursting is 4.5 kHz. The wavelength associated with this frequency, using frictional velocity as local variables is:

$$\lambda = u_\tau T = 0.00198 m$$

The average streak spacing is roughly 100 times the $\Delta\xi$ (corresponding to $\xi^+ = 1$). The ξ is the distance in the span-wise direction over the flap.

$$\ell \approx 100\, \Delta\xi$$

$$\ell = 100\left(\frac{\nu}{u_\tau}\right) = 167\, \mu m$$

The ratio of streak spacing to wavelength is:

$$\frac{\ell}{\lambda} = 11.9$$

So 12 actuators are optimum on this flap. However, due to the space and size reasons one-fourth of this number are installed i.e., three or four actuators. Nevertheless, the span-wise dimensions of the MEMS actuators should not exceed 30% of ℓ. The recommended spacing for MEM actuators is based on streaks spacing. The puff of air generated by a synthetic jet could trigger the bursting phenomenon more accurately and help in expediting the transition from laminar to turbulent flow.

REFERENCES

R. Viskanta, Heat transfer to impinging isothermal gas and flame jets, Exp. Thermal and Fluid Sciences , vol. 6, pp. 111–134, 1993.

H. Schlichting, Boundary Layer Theory, 7th ed. New York: McGraw-Hill, 1975.

E. Gutmark and C. Ho, Preferred modes and the spreading rates of jets, Phy. Fluids, vol. 26, no. 10, pp. 2932–2938, 1983.

T. Liu and J. P. Sullivan, Heat transfer and flow structures in an excited circular impinging jet, Int. J. Heat and Mass Transfer, vol. 17, pp. 3695–3706, 1996.

A. K. M. F. Hussain, Coherent structures - reality and myth, Phy. Fluids, vol. 26, no. 10, pp. 2816–2850, 1983.

H. T. Kim, S. J. Kline, W. C. Reynolds, The production of turbulence near a smooth wall in a turbulent boundary layer, Cambridge University Press, Volume 50, Issue 1, 15 November 1971, pp. 133–160.

13 Wakes and Separated Flows

Fluid flow is classified as external or internal, depending on whether the fluid is forced to flow over a surface or in a conduit. Internal and external flows exhibit very different characteristics. This chapter considers the external flows separation and its influence on the aerodynamic drag.

Learning outcomes: After finishing this chapter, you should be able to:

- Understand the dynamics of separated flows.
- Be able to calculate the location of separation for laminar boundary layer flows.
- Understand the use of drag coefficient.
- Understand difference between skin friction drag, shape drag, induced drag, and wave drag.

13.1 SEPARATION AND WAKE SHEAR LAYERS

Figure 13.1 shows the boundary layer separation over a curved surface. Initially, the boundary layer is laminar, but as the flow moves over the surface, the flow is accelerated. Then it is separated, leading to the formation of wake and separated flow region. In a turbulent flow, the turbulent mixing causes the separation point of a body in a flow to be displaced downstream. Thus, the back flow region in the wake of the body becomes considerably smaller. Related to this is a considerable reduction in the pressure drag.

13.2 LAMINAR FLOW (DP/DX $\neq$ 0) AND SEPARATION LOCATION PREDICTION

Consider the momentum integral equation for boundary layer flows:

$$\frac{\tau_o}{\rho U_\infty^2} = -\frac{\rho_o v_o}{\rho U_\infty} + \frac{d\theta}{dx} + \theta\left[\frac{1}{U_\infty}\frac{dU_\infty}{dx}(2+H)\right] \tag{13.1}$$

This equation is valid for steady, incompressible external flows only. The equation shows that the most realistic thickness of the boundary layer for flow analysis is not the geometric thickness (δ), rather it is the momentum thickness of the boundary layer (θ). It is therefore advisable that dimensionless pressure gradient is used for boundary layer analysis:

DOI: 10.1201/9781003315117-13

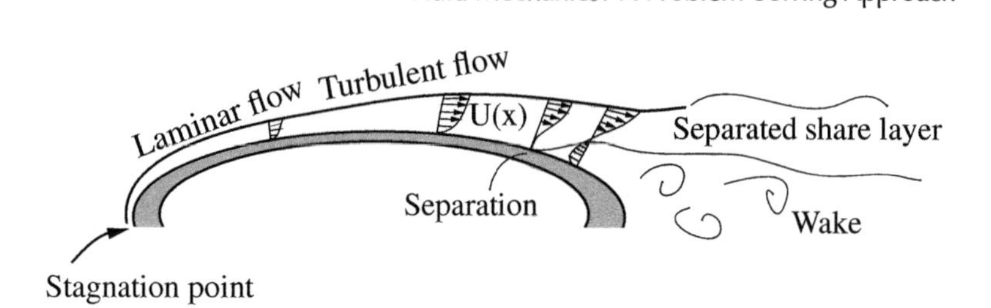

Figure 13.1 Boundary layer separation on a curved surface.

$$\lambda = \frac{\theta^2}{\nu}\left(\frac{dU_\infty}{dx}\right)$$

We define some parameters:
Shear thickness:

$$\delta_s = \frac{\mu U_\infty}{\tau_o}$$

Shear correlation:

$$S = \frac{\theta}{\delta_s} = \frac{\theta \tau_o}{\mu U_\infty} = \frac{C_f \mathrm{Re}_\theta}{2}$$

Assuming no suction and blowing, consider Eq. 13.1 and multiply both sides by $2\theta U_\infty/_\nu$ and adjust:

$$\frac{U_\infty}{\nu}\frac{d(\theta^2)}{dx} + \frac{2\theta^2 U_\infty}{\nu R}\frac{dR}{dx} = 2\left[S - (2+H)\frac{\theta^2}{\nu}\frac{dU_\infty}{dx}\right]$$

$$\frac{U_\infty}{\nu}\frac{d(\theta^2)}{dx} + \frac{2\theta^2 U_\infty}{\nu R}\frac{dR}{dx} = 2\left[S - (2+H)\lambda\right] = F(\lambda) \approx a - b\lambda$$

$$\frac{1}{R^2}\frac{d(\theta^2 U_\infty R^2)}{dx} - \frac{\theta^2}{\nu}\frac{dU_\infty}{dx} \approx a - b\left(\frac{\theta^2}{\nu}\frac{dU_\infty}{dx}\right)$$

$$y = \frac{\theta^2 U_\infty R^2}{\nu}$$

$$\frac{dy}{dx} + \frac{1}{U_\infty}\frac{dU_\infty}{dx}(b-1)y = aR^2$$

This is an ODE:

$$y' + py = Q$$

Solution of this ODE is known:

$$ye^{\int p dx} = \int Q\left[e^{\int p dx}\right] + C$$

Momentum thickness can be found through relation, if we know a and b:

$$\theta^2 = \frac{a\nu\int_0^x R^2 U_\infty^{b-1} dx}{U_\infty^b R^2}$$

A plot of $F(\lambda)$ for various flows shows that curves can merge for the accelerating flows.

$$F(\lambda) \approx 0.44 - 5.68\lambda$$

By examining the boundary layer data from experiments and exat solutions, Thwaites determined a relation between shear correlation S and λ. Kay et al. suggested the values:

$$F(\lambda) \approx 0.44 - 5.68\lambda \quad (13.2)$$

$$S(\lambda) = (\lambda + 0.09)^{0.62} \quad (13.3)$$

Figure 13.2 shows the shape factor and shear factor data vs. λ.

$$H(\lambda) = \left\{ \begin{array}{ll} 2.61 - 3.75\lambda + 5.42\lambda^2 & 0 \le \lambda \le 0.1 \\ \frac{0.0731}{0.15+\lambda} + 2.088 & -0.1 \le \lambda \le 0 \end{array} \right\}$$

$$\theta = \sqrt{\frac{a\nu\int_0^x R^2 U_\infty^{b-1} dx}{U_\infty^b R^2}} = \sqrt{\frac{0.44\nu\int_0^x R^2 U_\infty^{4.68} dx}{U_\infty^{5.68b} R^2}}$$

$$\theta = \frac{0.663}{RU_\infty^{2.84}} \left(\nu \int_0^x R^2 U_\infty^{4.68} dx \right)^{1/2}$$

$$\tau_o = \frac{\mu U_\infty}{\theta} (\lambda + 0.09)^{0.62}$$

Table 13.1 lists the values of λ, $S(\lambda)$, and H, used for Thwaites integral method.

Important:

Thwaites shows that the onset of laminar separation typically occurs when λ drops below the value of -0.09.

Schlichting outlines the case of stagnation in plane flow [Hiemenz flow] and finally concludes that the shape factor at stagnation is equal to the constant value of 2.21. From this, Thwaites λ is found to be 0.125 and $H = 2.21$.

For laminar bubble reattachment to occur, in engineering calculations it can be assumed that the following criteria must be satisfied:

$$\mathrm{Re}_{\delta^*} \ge 500$$

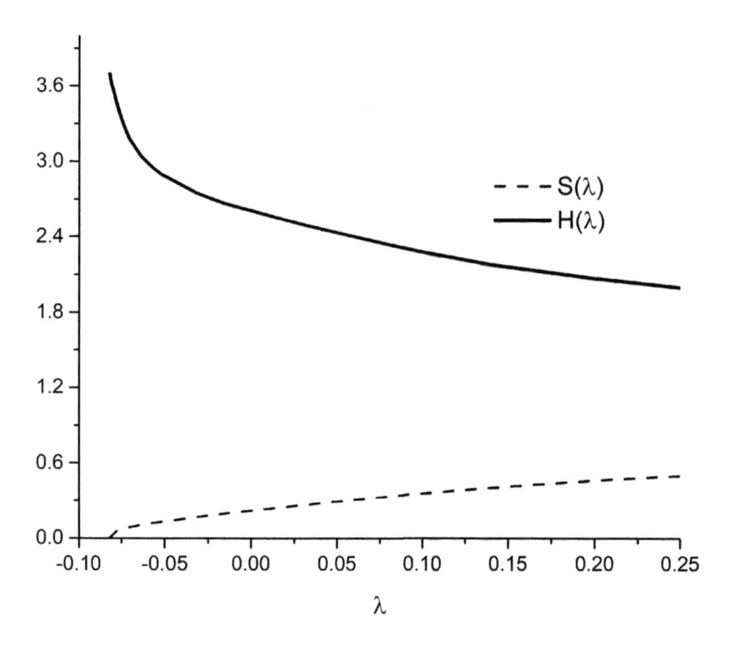

Figure 13.2 Shape factor and shear factor data vs. λ [source: Reference 1].

Example 13.1

Example Consider the flow over pipe with free stream velocity of 10 m/s as shown in Figure 13.3. Estimate the location of boundary layer separation using Thwaites integral method for the flow over a pipe of outer diameter 5 cm.

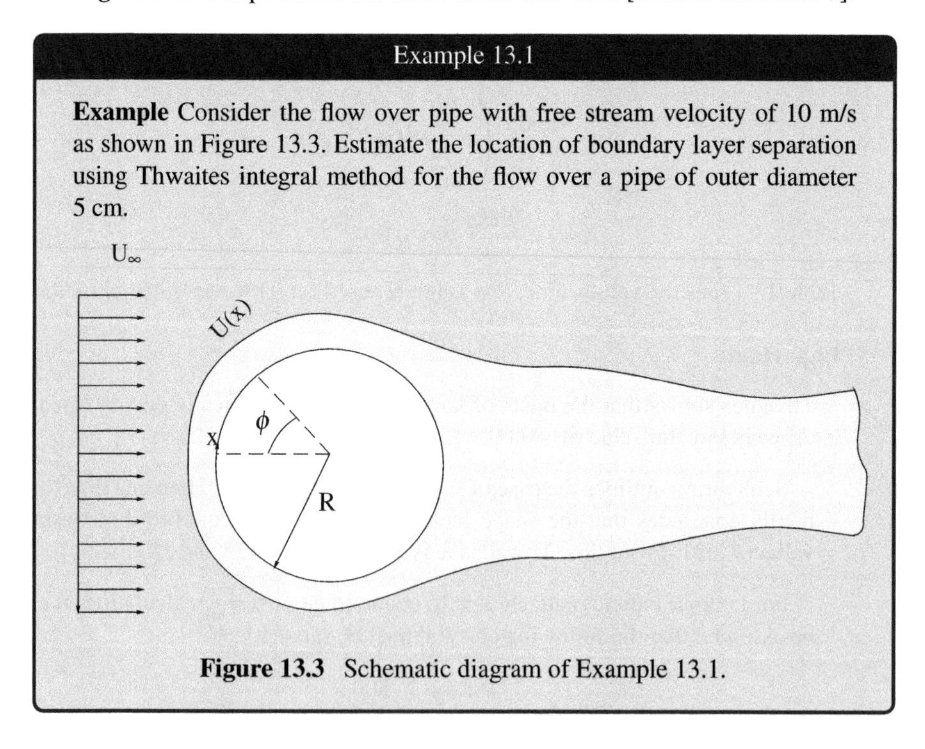

Figure 13.3 Schematic diagram of Example 13.1.

Solution As the velocity for the case of flow over a cylinder is known from potential flow theory that

$$U_\infty = U_o \cdot \sin\left(\frac{x}{R}\right)$$

where x is the coordinate at the outer wall of the pipe and U is the free stream velocity. Substituting ϕ we can cast the velocity in form:

$$U_\infty(\phi) = 2U_o \sin(\phi)$$

Using Thwaites integral method we can estimate the momentum thickness:

$$\theta = \frac{0.663}{RU_\infty^{2.84}} \left(\nu \int_0^x R^2 U_\infty^{4.68} dx \right)^{1/2}$$

Considering the substitution:

$$x = R\phi$$
$$dx = Rd\phi$$

we have the momentum thickness equation as:

$$\theta = \frac{0.663}{U_\infty^{2.84}} \left(\nu \int_0^x U_\infty^{4.68} dx \right)^{1/2}$$

$$= \frac{0.663}{\left[2U_o \sin\left(\frac{x1}{R}\right)\right]^{2.84}} \left(\nu \int_0^{x_1} \left[2U_o \sin\left(\frac{x}{R}\right)\right]^{4.68} dx \right)^{1/2}$$

Taking $\phi = 10^\circ$ we have:

$$\theta = \frac{0.663}{3.47^{2.84}} \left(1.5 \times 10^{-5} \int_0^{0.00436} \left[2U_o \sin\left(\frac{x}{R}\right)\right]^{4.68} dx \right)^{1/2}$$

Numerical integration of this equations yields a momentum thickness value of $\theta = 0.0383$ mm.

$$\lambda = \frac{\theta^2}{\nu} \left[\frac{dU(x)}{dx}\right]_{x=x_1} = 0.077$$

Now selecting the suitable correlation, we have shape factor value

$$H(\lambda) = \left\{ \begin{array}{ll} 2.61 - 3.75\lambda + 5.42\lambda^2 & 0 \le \lambda \le 0.1 \\ \frac{0.0731}{0.15+\lambda} + 2.088 & -0.1 \le \lambda \le 0 \end{array} \right\}$$

$$H = 2.35$$

Taking $\phi = 80^\circ$ repeating the above procedure we get:

$$\theta = 0.058mm$$
$$\lambda = 0.03164$$
$$H = 2.49$$

At angle $\phi = 103.2^\circ$ we have:

$$\lambda = -0.089$$
$$H = 3.29$$
$$\frac{C_f \mathrm{Re}_\theta}{2} = S = 0.00$$

The skin-friction coefficient has become zero, so at an angle of $\phi = 103.2^\circ$, the laminar boundary layer will separate from the cylinder.

According to Terril (1960) the exact location of separation is 1.823 radians, for laminar flow over cylinder, which corresponds to 104.3° (see Table 4.5, White (1991)). This shows that Thwaites method is under-predicting the separation location only by 1.247%.

13.3 DRAG IN FLUID

The part of the drag that arises due to surface friction is called the skin friction drag, whereas the drag that occurs because of the object's shape, which causes the pressure difference across a body, is called the pressure drag. In certain cases, the vortex shedding also induces a drag in the object, called the induced drag. If the object is experiencing a shockwave, there will be an added drag component, called the wave drag. Figure 13.4 shows schematically the pressure variation over a vertical flat plate. The drag experienced by the vertical flat plate is mainly shape drag.

From an incompressible point of view, in this section, we do not discuss wave drag. The bodies with extensive surface area experience considerable frictional drag. The friction drag is minimal for the flat surface that is normal to the flow, whereas the pressure drag is minimum for the surface parallel to the flow. In specific industrial and practical applications, the object's shape cannot be changed. However, by designing the wall/surface, the surface drag or skin friction drag can be significantly reduced. For example, a swimmer may use a swimsuit to reduce the drag associated with the surface. Many swimsuits that mimick the surface texture of a whale or a dolphin have already been designed, and such swimsuits have been banned for international competitions. The same strategy can be used for the drag reduction for ships, boats, torpedoes, etc. The pressure drag is proportional to an object's frontal or projected area. Therefore, it is very dominant for blunt objects or flat surfaces placed perpendicular to the flow. Designing a body over which the streak-lines are parallel is called a **streamlined design**. Usually, in such cases the separation of flow is

Table 13.1
λ, $S(\lambda)$, and H values

λ	$S(\lambda)$	H	λ	$S(\lambda)$	H
−0.082	0	3.7 (Separation)	−0.048	0.138	2.87
−0.0818	0.011	3.69	−0.04	0.153	2.81
−0.0816	0.016	3.66	−0.032	0.168	2.75
−0.0812	0.024	3.63	−0.024	0.182	2.71
−0.0808	0.03	3.61	−0.016	0.195	2.67
−0.0804	0.005	3.59	−0.008	0.208	2.64
−0.08	0.039	3.58	0	0.22	2.61 (Flat plate)
−0.079	0.049	3.52	0.016	0.244	2.55
−0.078	0.055	3.47	0.032	0.268	2.49
−0.076	0.001	3.38	0.048	0.291	2.44
−0.074	0.01	3.3	0.064	0.313	2.39
−0.072	0	3.23	0.08	0.333	2.34
−0.07	0.009	3.17	0.1	0.359	2.28
−0.068	0.094	3.13	0.12	0.382	2.23
−0.064	0.104	3.05	0.14	0.404	2.18
−0.06	0.113	2.99	0.2	0.463	2.07
−0.056	0.122	2.94	0.25	0.5	2
−0.052	0.13	2.9			

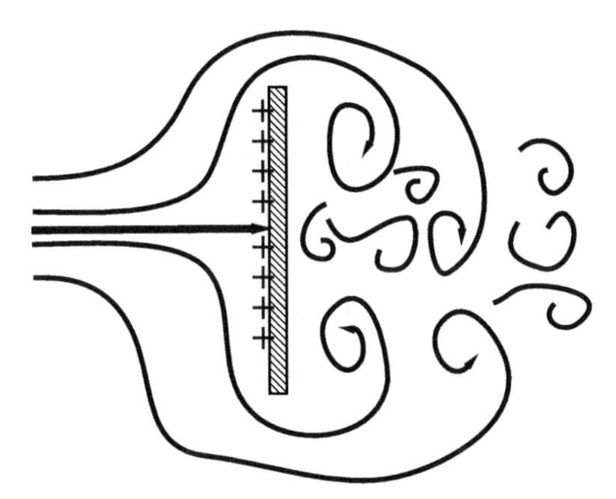

Figure 13.4 The drag experienced by the vertical flat plate is mainly shape drag.

Figure 13.5 Streaklines over the race car. (*Author: Bernardo Malfita, University of Stanford; Courtesy of Dr. Marian Muste, IIHR - Hydroscience & Engineering, University of Iowa. Source: Intl. Assoc. for Hydro-Environment Engr. and Research.*)

Table 13.2
Flow over Cylinder

Reynolds number range	Flow over cylinder
$0 < Re_d < 5$	No flow separation
$5 < Re_d < 45$	Vortex pair attached to cylinder
$45 < Re_d < 200$	Purely laminar, vortex street
$200 < Re_d < 3.8\times 10^5$	Formation of vortex layer, Vortex street superimposed with irregular frequencies
$Re_d > 3.8\times 10^5$	Turbulent flow

minimized while the flow traverses around the object. In common usage, a misnomer term "streamlined design" is used, which refers to an object's shape that has reduced pressure drag. Note that the streamlines are imaginary/hypothetical lines, and the concept is only valid for nonviscous flows. Figure 13.5 shows the flow streaklines over a race car in a testing facility.

13.3.1 DRAG OVER CYLINDRICAL OBJECTS

Experimentally, it has been found that different vortex structures form over the cylinder when free-stream flow passes over it. Different flow behavior, according to the Reynolds number, based on the cylinder diameter and the free-stream velocity are outlined in Table 13.2.

We may divide the flow cylinder into several regimes. Some of the flow regimes are depicted schematically in Figure 13.6.

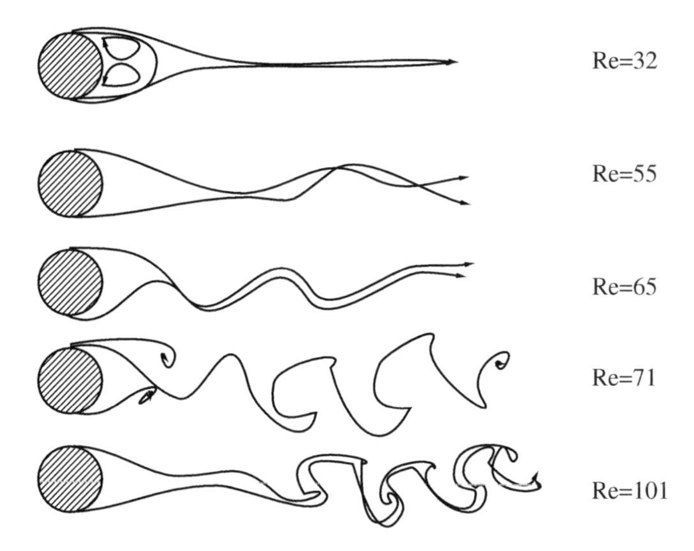

Figure 13.6 The vortex shedding behind a cylinder at different Reynolds numbers.

Regime I (Creeping flow) [Re $<$ 1]: In this regime, the Reynolds number is small, that is, less than one. We know that the Reynolds number is the ratio of inertia forces to viscous forces, so we could say that the viscous forces predominate in this flow. The flow pattern in this case is almost symmetric over the cylinder and the wake will be free from oscillation. This is one type of creeping flow.

Regime II [10 $<$ Re $<$ 1000]: As a Reynolds numbers increases beyond 10 to 1,000, small eddies are formed near the rear stagnation point and they will grow larger as a Reynolds number is increased. The pattern of flow in wake is called the von Karman vortex street or vortex trail. The wake in this regime is unsteady but periodic.

Regime III [1000 $<$ Re $<$ 5×10^5]: In this third regime, the point of separation stabilizes about 80° from the forward stagnation point. The wake is no longer characterized by large eddies, but the flow remains unsteady in wake. The boundary layer near the stagnation point to the point of separation is laminar and the shear stress in this interval is appreciable only in a thin layer near the surface. The drag coefficient value levels out to almost a constant value of one.

Regime IV [Re $\approx 5 \times 10^5$] With a Reynolds number near 5×10^5, the drag coefficient suddenly decreases to a value of 0.3. The point of separation will move past 90°. Close to the stagnation point on the cylinder the boundary layer will quickly go into the transition, and flow is predominantly controlled by inertia forces as the boundary layer traverse over the remaining cylinder body.

Figure 13.6 shows the vortex shedding behind cylinders at Reynolds numbers from $Re_D = 55 - 101$.

Drag and Lift Calculations:

Drag Force:

$$F_D = C_D \underbrace{\left(\frac{1}{2}\rho V^2\right)}_{Dynamic\ Pressure} A_{projected}$$

Lift Force:

$$F_L = C_L \underbrace{\left(\frac{1}{2}\rho V^2\right)}_{Dynamic\ Pressure} A_{planform}$$

13.3.2 D'ALEMBERT'S PARADOX

French mathematician Jean le Rond d'Alembert, in 1752, mathematically proved that according to inviscid potential flow theory, the drag is zero on a body moving with constant velocity relative to the fluid (see Figure 13.7). The simplistic exhibition of this paradox can be seen in the pressure distribution over a smooth cylinder as ideal

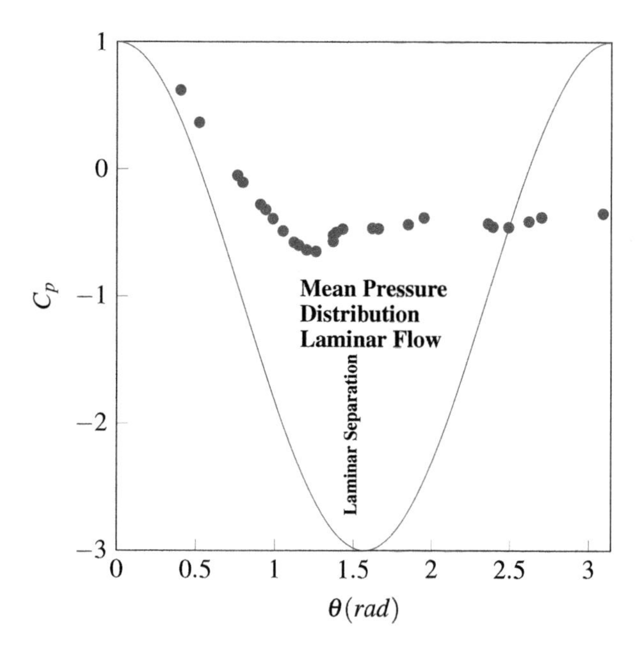

Figure 13.7 Pressure coefficient vs. angle over the cylinder surface. Shape drag over cylinder is not zero as predicted by irrotational flow theory. Experimental data for laminar flow is plotted alongside the curve showing laminar boundary layer separation around 82 degrees.

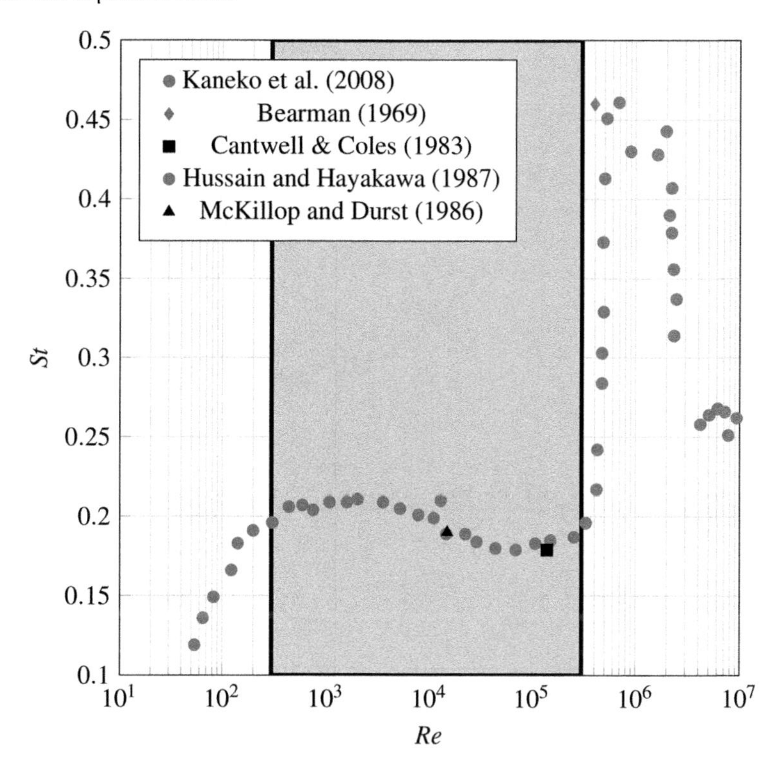

Figure 13.8 Strouhal number (vortex shedding frequency) dependence on Reynolds number.

or inviscid fluid moves over the cylinder. Geometers or mathematicians mostly did the analytical study of fluid mechanics, and thus engineers discredited it as it was far from their field of expertise. This zero drag was baffling, as mostly viscosity was ignored, and later Navier and Stokes independently highlighted the importance of viscosity.

13.3.3 VORTEX SHEDDING FREQUENCIES

Figure 13.8 shows that if a cylinder is placed in a flow, the frequency of the vortex shedding from the cylinder will be constant for a range of Reynolds number $\mathrm{Re} = 300 - 2 \times 10^5$. The frequency dedimensionalized as Strouhal number is almost constant for this range and it is roughly 0.2. This characteristic has been exploited to design the vortex flow meters, when flow vortex shedding behind the cylinder is measured via strain gauge or hot-wire and using following equation the velocity can be estimated.

$$U = \frac{D.f}{0.2}$$

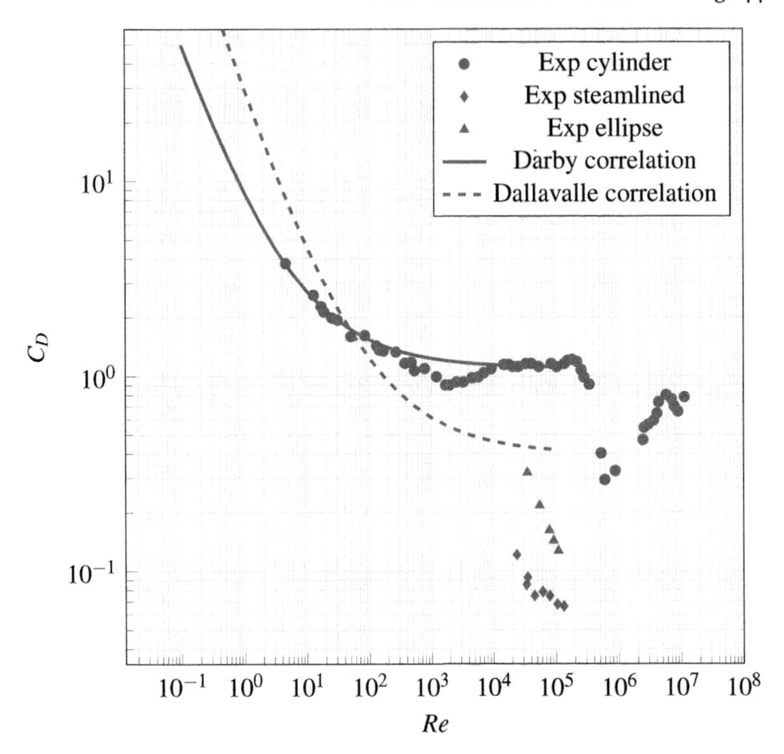

Figure 13.9 Drag coefficients for cylinders and other column-shaped bodies.

In case of periodic vortex shedding after Re = 3.5×10^6 (the super critical regime), the drag force could also oscillate in a sinusoidal manner:

$$F_D(t) = [C_D \sin(2\pi f t)] \frac{1}{2}\rho U^2 A$$

Lyn et al. (1995) found that the drag coefficient of a square cylinder is related with a circular cylinder as:

$$\frac{C_{Dsquare}}{C_{Dcircular}} \simeq 1.7$$

Wadell (1934) and Dallavalle (1948) proposed the equation for drag coefficient calculation:

$$C_D = \left(0.632 + \frac{4.8}{\sqrt{\text{Re}_D}}\right)^2$$

Darby proposed the equation for drag coefficient calculation:

$$C_D = \left(1.05 + \frac{1.9}{\sqrt{\text{Re}_D}}\right)^2 \qquad \text{Re} \leq 2 \times 10^5$$

Figure 13.9 shows the drag coefficients for cylinders and other column-shaped bodies.

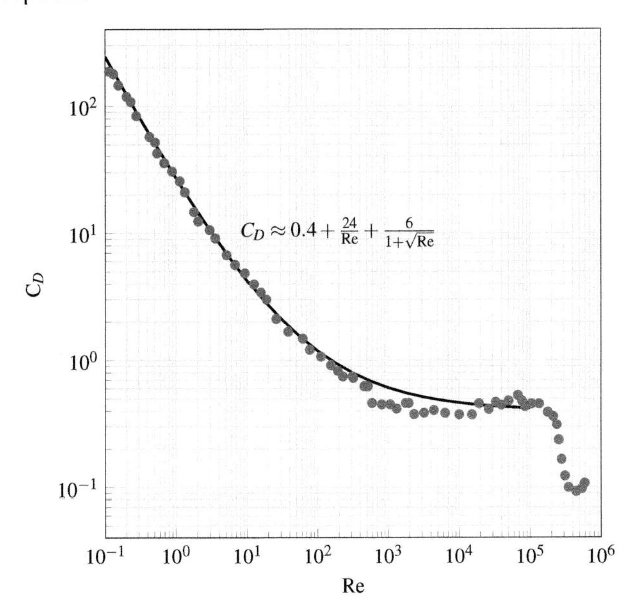

Figure 13.10 Comparison of experimental and empirical formulas for drag coefficients of a smooth sphere.

13.3.4 DRAG OVER SPHERE

The drag over a sphere is an interesting inquiry that has attracted the attention of many researchers over the years. Flow over a sphere has been divided into regimes:

$$\begin{aligned}
&\mathrm{Re} \sim 24 \; (Laminar\ Flow\ Separation) \\
&Re \le 2\times10^5 \; (Subcritical\ Flow) \\
&2\times10^5 \le Re \le 4\times10^5 \; (Critical\ Flow) \\
&4\times10^5 \le Re \le 10^6 \; (Supercritical\ Flow) \\
&Re \ge 10^6 \; (Transcritical\ Flow)
\end{aligned}$$

Figure 13.10 shows the drag coefficient over a sphere, and Figure 13.11 shows the experimental visualization of the flow separation over the sphere.

The following curve-fit formula for the (laminar-flow) data is provided by White (1993) and is plotted along with the experimental data.

$$C_D \approx 0.4 + \frac{24}{\mathrm{Re}} + \frac{6}{1+\sqrt{\mathrm{Re}}}$$

It is valid for the range $0 \le \mathrm{Re} \le 2\times10^5$.

The accuracy is $\pm$ 10% up to the drag crisis, $\mathrm{Re}_D \approx 250{,}000$, where the boundary layer on the sphere will become turbulent and the wake will be thin. This will lead to sudden drag reduction, and it may occur earlier with rough surfaces and with fluctuating free-stream velocity conditions.

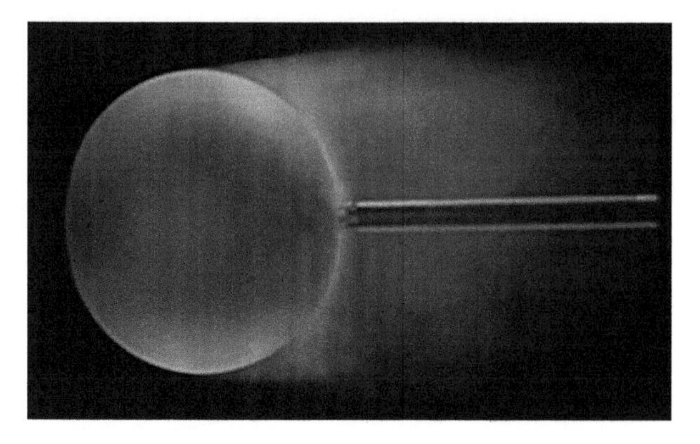

Figure 13.11 Laminar boundary layer separation over a sphere. (*Author: Hunter Rouse; Courtesy of Dr. Marian Muste, IIHR - Hydroscience & Engineering, University of Iowa. Source: Intl. Assoc. for Hydro-Environment Engr. and Research.*)

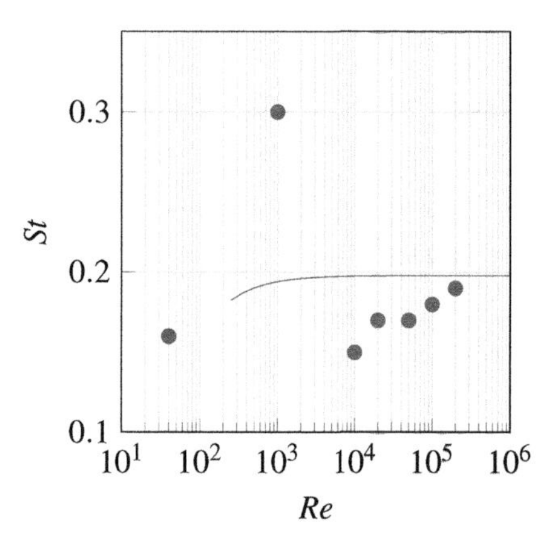

Figure 13.12 Strouhal number of sphere is a function of Reynolds number.

Taylor (1886–1975) proposed an empirical correlation for frequency of vortex shedding, which is valid for $250<\mathrm{Re}<2 \times 10^5$:

$$f = 0.198\left(\frac{U_\infty}{D}\right)\left[1-\left(\frac{19.7}{\mathrm{Re}}\right)\right]$$

$$St = \frac{f \cdot D}{U_\infty} = 0.198\left[1-\left(\frac{19.7}{\mathrm{Re}}\right)\right]$$

Figure 13.12 shows the experimental data with the Taylor's correlation. Table 13.3 lists the drag coefficient for rods and some 3D objects.

Table 13.3
Drag Coefficient for Rods and Some 3D Objects

Drag Coefficients for Rods			Drag Coefficient for 3D Objects		
Circular Rod	○	Cd=1.2 (Lam) Cd=0.3 (Turb)	Cube	■	C_D=1.05
Elliptic Rod	⬭	C_D=0.5	Disc	( )	C_D=20/Re
Equilateral Triangular Rod	→◁ →▷	C_D=1.5 C_D=2.0	Cone (60°)	◀	C_D=0.8
Semicircular Shell	→) → (	C_D=2.3 (Concave) C_D=1.2 (Convex)	Sphere	●	C_D= 24/Re (Lam) C_D=0.2 (Turb)
Square Rod	□ ▢	C_D=2.2 C_D=1.2	Hemisphere	→◗ →◖	C_D=22/Re C_D=1.2

Example 13.2

Example: Falling ball viscometer. A smooth sphere of radius R and density ρ_s is falling in a column filled with oil. Assuming that the sphere is falling without acceleration, find the viscosity of the oil. Assume that sphere is of aluminum 2.7 g/cm^3 and radius is 0.5 mm. The oil density is $\rho = 908\ \mathrm{kg/m^3}$ and the velocity of the falling sphere is 0.0366 m/s. Find the viscosity of the oil.

Solution We apply Newton's second law of motion.

$$\sum F_y = 0 \qquad \underbrace{+F_D}_{drag} + \underbrace{F_B}_{bouyant} - \underbrace{W}_{weight} = \cancel{ma_y}$$

$$C_D \frac{1}{2} \rho w^2 A + \gamma \forall - mg = 0$$

$$C_D \frac{1}{2} \rho w^2 \left(\pi R^2\right) + \rho g \left(\frac{4}{3} \pi R^3\right) - \rho_s g \left(\frac{4}{3} \pi R^3\right) = 0$$

$$C_D \frac{1}{2} \rho w^2 \left(\pi R^2\right) = g \left(\frac{4}{3} \pi R^3\right) (\rho_s - \rho) = 0$$

$$w = \sqrt{\frac{8Rg(\rho_s - \rho)}{3\rho C_D}}$$

Since viscosity is unknown, we cannot calculate Reynolds number. So by assuming Reynolds number, we calculate C_D and calculate velcity. The trial and error will be done until $u = 0.0366$ m/s is achieved. This gives us Re = 0.7, less than 1.

$$C_D \approx 0.4 + \frac{24}{\text{Re}} + \frac{6}{1 + \sqrt{\text{Re}}}$$

$$C_D = 38.46$$

$$w = \sqrt{\frac{8Rg(\rho_s - \rho)}{3\rho C_D}} = 0.0366 m/s$$

Using this velocity we calculate viscosity as:

$$\mu = \left(\frac{2wR\rho}{Re}\right) = 0.089 Pa \cdot s$$

Alternatively, one may approximate drag coefficient as:

$$C_D \approx \frac{24}{\text{Re}}$$

Substituting it into drag force relation, we get drag force as:

$$F_D = 3\pi\mu w D$$

This may help in quick calculations.

Example 13.3

Example Find the trajectory of a free-falling object that is experiencing free stream velocity of W∞.

Solution The resultant force balance, according to Newton's second law of motion, is

$$F_R = m \cdot a_z = m\frac{dw}{dt} = F_w - F_B - F_D$$

where a_z is acceleration in vertical z direction, w is the velocity component in vertical z direction, F_w is force due to weight, F_B the buoyant force, and F_D is the drag force. W_∞ is the free stream velocity experienced by the object falling.

$$m\frac{dw}{dt} = m \cdot g - \gamma \forall - C_D\left(\frac{\rho}{2}w^2\right)A$$

where $\forall$ is volume of the object.

$$\int_0^t dt = \frac{m\,du}{m \cdot g - \gamma\forall - C_D\left(\frac{\rho}{2}w^2\right)A}$$

$$w = W_\infty \tanh\left[\left(\frac{g \cdot m - \gamma\forall}{m \cdot W_\infty}\right)t\right]$$

Since velocity $w = \mathrm{dz}/\mathrm{dt}$, we can integrate the above equation and get

$$z = W_\infty \ln\left(\cosh\left(\frac{g \cdot m - \gamma\forall}{m \cdot W_\infty}\right)t\right)$$

$$z = \left(\frac{m \cdot W_\infty^2}{g \cdot m - \gamma\forall}\right)\ln\left(\cosh\left(\frac{g \cdot m - \gamma\forall}{m \cdot W_\infty}\right)t\right)$$

We now consider a limiting case of negligible volume ($\forall \approx 0$); we have

$$w = W_\infty \tanh\left[\left(\frac{g}{W_\infty}\right)t\right]$$

Figure 13.13 shows the velocity distributions for the free-falling object.

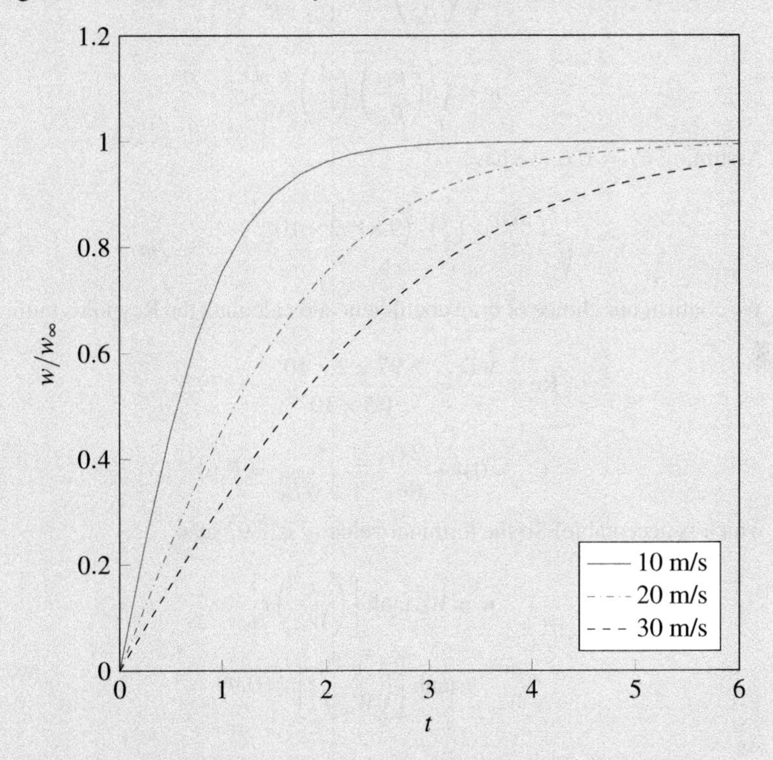

Figure 13.13 Free falling object with negligible volume at free stream velocity of 10 m/s (solid), 20 m/s (dash-dotted), and 30 m/s (dashed).

Example 13.4

Example A water droplet took the form of a sphere as it dropped in air. Find the velocity of the water droplet if the diameter is 2 mm. Ignore the buoyant force and take density of water $\rho_w = 1,000\ \text{kg/m}^3$ and density of air as $\rho_a = 1.22\ \text{kg/m}^3$. The kinematic viscosity of air is $1.5\times10^{-5}\ \text{m}^2/\text{s}$. Also find the position z of the droplet.

Solution The balance of air drag and weight of water droplet gives the velocity w of the falling droplet as:

$$C_D\left(\frac{1}{2}\rho_a w^2\right)A = m_w\cdot g$$

$$w = \sqrt{\frac{2m_w\cdot g}{\rho_a C_D A}} = \sqrt{\frac{2\cdot g\cdot\rho_w\forall}{\rho_a C_D A}}$$

$$w = \sqrt{\left(\frac{\rho_w}{\rho_a}\right)\frac{2\cdot g\left((4/3)\not{\pi}R^{\not{3}}\right)}{C_D\,\not{\pi}\not{R}^{\not{2}}}}$$

$$w = \sqrt{\left(\frac{\rho_w}{\rho_a}\right)\left(\frac{4}{3}\right)\frac{g\cdot D}{C_D}}$$

Assuming $C_D = 0.6$, we have

$$w = \sqrt{\frac{\left(\frac{1000}{1.22}\right)\cdot\left(\frac{4}{3}\right)\cdot\left(9.81\cdot 2\times10^{-3}\right)}{0.6}} = 5.97m/s$$

We confirm our choice of drag coefficient and calculate the Reynolds number

$$\text{Re} = \frac{wD}{\nu} = \frac{5.97\times2\times10^{-3}}{1.5\times10^{-5}} = 797$$

$$C_D = 0.4 + \frac{24}{Re} + \frac{6}{1+\sqrt{Re}} = 0.61$$

which is acceptable! So the terminal velocity is 5.97 m/s.

$$w = W_\infty \tanh\left[\left(\frac{g}{W_\infty}\right)t\right]$$

$$\frac{w}{W_\infty} = \tanh\left[\left(\frac{g}{W_\infty}\right)t\right] = 0.99$$

As

$$\tanh(2.65) = 0.99$$

we can have

$$\left[\left(\frac{g}{W_\infty}\right)t\right] = 2.65$$

This gives $t = 1.59$ s.
Now from trajectory equation we have

$$z = \left(\frac{m \cdot W_\infty^2}{g \cdot m - \gamma\forall}\right) \ln\left(\cosh\left(\frac{g \cdot m - \gamma\forall}{m \cdot W_\infty}\right)t\right)$$

$$z = \left(\frac{W_\infty^2}{g}\right) \ln\left(\cosh\left(\frac{g}{W_\infty}\right)t\right) = 3.57 \ln\left(\cosh\left(1.65\right)t\right)$$

$$z = 6.95m.$$

13.3.5 DRAG OVER SOME OTHER 3D OBJECTS

Figure 13.14 shows the drag coefficient distribution for cone vs. cone angle. Figure 13.15 gives the plot of drag coefficient for a circular rod parallel to flow, a cylindrical rod perpendicular to flow, a thin rectangular plate perpendicular to flow, and a square rod parallel to flow.

Ellipsoid of revolution of width D and length L:

$$C_D\big|_{Ellipsoid} = 0.44\left(\frac{D}{L}\right) + 0.016\left(\frac{L}{D}\right) + 0.016\sqrt{\left(\frac{D}{L}\right)}$$

Sphere of radius R in a duct of radius R_o:

$$C_D\big|_{Sphere^{\cdot}Duct} = \left[1 + 1.45\sqrt{\left(\frac{R}{R_o}\right)^9}\right]$$

Table 13.4
Drag Coefficient for the Rectangular Plate Facing Flow

Rex10^{-4}	Height/Thickness	St
3	3.8	0.131
1.9	1.62	0.128
1.3	1	0.132
2.1	1	0.13

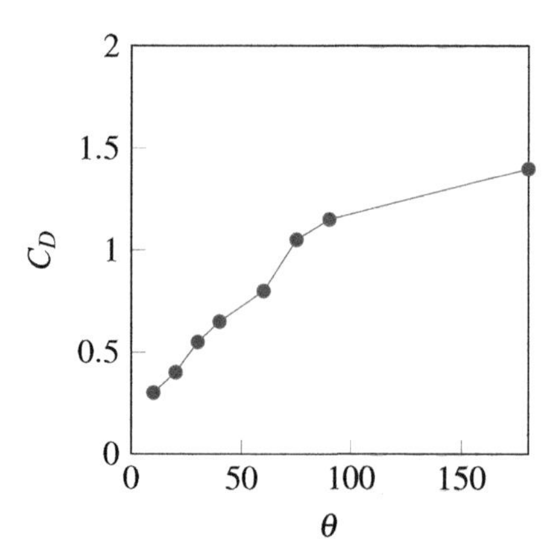

Figure 13.14 Drag coefficient vs. cone angle.

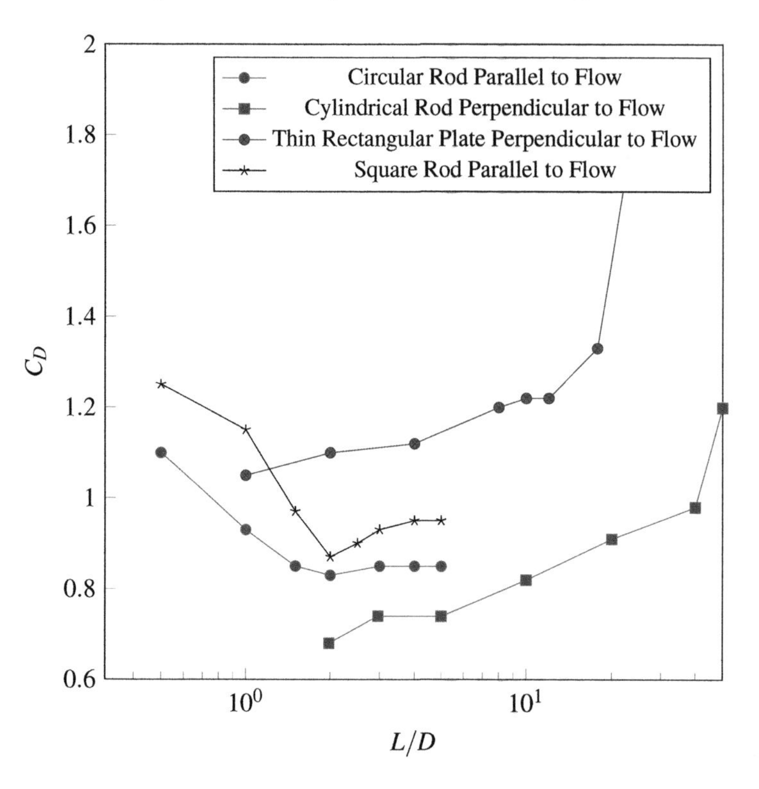

Figure 13.15 Drag coefficient for some common 3D bluff bodies (Re $\geq 10^4$).

13.3.6 DRAG MEASUREMENTS IN WIND TUNNEL

We can measure the drag by placing the model of the object in the wind tunnel. The rate of momentum loss in flow before and after the object gives us the force

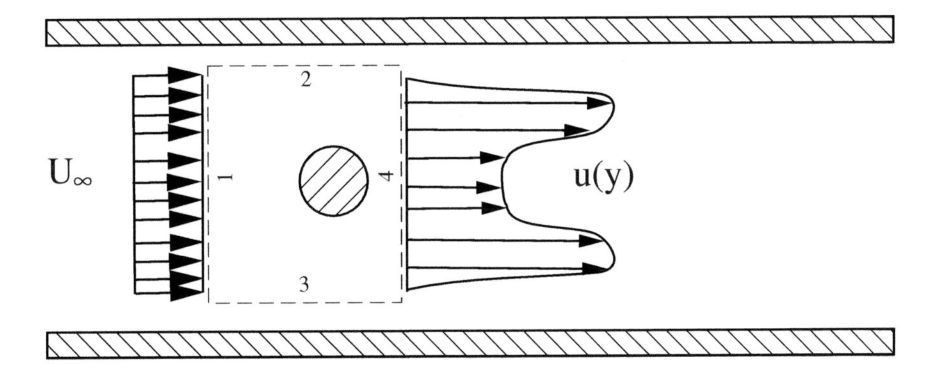

Figure 13.16 Drag can be estimated by measurement of velocity distributions in wake region. The dashed line represent the control surface. The faces of control volume are indicated by 1, 2, 3, and 4.

experienced by an object. Care must be taken as we need to estimate the displacement thickness (δ^*) and make sure that boundary layer is not disturbing the approaching uniform flow. Figure 13.16 shows the control volume around the bluff body to measure the drag in the wind tunnel.

If b and h are the width and height of the control volume, respectively, then we can write using law of conservation of mass:

$$+\rho U_\infty bh - \dot{m}_2 - \dot{m}_3 - \int [\rho u(y)]dA = 0$$

$$\dot{m}_2 + \dot{m}_3 = \rho U_\infty bh - \int [\rho u(y)]dA$$

$$\dot{m}_2 = \dot{m}_3 = \frac{1}{2}\left[\rho U_\infty bh - \int [\rho u(y)]dA_4\right]$$

$$\dot{m}_{top} = \frac{1}{2}\left[\rho U_\infty bh - \int [\rho u(y)]dA_4\right]$$

According to Reynolds transport theorem, the force can be estimated as:

$$\mathbf{F} = \int_{CS} \mathbf{V}\rho\mathbf{V}\cdot dA + \frac{\partial}{\partial t}\int_{CV} \mathbf{V}\rho d\forall \tag{13.4}$$

for steady state:

$$\mathbf{F} = \int_{CS} \mathbf{V}\rho\mathbf{V}\cdot dA$$

$$\mathbf{F} = -(U_\infty \rho U_\infty A)_1 + \dot{M}_2 + \dot{M}_3 + \int (u(y)\rho\, u(y))dA$$

where $\dot{M}$ is the rate of changed of momentum

$$\mathbf{F} = -(U_\infty \rho U_\infty bh)_1 + \int \left(u^2(y)\rho\right) dA + \dot{M}_2 + \dot{M}_3$$

$$\mathbf{F} = \int \left[\rho u^2(y)\right] dA - (\rho U_\infty U_\infty bh)_1 + \dot{M}_2 + \dot{M}_3$$

momentum from the face 3 and 4 can be estimated as:

$$\dot{M}_2 + \dot{M}_3 = \left[\rho U_\infty bh - \int [\rho u(y)] dA_4\right] (U_\infty) = U_\infty (2\dot{m}_{top})$$

so the drag on the object is:

$$\mathbf{F} = \int \left[\rho u^2(y)\right] dA - (\rho U_\infty U_\infty bh)_1 + U_\infty (2\dot{m}_{top})$$

This shows that if we can accurately measure the velocity behind the model then this can give us the drag experienced by the object.

Example 13.5

Example Find the total drag on an elliptic cylinder with characteristic thickness h, as shown in Figure 13.17. Half of the velocity profile downstream of the elliptic cylinder is represented by the distribution:

$$u(y) = \frac{U_\infty}{3}\left[1 + \left(\frac{y}{h}\right)\right] \quad 0 \leq y \leq 3h$$

where h is the thickness of the cylinder and b is the length of the cylinder.

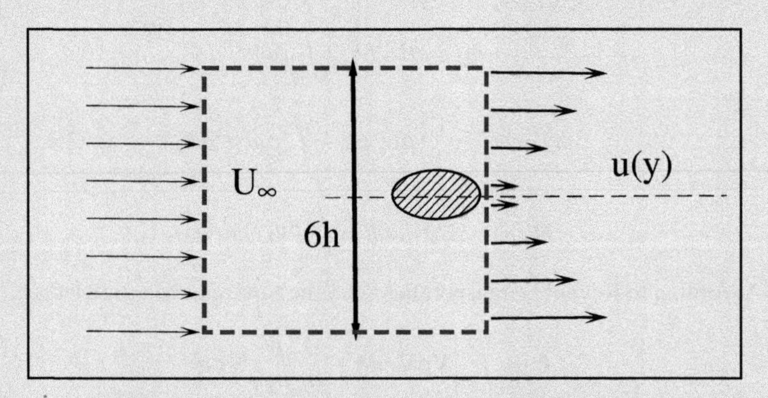

Figure 13.17 Schematic diagram of Example 13.5.

Solution We first find the mass going out of the top or bottom boundaries of control volume. Using the equation developed in the section above, we have

$$\dot{m}_{top} = \frac{1}{2}\left[\rho U_\infty bh - \int \rho u(y)\, dA\right]$$

$$\dot{m}_{top} = \frac{1}{2}\left[\rho U_\infty b(6h) - 2\int_0^{3h} \rho u(y)\, b\, dy\right]$$

$$\dot{m}_{top} = \frac{1}{2}\rho U_\infty b\, h$$

The drag force is:

$$\mathbf{F} = \int \left[\rho u^2(y)\right] dA - (\rho U_\infty U_\infty bh)_1 + U_\infty(2\dot{m}_{top})$$

$$\mathbf{F} = 2\int_0^{3h} \rho u^2(y)\, b{\cdot}dy - (\rho U_\infty U_\infty b\cdot 6h) + U_\infty(2\dot{m}_{top})$$

$$\mathbf{F} = \frac{-1}{3}\rho U_\infty{}^2 bh$$

The negative sign indicates that this force is acting opposite to the main flow direction.

13.3.7 DRAG ESTIMATION USING WAKE PARAMETERS

Some researchers have related the drag force with the spread of wake region. They related the maximum change in velocity in wake region with the momentum thickness and the distance from the object. Sreenivasan and Narasimha (1982) proposed the following relations:

$$\Delta u_{\max} = 1.63 U_\infty \sqrt{\frac{\theta}{x}}$$

It is found that the spreading of wake is similar to jet spreading, so taking an analogy with jet parameters, the wake half-width is introduced as

$$y_{1/2} \approx 0.3\sqrt{x\cdot\theta}$$

where θ is the wake momentum thickness and x is the distance from the object. Note that the velocity distribution in the wake region is not self-similar unless it is several hundred times the characteristic length of the object.

The drag force can be calculated as:

$$\mathbf{F} = \rho U_\infty^2 \theta = C_D \frac{\rho U_\infty^2}{2} A$$

Example 13.6

Example The column in an offshore platform has a diameter of 6 m. The ocean currents are 0.777 knots. Find the drag experienced by the column if in the wake region at $x = 1,000$ m, $\Delta u_{max} = 0.03$ m/s. Also, find the minimum width of the wake region at this location. Take sea water density as 1023.6 kg/m^3.

Solution The free stream velocity is $U_\infty = 0.77$ knots $= 0.4$ m/s.

$$\Delta u_{max} = 1.63 U_\infty \sqrt{\frac{\theta}{x}}$$

$$0.03 = 1.63 \times 0.4 \sqrt{\frac{\theta}{100}}$$

$$\theta = 1.32m.$$

$$\mathbf{F} = \rho U_\infty^2 \theta = 1023.6(kg/m^3) \times (0.4m/s)^2 \times 1.32m = 216.057N$$

$$y_{1/2} = 0.3\sqrt{x \cdot \theta}$$

$$y_{1/2} = 10.89m$$

The width of the wake will not be less than $2y_{1/2} = 2 \times 10.89 = 21.79$ m.

13.4 FORCES OVER AIRFOIL

Airfoil are the cross-section of aircraft wings. The Wright Flyer or Kitty Hawk, Flyer I, or the 1903 Flyer was the first plane invented at the beginning of 20th century. Old planes were one- or two-man aircrafts (see Figure 13.18). Airfoils are also used in helicopters (see Figure 13.22). The chord line is the straight line joining the leading and trailing edges of an airfoil (Abbott and von Doenhoff, 1959). Lift and drag coefficients for airfoil depend on Reynolds number and angle of attack between the chord line and the free stream flow. When the angle of attack is positive, air flowing toward the wing is split into two sections; air flowing above the wing is forced to narrow down, while the air flowing beneath the wing surface is expanded. As a result, the flow velocity increases at the upper section of the wing, and the pressure reduces. On the other hand, the flow velocity drops at the lower section of the wing, and the pressure grows. The pressure difference between lower and upper wing sections stimulates the generation of the well-known aerodynamic force called lift force. The lift force generated by an airfoil is mainly caused by the angle of attack variation, which is the angle formed by the chord of the airfoil and the wind flow direction representing the relative motion between the aircraft body and the atmospheric air. When the airfoil is tilted at a suitable angle, it deflects the incoming air airstream for fixed-wing planes forcing them to flow below it, thus, creating a force beneath the airfoil in the direction perpendicular to the airstream deflection.

Figure 13.18 Old planes were made for one man.

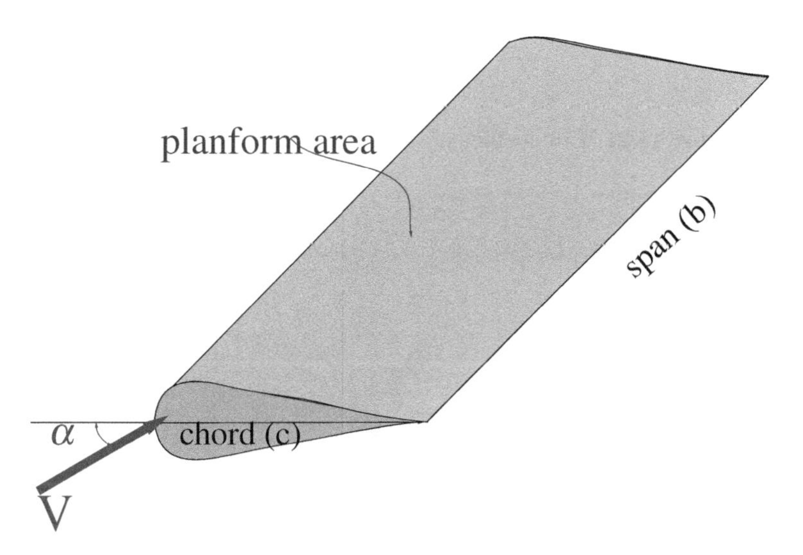

Figure 13.19 Airfoil geometry nomenclature.

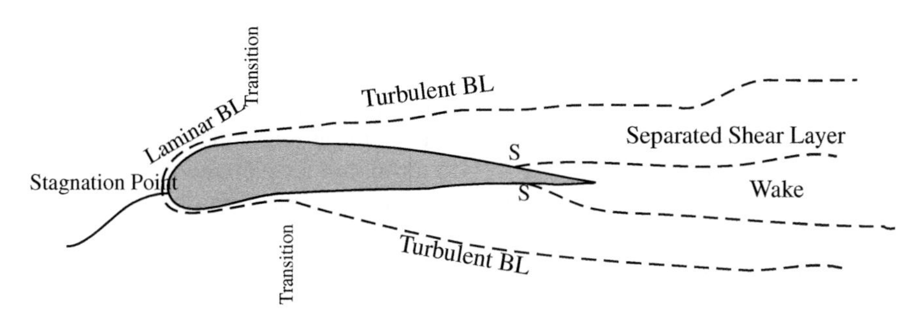

Figure 13.20 Viscous flow around an airfoil.

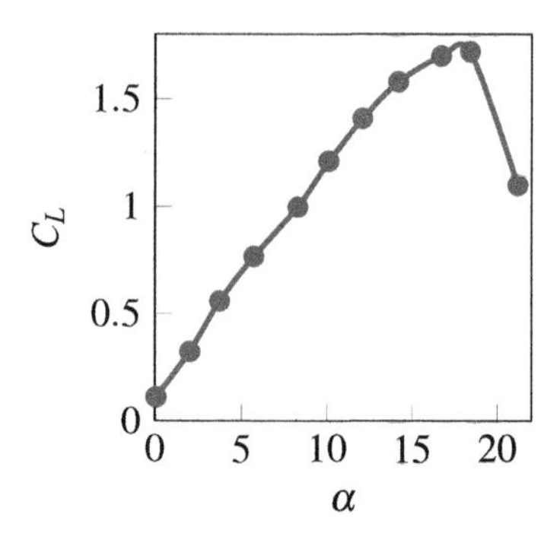

Figure 13.21 Lift coefficient vs. angle of attack α for NACA 23015.

Figure 13.22 Blade used in the helicopter.

Figure 13.19 show the viscous flow around an airfoil. Figure 13.20 shows the distribution of lift coefficient vs. angle of attack α for NACA 23015.

Nomenclature of Airfoil Series by NACA

4-digit Series: The first digit informs about the maximum camber (m) in the chord's percentage (airfoil length), the second digit shows the position of the maximum camber in tenths of the chord, and the last two numbers give the information on maximum thickness (t) of the airfoil in the percentage of a chord. For example, the NACA 2415 airfoil has a maximum thickness of 15%, with a camber of 2% located 40% back from the airfoil leading-edge (or 0.4c).

5-digit Series: The NACA Five-Digit Series uses the same thickness forms as the Four-Digit Series, but the mean camber line is defined differently, and the naming convention is more complex. When multiplied by 3/2, the first digit yields the design lift coefficient (cl) in tenths. When divided by 2, the next two digits give the maximum camber position in tenths of a chord. The final two digits again indicate the maximum thickness (t) in the percentage of a chord. For example, the NACA 23012 has a maximum thickness of 12%, a design lift coefficient of 0.3, and a maximum camber located 15% back from the leading edge.

NACA 1-Series or 16-Series: The 1-Series was developed based on airfoil theory in the 1930s. Example NACA 16-212. The first digit, 1, shows the series. The 6 specifies the location of minimum pressure in tenths of the chord, i.e., 60% back from the leading edge. Following a dash, the first digit shows the design lift coefficient in tenths (0.2), and the final two digits specify the maximum thickness in tenths of a chord (12%).

NACA 6-Series: The airfoils that maximized the region over which the airflow remains laminar. By so doing, the drag over a small range of lift coefficients can be substantially reduced. NACA $64_1 - 212$, $a = 0.6$.

In this example, 6 denotes the series and shows that this family is designed for greater laminar flow than the Four- or Five-Digit Series. The second digit, 4, is the location of the minimum pressure in tenths of the chord (0.4c). The subscript 1 shows that low drag is maintained at lift coefficients 0.1 above and below the design lift coefficient (0.2) specified by the first digit after the dash in tenths. The final two digits specify the thickness in the chord's percentage, 12%. The fraction set by $a = 0.6$ shows the percentage of the airfoil chord over which the pressure distribution on the airfoil is uniform, with 60% chord in this case. If not specified, the quantity is assumed to be 1, or the distribution is constant over the entire airfoil.

The lift and drag coefficients for an airfoil are Reynolds number and angle of attacks functions. Flow on the upper surface must speed up sharply around the nose of the airfoil. The minimal pressure becomes lower as the angle of attack increases, and its location moves forward on the upper surface. When the separation point is reaches

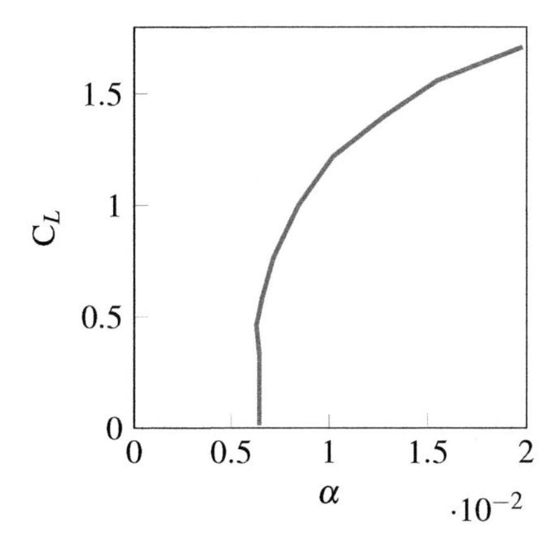

Figure 13.23 Lift-drag polar for airfoil NACA 23015. $C_{L,max} = 1.72$. Data from Abbott and von Doenhoff.

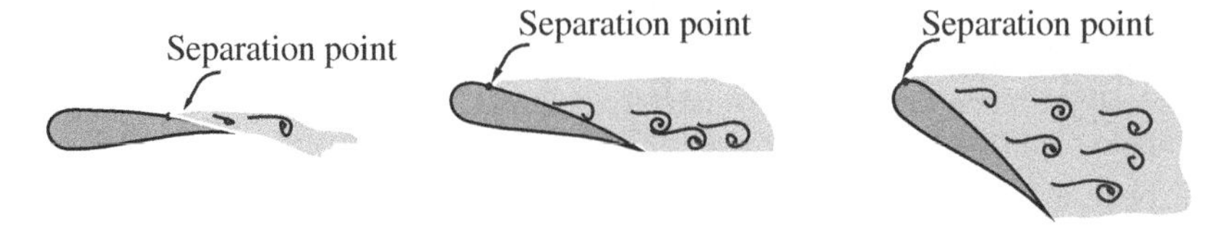

Figure 13.24 Flow over airfoil and separation point movement.

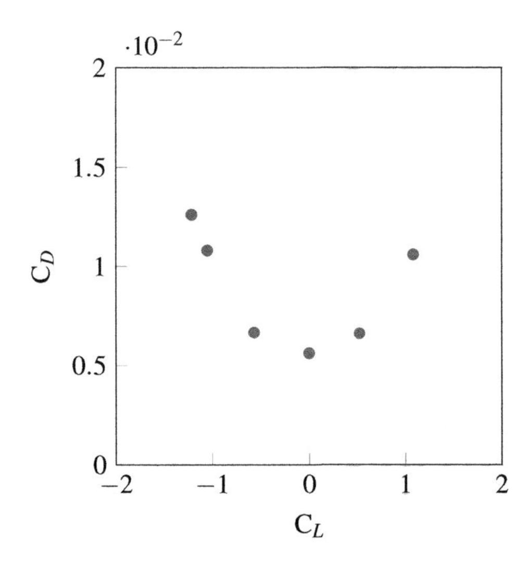

Figure 13.25 C_D vs. C_L for NACA 0012 airfoil (Abbott and Von Deonhoff data).

close to the leading edge, the entire upper surface is immersed in the wake. In that case the flow will separate from the entire upper surface. For this condition, the pressure on the upper surface is roughly the same as that for upstream pressure, which is clear in Figure 13.24. This results in a decrease in the lift force and the condition is called a stall. Aircraft with laminar-flow regions are designed for cruising in the low-drag region. Movement of the minimum pressure point and accentuation of the adverse pressure gradient handle the rapid rise in drag coefficient for the laminar-flow section. The rapid rise in drag coefficient is provoked by the early transition from laminar to turbulent boundary-layer flow on the upper surface. Figure 13.25 shows the C_D vs. C_L for NACA 0012 airfoil, and Figure 13.26 shows the effect of flaps on the lift coefficients of an airfoil.

13.4.1 SEPARATION BUBBLE

Some airfoils have large upper-surface curvature, which could cause the laminar boundary layer separation when the airfoil is at position at a moderate angle of attack. If the laminar boundary layer separates then the departed shear layer will go into transition and reattach again to the surface as a turbulent boundary layer.

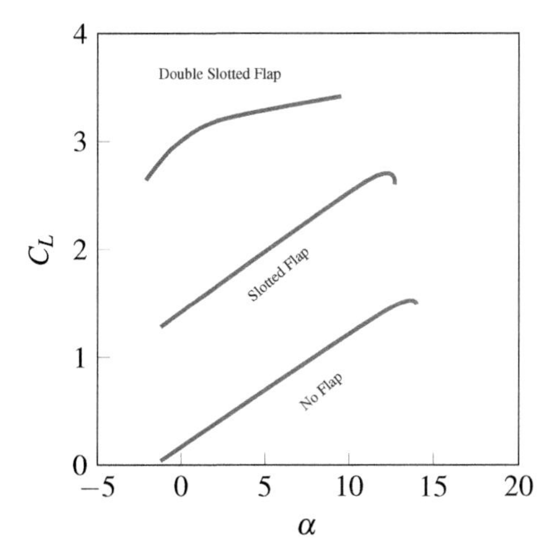

Figure 13.26 Effect of flaps on the lift coefficients of an airfoil.

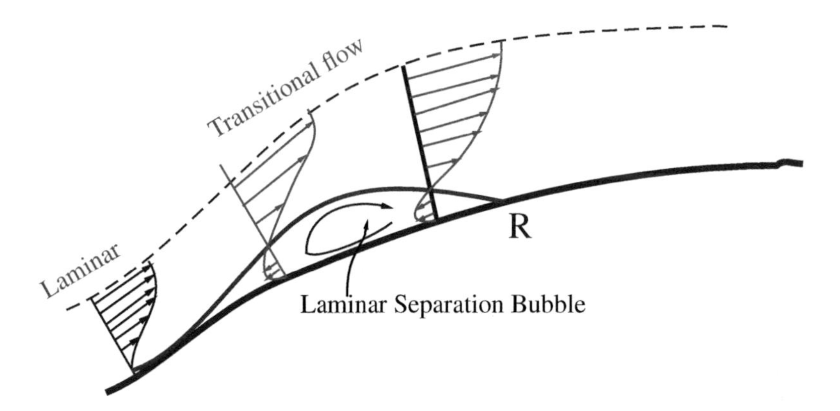

Figure 13.27 Dynamics of a separation bubble over an airfoil. R is the reattachment point.

The phenomenon is depicted in the Figure 13.27. However, a bubble of fluid will be trapped underneath the separated shear layer between the separation and reattachment points. Two bubble types have been discovered. One is called the short bubble, whose size is on the order of 1% of the chord-length or 100 times the displacement thickness at separation point. Another one is a long bubble, which could have a size the order of magnitude of 10,000 of displacement thicknesses. In case of airfoils, the sudden change in curvature is sometimes needed, which could cause the formation of separation bubble and thus leads to a very detrimental stall characteristics.

The short bubble points to a bubble confined between the limits of separation and reattachment and set into circulating motion. This short bubble breaks down when the bubble shrinks, for example, because of the increment of attack angle. Thus, the flow separates wholly from the surface without reattachment. Although the length of the bubble is short, e.g., with an airfoil, it is in the order of 1% of the chord length and does not substantially affect the pressure distribution. When the bubble breaks down, a stall occurs, leading to a sudden loss of lift and surge of drag. The long bubble whose length is long (2 or 3% of chord-length) likewise breaks down, but does not lead to a complete separation of flow; instead, the separated flow passes over the body surface and reattaches further downstream and, finally, may reattach to the trailing edge. The pressure distribution varies with the long bubble, but the discontinuity in a lift is meager.

Owen-Klanfer proposed a useful criterion to judge whether a short bubble or long bubble will form. Their criterion suggest that the Reynolds number based on displacement-thickness plays an important role in the bubble formation. As a boundary layer grows over the surface, the displacement thickness will increase.

$$Re_{\delta^*} = \frac{U_\infty \delta^*}{\nu} < 400$$

$$Re_{\delta^*} < 400 \quad (Long\ bubble)$$

$$Re_{\delta^*} > 550 \quad (Short\ bubble)$$

where δ^* is boundary layer displacement thickness at the separation.

The bubbles are very problematic for the flights with thin-airfoils. The length of long bubble rapidly increase with increasing incidence of the flow, leading to a continuous reduction of the leading-edge suction capabilities. The occurrence of the reattachment is incumbent upon the Reynolds number. If the Reynolds number is larger than critical Reynolds number (i.e. $Re = 400$), then flow becomes unstable, circulating motion is energized, and flow reattaches itself back to the surface.

The separated flows may happen due to deflected flaps, bomb bays, open cockpit, escape hatches, spoiler control, over-expanded rocket nozzle, leeward side of an object inclined at a large angle of attack, the surface of a ship's hull, etc. In most cases, the vortex shedding is unsteady, and the experimental study is not possible in many cases. To investigate this complex flow often simulations are conducted to understand the dynamics of this complex phenomenon.

Active or passive control of boundary layer. Important for control of lift and drag forces, heat, mass transfer, etc.

Suction

Blowing

Excitation at different frequencies

Adding additives

Devices like actuators, MEMS, etc. introduce the jets in air stream.

A very effective method of avoiding boundary-layer separation is suction. In this case, the fluid in the boundary layer is sucked into the interior of the body through small slits or pores in the wall of the body in the backflow region. If the suction is sufficiently strong, the accumulation of decelerated fluid is avoided, and the boundary-layer separation can be prevented.

The separation of the boundary layer can also be prevented by tangential blowing into the boundary layer. A wall jet blown into the boundary layer through a slit in the contour parallel to the main flow direction can supply enough kinetic energy to the boundary layer to prevent separation. According to this principle, for example, the maximum lift of a wing can be greatly increased, although at the expense of a large drag.

There exists a critical value of $\mathrm{Re}_\theta = 300$ above which the occurrence of sudden stalls takes place due to reseparation of boundary layer, and below $\mathrm{Re}_\theta = 300$ the occurrence of sudden stalls can take place because of bubble breakdown.

13.5 BOUNDARY LAYER CONTROL

For drag reduction, a non-Newtonian fluid can be injected into the boundary layer region via tiny surface pores or slots. It has been found experimentally that some fish species use a similar technique by injecting slime from their skin (for example Pacific barracuda or *Sphyraena argentia*). This helps not only in drag reduction and lubrication between scales, but it also prevents surface infection.

Example 13.7

Example Consider the laminar boundary layer flow over a flat plate with blowing of fluid at velocity $v_w > 0$. The flow inside the boundary layer is following the velocity distribution:

$$\frac{u}{U_\infty} = 2\left(\frac{y}{\delta}\right) - \left(\frac{y}{\delta}\right)^2$$

Solution Using the momentum thickness definition we can get:

$$\theta(x) = \frac{2}{15}\delta(x)$$

and from Newton's law of viscosity:

$$\tau_o(x) = \frac{2\mu U_\infty}{\delta(x)}$$

For a flat plate with non-accelerating flow over it, we may use the reduced form of momentum integral equation:

$$\frac{2\mu \cancel{U_\infty}}{\delta}\left(\frac{1}{\rho U_\infty^{\cancel{2}}}\right) = -\frac{\upsilon_o \rho_o}{\rho U_\infty} + \frac{d}{dx}\left[\frac{2}{15}\delta(x)\right]$$

$$\frac{2\mu}{\rho\delta U_\infty} = -\frac{\upsilon_o}{U_\infty} + \frac{d}{dx}\left[\frac{2}{15}\delta(x)\right]$$

Assuming that $\delta = Bx^{1/2}$ and $\upsilon_o = Kx^{-1/2}$ will lead to equation:

$$\frac{2\nu}{B(x^{1/2})U_\infty} = -\frac{K(x^{-1/2})}{U_\infty} + \frac{d}{dx}\left[\frac{2}{15}B(x^{1/2})\right]$$

$$\frac{2\nu}{U_\infty B\sqrt{x}} - \frac{1}{15}\frac{B}{\sqrt{x}} + \frac{K}{\sqrt{x}U_\infty} = 0$$

and quadratic equation in terms of B can be cast as:

$$\frac{-30\nu + B^2U_\infty - 15KB}{U_\infty} = 0$$

The solution of this equation is:

$$B = \pm\frac{1}{2}\frac{15K + \sqrt{225K^2 + 120U\nu}}{U}$$

Now we need to investigate it carefully as we have two solutions.
No blowing condition: If we set $K = 0$, we can have no blowing condition and the solution for this case $B = B_o$.

$$B_o = \frac{30\nu}{U_\infty}$$

$$\delta_o = B_o\sqrt{x}$$

$$\delta_o(x) = \sqrt{30}\sqrt{\frac{x\nu}{U_\infty}}$$

This is exactly the solution if we solve the momentum integral equation with reasonable profile with no suction and blowing case. Thus the B factor can now be written as

$$B = \frac{1}{2}\frac{15K + \sqrt{225K^2 + 120U\,\nu}}{U_\infty}$$

$$C_f = \frac{2\cdot\tau_o(x)}{\rho\cdot U_\infty^2} = \frac{\mu U_\infty}{\rho U_\infty\left(15K + \sqrt{225K^2 + 120U_\infty\nu}\right)\sqrt{x}}$$

$$Cf = \frac{8\mu}{\left(\frac{15v_w^*\sqrt{Re}}{U_\infty} + \sqrt{\frac{225v_w^{*2}Re}{U^2} + 120U_\infty\nu}\right)\sqrt{x}\rho}$$

where $\mathrm{Re} = U_\infty x/\nu$ and $v_w^* = K/\sqrt{U_\infty\nu}$

PROBLEMS

13P-1 A radar is installed to track the enemy aircraft in the region where air speeds are varying from 20 m/s to 30 m/s. The radar is approximated as a hemisphere facing flow with drag coefficient of 1.42. If the diameter of the radar is 30 m, find the mean drag force experienced by the radar if it is made up of solid metal sheets.

13P-2 An array composed of 30 solar troughs is installed in an open field. Considering it as flow facing C-section type geometry ($C_D = 2.30$), what is the drag experienced by the array of solar troughs (length = 10 m, width = 1 m), if a sandstorm is approaching them with wind speeds of 20 miles per hour? Ignore the sand particle related drag.

13P-3 A sphere of diameter 0.1 mm is descending in an oil bath of specific gravity 5.9. The viscosity of oil is 150 cP (centi Poise). Find the terminal velocity of the sphere. The density of steel is 7.85 g/cm^3.

13P-4 A researcher is looking for the most suitable position for positioning third row of wind turbines. The turbine wake causes $\Delta u_{max} = 0.8$ m/s, and the free stream velocity is 12 m/s. The drag force experienced by one turbine is 500 N. The turbine rotor diameter is 120 m. What L value would you propose?

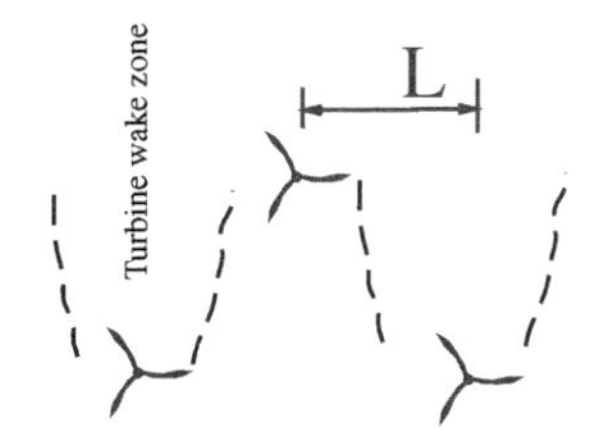

Figure 13.28 Figure of problem 13P-4.

13P-5 The wind blowing over a power-line tower is expressed by relation:

$$u(z) = 26.8\left(\frac{z}{9.114}\right)^{4/25}$$

where z is the vertical height. The height of tower is 22 m. The towers are holding four power conductor cables of 2 cm diameter each. Find the drag on the cables.

13P-6 Calculate the terminal velocity of a 15-μm diameter solid spherical inclusion particle in a stagnant molten steel. The density of inclusion particle is 2.7×10^3 kg/m^3; density of molten steel is 7.1×10^3 kg/m^3; viscosity of molten steel is 5.5×10^{-3} Pa · s. Take drag coefficient is 0.2.

13P-7 A fishery company in Brunei is designing a fish farm using the water of Brunei river, flowing at 3 m/s. The farm will be located at one section of the river and it is

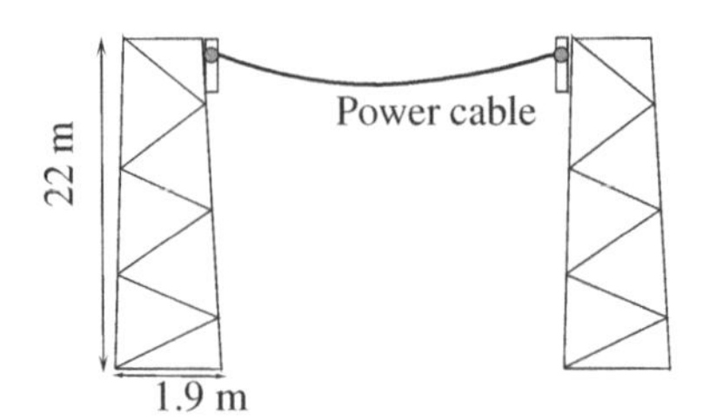

Figure 13.29 Figure of problem 13P-5.

planned that three different species of fishes will be harvested in compartments with an in-line arrangement, i.e., water from compartment 1 will flow into compartment 2 and so on. The three compartment of fishes will be separated by a fish, net and following is the distribution of fishes in each compartment.

Compartment 1: 10 fishes of species one, $C_D = 0.1$; the lengthwise cross-sectional area of fish is 0.2 m^2.

Compartment 2: 20 fishes of species two, $C_D = 0.02$; the lengthwise cross-sectional area of fish is 0.01 m^2.

Compartment 3: 15 fishes of species three, $C_D = 0.12$; the lengthwise cross-sectional area of fish is 0.5 m^2.

All compartments are of equal sizes. The flow is entering the compartment in area of size 4 m $\times$ 4 m.
(i) Do proper mass balance to estimate the mass flow rate entering the fish farm.
(ii) Find the maximum drag experienced by this fish farm. Ignore the pressure drop across the fish net.

13P-8 A cube-shaped house of sides of 10 ft, in a water village, is built over a wooden platform support with four masts of wood logs, going all the way deep into the river bed. The cylindrical mast diameter is 60 cm and length is 10 m. The river speed is 4 m/s and the air speed is 5 m/s. Assuming 60% of mast length in water, calculate (i) the total aerodynamic drag experienced by all four masts. (ii) The moment at the base of a mast. Take necessary assumptions.

13P-9 The cross-section of the wing of an aircraft is based on NACA0012, whose maximum thickness is 12% at location where chord length is 30% starting from leading edge. The length of the wing is 7 m and the chord length is 1.2 m. The plot of drag and lift coefficients are given below. Assume angle of attack of 5 and 10 degrees, with aircraft flying at a cruising speed of 1050 km/h, calculate the drag and lift forces.

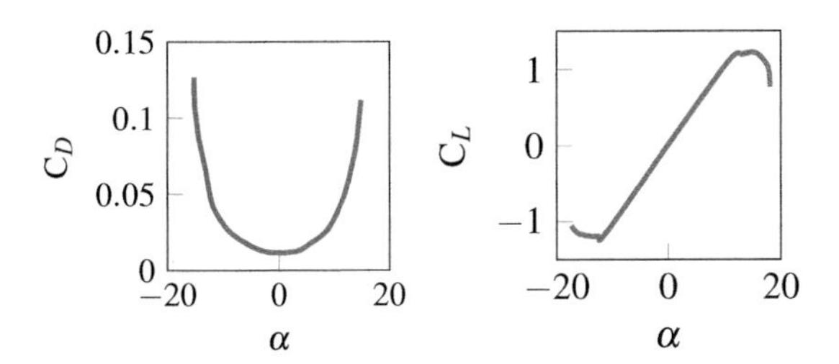

Figure 13.30 Figure of problem 13P-9.

13P-10 Estimate the drag on a triangular plate shown in Figure 13.31 below, which is placed parallel to the air flowing at 10 m/s and 20°C. What is the skin friction drag for the orientations shown in the figure?

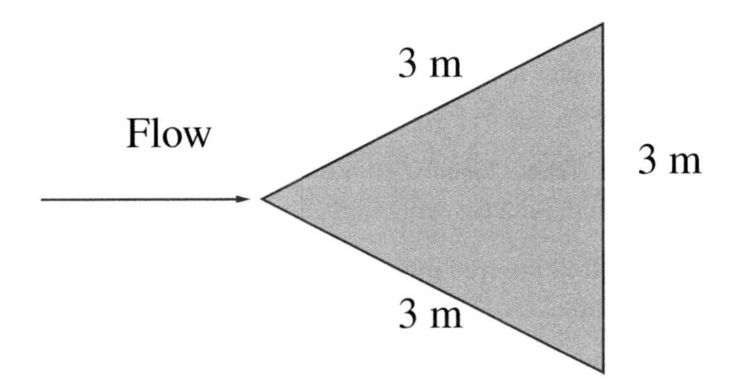

Figure 13.31 Figure of problem 13P-10.

13P-11 A sphere is dropped in oil. Find the terminal velocity of the sphere if the diameter is 0.2 mm, 2 mm, and 20 mm. The oil viscoity is 0.23 Pa · s. The specific gravity of the sphere is 8, and the specific gravity of oil is 0.7. The density of water is 998 kg/m^3.

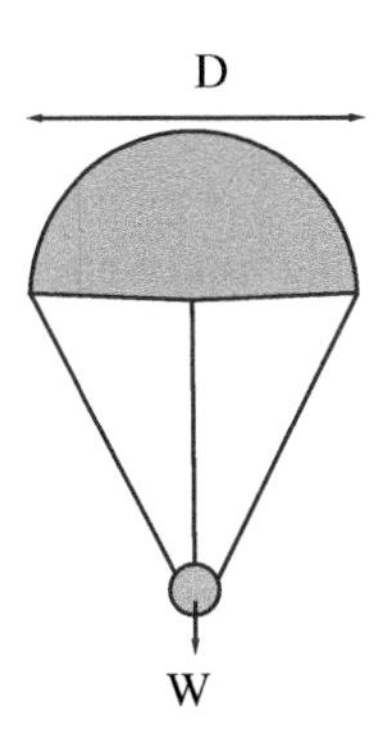

Figure 13.32 Figure problem 13P-12.

13P-12 It is planned that a parachute be used to transfer goods packed in wooden boxes. A box is 50 kg and the box can tolerate a maximum impact speed of 7 m/s. Find the diameter of the parachute whose drag coefficient is 1.33. What is the speed of descent at an altitude where the density of air is 1.00 kg/m^3?

REFERENCES

B. Thwaites, Approximate calculation of the laminar boundary layer, Aeronautical Quarterly, vol. 1, no. 3, pp. 245–280, 1949.

F. M. White, Viscous Fluid Flow, 2nd. ed. McGraw-Hill Publishers, USA, 1991.

W. M. Kays, M. E. Crawford, and B. Weigand, Convective Heat and Mass Transfer. McGraw-Hill, USA, 2004.

R. Viskanta, Heat transfer to impinging isothermal gas and flame jets, Exp. Thermal and Fluid Sciences, vol. 6, pp. 111–134, 1993.

C. O. Popiel and O. Trass, Visualisation of a free and impinging round jet, Exp. Thermal and Fluid Sciences, vol. 4, pp. 253–264, 1991.

E. Gutmark and C. Ho, Preferred modes and the spreading rates of jets, Phy. Fluids, vol. 26, no. 10, pp. 2932–2938, 1983.

T. Liu and J. P. Sullivan, Heat transfer and flow structures in an excited circular impinging jet, Int. J. Heat and Mass Transfer, vol. 17, pp. 3695–3706, 1996.

Ron Darby, Chemical Engineering Fluid Mechanics. Marcel Dekker, Inc, USA, 2001.

J. M. Dallavalle, Micrometrics, 2nd ed. Pitman, USA, 1948.

I. H. Abbott and A. E. von Doenhoff, Theory of Wing Sections, Including a Summary of Airfoil Data. New York: Dover, 1959.

S. Kaneko, T. Nakamura, F. Inada and M. Kato, Flow Induced Vibrations, Elsevier Science, USA, 2008.

P. W. Bearman, On vortex shedding from a circular cylinder in the critical Reynolds number regime, J. Fluid Mech, vol. 37, pp. 577–585, 1969.

B. J. Cantwell, and D. Coles, An experimental study of entrainment and transport in the turbulent near wake of a circular cylinder, J . Fluid Mech., vol. 135, pp. 321–374, 1983.

A. K. M. F. Hussain and M. Hayakawa, Eduction of large-scale organized structures in a turbulent plane wake, J . Fluid Mech., vol. 180, pp. 193–229, 1987.

D. A. Lyn, S. Einav, W. Rodi, and J. H. Park, A laser Doppler velocimetry study of ensemble-averaged characteristics of the turbulent near wake of a square cylinder, J. Fluids and Structures, vol. 304, pp. 285–319, 1995.

Robert D. Blevins, Applied Fluid Dynamics Handbook, Van Nostrand Reinhold Company, USA, 1984.

K. R. Sreenivasan and R. Narasimha, Equilibrium parameters for two-dimensional turbulent wakes, J. Fluids Engr., vol. 104, no. 2, p. 167, 1982.

P. R. Owen and L. Klanfer, On the Laminar Boundary Layer Separation from the Leading Edge of a Thin Airfoil, RAE Rept. Aero. 2508, 1953; reissued as CP No. 220, 1955.

W. T. Evans and K. W. Mort, Analysis of Computed Flow Parameters for a Set of Sudden Stalls in Low-speed Two-dimensional Flow, NACA TN D-85, 1959.

M. S. Olcmen and R. L. Simpson, Perspective: On the near wall similarity of three-dimensional turbulent boundary layers. J. Fluids Engr., vol. 114, p. 487, 1992.

14 Waves and Tsunamis

Flows with a free surface have the possibility to distort into a variety of forms or shapes. At one position in oscillatory ocean flow, some liquid climbs above the mean level and then dwindles below it. In doing so, it appears to move over the surface of the liquid. But what actually moves over the surface is only the form/shape of the disturbance. This chapter will focus on the disturbances moving over a liquid's free surface repeatedly.

Learning outcomes: After finishing this chapter, you should be able to:

- Classify different waves that appear in ocean and rivers.
- Calculate the wavelength of different waves.
- Estimate the extractable energy from the waves.
- Estimate the drag forces due to the waves.
- Understand the dynamics and speed of a Tsunami wave.

A wave is transmission of energy through matter. The major parts of a wave are the crest (the highest point), the trough (the lowest point), the height (distance from trough to crest), the wavelength (distance between identical points on two waves, typically crest to crest), and the period (the time it takes for the same spot on two consecutive waves to pass the same point).

The ratio of a wave height to wavelength can tell us some information about the wave, for example, if it is about to break. Speed of wave is wavelength/period. There is more than one type of wave. The most important for the hydropower and tsunami protection are the shallow-water waves and the deep-water waves. These waves are defined by their wavelength compared to the depth of the water in which they occur. A deep-water wave occurs when water depth is greater than or equal to 1/2 of its wavelength. A shallow-water wave occurs when water depth is less than or equal to 1/20 of its wavelength. As waves move from deep water to shallow water, they come into contact with the ocean floor. This causes the wavelength to decrease and wave height to increase. Once the wave height to wavelength ratio surpasses 1:7, the wave breaks.

14.1 OSCILLATORY WAVES

Free surface flows distort into a variety of patterns (see Figure 14.1). In this section, we indulge in analyzing the disturbances that periodically move over the free surface, as happens in wind-generated ocean waves. In such cases, at any one location on the surface, the water-free surface is undulating at the mean, i.e., the water rises above the mean level and then subsides below it, and thus gives an appearance of a traveling

DOI: 10.1201/9781003315117-14

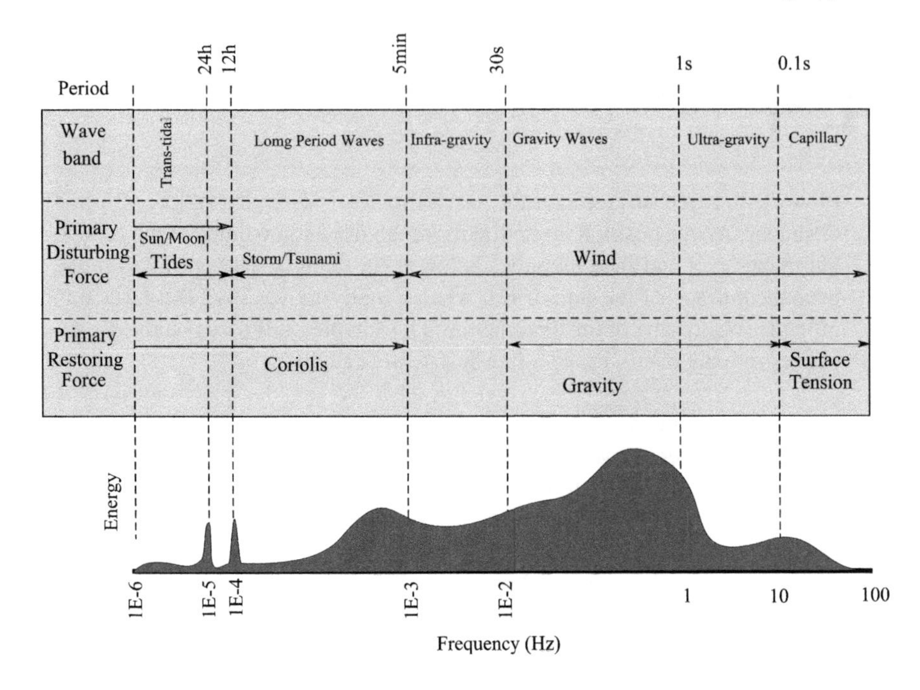

Figure 14.1 Wave types with principal causes and energy distribution.

wave to the onlookers. A similar phenomenon can be observed in a grain field when wind passes over it; to an onlooker, the waves travel across the upper surface of the grain field. However, in reality, the stalks of the grain field are not moving; instead, they oscillate about a mean position. What travels over the free surface is the pattern of that disturbance. There will be no net movement of the liquid itself. The object floating on the surface like a leaf will move forward with the wave's crest, but it returns close to its original position in the next trough.

With wind-generated ocean waves, the major movement is toward the shallow water close to the coastline, but there will be no net flow of sea water toward the coast except the small flow that occurs because of the rise and fall of tides. We shall determine first the velocity with which a wave moves over the free surface.

A disturbance at some point in a liquid generally will give rise to three types of waves, the short surface-tension waves called the capillary waves and longer gravity waves. We now consider the irrotational flow model of the water wave movement. The fluid body is considered as a bed of water reaching the floor. The depth is taken as h, and wave movement over and below the stationary surface of water is taken as η. The peak to peak distance is called the wavelength λ in this analysis. The co-ordinate system is 2D Cartesian with origin fixed at stationary surface level with z taken as vertical direction and x as forward flow direction. It is assumed that water is infinitely available in x direction and its bed height will remain constant (see Figure 14.2).

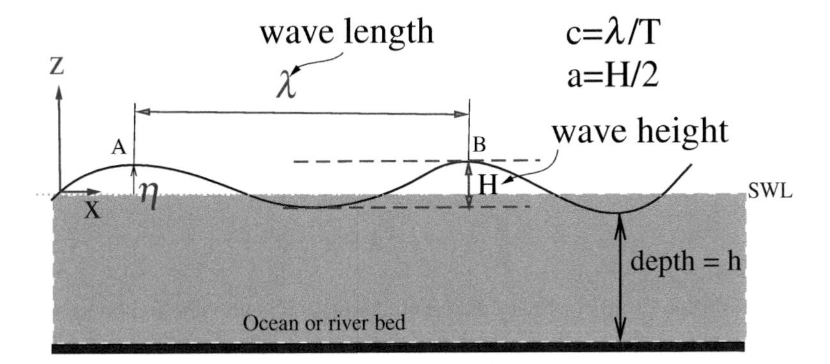

Figure 14.2 Schematic diagram of wave movement in an ocean or river. SWL is the stationary water level. η is the variable indicating the wave position over and above the SWL. The amplitude of wave is indicated by a, which is small compared to wavelength. The H is height of the wave, and λ is the wavelength from crest A to B. c is the wave speed. Note that at $x = \pi, \eta = a$.

The viscous effects are ignored and there is no salinity gradients in the flow field that may influence the density of the fluid. The wind shear is considered as negligible and thus is ignored in the analysis. We assume that wave shape and speed of propagation (c) are constant. The velocity c is called the phase velocity because, regardless of the shape of the wave, the points of the same phase or equal η move at this velocity on the surface. The celerity can also be defined as speed at which the wave crest or trough moves. Celerity is used in the context of waveform movement where no mass is in motion.

We start with the simplest model of motion, so we assume that the stream function of the wave at the free surface is resembling the sinusoidal wave like:

$$\psi(z,x) = cz + f(z)\sin(mx)$$

where $m = 2\pi/\lambda$ is called the wave number, and the cz term appears as the uniform velocity, c is applied to the whole field to make the flow appear steady according to a moving observer. The velocity c is called the phase velocity, or celerity, as irrespective of the shape of the wave, it refers to the fluid particles in same phase (i.e., of equal η) which move at this velocity. Such a type of wave is called the dispersive wave if the phase velocity is a function of the wavelength λ. We assume that h remains constant and the wave can be modeled as a two-dimensional phenomenon. We also assume that wavelength and period are not changing with x. The Coriolis force and vorticity due to Earth's rotation are ignored.

According to irrotational flow, a stream function should satisfy the following condition:

$$\frac{\partial^2 \psi(z,x)}{\partial x^2} + \frac{\partial^2 \psi(z,x)}{\partial z^2} = 0$$

Substituting $\psi(z,x)$ into the above equation leads to form:

$$\left(\frac{d^2}{dz^2} f(z)\right) \sin(mx) - f(z)\sin(mx)\, m^2 = 0$$

We simplify it by dividing whole equation by sin($m\,x$):

$$\frac{d^2}{dz^2} f(z) - f(z)\, m^2 = 0$$

The solution of this ordinary differential equation is already known as:

$$f(z) = A\sinh(B + mz)$$

Based on this solution we update the stream function expression as:

$$\psi(x,z) = cz + A\sinh(B + mz)\sin(mx)$$

where A and B are constants, which we will estimate by applying the boundary conditions.

Floor condition: At the bed floor, where $z = -h$, there is no flow there and we have

$$\cancel{\psi(x,-h)} = \cancel{c(-h)} + A\sinh(B + m(-h))\sin(mx)$$

This leads to

$$\begin{aligned} &\sinh(B + m(-h)) = 0 \\ &(B + m(-h)) = 0 \\ &B = mh \end{aligned}$$

This gives:

$$\psi(x,z) = cz + A\sinh(mh + mz)\sin(mx)$$

Free surface condition: We designate the free-surface streamline that the above equation shows and this streamline must cross the origin (x, η) location.

$$\psi(x,z) = cz + A\sinh(mh + mz)\sin(mx)$$

$$\psi(x,\eta) = c\eta + A\sinh(mh + m\eta)\sin(mx)$$

We set the free surface streamline as zero:

$$\cancel{\psi(x,\eta)}^{zero} = c\eta + A\sinh(mh + m\eta)\sin(mx)$$

Since $\eta \ll h, \lambda$ so we can approximate:

$$\eta \simeq \frac{-A\sinh\left(mh + \cancel{m\eta}^{ignored}\right)\sin(mx)}{c}$$

$$\boxed{\eta \simeq \frac{-A\sinh(mh)\sin(mx)}{c}}$$

According to irrotational flow, the velocity components can be calculated using stream function so:

$$Vel^2 = \left(\frac{\partial \psi(z,x)}{\partial x}\right)^2 + \left(\frac{-\partial \psi(z,x)}{\partial z}\right)^2$$

$$Vel^2 = -A^2 \cos(mx)^2 m^2 + c^2 + 2cA\cosh(mh + m\eta)\, m \sin(mx)$$
$$+A^2 \cosh(mh + m\eta)^2 m^2$$

We can take A from the η definition:

$$A = -\frac{c\eta}{\sin(mx)\cdot\sinh(mh)}$$

Since $x = \pi, \eta = a$ we have

$$A = \frac{-ca}{\sinh(mh)\underbrace{\cos(m\pi)}_{1}}$$

$$A = \frac{-ca}{\sinh(mh)}$$

$$A = -c\cdot a\cdot \text{cosech}(mh)$$

$$Vel^2 = c^2 - \frac{2c^2\eta\cosh(mh + m\eta)m}{\sinh(mh)}$$

$$\underbrace{-\frac{c^2 cos(mx)^2\eta^2 m^2}{(\sin(mx)^2\sinh(mh)^2)} + \frac{c^2\cosh(mh + m\eta)^2\eta^2 m^2}{(\sin(mx)^2\sinh(mh)^2)}}_{\eta^2 m^2 \approx 0}$$

We neglect the term $m^2\eta^2$ so the equation is

$$Vel^2 = c^2 - \frac{2c^2\eta\cosh(mh + m\eta)m}{\sinh(mh)}$$

$$Vel^2 = c^2\left[1 - \frac{2m\,\eta\cosh(mh + m\eta)}{\sinh(mh)}\right]$$

$$Vel^2 = c^2\left[1 - \frac{2m\,\eta\cosh(mh + \cancel{m\eta}^{\eta \ll h})}{\sinh(mh)}\right]$$

$$Vel^2 = c^2\left[1 - \frac{2m\,\eta\cosh(mh)}{\sinh(mh)}\right]$$

$$Vel^2 = c^2\left[1 - \frac{2m\,\eta}{\tan h(mh)}\right]$$

14.2 PRESSURE ON THE SURFACE

The pressure at the liquid's free surface can be influenced by surface tension (σ).

$$\Delta F_{\sigma} = \underbrace{\frac{d}{dx}(\sigma \sin\theta)}_{surface\ tension} \underbrace{\delta x}_{area}$$

The upward force due to rate of increase of vertical momentum of the fluid is:

$$+\uparrow F = \frac{d}{dx}(\sigma \sin\theta)\delta x + p_{gauge}\delta x$$

Force immediately below the free surface is

$$\lim_{\delta\eta \to 0}\left[\frac{d}{dx}(\sigma \sin\theta)\delta x + p_{gauge}\delta x\right] = \lim_{\delta\eta \to 0} F$$

momentum term or force tends to zero as $\delta\eta \to 0$

$$-p_{gauge} = \frac{d}{dx}(\sigma \sin\theta) \approx \sigma\frac{d}{dx}(\sin\theta)$$

We now introduced the sin(θ) as following function:

$$\sin\theta = \frac{\zeta}{\sqrt{1+\zeta^2}}$$

where $\zeta = d\eta/dx$.

$$\eta = -\frac{A\sinh(mh)\sin(mx)}{c}$$

$$\frac{d\eta}{dx} = -\frac{A\sinh(mh)\cos(mx)m}{c}$$

Substituting A into the above equation leads to

$$\frac{d\eta}{dx} = \left(\frac{\eta\sinh(mh)\cos(mx)m}{\sinh(m\cdot(h+\eta))\cdot\sin(mx)}\right)$$

Trigonometrical relation can help in simplification,

$$\sinh(m\cdot(h+\eta)) = \sinh(mh)$$

$$\frac{d\eta}{dx} = \left(\frac{\eta\cancel{\sinh(mh)}\cos(mx)m}{\cancel{\sinh(mh)}\cdot\sin(mx)}\right)$$

$$-p_{gauge} = \frac{d}{dx}(\sigma \sin\theta) \approx \sigma\frac{d}{dx}\left(\frac{\zeta}{\sqrt{1+\zeta^2}}\right)$$

We here introduce $k = \eta^2 m^2$ for simplification purposes

$$\frac{d\eta}{dx} = \eta m \cot(mx)$$

$$p = -\sigma\eta m^2 \left(-1 - \cot(mx) \left(\frac{1}{\sqrt{1 + k\cot(mx)^2}} - \frac{k\cot(mx)^2}{\left(1 + k\cot(mx)^2\right)^{3/2}} \right)^2 \right)$$

The equation will reduced to the form

$$p = -\sigma\eta m^2 \left(-1 - \frac{1}{\tan(mx)} \right)$$

For large x, $\tan(\infty) = \infty$ so

$$p = -\sigma\eta m^2 \left(-1 - \frac{1}{\infty} \right) \approx \sigma\eta m^2$$

We can apply the Bernoulli's equation on free surface streamline and η:

$$p + \frac{\rho}{2}V^2 + \rho g z = c$$

$$\sigma\eta m^2 + \frac{\rho}{2}c^2 \left[1 - \frac{2m\,\eta}{\tanh(mh)} \right] + \rho g z = \frac{\rho}{2}c^2$$

$$\sigma\eta m^2 + \frac{\rho}{2}c^2 \left[1 - \frac{2m\,\eta}{\tanh(mh)} \right] + \rho g z = \frac{\rho}{2}c^2$$

$$\frac{\sigma\eta m^2}{\frac{\rho}{2}} + \frac{\frac{\rho}{2}c^2}{\frac{\rho}{2}} \left[1 - \frac{2m\,\eta}{\tanh(mh)} \right] + \frac{\rho g \eta}{\frac{\rho}{2}} = \frac{\frac{\rho}{2}c^2}{\frac{\rho}{2}}$$

$$\frac{g}{m} + \frac{\sigma m}{\rho} = c^2 \left[\frac{1}{\tanh(mh)} \right]$$

$$\boxed{c^2 = \left(\frac{g}{m} + \frac{\sigma m}{\rho} \right) \tanh(mh)}$$

Substituting $m = (2\pi/\lambda)$, we have

$$\boxed{c^2 = \left(\frac{g}{(2\pi/\lambda)} + \frac{\sigma(2\pi/\lambda)}{\rho} \right) \tanh\left(\frac{2\pi h}{\lambda} \right)} \tag{14.1}$$

The waves that follow this relation are called Airy waves after Sir George B. Airy (1801–1892), who first mathematically analyzed them. Wave propagation is considered as a dispersive type when the phase velocity depends on the wavelength. In such waves, the dispersion might occur as the waves can split up and the components of

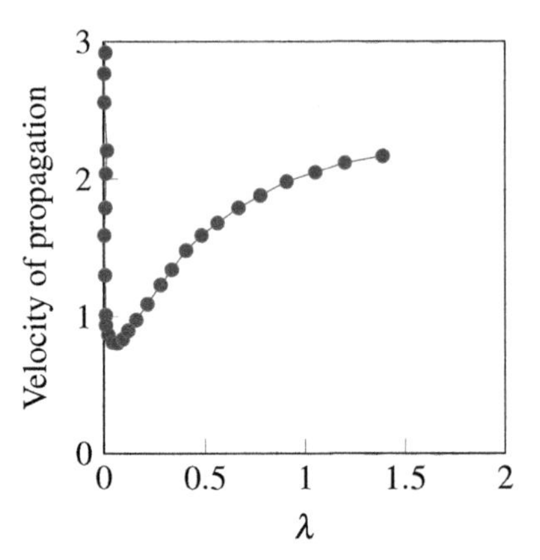

Figure 14.3 Typical capillary wave behavior.

different wavelengths will form, which will move at different velocities and thus become separated from the original wave.

Capillary waves:

Waves whose characteristics are governed principally by surface tension are known as capillary waves, and their propagation velocity is defined by Eq. 14.1. When λ is small compared with h then

$$\tanh\left(\frac{2\pi h}{\lambda}\right) \to 1$$

$$c^2 = \left(\underbrace{\frac{g\lambda}{(2\pi)}}_{negligible} + \frac{\sigma 2\pi}{\rho\lambda}\right) \approx \frac{\sigma 2\pi}{\rho\lambda}$$

Figure 14.3 shows the plot of c against λ for water depth (h) of 0.5 in. and 2.0 in. This figure shows that waves cannot exist with velocity of propagation less than approximately 0.75 foot per second. Disturbances of wave length less than that corresponding to the minimum wave velocity are termed capillary waves since they depend primarily on the surface tension of the fluid. Propagation velocities greater than this minimum velocity correspond to shorter capillary waves and longer gravity waves.

Negligible surface tension: If the effect of surface tension on the phase velocity is negligible then Eq. 14.1 will be reduced to

$$\frac{g}{(2\pi/\lambda)} \gg \frac{\sigma(2\pi/\lambda)}{\rho}$$

$$\frac{g\lambda^2}{2\pi} \approx \frac{2\pi\sigma}{\rho}$$

leading to result

$$\lambda \approx 2\pi\sqrt{\left(\frac{\sigma}{\rho g}\right)}$$

The effect of surface tension on the phase velocity is negligible if

$$\frac{2\pi\sigma}{\rho\lambda} \ll \frac{g\lambda}{2\pi}$$

or

$$\lambda \gg 2\pi\sqrt{\left(\frac{\sigma}{\rho g}\right)}$$

For water, $\lambda \approx 17$ mm which gives $c = 0.23$ m/s and the surface tension effects can be ignored, if

$$\lambda \gg 17$$

Deep water waves: Deep water waves are found in water deeper than half of their wavelength $((h/\lambda\ ) > (1/2))$. For $\lambda >$ h/2 the $\tanh\left(\frac{2\pi h}{\lambda}\right)$ term can be unity, and velocity of propagation will become

$$c = \sqrt{\left(\frac{g}{(2\pi/\lambda)} + \frac{\sigma(2\pi/\lambda)}{\rho}\right)}$$

which can be reduced further as:

$$c = \sqrt{\left(\frac{g}{(2\pi/\lambda)}\right)}$$

Gravity waves: Waves whose properties are primarily determined by gravitational effects are referred to as gravity waves, represented by:

$$c^2 = \left(\frac{g}{(2\pi/\lambda)} + \underbrace{\frac{\sigma(2\pi/\lambda)}{\rho}}_{negligible}\right)\tanh\left(\frac{2\pi h}{\lambda}\right)$$

$$c^2 = \left(\frac{g}{(2\pi/\lambda)}\right)\tanh\left(\frac{2\pi h}{\lambda}\right)$$

Shallow water waves: If $\lambda \gg h$ (shallow-water wave or waves of long wavelength) then

$$c^2 = \left(\frac{g}{(2\pi/\lambda)}\right)\tanh\left(\frac{2\pi h}{\lambda}\right)$$

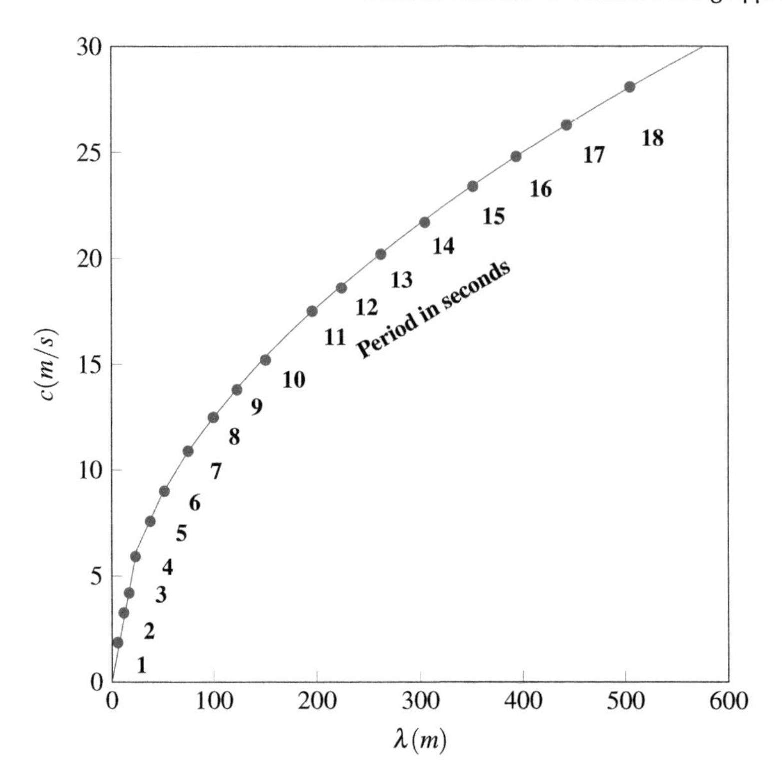

Figure 14.4 Celerity vs. wavelength (valid for deep water waves). Plot of $c = g\lambda/2\pi$.

$$c^2 \approx \left(\frac{g\lambda}{(2\pi)}\right)\left(\frac{2\pi h}{\lambda}\right) = gh$$

Shallow water waves are defined as waves in water shallower than 1/20 their wavelength ($(h/\lambda) < (1/20)$).

Figure 14.4 shows the celerity vs. wavelength (valid for deep water waves). The figure provides a quick estimate of the unknown quantities.

Example 14.1

Example Gravity waves on water with a mean depth of 3 m have a period of 10 s. What is the wavelength?

Solution

$$c^2 = \left(\frac{g}{(2\pi/\lambda)}\right)\tanh\left(\frac{2\pi h}{\lambda}\right)$$

since

$$c = \frac{\lambda}{T}$$

where T is time period and λ is the wavelength. We have

$$\lambda - \left(\frac{gT^2}{2\pi}\right)\tanh\left(\frac{2\pi h}{\lambda}\right) = 0$$

This is a transcendental equation and can be solved via trial and error. As a first approximation, we can assume a small value for

$$\tanh\left(\frac{2\pi h}{\lambda}\right) = 0.75$$

This can also be solved by trial and error, which gives an answer $\lambda = 53.15$.

Example 14.2

Example The wave characters are being investigated in a water channel. The oscilloscope is being used to measure the water's free surface wavelength and frequency. It is noted that wavelength is 0.813 m and frequency is 1.3 Hz. find the celerity of the wave if the depth of the channel is 0.6 m.
Solution Since $h/\lambda = 0.6/0.813 = 0.73$ the wavelength is deep water wave ($h/\lambda \geq 0.5$). For such waves, the celerity (c) is :

$$c^2 = \frac{g\lambda}{2\pi}$$

$$c = 1.12m/s$$

14.3 ABSOLUTE VELOCITY COMPONENTS

So far we have discussed the phase velocity and relative flow velocities. We will now derive the expression for u and w components of the absolute velocity.

$$u = c - \frac{\partial \psi}{\partial z}$$

and

$$w = \frac{\partial \psi}{\partial z}$$

The stream function we have already defined as:

$$\psi(x,\eta) = c\eta + A\sinh(mh + m\eta)\sin(mx)$$

so u and w are

$$u = -A \cdot m \cdot \cosh\left(m \cdot (h+z)\right) \cdot \sin\left(m\left(x - c \cdot t\right)\right)$$

$$w = A \cdot m \cdot \sinh\left(m\left(h+z\right)\right) \cos\left(m\left(x - ct\right)\right)$$

respectively. Since $A = -ca$ cosech(mh) we may write the u and w as:

$$u = \frac{c \cdot a \cdot m \cdot \cosh\left(m\left(h+z\right)\right) \sin\left(m\left(x - ct\right)\right)}{\sinh\left(mh\right)}$$

$$w = -\frac{c \cdot a \cdot m \cdot \sinh\left(m\left(h+z\right)\right) \cos\left(m\left(x - ct\right)\right)}{\sinh\left(mh\right)}$$

$$u = \frac{\pi H}{T}\left\{\frac{\cosh\left(m\left(h+z\right)\right)}{\sinh\left(mh\right)}\right\} \sin\left(\theta\right)$$

where $\theta = m\left(x - ct\right)$ is independent of z and $a = H/2$.

For gravity waves $h \gg \eta$, and we can further cast the u equation into simpler form. As the following term

$$\lim_{h \to \infty} \frac{\cosh(mh + mz)}{\sinh(mh)} = \frac{(\infty)}{\infty} = \textit{undefined}$$

is undefined when h tends to ∞ we will apply the L'Hospital rule.

$$\lim_{h \to \infty} \left[\frac{\sinh(mh + mz)}{\cosh(mh)}\right]$$

Using double angle rule we can expand the numerator and write further as

$$\frac{\sinh(mh + mz)}{\cosh(mh)} = \frac{\sinh(mh) * \cosh(mz) + \cosh(mh) * \sinh(mz)}{\cosh(mh)}$$

$$\frac{\sinh(mh + mz)}{\cosh(mh)} = \tanh(mh) \cdot \cosh(mz) + \sinh(mz)$$

Now applying limit

$$\lim_{h \to \infty} \left[\tanh(mh) \cdot \cosh(mz) + \sinh(mz)\right] \approx \cosh(mz) + \sinh(mz) = e^{mz}$$

so the more simplified form of u for the gravity waves is

$$u = \frac{\pi H}{T}\left\{\frac{\cosh\left(m\left(h+z\right)\right)}{\sinh\left(mh\right)}\right\} \sin\left(\theta\right) = \frac{\pi H}{T}\left\{e^{mz}\right\} \sin\left(\theta\right)$$

$$u = \frac{\pi H}{T}\left\{e^{mz}\right\} \sin\left(m\left(x - ct\right)\right)$$

14.4 WAVE MOVEMENT

We now have velocities from the Airy wave theory as:

$$u = -A \cdot m \cdot \cosh(m(h+z))\sin(-mx+mct)$$
$$w = Am\sinh(m(h+z))\cos(-mx+mct)$$

For the sake of simplification, we introduced $\theta = m(h+z)$ here, and equations can be cast into form

$$u = -A \cdot m \cdot \cosh(\theta)\sin(-mx+mct)$$
$$w = Am\sinh(\theta)\cos(-mx+mct)$$

We can find the displacement **x** and **z** by integrating these equations, which gives us the result

$$\mathbf{x} = -\frac{A\cosh(\theta)\cos(-mx+mct)}{c}$$
$$\mathbf{z} = \frac{A\sinh(\theta)\sin(-mx+mct)}{c}$$

We now take the square of **x** and divide it by $\left(\left(\frac{A}{c}\right)\cdot\cosh(\theta)\right)^2$ and we square **z** and divide it by$\left(\left(\frac{A}{c}\right)\cdot\sinh(\theta)\right)^2$ and add the two terms together. With a little rearrangement, these terms can be written in a combined form as:

$$m^2\sin(-mx+mct)^2c^2 + m^2\cos(-mx+mct)^2c^2$$

dividing the whole equation by m^2c^2 will give us

$$\sin^2(m(ct-x)) + \cos^2(m(ct-x))$$

which is an equation of a circle in new horizontal $\left(\frac{A}{c}\right)\cdot\cosh(\theta)$ and vertical coordinate $\left(\frac{A}{c}\right)\cdot\sinh(\theta)$. Trigonometrical relations can prove that

$$\sin^2(\theta) + \cos^2(\theta) = 1$$

so the wave displacement is like equation of a circle or an orbit. Figure 14.5 shows the schematic diagram of the elliptic movement of the particle.

From this derivation we can conclude that

1. As energy travels through matter, the energy is transmitted to adjacent matter.
2. As energy moves through matter, the matter moves and then returns to its original position.

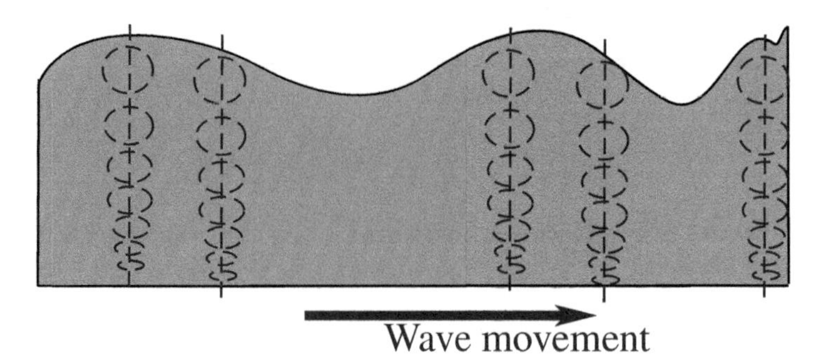

Figure 14.5 Ocean waves transfer energy, and internally in a body of water, the wave moves in circles, also called orbital waves. A slight forward movement is observed, which is called the Stoke drift. This drift phenomenon is named after George Gabriel Stokes, who derived expression for the wave drift in his 1847 study of water waves.

14.5 THE WAVE KINETIC ENERGY

The wave kinetic energy is defined as:

$$KE(x,z) = \frac{\rho}{2} \cdot \left(u^2 + w^2\right)$$

$$KE(x,z) =$$

$$\frac{1}{2}\rho \left(A^2 m^2 \cosh(m(h+z))^2 \sin(m(x-ct))^2 + A^2 m^2 \sinh(m(h+z))^2 \cos(m(x-ct))^2\right)$$

We integrate this twice. First from depth to the instantaneous location η and then across wave length

$$KE(x,z) = \int_0^{\lambda} \left(\int_{-h}^{\eta} K(x,z)\, dz\right) dx$$

$$KE(x,z) = \frac{1}{8}\rho A^2 m \ \sinh(2mh)\lambda$$

Now since $A = -ca$ cosech(mh) and

$$c = \sqrt{\left(gm + \frac{g}{m}\right)\tanh(mh)}$$

we have

$$KE(x,z) = \frac{1}{8}\frac{a^2\rho\left(\frac{\sigma m^2}{\rho} + g\right)\tanh(mh)\sinh(2mh)\lambda}{\sinh(mh)^2}$$

$$KE(x,z) = \frac{\lambda}{8}a^2\rho\left(\frac{\sigma m^2}{\rho} + g\right)\left[\underbrace{\frac{\tanh(mh)\sinh(2mh)}{\sinh(mh)^2}}_{2}\right]$$

The square bracket term is 2 according to trigonometry so we may write

$$KE\,(x,z) = \frac{2\lambda}{8} a^2 \rho \left(\frac{\sigma m^2}{\rho} + g \right)$$

$$KE\,(x,z) = \frac{\lambda}{4} a^2 \rho \left(\frac{\sigma m^2}{\rho} + g \right)$$

14.6 RATE OF ENERGY OF WAVE

Power extraction from the tides dates back to medieval tide mills. Most designs have some core elements like a mechanical device or turbine, which extracts energy from waves and then passes it to a generator. The most significant recent developments are the La Rance Tidal Power Station (Brittany, France), which was opened in 1966, and the Siwha Lake Tidal Power Station (Gyeonggi Province, South Korea), which was opened in 2011. In practice, turbines of different sizes are used, which can be between 0.001 and 100 m. Also, they must operate across a range of flow speeds, with the power capping mechanisms to protect the drivetrain mechanical and electrical systems from the unsteady flows or extreme wave loading conditions.

We can now apply the Bernoulli's equation to at two stations on the wave. The wave is moving with speed c and the fluid particles have speed u and w. So we may write as:

$$p + \frac{1}{2}\rho \left\{ (u-c)^2 + w^2 \right\} = \frac{1}{2}\rho (c)^2$$

$$p + \frac{1}{2}\rho \left\{ (u^2 - 2uc + c^2) + w^2 \right\} = \frac{1}{2}\rho (c)^2$$

$$p + \frac{1}{2}\rho \left\{ u^2 + w^2 \right\} = \cancel{\frac{1}{2}\rho (c)^2} - \cancel{\frac{1}{2}\rho (c)^2} + \frac{1}{2}\rho (2uc)$$

$$\frac{E}{\forall} = p + \frac{1}{2}\rho \left\{ u^2 + w^2 \right\} = \frac{1}{2}\rho (2uc) = \rho uc$$

where $\forall$ is volume. The L.H.S. shows that total energy in an irrotational wave is ρuc. We now take mean energy to estimate the power in a single wavelength across the sea/river depth, which leads to the result:

$$\left(\frac{Power}{width} \right)_{\lambda} = \left(\frac{\rho c}{2} \right) u \; dz = \left(\frac{\rho c}{2} \int_{-h}^{\eta} u^2 dz \right)$$

$$= \frac{1}{4} \cdot \Omega \cdot \Xi$$

where Ω and Ξ are defined as:

$$\Omega = \left\{ \rho c^3 a^2 m \,\mathrm{cosech}^2\,(mh) \sin^2 [m(x-ct)] \right\}$$

$$\Xi = \left\{ \cosh\,(m\,\eta + m\,h) \sinh\,(m\,\eta + mh) + m\,\eta + mh \right\}$$

This is the expression for average power of the wave. Since for deep water waves $\eta \ll h$, we can ignore $m\eta$ terms leading to form

$$\frac{1}{4}\left\{\rho c^3 a^2 m \sin^2[m(x-ct)]\right\} \cdot \left\{\underbrace{\frac{\cosh(m\,h)\sinh(mh)+mh}{\sinh^2(mh)}}_{1\ for\ h\to\infty}\right\}$$

The trigonometric functions depending on h will be unity for very large h values, so we can write power per unit width as:

$$\frac{1}{4}\left\{\rho c^3 a^2 m \sin^2[m(x-ct)]\right\}$$

Now, since power is maximized when $\sin^2[m(x-ct)] = 1$, the maximum power per unit width that can be extracted from the wave will be

$$= \frac{1}{4}\left\{\rho c^3 a^2 m\right\}$$

It is already proved for deep-water waves that $c^2 = g\lambda/2\pi$, so now substitute the wave variables into parameters:

$$\begin{aligned} c &= gT/2\pi \\ a &= H/2 \\ m &= 2\pi/\lambda \\ \lambda &= c\cdot T = gT^2/2\pi \end{aligned}$$

and rearranging we get

$$\frac{Power}{width} = \frac{\rho a^2}{8}\left(\frac{g^2 T}{\pi}\right)$$

This expression is valid for deep-water waves.

Example 14.3

Example In an ocean, the waves bearing the period of 10 s and height of 7 m are approaching an inline array of hydroturbines. An array of turbines is installed parallel to the shore in water at a depth of 40 m. Estimate the mean power per unit array width that can be extracted if the array has a cumulative efficiency of 60%. Take the density of sea water as 1023 kg/m^3.

Solution We first need to confirm that the waves are deep-water waves. For that we calculate the wavelength. It is given that $T = 10$ s, $H = 7$ m and the depth $h = 40$ m.

$$\lambda - \frac{g\cdot T^2}{2\cdot\pi}\cdot\tanh\left(\frac{2\cdot\pi\cdot h}{\lambda}\right) = 0$$

This gives $\lambda = 78.9$ m. For the deep-water wave, the necessary condition is $h \geq \lambda/2$, which is satisfied as $40 \geq 39.45$.
We can use the power relation:

$$\frac{Power}{width} = \frac{\rho a^2}{8}\left(\frac{g^2 T}{\pi}\right)$$

As amplitude is H/2, we have $a = 7/2 = 3.5$ m.

$$\frac{Power}{width} = 0.6\frac{\rho a^2}{8}\left(\frac{g^2 T}{\pi}\right) = 0.6\frac{1023 \times (3.5)^2}{8}\left(\frac{9.81^2 \times 10}{\pi}\right)$$

$$\frac{Power}{width} = 0.6 \times 480kW/m = 288kW/m$$

14.7 DRAG IN OSCILLATORY FLOWS

Morison et al. (1950) gave a semi-empirical heuristic formula for the oscillatory flow force calculations. The Morison equation is the sum of two force components: first, an inertia force in phase with the local flow acceleration and second the drag force proportional to the square of the instantaneous flow velocity.

$$F = \frac{1}{2}\rho C_D A u\,|u| + C_m \rho \forall \left(\frac{du}{dt}\right)$$

In case the body moves as well, with velocity $v(t)$, the Morison equation is written as:

$$F = \underbrace{\rho \forall \left(\frac{du}{dt}\right)}_{Froude-Krylov\ force} + \underbrace{C_a \rho \forall \left(\frac{du}{dt} - \frac{dv}{dt}\right)}_{Hydodynamic\ mass\ force} + \underbrace{\frac{1}{2}\rho C_D A (u-w)\,|u-w|}_{drag\ force}$$

The Morison equation is an empirical equation, so the values of drag and inertia coefficients are determined experimentally in a number of ways. The C_D and C_a are dimensionless quantities and the most widely used values are 0.7 and 2.0, respectively. The API recommended $C_D = 0.65$ and $C_a = 1.6$, for smooth or $C_D = 1.05$ and $C_a = 1.2$ for rough surfaces in case of marine growth.

Following are the wave speed functions, which are also proposed in the literature:

$$u = \left[\frac{\pi H}{T}\cos\left(\frac{2\pi}{\lambda}x - \frac{2\pi}{T}t\right)\right]\left[\frac{\cosh(z+h)}{\sinh(\frac{2\pi}{\lambda}h)}\right] \quad 0.05\lambda < h < 0.5\lambda \quad [Intermediate\ waves]$$

$$u = \left[\frac{\pi H}{T}\cos\left(\frac{2\pi}{\lambda}x - \frac{2\pi}{T}t\right)\right]\left[e^{mz}\right] \quad h \geq 0.5\lambda \quad [Deep\ water\ waves]$$

$$u = \left[\frac{\pi H}{T}\cos\left(\frac{2\pi}{\lambda}x - \frac{2\pi}{T}t\right)\right]\left(\frac{\lambda}{2\pi h}\right) \quad h \leq 0.5\lambda \quad [Shallow\ water\ waves]$$

Note that the wave here is modeled as a cosine wave instead of the sine wave representation as done earlier in this chapter; this is a matter of taste.

Example 14.4

Example A vertical pole in an offshore structure has diameter of 1.5 m. It is installed at a location in the sea where the depth of sea is 200 m and the mean sea currents are negligible. The pole is designed to withstand the drag caused by waves of heights of 7 m and a period of 10 s. Find the wave-induced drag and inertia force drag on the pole. Take $\rho = 1023$ kg/m^3, $C_D = 1.3$ and $C_M = 2$.

Solution It is given that $T = 10$ s and $h = 200$ m, so we first calculate the

$$\lambda - \left(\frac{gT^2}{2\pi}\right)\tanh\left(\frac{2\pi h}{\lambda}\right) = 0$$

This is a transcendental equation and can be solved via trial and error. As a first approximation we can assume a small value for

$$\tanh\left(\frac{2\pi h}{\lambda}\right) = 0.75$$

This can also be solved by trial and error, which gives $\lambda = 156.130$ m.
The ratio of $D/\lambda = 1.5/156.130 = 0.0096 < 0.2$
Morison equation for small structures and mean currents is applicable.

$$F = \frac{1}{2}\rho C_D A u\,|u| + C_m \rho \forall \left(\frac{du}{dt}\right)$$

The depth of ocean at this location is very large. For the assumption of deep water waves, the condition is:

$$Depth \geq \frac{\lambda}{2}$$

$$200m > \left(\frac{156}{2} = 78\right)$$

hence the waves are deep-water waves.
It is given that $H = 7$ m and $T = 10$ s, so we can write velocity and accelerations:

$$u = \frac{\pi H}{T} e^{mz} \sin\left(mx - \frac{2\pi t}{T}\right)$$

$$\frac{du}{dt} = \frac{2\pi^2}{T^2} H e^{mz} \cos\left(mx - \frac{2\pi t}{T}\right)$$

where $m = 2\pi/\lambda = 0.0402$. We first compute the inertia force term as below:

$$F_{inertia} = C_m \rho \forall \left(\frac{du}{dt}\right) = C_m \rho \left[\frac{\pi}{4} D^2 dz\right] \frac{2\pi^2}{T^2} H e^{mz} \cos\left(mx - \frac{2\pi t}{T}\right)$$

$$F_{inertia} = C_m \rho \left(\frac{2\pi^2}{T^2}\right) \left(\frac{\pi}{4} D^2\right) H \cos\left(mx - \frac{2\pi t}{T}\right) [e^{mz} dz]$$

Since the volume $\forall$ contains the z, we need to integrate this equation.

$$F_{inertia} = C_m \rho \left(\frac{2\pi^2}{T^2}\right) \left(\frac{\pi}{4} D^2\right) H \cos\left(mx - \frac{2\pi t}{T}\right) \int_{-depth}^{0} [e^{mz} dz]$$

$$F_{inertia} = C_m \rho \left(\frac{2\pi^2}{T^2}\right) \left(\frac{\pi}{4} D^2\right) H \cos\left(mx - \frac{2\pi t}{T}\right) \left[\frac{1}{k}(1 - e^{-m(depth)})\right]$$

$$F_{inertia} = 124 \cos\left(mx - \frac{2\pi t}{T}\right) \quad kN$$

The force is maximum when $\cos\left(mx - \frac{2\pi t}{T}\right) = 1$, so maximum inertia force is 124 kN.

Wave-induced drag force: We will now calculate the drag force as follows:

$$F_{drag} = \frac{1}{2} \rho C_D (D.dz) u |u|$$

Note the projected area in terms of z in the above equation, so we need to integrate this equation over the pole.

$$F_{drag} = \frac{\rho}{2} C_D(D) \int_{-depth}^{0} u|u| \; dz$$

We have velocity as:

$$u = \frac{\pi H}{T} e^{mz} \sin\left(mx - \frac{2\pi t}{T}\right)$$

$$F = \frac{1}{2} \rho C_D (D.dz) \left[\frac{\pi H}{T} e^{mz} \sin\left(mx - \frac{2\pi t}{T}\right) \left| \frac{\pi H}{T} e^{mz} \sin\left(mx - \frac{2\pi t}{T}\right) \right| \right]$$

$$F = \frac{1}{2} \rho C_D (D) \left(\frac{\pi H}{T}\right)^2 \sin\left(mx - \frac{2\pi t}{T}\right) \left| \sin\left(mx - \frac{2\pi t}{T}\right) \right| \left(\int_{-depth}^{0} e^{mz} dz \right)$$

The maximum force will occur when

$$\sin\left(mx - \frac{2\pi t}{T}\right) \approx 1$$

so the maximum drag is

$$F = \frac{1}{2}\rho C_D(D)\left(\frac{\pi H}{T}\right)^2 \left[\frac{1}{2m}(1 - e^{-2m(depth)})\right]$$

Drag force = 59.93 kN.

14.8 SHIP RESISTANCE

Flow in rivers and oceans is mostly turbulent, with some exceptions. Figure 14.6 shows NASAs Earth Observatory photograph of eastern Russia, including the laminar and turbulent flow patterns.

Figure 14.6 Strong tides create laminar and turbulent flow patterns in waters off far eastern Russia. *(Source: NASAs Earth Observatory)*

Flow types have a strong effect on the drag experienced by the ships. Consider a ship moving in water having a density as ρ and viscosity as μ. The drag on the ship can be calculated as:

$$F_D = C_W\left(\frac{\rho}{2}V^2\right)S_w$$

where S_w is the wetted surface area of the ship, which can be estimated from formula

$$S_w = 1.025L_{pp}\left[B \cdot C_B + 1.7Dr\right]$$

where B is the waterline breadth of the hull,
L_{pp} Length between perpendiculars and it is the length of ship which is in water, so it is less than the total length of the ship,
C_B is the block coefficient. Some values are tabulated in Table 14.1.
Dr is the draft of the ship.

C_B = (underwater submerged volume) / (length x breadth x draft of the vessel)

C_W is the drag coefficient due to water, which can be further synthesized into the following components:

$$C_W = C_R + C_f + \Delta C_\varepsilon$$

Table 14.1
Vessel - Block Coefficient

Vessel	Block Coefficient
Ferry	0.5–0.65
Dry cargo	0.6–0.75
Container/Ro-Ro/Passenger	0.65–0.7
Tanker/Bulk	0.72–0.85

Table 14.2
Residual Coefficient

Froude Number	Residual Coefficient
0.15	5.30E-04
0.2	7.20E-04
0.25	8.70E-04
0.3	1.53E-03
0.35	2.00E-03

Here,

C_R is a function of Froude number and is called the residual coefficient (see Table 14.2),

C_f is a function of Reynolds number and it indicates the skin friction coefficient, and

ΔC_ε represents the friction arisen due to roughness of the outer ship walls.

14.9 TSUNAMI

The word *Tsunami* comes from the Japanese characters for harbor *tsu* and wave *nami*. A tsunami is a series of waves caused by earthquakes or undersea volcanic eruptions in the depths of the ocean, or from icebergs falling into the ocean.

- **Tsunami** waves do not dramatically increase in height. But as the waves travel inland, they build up to higher and higher heights as the depth of the ocean decreases. The speed of tsunami waves depends on ocean depth rather than the distance from the source of the wave. Tsunami waves may travel as fast as jet planes over deep waters, only slowing down when reaching shallow waters. In deep ocean, the observer may not perceive the movement of a Tsunami wave,

Figure 14.7 Tsunami wave generated in a model testing facility. (*Hubert Chanson, Civil Engineering, The University of Queensland, Australia; Source: Intl. Assoc. for Hydro-Environment Engr. and Research.*)

as its height can be one meter. A tsumani takes from a few minutes to hours to reach the coast. It's wavelength is in order of kilometer. There is a sudden rise in the water level at the coast as the wave reaches the coast and it forms a wall of water. This is actually a wave front formed as waves merged into each other. Tsumani waves should not be called a tidal wave.

- **Seiches** are standing waves with longer periods of water-level oscillations (typically exceeding periods of three or more hours). They can be generated due to storms and strong winds in lakes and large standing bodies of water. This water swishing back and forth is called a seiche and can cause flooding and damage along the shoreline of a lake.
- **Meteotsunamis** are progressive waves limited to the tsunami frequency band of wave periods (two minutes to two hours). A meteotsunami is a tsunami wave that is generated due to weather conditions and not due to an earthquake. Meteotsunamis have been observed to reach heights of 6 ft or more. They occur in many places around the world, including the Great Lakes, Gulf of Mexico, Atlantic Coast, and the Mediterranean and Adriatic Seas.

Figure 14.7 shows the Tsunami run-up onto a 1:10 slope with an offshore still water depth of 0.150 m in a hydraulic model testing facility.

Close to shores the tsunami wave is a shallow water wave, so we can use the shallow water wave relations.

Speed of the tsunami wave: The celerity of a tsunami wave can be estimated from the formula:

$$c = \sqrt{g \cdot h} \tag{14.2}$$

where g is acceleration due to gravity and h is depth of origin of disturbance.

Time taken by Tsumani to hit the coast is:

$$t = s/c \tag{14.3}$$

where s is displacement from point of origin to coast. These are the rough calculations to estimate the time available for evacuation.

An eruption in 1792 in Japan created waves that were several hundred feet high.

On December 26, 2004, waves triggered by an earthquake caused at least 230,000 deaths in Asian countries, including Indonesia, Sri Lanka, India, Maldives, and Thailand.

On September 29, 2009, a tsunami caused substantial damage and loss of life in American Samoa, Samoa, and Tonga. The tsunami was generated by a large earthquake in the Southern Pacific Ocean.

The March 11, 2011, magnitude 9.0 Honshu, Japan earthquake (38.322 N, 142.369 E, depth 32 km) generated a tsunami observed over the Pacific region and caused tremendous local devastation. This is the fourth largest earthquake in the world and the largest in Japan since instrumental recordings began in 1900. The disaster resulted in economic losses of US 360 billion dollar.

On January 15, 2022, the Hunga-Tonga-Hunga-Ha'apai volcano, about 30 kilometers southeast of Tonga's Fonuafo'ou island, erupted and created a tsunami wave.

Example 14.5

Example An earthquake off the coast of Hawaii at a depth of 100 ft causes a tsunami wave. Estimate how long it will take for this wave to reach New Zealand. The distance from the earthquake location in deep sea to New Zealand is 7,413 km.

Solution We can calculate the celerity of the wave as:

$$c = \sqrt{g \cdot h} = \sqrt{9.81 \times 30.48} = 17.29 m/s$$

The time the tsunami wave will take to reach New Zealand is

$$t = \frac{s}{c} = \frac{7413}{17.29} = 4.9 days$$

PROBLEMS

14P-1 A ship 125 m long (at the water-line) and having a wetted surface area of 4,000 m^2 is to be driven at 12 m/s in sea water. A model ship of 1/20th scale is to be tested to determine its resistance. Find the total resistance on the prototype. (Note: use dimensional analysis.)

14P-2 An earthquake hypocenter is at a depth of 29 km. The nearest coast is at 100 km. How long does it take for the Tsunami wave to hit the coast? [Answer: $t = 3.11$ minutes]

14P-3 A ripple tank contains a liquid of density of 900 kg/m^3 to a depth of 4 mm. Waves of length 7.5 mm are produced by a reed vibrating at 25 Hz. Determine the surface tension of the liquid.

REFERENCES

B. Massey and J. Ward-Smith, Mechanics of Fluids, Taylor and Francis, USA, 2006.

J. R. Morison, M. P. O'Brien, J. W. Johnson, and S. A. Schaaf, The force exerted by surface waves on piles, Petroleum Transactions, American Institute of Mining Engineers, vol. 189, pp. 149–154.

H. O. Kristensen and M. Luetzen, Prediction of Resistance and Propulsion Power of Ships, Project no. 2010-56, Emissionsbeslutningsstttesystem, Work Package 2, Report no. 04, May 2013.

15 Channel Flow

If a liquid-free surface is exposed to an open atmosphere and bounded by sidewalls only, then such a flow is classified as an open channel flow. In these flows, the free surface is subjected most of the time only to atmospheric pressure. We know that atmospheric pressure is almost constant across the globe, and therefore, the open channel flow is driven by the component of the weight of the fluid body. Natural streams and rivers, artificial canals, irrigation ditches, and flumes are examples of open channel flows.

Learning outcomes: After finishing this chapter, you should be able to:

- Classify the channel flows as uniform or nonuniform flows.
- Calculate the frictional head loss for flow through a channel.
- Estimate the height of hydraulic jump.
- Understand the role of Froude number in open channel designs.

If a liquid is flowing in a channel where part of the free surface is exposed to an open atmosphere, such a flow is called an open channel flow. Examples of open channel flows are river flows, artificial channels, or canals used for irrigation. They might be carrying water or sewerage water, etc. The flow in an open channel can be uniform or nonuniform. A uniform flow means that the flow velocity is independent of the downstream direction. This implies that the velocity profile does not change with the distance. In the cases of a nonuniform flow, the velocity will change from one location to another, and thus an observer may view the variation in the channel depth. For all cases discussed in this chapter, we assume that the density of the liquid flowing in the channel is not changing with the distance. Therefore, according to the continuity equation, the flow should accelerate in a narrow channel, and the velocity will be reduced when flowing in a broad or wide channel. We also assume that the flow is moving steadily and thus is independent of time. Figure 15.1 schematically depicts flow in an open channel.

The flow in an open channel is said to be uniform if the velocity of the liquid does not change either in magnitude or direction from one section to another in the channel's part under consideration. The uniform flow condition is achieved only if the cross-section of the channel does not change along the length of the channel. Here, the depth of the liquid will be unchanged; hence, the uniform flow is ensured as per the continuity equation. The mean velocity and the velocity profile will be the same at all the cross-sections. The flow in which the liquid surface height is varied, because of variations in channel depth, is nonuniform. This change in depth can be

DOI: 10.1201/9781003315117-15

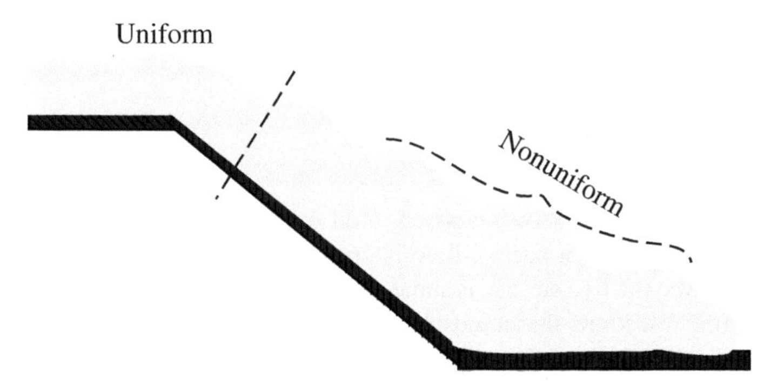

Figure 15.1 Uniform and nonuniform open channel flows.

rapid or gradual, and hence the flow is classified as rapidly varied or gradually varied flow.

15.1 THE DIMENSIONLESS PARAMETERS

The major driving force in an open channel flow is gravity, unlike the pressure difference, which is the primary driving force in the wall-enclosed pipe flows. The relevant parameters are the velocity of the fluid, the characteristic length of the channel, the fluid's viscosity, and the fluid's density. When we execute the standard Buckingham Pi theorem, we can obtain dimensionless parameters, namely the Froude and Reynolds numbers.

$$Fr = \frac{u}{\sqrt{g\hbar}}$$

where $\hbar$ is channel depth. The denominator is defined as inertial wave celerity

$$c_i = \sqrt{g\hbar}$$

So Froude number is defined as ratio of the the depth-averaged velocity to the wave celerity. The Froude number is an incompressible analogue of Mach number, which is used for compressible flows.

A shallow-water wave occurs when water depth is less than or equal to 1/20 of its wavelength. We have derived the equation for celerity of shallow water wave:

$$c^2 \approx gh$$

This is the speed at which a surface disturbance travels through a liquid in a channel.

The Reynolds number would help us in identifying the flows as laminar or turbulent flows. It is defined as

$$\mathrm{Re} = \frac{uR_\hbar}{\nu}$$

The characteristic length is the channel hydraulic radius, defined as

$$R_{\hbar} = \frac{B\hbar}{B + 2\hbar}$$

where B is the width of the channel and $\hbar$ is the depth of the liquid in the channel. Note that the flow is considered laminar if $\mathrm{Re} < 500$ in open channel flows and turbulent if $\mathrm{Re} > 750$.

If the Reynolds number is defined directly using channel depth instead of the hydraulic radius, then it is also defined in literature as

$$\mathrm{Re}_2 = \frac{u\hbar}{\nu}$$

In this case, the range of critical Reynolds number is suggested as:

$$1000 < \mathrm{Re}_2 < 3000$$

15.2 ENERGY GRADIENT LINE

From Bernoulli's equation we have energy in streamlines

$$E = z + \frac{V^2}{2g}$$

Substituting V formula into the above equation

$$E = z + \left(\frac{Q}{A}\right)^2 \frac{1}{2g} = z + \frac{Q^2}{(zB)^2}\frac{1}{2g}$$

Now for minimum z position, we have

$$\left(\frac{dE}{dz}\right)_{Q=const} = 0$$

This indicates that

$$z = \left(\frac{Q^2}{gB^2}\right)^{1/3}$$

is the minimum depth of the channel as

$$\left(\frac{d^2E}{dz^2}\right)_{Q=const} > 0$$

The corresponding minimum energy is

$$E_{\min} = \frac{3}{2} z_{\min}$$

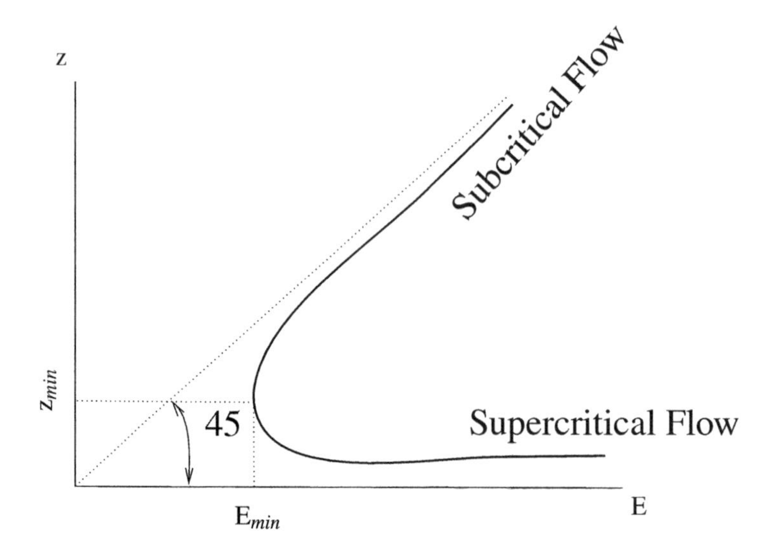

Figure 15.2 Specific energy diagram.

where

$$z_{\text{min}} = \left(\frac{Q^2}{gB^2}\right)^{1/3}$$

is the critical depth of the channel.

The specific energy can be rewritten in dimensionless form as

$$\frac{E}{z_{\text{min}}} = \frac{z}{z_{\text{min}}} + \frac{1}{2}\left(\frac{z_{\text{min}}}{z}\right)^2$$

The critical speed is

$$V_c = \left(g\frac{Q}{B}\right)^{1/3}$$

If the flow conditions are like z_{min} and V_c are present then the mean specific energy is minimum, and such flow conditions are called the critical flow conditions.

The flow is classified as subcritical if $\text{Fr} < 1$, and in such flow surface disturbances, can move in both upstream and downstream directions. The flow is classified as supercritical for $\text{Fr} > 1$, and in this case disturbances can only move in the downstream direction. For $\text{Fr} = 1$, the flow is said to be in a critical state.

The specific energy diagram is shown in Figure 15.2. Hydraulic structures like gates and weirs that cause critical flow are used as control sections for open channel flows. These control sections can be the spillway crest, sluice gate, over-fall, weir, etc. The flow rates at these stations can be recorded simply by measuring the critical flow depth.

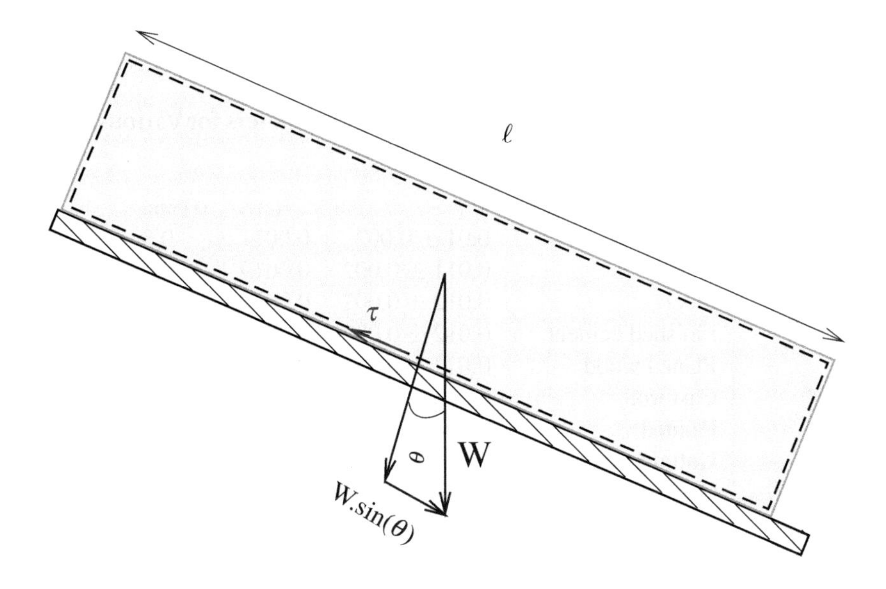

Figure 15.3 The flow in an inclined channel. The dashed rectangular box is the control volume used for the analysis.

15.3 PRESSURE LOSS IN OPEN CHANNEL FLOWS

Figure 15.3 shows the flow inside an inclined channel under the action of gravity. The shear stress at the base of the channel is considered here as a major opposing force to the flow and the friction from side walls is ignored.

The balance of forces gives

$$\tau(P\cdot\ell) = W\sin(\theta)$$

where P is perimeter and ℓ is the length of the channel. For small angles of inclination

$$\sin(\theta) \approx \tan(\theta) \approx S$$

$$\tau(P\cdot\ell) = W\cdot S$$

$$\tau = W\cdot S\frac{1}{P\cdot\ell} = (\gamma Vol)\cdot\frac{S}{P\cdot\ell} = (\gamma\cdot A\cdot\ell)\cdot\frac{S}{P\cdot\ell}$$

$$\tau = \frac{\gamma AS}{P} = \gamma SR_\hbar$$

$$R_\hbar = \frac{A}{P}$$

$$\tau = K\left(\frac{\rho V^2}{2}\right)$$

Table 15.1
Manning Coefficient and Average Roughness Parameters for Various Channel Surfaces

Material	n	(ε) ft	(ε) mm
Glass	0.01 ± 0.002	0.0011	0.3
Brass	0.011 ± 0.002	0.0019	0.6
Steel	0.012 ± 0.002	0.0032	1
Finished cement	0.012 ± 0.002	0.0032	1
Planed wood	0.012 ± 0.002	0.0032	1
Cast iron	0.013 ± 0.003	0.0051	1.6
Painted	0.014 ± 0.003	0.008	2.4
Unfinished cement	0.014 ± 0.002	0.008	2.4
Clay tile	0.014 ± 0.003	0.008	2.4
Riveted	0.015 ± 0.002	0.012	3.7
Brickwork	0.015 ± 0.002	0.012	3.7
Asphalt	0.016 ± 0.003	0.018	5.4
Corrugated metal	0.022 ± 0.005	0.12	37
Rubble masonry	0.025 ± 0.005	0.26	80

$$K\left(\frac{\rho V^2}{2}\right) = \gamma S R_\hbar$$

$$V = \sqrt{\left(\frac{2}{\rho K}\gamma\right) S R_\hbar} = C\sqrt{S R_\hbar}$$

French hydraulics engineer, Antoine de Chezy, in 1775, proposed the formula to calculate the friction coefficient:

$$Q = C \cdot A \cdot \sqrt{S \cdot R_\hbar}$$

where S is the slope of the EGL line, $R_\hbar$ is the hydraulic radius of the channel, A is area of the channel.

Robert Manning, an Irish hydraulic engineer found that velocity is proportional to $V \propto R_\hbar{}^{2/3}$ instead of the square-root of hydraulic radius. He therefore proposed, in 1885, a corrected version of the Chezy formula, now known as the Manning formula.

$$Q = \frac{\varphi A R_\hbar^{2/3}}{n}\sqrt{S}$$

where φ is 1 for SI units and 1.486 for English/US Customary units.

According to Darcy-Weisbach equation

$$h_f = f\left(\frac{L}{D}\right)\frac{V^2}{2g}$$

Table 15.2
Manning Coefficient for Natural Channels

Straight gravel beds	0.025
Gravel beds with large boulders	0.04
Straight Earth with some grass	0.026
Winding Earth with no vegetation	0.03
Winding Earth with weedy banks	0.05
Earth with overgrown/weedy base	0.08

$$\frac{h_f}{L} = \frac{f}{4R_\hbar}\frac{V^2}{2g} \equiv S$$

$$V = \sqrt{\frac{8gSR_\hbar}{f}}$$

The friction factor f can be estimated from the pipe flow correlation used in the Moody diagram as a first approximation. The $(\varepsilon/4R_\hbar)$ can be used as relative roughness.

Example 15.1

Example Water is flowing in a trapezoidal cross-section, rubble masonry channel that is inclined at a slope of 3 degrees. The height (z) of the water in the channel is 0.7 m and the bottom base of the channel is 0.5 m. The walls of the channel are angled at $\theta = 45°$. Determine the flow rate of water through this channel whose cross-section is shown in Figure 15.4.

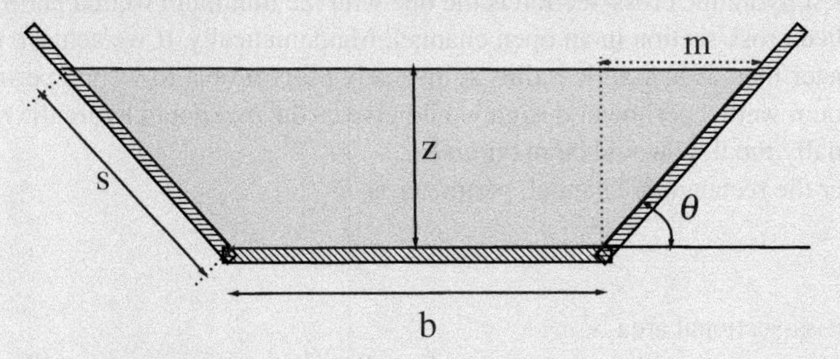

Figure 15.4 Schematic diagram of Example 15.1.

Solution The trapezoidal cross-section can be converted into two triangles and one rectangle. The hydraulic radius is

$$s = \frac{z}{\sin(\theta)} = 0.99m$$

$$m = \frac{z}{\tan(\theta)} = 0.7m$$

$$A = \frac{z^2}{\tan(\theta)} + zb = 0.84m^2$$

$$P = b + \frac{2z}{\sin(\theta)} = 2.48m$$

$$R_h = \frac{\frac{z^2}{\tan(\theta)} + zb}{b + \frac{2 \cdot z}{\sin(\theta)}} = 0.338m$$

where for small angles slope is $S = \sin(\theta) = 0.052$. For rubble masonry channel $n = 0.025 + 0.005 = 0.03$ is taken.

Using Manning's equation we can calculate the volume flow rate

$$Q = \frac{\varphi A R_h^{2/3}}{n} \sqrt{S} = 3.11m^3/s$$

The absolute viscosity of water at 20°C is $\mu = 8.91 \times 10^{-4}$ Pa · s and $\rho = 997$ kg/m^3. The Reynolds number is

$$\text{Re} = \frac{VR_h}{\nu} = 1.408 \times 10^6$$

as Re > 750, the flow is turbulent.

15.4 BEST HYDRAULIC CROSS-SECTION

The best hydraulic cross-section is the one with the minimum wetted perimeter for a specified cross-section in an open channel. Mathematically, if we achieve minimum perimeter then as hydraulic radius is inversely proportional to wetted perimeter, the minimum wetted perimeter design would give us the maximum hydraulic radius and eventually the flow would be maximized.

For the rectangular channel, perimeter is

$$Peri = B + 2z$$

and cross-sectional area is

$$A_c = Bz$$

In terms of area, perimeter is

$$Peri = \frac{A_c}{z} + 2z$$

Table 15.3
Geometric Properties of Common Open-Channel Shapes

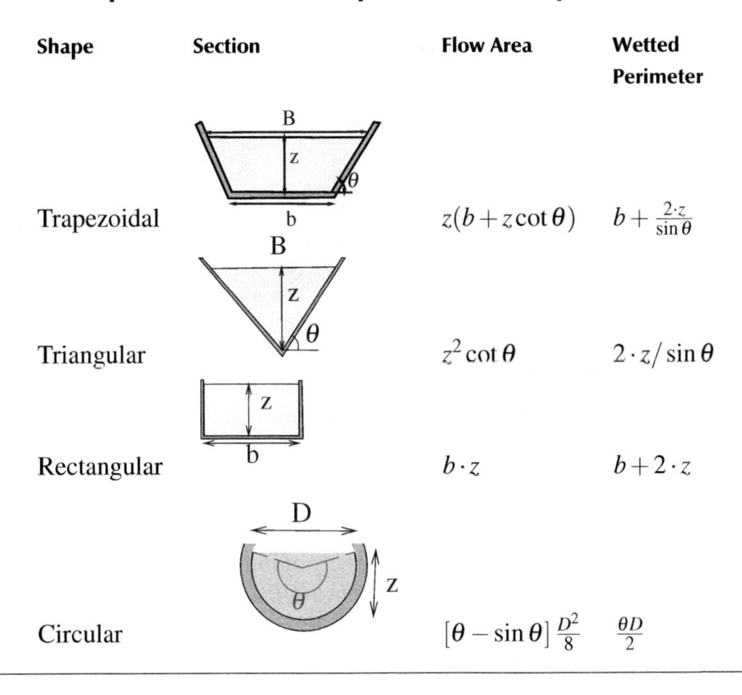

Shape	Section	Flow Area	Wetted Perimeter
Trapezoidal		$z(b+z\cot\theta)$	$b+\frac{2\cdot z}{\sin\theta}$
Triangular		$z^2\cot\theta$	$2\cdot z/\sin\theta$
Rectangular		$b\cdot z$	$b+2\cdot z$
Circular		$[\theta-\sin\theta]\frac{D^2}{8}$	$\frac{\theta D}{2}$

Differentiating this perimeter with z and setting it to zero gives us the best perimeter

$$\frac{d(Peri)}{dz}=\frac{-A_c}{z^2}+2=\frac{-Bz}{z^2}+2=\frac{-B}{z}+2$$

$$z=\frac{B}{2}$$

15.5 HYDRAULIC JUMP

In open channels and dam spillways, the flow may go into transition from a rapid flow to a slow-moving flow. This phenomenon is called a hydraulic jump (see Figure 15.5). This sudden change in flow would cause the rapid rise of liquid surface. Figure 15.6 shows the schematic representation of the formation of a hydraulic jump. The flow velocity before the hydraulic jump is V_1 and after the jump is V_2. The level of liquid above the ground before the hydraulic jump is $\hbar_1$, and the level of liquid above the ground after the hydraulic jump is $\hbar_2$.

From continuity equation

$$V_1\hbar_1 B=V_2\hbar_2 B$$

Figure 15.5 Hydraulic jump at the end of a spillway. (*Courtesy of Dr. Marian Muste, IIHR-Hydroscience & Engineering, University of Iowa. The Fluids Laboratory is an integrated learning environment developed by IIHR-Hydroscience & Engineering for the College of Engineering of The University of Iowa; Source: Intl. Assoc. for Hydro-Environment Engr. and Research.*)

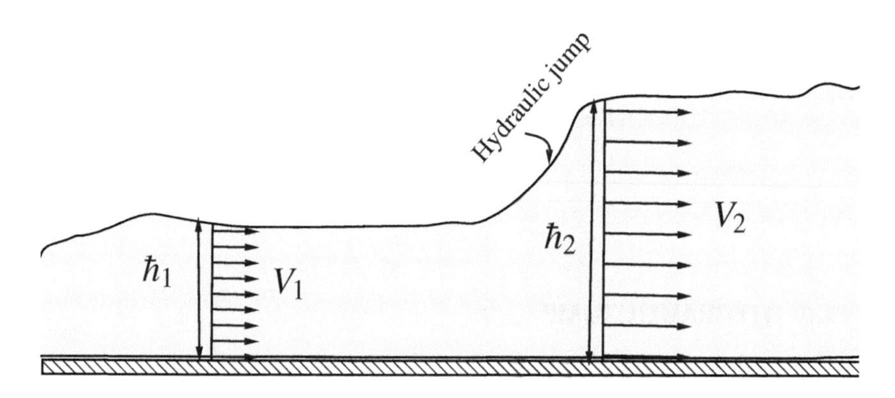

Figure 15.6 Schematic representation of the formation of hydraulic jump.

We now apply the Bernoulli's equation on the phenomenon of hydraulic jump.

$$\frac{p_2}{\gamma}+\frac{V_2^2}{2g}+z_2+h_f=\frac{p_1}{\gamma}+\frac{V_1^2}{2g}+z_1$$

For simplified model we ignore the friction loss:

$$\hbar_2 - \hbar_1 = \frac{V_1^2}{2g}\left[1 - \left(\frac{V_2}{V_1}\right)^2\right]$$

From continuity equation

$$V_1\hbar_1 B = V_2\hbar_2 B$$

$$\frac{V_2}{V_1} = \frac{\hbar_1 B}{\hbar_2 B} = \frac{\hbar_1}{\hbar_2}$$

$$\hbar_2 - \hbar_1 = \frac{V_1^2}{2g}\left[1 - \left(\frac{\hbar_1}{\hbar_2}\right)^2\right]$$

French engineer Jean-Baptiste Blanger (1828) derived this equation. We can also cast it into dimensionless form

$$\frac{\hbar_2}{\hbar_1} - 1 = \frac{V_1^2}{2g\hbar_1}\left[1 - \left(\frac{\hbar_2}{\hbar_1}\right)^{-2}\right]$$

$$Fr_1 = \frac{V_1^2}{g\hbar_1}$$

$$\left(\frac{\hbar_2}{\hbar_1}\right) = 1 + \frac{1}{2}Fr_1\left[1 - \left(\frac{\hbar_2}{\hbar_1}\right)^{-2}\right]$$

When compared with experiments, this equation gives a bit deviated results. Therefore, Charles Bresse, in 1860, proposed another equation, which gives results close to experiments. We now derive Bresse equation for hydraulic jump, using the momentum balance:

$$\rho Q(V_2 - V_1) = \Delta(\gamma\bar{z}\cdot(A)) - F_f$$

where F_f is channel friction force and $\bar{z}$ is the centroid of an area. Note that the pressure force is written as

$$F_{pressure} \approx P\cdot A = \gamma\bar{z}\cdot(A)$$

Ignoring friction, we have

$$Q(V_2 - V_1) = g\left[\bar{z}_1 A_1 - \bar{z}_2 A_2\right]$$

Since $Q = A\cdot V$ we can cast this equation into form

$$\frac{Q^2}{gA_1} + \bar{z}_1 A_1 = \frac{Q^2}{gA_2} + \bar{z}_2 A_2$$

This combination of terms is called the specific force or momentum function

$$F_s = \frac{Q^2}{gA} + \bar{z}A$$

$$\frac{Q^2}{gA_1} - \frac{Q^2}{gA_2} = \bar{z}_2 A_2 - \bar{z}_1 A_1$$

$$\frac{Q^2}{g}\left[\frac{1}{A_1} - \frac{1}{A_2}\right] = \bar{z}_2 A_2 - \bar{z}_1 A_1$$

$$\frac{Q^2}{g}\left[\frac{A_2 - A_1}{A_1 A_2}\right] = \bar{z}_2 A_2 - \bar{z}_1 A_1$$

where $\bar{z}$ is the depth of the centroid of flow area A.

Assuming a rectangular channel, so, $A = B\hbar$ and $\bar{z} = \hbar/2$, we can cast the equation in form

$$\frac{Q^2}{g}\left[\frac{\hbar_2 - \hbar_1}{B\hbar_1\hbar_2}\right] = \left(\frac{\hbar_2}{2}\right)(B\hbar_2) - \left(\frac{\hbar_1}{2}\right)(B\hbar_1)$$

$$\frac{Q^2}{g}\left[\frac{\hbar_2 - \hbar_1}{\hbar_1\hbar_2}\right] = \frac{B^2}{2}(\hbar_2^2 - \hbar_1^2)$$

$$\frac{Q^2}{g}(\hbar_2 - \hbar_1) = \frac{B^2}{2}(\hbar_1\hbar_2)(\hbar_2^2 - \hbar_1^2)$$

Introducing $Q = B \cdot \hbar_1 \cdot V_1$ and the Froude number $\mathrm{Fr}_1 = V_1/(g\hbar_1)^{1/2}$ we have

$$-2Fr_1^2 + \frac{\hbar_2}{\hbar_1} + \left(\frac{\hbar_2}{\hbar_1}\right)^2 = 0$$

The solution of this equation is

$$\frac{\hbar_2}{\hbar_1} = \frac{1}{2}\left[-1 + \sqrt{1 + 8Fr_1^2}\right]$$

Using continuity equation in form of V_2 gives

$$Fr_2 = \frac{V_2}{\sqrt{g\hbar_2}} = \frac{V_1}{\sqrt{g\hbar_1}}\left(\frac{z_1}{z_2}\right)^{3/2} = \frac{2^{3/2}Fr_1}{\left(-1 + \sqrt{1 + 8Fr_1^2}\right)^{3/2}}$$

Table 15.4 lists the types of hydraulic jump that can be encountered in an open channel. Figure 15.7 shows the types of hydraulic jumps in horizontal rectangular channels.

We can estimate the head loss from specific energy change:

$$h_f = \Delta E = \left(\hbar + \frac{Q^2}{2gA^2}\right)_1 - \left(\hbar + \frac{Q^2}{2gA^2}\right)_2$$

Table 15.4
Types of Hydraulic Jump

Froude Number	Jump Type	
1–1.7	Undular jump	No hydraulic jump
1.7–2.5	Weak jump	Low energy loss
2.5–4.5	Oscillating jump	Wavy free surface
4.5–9	Steady jump	45–70% of energy dissipation
9	Strong/Rough jump	85% of energy dissipation

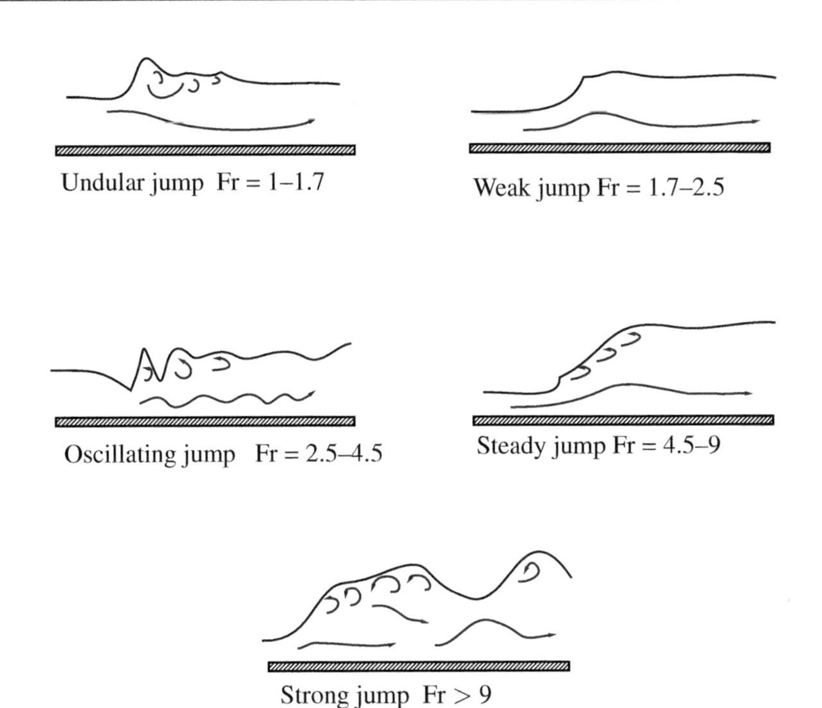

Figure 15.7 Types of hydraulic jumps in a horizontal rectangular channel.

$$h_f = (\hbar_1 - \hbar_2) + \frac{Q^2}{2g}\left[\frac{A_2^2 - A_1^2}{A_1{}^2 A_2{}^2}\right]$$

$$h_f = (\hbar_1 - \hbar_2) + \frac{Q^2}{2g}\left[\frac{B^2(\hbar_2^2 - \hbar_1^2)}{B^4 \hbar_1{}^2 \hbar_2{}^2}\right]$$

$$h_f = (\hbar_1 - \hbar_2) + \frac{Q^2}{2gB^2}\left[\frac{(\hbar_2^2 - \hbar_1^2)}{\hbar_1{}^2 \hbar_2{}^2}\right]$$

From the momentum equation

$$\frac{Q^2}{gB^2} = \frac{\hbar_1 \hbar_2}{2}(\hbar_1 + \hbar_2)$$

$$h_f = \frac{(\hbar_2 - \hbar_1)^3}{(4 \cdot \hbar_1 \cdot \hbar_2)}$$

Example 15.2

Example At a distance downstream of a control gate, a hydraulic jump is formed in a channel 3 m wide. If the flow depth just downstream of the gate is 2 m and the outlet discharge is 170 m^3/s, assuming there are no losses in the flow through the gate, find

(i) Flow depth downstream of the jump;

(ii) Head losses in the jump.

Solution The discharge flow rate is

$$V_1 = \frac{Q}{A_1} = \frac{170}{B \cdot \hbar_1} = \frac{170}{(3m)(2m)} = 28.33\ m/s$$

The Froude number is defined as

$$Fr = \frac{V_1}{\sqrt{g\hbar_1}} = 6.39$$

Using Bresse equation, we have

$$\frac{\hbar_2}{\hbar_1} = \frac{1}{2}\left[-1 + \sqrt{1 + 8Fr_1^2}\right] = 8.55$$

$$\hbar_2 = 17.11m$$

Head loss: From Bernoulli's equation we have

$$h_f = (\hbar_1 - \hbar_2) - \frac{{V_2}^2}{2g} + \frac{{V_1}^2}{2g} = 25.23\ m$$

The change in the specific energy at two stations can also be used to give the head loss

$$h_f = \Delta E = \left(\hbar + \frac{Q^2}{2gA^2}\right)_1 - \left(\hbar + \frac{Q^2}{2gA^2}\right)_2$$

$$h_f = 25.23m$$

Head loss can also be computed from a handy formula:

$$h_f = \frac{(h_2 - h_1)^3}{(4 \cdot h_1 \cdot h_2)}$$

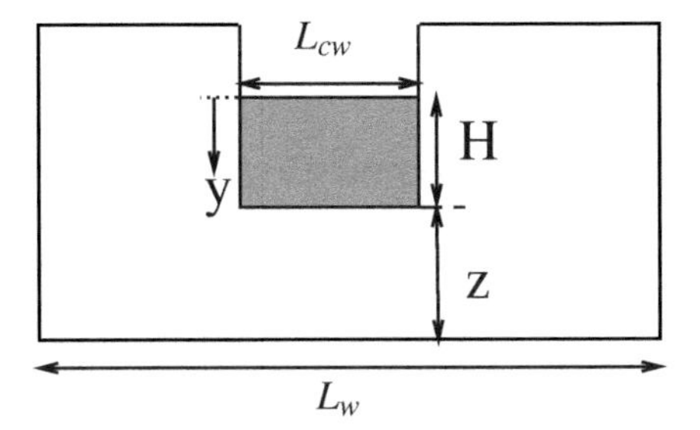

Figure 15.8 The rectangular wier.

15.6 WIERS

The purpose of weirs is to measure water flow rates in open channels with a free liquid surface. A weir is also used in dams over which the liquid is made to flow. Weirs are classified according to the form of their notch or opening as follows:

(i) Rectangular notch,

(ii) The V or triangular notch, and

(iii) Special notches (like the trapezoidal, hyperbolic, and parabolic notches) are designed to have a constant discharge coefficient.

15.6.1 RECTANGULAR WIER

According to Bernoulli's equation the velocity of the flow is $\sqrt{2gy}$. We can measure the flow rate as

$$Q = \int_A u\,dA = \int_0^H \left(\sqrt{2gy}\right) L_{cw} dy$$

where L_{cw} is crest width and H is the water level height from crest till free surface.

$$Q_{ideal} = \frac{2\sqrt{2g}}{3} L_{cw} H^{3/2}$$

This is the ideal volume flow rate. The discharge coefficient (see Table 15.5) and adjusted corrected crest width are introduced in this relation:

$$Q_{actual} = c_d \cdot L_{acw} \cdot \left(\frac{2\sqrt{2g}}{3} {H_a}^{3/2} \right)$$

where $L_{acw} = L_{cw} + \zeta$ is adjusted crest width, and H_a is adjusted weir head $= H +$ 0.003 ft or $H + 0.9$ mm. The values of ζ can be taken from Table 15.6.

Table 15.5
Coefficient of Discharge (c_d) at Various Crest Length-to-Channel Width Ratios

L_{cw}/L_c	0	0.2	0.4	0.6	0.7	0.8	0.9	1
H/z								
0	0.587	0.588	0.59	0.593	0.595	0.597	0.599	0.602
0.15	0.587	0.587	0.593	0.604	0.611	0.619	0.629	0.64
1	0.586	0.586	0.595	0.614	0.627	0.642	0.659	0.679
1.5	0.585	0.585	0.598	0.624	0.643	0.664	0.689	0.718
2	0.584	0.583	0.6	0.635	0.659	0.687	0.719	0.756
2.5	0.584	0.582	0.603	0.645	0.674	0.709	0.749	0.795
3	0.583	0.581	0.605	0.655	0.69	0.732	0.779	0.834

Table 15.6
Adjustment for Crest Length ζ

L_{cw}/L_c	0	0.2	0.4	0.6	0.7	0.8	0.9	1
ft	0.007	0.008	0.009	0.012	0.013	0.014	0.013	0.005
m	0.0021	0.0024	0.0027	0.0037	0.004	0.0043	0.004	0.0015

Example 15.3

Example Water flows in a 9-ft wide channel. At the end of the channel a rectangular wier with a width of 3.6 ft is installed. The wier crest height from channel base is 4 ft. The water rises 5 ft above the crest of the wier. Estimate the volume flow rate.

Solution Using Table 15.5 the discharge coefficient can be selected or interpolated.

$$\frac{L_{cw}}{L_c} = \frac{\textit{crest width}}{\textit{channel width}} = \frac{3.6}{9} = 0.4$$

$$\frac{H}{z} = \frac{\textit{water height}}{\textit{crest height}} = \frac{5}{4} = 1.25$$

From the table we take $c_d = 0.59$.

$$L_{acw} = L_{cw} + \zeta$$

From Table 15.6 we take ζ. For crest length/channel width = 0.4, adjusted crest width is −0.009.

$$L_{acw} = L_{cw} + \zeta = 3.6 + (-0.009) = 3.591$$

$$H_a = H + 0.003 = 5 + 0.003 = 5.003$$

$$Q_{actual} = c_d \cdot L_{acw} \cdot \left(\frac{2\sqrt{2g}}{3} {H_a}^{3/2}\right) = 0.59 \times 3.591 \left(\frac{2\sqrt{2g}}{3}(5.003)^{3/2}\right)$$

$$= 219.27 ft^3/s$$

PROBLEMS

15P-1 The water in a dam has a level of $H = 2$ m, which reached over the crown, or crest of an overflow section of the dam. Above the overflow crown the constant critical water level is h_k. The speed v changes as the water goes out. Assuming frictionless flow of water, find the minimum height h_k, the minimum velocity, and the minimum volume flow Q that flows off over the weir with width of $b = 100$ m.

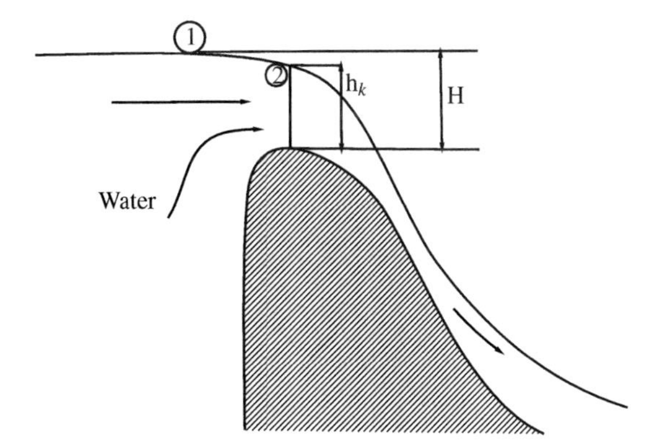

Figure 15.9 Figure of problem 15P-1.

15P-2 A channel of symmetrical trapezoidal section, 1,000 mm deep, carries water at a depth of 600 mm. The channel's top and bottom widths are 1.8 m and 600 mm respectively. If the channel slopes uniformly at 1 in 2600 and Chezys coefficient is 60 $m^{1/2}$/s, calculate the steady rate of flow in the channel.

15P-3 An open channel with bed of rubble masonry is of trapezoidal section, 3 m wide at the base and having sides inclined at 60° to the horizontal, and has a bed slope of 1 in 600. It is found that when the rate of flow is 1.3 m^3/s, the depth of water in the channel is 400 mm. Using Mannings formula, calculate the rate of flow when the depth is 500 mm.

15P-4 Consider a 3.45-m-wide rectangular channel with a brickwork bed and slope of 0.00035 and Manning's roughness factor of 0.012. A weir is placed in the channel, and the depth upstream of the weir is 1.3 m for a discharge of 7.66 m^2/s. Determine whether a hydraulic jump forms upstream of the weir.

15P-5 Consider a 2.45-m-wide rectangular channel with an asphalt bed and slope of 8 ft/mile. The discharge flow is 30 cfs. What will be the depth and width for the most economical channel?

REFERENCES

B. S. Massey, J. Ward-Smith, Mechanics of Fluids, 9th Edition, CRC Press, London UK, 2012.

H. Chanson, The Hydraulics of Open Channel Flow: An Introduction, Second Edition, Elsevier Butterworth-Heinemann, UK, 2004.

16 Compressible Flows

In the previous chapters, we focused our attention on the flows with negligible density and pressure variations. This chapter will consider the flows that involve a significant variation in density because of pressure and temperature change. Such flows are called compressible flows, and they have frequently been encountered in nozzles, compressors, and combustion chambers. Compressible flows are also significant in rocket, missile, and space shuttle designs. We will develop the general relationship relevant for the compressible gases, considering them as an ideal gas that work below the critical temperature and pressure. In this chapter, we first introduce the acoustic wave, which is a slight pressure disturbance wave. After that, we will discuss the isentropic flow relations valid for compressible flows, followed by the normal shockwave relations and the nozzle flows. We will also discuss the flow in heated and unheated ducts.

■■■■■■

Learning outcomes: After finishing this chapter, you should be able to:

- Understand the difference between a shockwave and an acoustic wave
- Calculate the location of the shockwave and the pressure or temperature change across the shockwave.
- Understand the classification of the flows based on the Mach number.
- Formulate the compressible flow relations for frictional duct flows
- Formulate the relations for compressible flow through duct with heat transfer.

In high-speed gas flows there are significant density changes, and the Bernoulli's equation is no longer valid. Some highly compressible flows are flows in jet engines and rocket motor nozzles, flow through compressors and turbine blade passages, steam turbine control valves and nozzles, and gas welding torches.

16.1 MOVEMENT OF SMALL PRESSURE DISTURBANCE

The speed of sound, or the posted speed, is the speed at which an infinite chasm of a small pressure wave travels in the matter. Here we will develop the relation for the acoustic speed moment in the body of a liquid or gas. The small pressure wave may happen due to a small change in the local pressure. In order to derive the relation for the speed of sound in medium, we consider the movement of a small pressure wave in an enclosed duct. Let's say a certain disturbance in the flow is moving, causing a change in pressure, temperature, density, and enthalpy of the fluid. We consider the control volume around this wavefront as shown in the Figure 16.1. The duct has area A, density ρ, pressure P, and velocity V, originally. After the generation of a

DOI: 10.1201/9781003315117-16

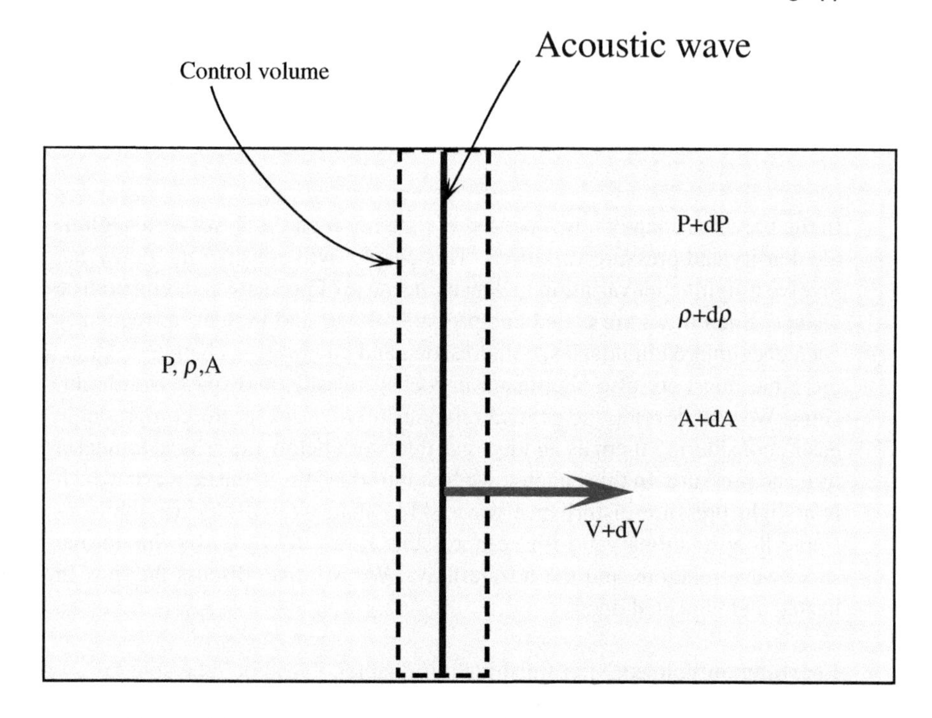

Figure 16.1 The movement of a small disturbance in the medium.

certain disturbance, the area remains A. However, the density increases to $\rho + d\rho$, and the velocity becomes $V + dV$. Refer to the balance of the mass across the control volume.

Figure 16.1 shows the movement of small disturbance in the medium. Applying the continuity equation, before and after the acoustic wave, we have:

$$\rho AV = (\rho + d\rho)(V + dV)A$$

$$\rho \cancel{A} V = (\rho + d\rho)(V + dV)\cancel{A}$$

$$\rho V = (\rho + d\rho)(V + dV)$$

Note that the area is constant in the duct. Therefore, the area A on both sides of the equity will be canceled out. Also, we consider that the infinitesimal pressure change multiplied by the infinitesimal density rise will be a very small quantity, so we shall ignore it for the rest of the analysis.

$$\cancel{\rho V} = \cancel{\rho V} + \rho dV + V d\rho + \underbrace{\cancel{d\rho dV}}_{small}$$

$$0 = \rho dV + V d\rho$$

$$\rho dV = -V d\rho$$

We now apply Newtons second law of motion on the pressure disturbance. Applying Newton's second law of motion we can have:

$$\sum F = ma = m\frac{\Delta V}{\Delta t} = \dot{m}\Delta V$$

$$pA - (p + dp)A = (\rho AV)\Delta V$$

$$pA - pA - Adp = (\rho AV)(V + dV - V)$$

$$-Adp = (\rho AV)dV$$

$$-dp = (\rho V)dV$$

$$-dp = -Vd\rho \cdot V = -V^2 d\rho$$

$$\frac{dp}{d\rho} = V^2$$

The speed inside the liquid or gas is a function of density. Thermodynamically, we can write:

$$p\rho^{-k} = const$$

Differentiation can give us:

$$\frac{dp}{d\rho} = \frac{kp}{\rho}$$

so the acoustic velocity is:

$$V = a = \sqrt{\frac{kp}{\rho}} \tag{16.1}$$

which is usually represented by letter a:

$$a = \sqrt{k\left(\frac{\partial p}{\partial \rho}\right)_T}$$

According to ideal gas relation, the pressure of gas is related with density and absolute temperature as follows:

$$p = \rho RT$$

The universal gas constant divided by the molecular weight is called R in the thermodynamics literature.

Substitution into Eq. 16.1 gives us:

$$V = \sqrt{kRT}$$

It is customary to indicate the acoustic speed by either "c" or "a" as:

$$a = \sqrt{kRT}$$

Mach number: The ratio of local gas velocity V to local speed of sound a is known as Mach number, M, after the Austrian/Czech physicist Ernst Mach (1838–1916), who investigated the flow a round rifle bullets

$$M = \frac{V}{a}$$

where a is the acoustic speed. The Mach number can also be defined as a ratio of inertia force to the elastic or compressibility force as:

$$F_{elastic} = p \cdot A = \underbrace{(\rho RT)}_{Ideal\ gas} \cdot L^2$$

Note that R is constant, so in terms of meaningful variables, elastic force is

$$F_{elastic} \approx \rho T L^2$$

Earlier the inertia force was defined as:

$$F_{inertia} \simeq \rho L^2 u^2$$

$$Ca = M^2 = \frac{F_{inertia}}{F_{elastic}} \equiv \frac{\rho L^2 u^2}{\rho L^2 T} = \frac{u^2}{T}$$

where Ca is called Cauchy number.

The Cauchy number owes its name to Augustin-Louis Cauchy (1789–1857), a French engineer mathematician and hydrodynamicist, and the Mach number is most commonly used for compressible flows, particularly in the fields of gas dynamics and aerodynamics.

$$M \propto u/\sqrt{T}$$

If the properties like specific heats are constant, the Mach number is related with velocity and temperature of the medium.

The Mach number can also be interpreted as a ratio of time necessary for a pressure signal to travel a distance to the time required by a fluid particle to travel the same distance.

We have discussed in Chapter 1 the bulk modulus of compressibility

$$\beta = \left[\frac{\Delta P}{\Delta\rho/\rho}\right]_T$$

$$\Delta P = \beta\left(\frac{\Delta\rho}{\rho}\right)$$

$$\left(\frac{\Delta P}{\Delta\rho}\right) = \left(\frac{\beta}{\rho}\right)$$

The acoustic speed is related to β as

$$a^2 = k\left(\frac{\Delta P}{\Delta\rho}\right) = k\left(\frac{\beta}{\rho}\right)$$

$$\beta = \frac{\rho \cdot a^2}{k}$$

From Bernoulli's equation we can approximate a relation between velocity and pressure difference:

$$\Delta P \simeq \frac{1}{2}\rho V^2$$

Substituting the pressure variation we have

$$\beta \left(\frac{\Delta \rho}{\rho}\right) \simeq \frac{1}{2}\rho V^2$$

$$\left(\frac{\rho \cdot a^2}{k}\right)\left(\frac{\Delta \rho}{\rho}\right) \simeq \frac{1}{2}\rho V^2$$

$$\frac{k}{2}\frac{V^2}{a^2} = \left(\frac{\Delta \rho}{\rho}\right)$$

$$\frac{k}{2}M^2 = \left(\frac{\Delta \rho}{\rho}\right)$$

Assuming that the flow remains incompressible as long as $\frac{\Delta \rho}{\rho}$ is 0.06, leads to Mach number as:

$$M \approx \sqrt{\frac{2}{k}\left(\frac{\Delta \rho}{\rho}\right)} \approx 0.29$$

The value of $M < 0.3$ is generally considered as the limit beyond which flow cannot be considered as incompressible.

$$M \leq 0.3 \quad (Incompressible)$$
$$M > 0.3 \quad (Compressible\ flow)$$
$$0.8 < M < 1.2 \quad (Transonic\ flow)$$
$$M = 1 \quad (Sonic\ flow)$$
$$M > 5 \quad (Hypersonic\ flow)$$

Figure 16.2 shows the distribution of c_p as a function of temperature.

The equation of state for liquids can be developed by correlating the property data. The Murnaghan-Tait equation of state for water is

$$p = \frac{\kappa_w}{n}\left[\left(\frac{\rho}{\rho_o}\right)^n - 1\right] + p_o$$

where

$$p_o = 101\ kPa, \rho_o = 1000\ kg/m^3, \kappa_w = 2.15GPa, n = 7.15$$

and this gives the speed of sound

$$a^2 = \frac{dp}{d\rho} = \frac{\kappa_w}{\rho}\left(\frac{\rho}{\rho_o}\right)^n$$

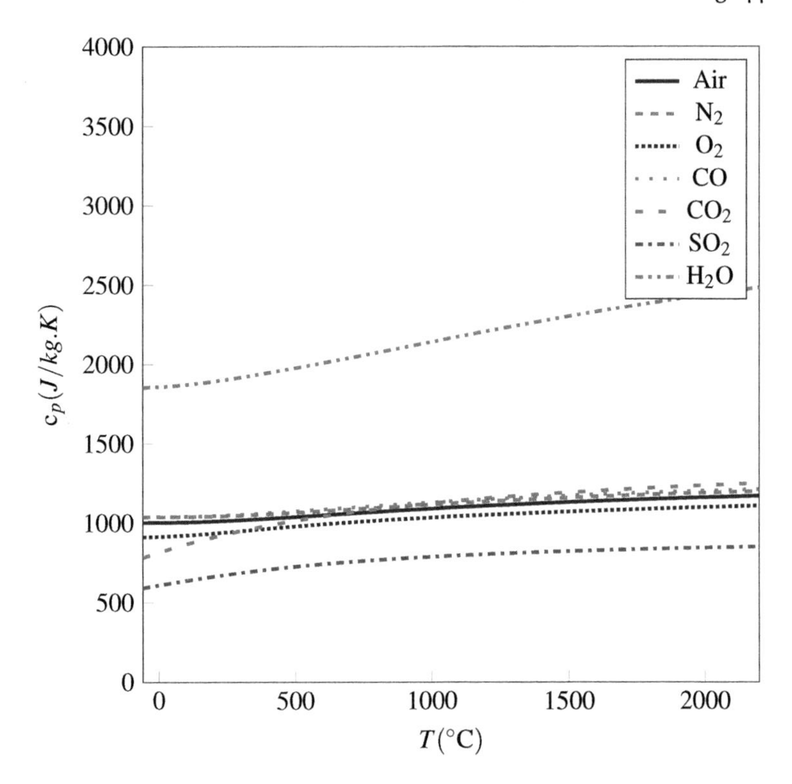

Figure 16.2 Specific heat of some ideal gases at constant pressure: air, N_2, O_2, CO, CO_2, SO_2, H_2O.

The Laval number (La), named after Swedish engineer Karl Gustaf Patrik de Laval (1845–1913), is defined as the ratio of the local flow velocity to the critical speed of sound. Also indicated as M^*.

$$La = \frac{u}{a^*}$$

Here, a^* is the critical acoustic speed of flow.

The Knudsen number can be related to Mach number and Reynolds number as (Sharipov, 2003):

$$Kn = \frac{M}{\mathrm{Re}}\sqrt{\frac{\gamma\pi}{2}}$$

$\boldsymbol{Kn \leq 10^{-2}}$ The flow is considered continuous. All equations written throughout this book are of this case.

$\boldsymbol{10^{-2} < Kn \leq 0.1}$ The characteristic scale is about 10 μm and this condition is valid for microfluidics.

$0.1 < Kn \leq 10$ This is a transition zone between the previous case and the following case.

$Kn > 10$ The mean free path is of the same order of magnitude as the characteristic scale. The modeling for this zone is done by kinetic molecular theory.

Example 16.1

Example Find the speed of sound in water and air at 20°C.
Solution For air:

$$a^2 = k \cdot R \cdot T = 1.4 \times 287 \times (20 + 273)$$

$a = 343$ m/s.
At 20°C, $\rho = 998.2 kg/m^3$ so using Murnaghan-Tait equation we have speed of sound through water as

$$a^2 = \frac{dp}{d\rho} = \frac{\kappa_w}{\rho}\left(\frac{\rho}{\rho_o}\right)^n$$

$a = 1457.2$ m/s
This shows that

$$a_{water} > a_{air}$$

Gas models

Ideal Gas is a gas which follows the relation

$$p = \rho RT$$

Many gases follow this relation at small pressure and temperature, including air.
Perfect Gas is an idealized gas that has constant specific heats c_p and c_v.

16.2 MOVEMENT OF LARGE DISTURBANCE IN FLOW

Consider small or weak pressure waves, issuing from a point source. In gases, three waves would create the zone of influence three-dimensionally like a spherical zone of influence. In 2D plane we can show disturbances as a circle. The speed of movement of these disturbances at acoustic speed or speed of sound, a. As the waves moves toward the left they will start merging into each other and will create two distinct zones. One is where the disturbances will have an influence, called the zone of action. A another zone on the left is where the disturbances are not felt; it is called the zone of silence. The accumulation of multiple pressure pulses or waves will form a plane tangent to the spherical zone of influence. This plane is called a Mach wave in gas

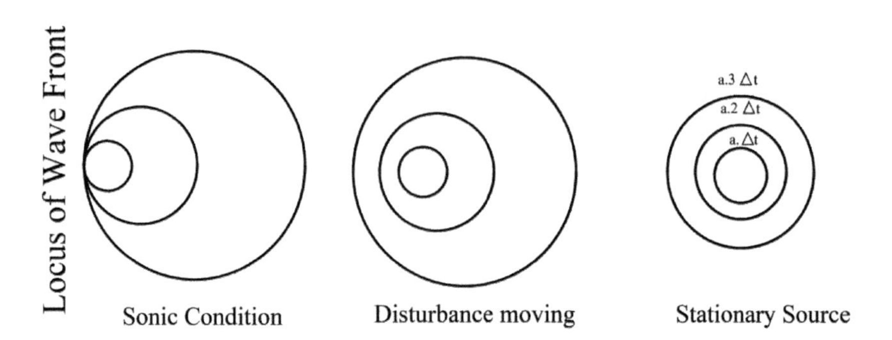

Figure 16.3 The movement of disturbance in a medium at $M \leq 1$.

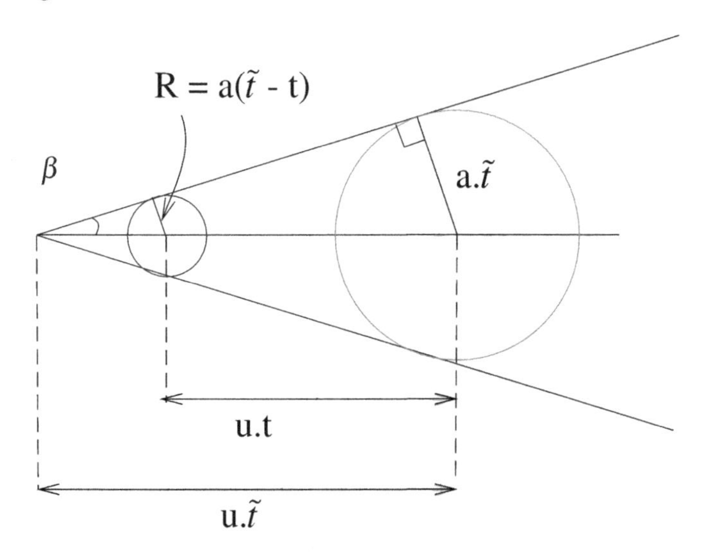

Figure 16.4 The movement of disturbance in a medium at $M > 1$ ($u \cdot \tilde{t} > a \cdot \tilde{t}$). A Mach cone will form with angle β with cone of oblique shock.

dynamics literature. Figure 16.3 shows the movement of disturbance in a medium at $M \leq 1$.

We can have another situation in which the disturbance is moving at a speed higher than the acoustic speed. In that case, a three-dimensional conical zone of influence will be formed. In Figure 16.5 a military aircraft is moving to the right of observer. The atmospheric moisture condenses as the temperature drops drastically, giving us the visualization of the shockwave. This cone is called the Mach cone in the gas dynamics literature.

The disturbance is moving three-dimensionally in medium like a sphere, and its growth is causing formation of spherical disturbances of radius $R = a(\tilde{t} - t)$ (see Figure 16.4). The disturbance, like a supersonic jet or a missile, is moving at speed

Figure 16.5 The movement of a military aircraft at a speed higher than the acoustic speed.

u, far greater than the speed at which the sphere is moving in the medium. The β is the cone half angle in a 2D plane. Note that the type of shockwave that is formed is an oblique shockwave. Figure 16.5 shows the movement of a military aircraft at a speed higher than the acoustic speed.

16.3 ABOVE EARTH

Above the Earth up to 7 miles is considered the troposphere. From 7 to 31 miles is the stratosphere, and from 31 to 50 miles is the mesosphere. From 50 to 440 miles is the thermosphere. The average altitude of the space shuttle is around 260 miles, and the maximum altitude the space shuttle has reached is 385 miles. Space starts somewhere around a hundred kilometers from the surface of the Earth, which is referred to as the Karman line. Figure 16.6 shows the different layers in atmospheres of planet Earth.

Table A.4 lists the air temperature and pressure as functions of the altitude.

16.4 COMPRESSIBLE FLOW MASS, MOMENTUM, AND ENTHALPY BALANCE

Mass flux:

$$G = \frac{\dot{m}}{A} = \rho_1 V_1 = \rho_2 V_2$$

$$\rho_1 A_1 V_1 = \rho_2 A_2 V_2$$

Momentum balance:

$$(p_1 - p_2)A = \dot{m}(V_2 - V_1)$$

$$(p_1 - p_2)\cancel{A} = \rho_1 \cancel{A} V_1 (V_2 - V_1)$$

$$p_1 - p_2 = \rho_1 V_1 (V_2 - V_1) = \rho_2 V_2 (V_2 - V_1)$$

$$(p_1 - p_2) = \rho_1 (V_1 V_2 - V_1^2)$$

Figure 16.6 The different layers in the atmosphere of our planet. (*This photo shows a crescent Moon hovering above an orbital sunset as the ISS passed over the Pacific Ocean east of New Zealand. Source: NASAs Earth Observatory.*)

$$\frac{(p_1 - p_2)}{\rho_2} = V_2^2 - V_1V_2$$

$$(p_1 - p_2)\left[\frac{1}{\rho_1} + \frac{1}{\rho_2}\right] = V_2^2 - V_1^2$$

Energy balance: In case of high speed gas flow without heat and work interaction we have:

$$\frac{V_1^2}{2} + h_1 = \frac{V_2^2}{2} + h_2 = h_o$$

$$\boxed{h_o = \frac{V^2}{2} + \left(\frac{k}{k-1}\right)\frac{p}{\rho}} \tag{16.2}$$

To analyze this problem we start from Eq. 16.10, and we will write the total enthalpy as:

$$h_o = \frac{V^2}{2} + \left(\frac{k}{k-1}\right)\frac{p}{\rho}$$

$$c_pT_o = \frac{V^2}{2} + \left(\frac{k}{k-1}\right)\frac{p}{\rho}$$

From ideal gas equation, we have $p = \rho RT$, so we adjust the LHS as:

$$\left(\frac{c_p}{R}\right)\frac{p_o}{\rho_o} = \frac{V^2}{2} + \left(\frac{k}{k-1}\right)\frac{p}{\rho}$$

As $c_p - c_v = R$ we can easily prove that

$$\frac{k-1}{k} = \frac{R}{c_v k} = \frac{R}{c_p}$$

Substituting it into the above relation gives:

$$\left(\frac{k-1}{k}\right)\frac{p_o}{\rho_o} = \frac{V^2}{2} + \left(\frac{k}{k-1}\right)\frac{p}{\rho} \tag{16.3}$$

For isentropic conditions, the following relation is valid

$$\frac{p}{\rho^k} = \frac{p_o}{\rho_o^k}$$

We use this to eliminate ρ in Eq. 16.3.

$$\frac{V^2}{2} + \left(\frac{k}{k-1}\right)\left(\frac{p_o}{\rho_o}\right)\left(\frac{p}{p_o}\right)^{(k-1)/k} = \left(\frac{k}{k-1}\right)\frac{p_o}{\rho_o}$$

$$\frac{V^2}{2} = \left(\frac{k}{k-1}\right)\frac{p_o}{\rho_o} - \left(\frac{k}{k-1}\right)\left(\frac{p_o}{\rho_o}\right)\left(\frac{p}{p_o}\right)^{(k-1)/k}$$

$$\frac{V^2}{2} = \left(\frac{k}{k-1}\right)\left[\frac{p_o}{\rho_o} - \left(\frac{p_o}{\rho_o}\right)\left(\frac{p}{p_o}\right)^{(k-1)/k}\right]$$

This is called Saint-Venant equation and it's Bernoulli's equation for the compressible flows. The equation shows that the dynamic pressure is no longer equal to difference of stagnation and static pressure in compressible flows.

Example 16.2

Example A venturi meter with 500 mm at inlet and 120 mm at throat is used to measure the air flow rate. The temperature and pressure at inlet are 150 kPa and 19°C. The pressure recorded at throat is 120 kPa. The coefficient of discharge for venturi meter c_d is 0.97. Find the flow rate of air if $k = 1.4$ and $R = 287$ kJ/kg · K.

Solution The flow is compressible through the venturi meter and we use the compressible Bernoulli's equation or Saint-Venant equation for the flow measurement.

$$\left(\frac{k-1}{k}\right)\left[\frac{p_1}{\rho_1} - \frac{p_2}{\rho_2}\right] = \frac{u_2^2}{2} - \frac{{u_1}^2}{2}$$

$$\left(\frac{k-1}{k}\right)\left(\frac{p_1}{\rho_1}\right)\left[1 - \frac{p_2/\rho_2}{p_1/\rho_1}\right] = \frac{{u_1}^2}{2}\left[\left(\frac{u_2}{u_1}\right)^2 - 1\right]$$

$$\left(\frac{k-1}{k}\right)\left(\frac{p_1}{\rho_1}\right)\left[1-r^{(k-1)/k}\right]=\frac{u_1{}^2}{2}\left[\left(\frac{u_2}{u_1}\right)^2-1\right]$$

For isentropic flow we have

$$\frac{\rho_2}{\rho_1}=\left(\frac{p_2}{p_1}\right)^{1/k}$$

$$\rho_2=\rho_1\left(\frac{p_2}{p_1}\right)^{1/k}$$

$$\frac{p_2}{\rho_2}=\frac{p_1}{\rho_1}\left(\frac{p_2}{p_1}\right)^{(k-1)/k}$$

$$r=\frac{p_2}{p_1}$$

$$\frac{p_2}{\rho_2}=\frac{p_1}{\rho_1}r^{(k-1)/k}$$

$$\frac{p_2/\rho_2}{p_1/\rho_1}=r^{(k-1)/k}$$

$$\frac{\rho_1}{\rho_2}=\left(\frac{p_1}{p_2}\right)^{1/k}=\left(\frac{1}{r}\right)^{1/k}$$

The mass flow rate is

$$\rho_1 A_1 u_1=\rho_2 A_2 u_2$$

$$u_2=\left(\frac{A_1}{A_2}\right)\left(\frac{\rho_1}{\rho_2}\right)u_1$$

$$\frac{u_2}{u_1}=\left(\frac{A_1}{A_2}\right)\left(\frac{\rho_1}{\rho_2}\right)$$

Substituting the isentropic relations and velocities ratio into Saint-Venant equation we have

$$\left(\frac{k-1}{k}\right)\left(\frac{p_1}{\rho_1}\right)\left[1-r^{(k-1)/k}\right]=\frac{u_1{}^2}{2}\left[\left(\frac{A_1}{A_2}\right)^2\left(\frac{1}{r}\right)^{2/k}-1\right]$$

$$u_{1,ideal}=\sqrt{\frac{2\left(\frac{k-1}{k}\right)\left(\frac{p_1}{\rho_1}\right)\left[1-r^{(k-1)/k}\right]}{\left[\left(\frac{A_1}{A_2}\right)^2\left(\frac{1}{r}\right)^{2/k}-1\right]}}$$

$$\dot{m}_{actual} = c_d \cdot \rho_1 \cdot A_1 \sqrt{\frac{2\left(\frac{k-1}{k}\right)\left(\frac{p_1}{\rho_1}\right)\left[1 - r^{(k-1)/k}\right]}{\left[\left(\frac{A_1}{A_2}\right)^2\left(\frac{1}{r}\right)^{2/k} - 1\right]}}$$

$$\dot{m}_{actual} = c_d \cdot \left(\frac{p_1}{R \cdot T_1}\right) \cdot A_1 \sqrt{\frac{2\left(\frac{k-1}{k}\right)\left(\frac{p_1}{\rho_1}\right)\left[1 - r^{(k-1)/k}\right]}{\left[\left(\frac{A_1}{A_2}\right)^2\left(\frac{1}{r}\right)^{2/k} - 1\right]}}$$

$$\dot{m}_{actual} = 0.911 kg/s$$

16.5 ONE-DIMENSIONAL ISENTROPIC FLOW

In many engineering devices, the flow quantities mainly vary in the flow direction. Thus these flows are classified as one-dimensional flows. The flow through devices, like nozzles, diffusers, and turbine blade passages, can be included in this category.

$$\rho AV = (\rho + d\rho)(V + dV)(A + dA)$$

$$\rho AV = (V\rho + \rho dV + Vd\rho + d\rho dV)(A + dA)$$

$$\rho AV = V\rho A + \rho AdV + VAd\rho + \cancel{Ad\rho dV} + V\rho dA + \cancel{\rho dVdA} + \cancel{Vd\rho dA} + \cancel{d\rho dVdA}$$

$$\rho AV = V\rho A + \rho AdV + VAd\rho + V\rho dA$$

Divide the whole equation by ρAV:

$$\frac{\rho AV}{\rho AV} = \frac{V\rho A}{\rho AV} + \frac{\rho AdV}{\rho AV} + \frac{VAd\rho}{\rho AV} + \frac{V\rho dA}{\rho VA}$$

$$\frac{dV}{V} + \frac{d\rho}{\rho} + \frac{dA}{A} = 0 \tag{16.4}$$

Considering the Bernoulli's equation:

$$p + \frac{1}{2}\rho V^2 = const$$

$$dp + \frac{1}{2}2VdV\rho = 0$$

$$\frac{dp}{\rho} + VdV = 0$$

We now substitute into this equation the relation of acoustic wave:

$$\frac{dp}{d\rho} = V^2$$

$$dp = V^2 d\rho$$

This gives us:

$$\frac{dp}{\rho}+VdV=\frac{V^2 d\rho}{\rho}+VdV=\frac{a^2 d\rho}{\rho}+VdV$$

$$\frac{a^2 d\rho}{\rho}+VdV=0$$

so,

$$\frac{d\rho}{\rho}=\frac{-1}{c^2}VdV$$

We substitute this into Eq. 16.4 :

$$\frac{dV}{V}+\frac{d\rho}{\rho}+\frac{dA}{A}=0$$

$$\frac{dV}{V}+\frac{1}{c^2}VdV+\frac{dA}{A}=\frac{dV}{V}-\frac{1}{c^2}\frac{V^2}{V}dV+\frac{dA}{A}=\frac{dV}{V}(1-M^2)+\frac{dA}{A}=0$$

Rearranging, we get:

$$\boxed{\frac{dA}{A}=\frac{dV}{V}(M^2-1)}$$

For isentropic flows, flowing is valid:

$$p_1{\upsilon_1}^k=p_2{\upsilon_1}^k$$

$$\frac{p_1}{\rho_1^k}=\frac{p_2}{\rho_2^k}$$

The pressure can be related with pressure and density of reservoir conditions indicated by subscript o. So, thermodynamically:

$$p=\left(\frac{p_o}{\rho_o^k}\right)\rho^k$$

The differential of pressure will be:

$$dp=\left(\frac{p_o}{\rho_o^k}\right)\left(k\rho^{k-1}\right)d\rho$$

$$\frac{dp}{\rho}=\left(\frac{kp_o}{\rho_o^k}\right)\rho^{k-2}d\rho$$

From differential form of Bernoulli's equation we have:

$$VdV+\frac{dp}{\rho}=0$$

We can substitute dp into this equation and integrate it.

$$\int VdV+\int\left(\frac{kp_o}{\rho_o^k}\right)\rho^{k-2}d\rho=0$$

$$\frac{V^2}{2}+\left(\frac{kp_o}{\rho_o^k}\right)\left(\frac{\rho^{k-2+1}}{k-2+1}\right)=const$$

$$\frac{V^2}{2}+\left(\frac{k}{k-1}\right)\left(\frac{p_o}{\rho_o^k}\right)\left(\frac{\rho^k}{\rho}\right)=const \quad (16.5)$$

Stagnation flow: If the gas stream is stopped and the velocity goes to zero, then the pressure and temperature will rise at the stagnation point. If V_o is the reservoir velocity, then it will become zero, leading to equation:

$$\frac{V_o^2}{2}+\left(\frac{k}{k-1}\right)\left(\frac{p_o}{\rho_o^k}\right)\left(\frac{\rho^k}{\rho}\right)=const$$

$$\frac{(0)^2}{2}+\left(\frac{k}{k-1}\right)\left(\frac{p_o}{\rho_o^k}\right)\left(\frac{\rho_o^k}{\rho_o}\right)=const$$

$$\frac{(0)^2}{2}+\left(\frac{k}{k-1}\right)p_o\upsilon_o=const$$

This gives the expression for constant as

$$\left(\frac{k}{k-1}\right)RT_o=const$$

Substituting the expression for constant of integration into the Eq. 16.5 gives:

$$\frac{V^2}{2}+\left(\frac{k}{k-1}\right)\left(\frac{p_o}{\rho_o^k}\right)\left(\frac{\rho^k}{\rho}\right)=\left(\frac{k}{k-1}\right)RT_o$$

Note that

$$p=\left(\frac{p_o}{\rho_o^k}\right)\rho^k$$

$$\frac{V^2}{2}=\left(\frac{k}{k-1}\right)\left[RT_o-\left(\frac{p}{\rho}\right)\right]=\left(\frac{k}{k-1}\right)[RT_o-p\upsilon]$$

$$\frac{V^2}{2}=\left(\frac{k}{k-1}\right)[RT_o-RT]$$

$$\frac{V^2}{2}=\left(\frac{k}{k-1}\right)R[T_o-T]$$

Substituting acoustic speed $a^2=kRT$ into this relation leads to form

$$\frac{V^2}{2}\frac{a^2}{a^2}=\frac{k}{k-1}R(T_o-T)$$

$$\frac{M^2}{2}=\frac{k}{k-1}\frac{R(T_o-T)}{a^2}$$

$$\frac{M^2}{2} = \frac{\cancel{k}}{k-1}\frac{\cancel{R}(T_o - T)}{\cancel{k}\cancel{R}T} = \frac{1}{(k-1)}\left[\frac{T_o}{T} - 1\right]$$

$$\frac{M^2}{2}(k-1) + 1 = \frac{T_o}{T}$$

$$\boxed{\frac{T_o}{T} = 1 + \frac{(k-1)}{2}M^2} \tag{16.6}$$

$$\boxed{\frac{p_o}{p} = \left(\frac{T_o}{T}\right)^{k/(k-1)} = \left(1 + \frac{(k-1)}{2}M^2\right)^{k/(k-1)}} \tag{16.7}$$

$$\boxed{\frac{\rho_o}{\rho} = \left(\frac{T_o}{T}\right)^{1/(k-1)} = \left(1 + \frac{(k-1)}{2}M^2\right)^{1/(k-1)}} \tag{16.8}$$

The subscript with o indicates the stagnation conditions.

The sonic condition: Often in high speed flow at the location of minimum area, the flow will attain the acoustic speed, and Mach number will become one.

For air with $k = 1.4$, these relations will have the following form

$$\frac{T_o}{T^*} = 1 + \frac{(k-1)}{2} = \frac{k+1}{2} = \frac{6}{5} = 1.2$$

$$\frac{p_o}{p^*} = \left(\frac{T_o}{T^*}\right)^{k/(k-1)} = \left(\frac{6}{5}\right)^{k/(k-1)} = \left(\frac{6}{5}\right)^{7/2} = 1.8929$$

$$\frac{\rho_o}{\rho^*} = \left(\frac{T_o}{T}\right)^{1/(k-1)} = \left(\frac{6}{5}\right)^{1/(k-1)} = \left(\frac{6}{5}\right)^{5/2} = 1.5774$$

Table 16.1 lists the k values for some gases.

Table 16.1

Ratio of Flow Quantities of Some Gases at Minimum Area Under Choking Condition

	Superheated steam	**Combustion gases**	**Air**	**Monoatomic gases**
	$k = 1.3$	$k = 1.33$	$k = 1.4$	$k = 1.667$
p^*/p_o	0.5457	0.5404	0.5283	0.4871
T^*/T_o	0.8696	0.8584	0.8333	0.7499
$\rho*/\rho\dot{}o$	0.6276	0.6295	0.634	0.6495

Mass balance at the sonic state: The sonic condition in the following sections is indicated by $*$. From the continuity equation we have:

$$\rho AV = \rho^* A^* V^*$$

$$\frac{A}{A^*} = \frac{\rho^*}{\rho}\frac{V^*}{V}$$

$$V^* = a^* = \sqrt{kRT^*}$$

$$V = M \cdot a = M\sqrt{kRT}$$

$$\frac{V^*}{V} = \frac{\sqrt{kRT^*}}{M\sqrt{kRT}} = \frac{1}{M}\sqrt{\frac{T^*}{T}}$$

$$\frac{V^*}{V} = \frac{1}{M}\sqrt{\frac{T^*T_o}{TT_o}} = \frac{1}{M}\sqrt{\frac{T^*}{T_o}}\sqrt{\frac{T_o}{T}}$$

$$\frac{V^*}{V} = \frac{1}{M}\sqrt{\frac{2}{k+1}}\sqrt{\left(1+\frac{(k-1)}{2}M^2\right)}$$

$$\boxed{\frac{V^*}{V} = \frac{1}{M}\left[\frac{1+\left(\frac{k-1}{2}\right)M^2}{\left(\frac{k+1}{2}\right)}\right]^{1/2}} \tag{16.9}$$

$$\frac{\rho^*}{\rho} = \frac{\rho^*}{\rho_o}\frac{\rho_o}{\rho} = \left[\frac{1+[(k-1)/2]M^2}{(k+1)/2}\right]^{1/(k-1)}$$

$$\frac{A}{A^*} = \frac{1}{M}\left[\frac{1+[(k-1)/2]M^2}{(k+1)/2}\right]^{\frac{(k+1)}{2(k-1)}}$$

For air with $k = 1.4$, the equation will reduced to form

$$\boxed{\frac{A}{A^*} = \frac{1}{M}\left[\frac{5+M^2}{6}\right]^3}$$

The equation is plotted in Figure 16.7.

Using Saint-Venant equation we can investigate the thermodynamically possible and impossible states for compressible flow through converging nozzles, venturi meters, and similar devices.

$$\frac{\dot{m}}{A\sqrt{2p_o\rho_o}} = \sqrt{\left(\frac{k}{k-1}\right)\left(\frac{p}{p_o}\right)^{2/k}\left[1-\left(\frac{p}{p_o}\right)^{\frac{k-1}{k}}\right]}$$

for $k = 1.4$

$$\frac{\dot{m}}{A\sqrt{2p_o\rho_o}} = \sqrt{3.5\left(\frac{p}{p_o}\right)^{1.428}\left[1-\left(\frac{p}{p_o}\right)^{0.2857}\right]}$$

$$\frac{\dot{m}\sqrt{RT_o}}{Ap_o\sqrt{7}} = \sqrt{\left(\frac{p}{p_o}\right)^{1.428}\left[1-\left(\frac{p}{p_o}\right)^{0.2857}\right]}$$

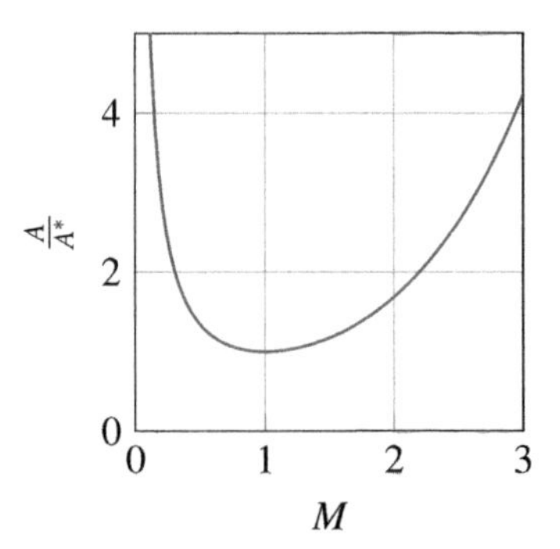

Figure 16.7 The plot shows that at minimum area the Mach number is unity (if flow is chocked already). For a further increase in Mach number an increase in area is required.

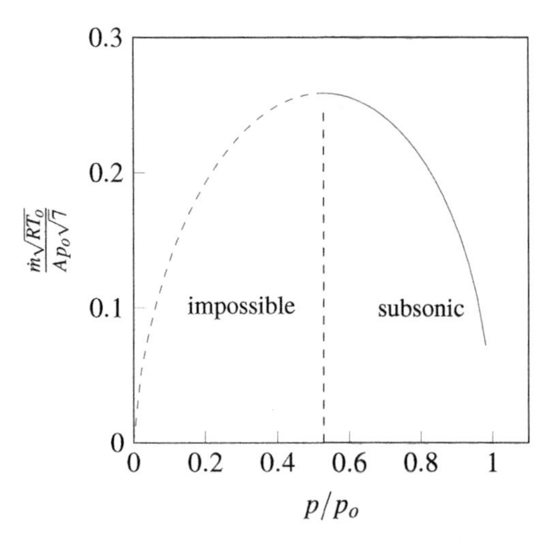

Figure 16.8 The mass flow rate factor for air at any section of the duct as long as the velocity at the throat is subsonic. The peak in flow rate corresponds to sonic conditions when $p^*/p_o = 0.528$.

This equation holds for any section and is applicable as long as the velocity at the throat is subsonic. The equation is plotted in Figure 16.8.

We consider again the continuity equation:

$$\dot{m} = \rho AV$$

Using ideal gas and isentropic relation

$$p = p_o\left(\frac{T}{T_o}\right)^{k/(k-1)}$$

we have

$$\dot{m} = \left(\frac{p}{RT}\right) A(aM)$$

$$\dot{m} = A \cdot M \left(\frac{1}{RT}\right) p_o \left(\frac{T}{T_o}\right)^{k/(k-1)} (\sqrt{kRT})$$

rearrangement gives:

$$\dot{m} = A \cdot p_o \left(\frac{\sqrt{k}}{\sqrt{RT_o}}\right) M \left(\frac{T}{T_o}\right)^{(k+1)/2(k-1)}$$

substituting the Isentropic relation for temperature

$$\frac{T}{T_o} = \left[1 + \frac{k-1}{2} M^2\right]^{-1}$$

we have

$$\dot{m} = A \cdot p_o \left(\frac{\sqrt{k}}{\sqrt{RT_o}}\right) M \left(1 + \frac{k-1}{2} M^2\right)^{-(k+1)/2(k-1)}$$

We now differentiate this relation with Mach number and set it to zero ($d\dot{m}/dM = 0$) to get the maximum mass flow rate condition. This will give us

$$M^2 = \frac{1}{\left(\frac{k-1}{2}\right)\left[2\left(\frac{k+1}{2(k-1)}\right) - 1\right]}$$

Simplification will prove that $M = 1$ for the maximum mass flow rate condition.

The choking condition: Since the mass flow rate is depending on area, the location of minimum area will dictate the maximum mass flow rate achieved, so

$$\dot{m}_{\text{max}} = \rho^* A^* V^* = \rho_o \left(\frac{2}{k+1}\right)^{1/(k-1)} A^* \sqrt{\frac{2kRT_o}{k+1}}$$

Replacing ρ_o by ideal gas relation p_o/RT_o:
For air $k = 1.4$, this equation will take the form:

$$\boxed{\dot{m}_{\text{max}} = \frac{A^* p_o}{\sqrt{RT_o}} \sqrt{k \left(\frac{2}{k+1}\right)^{\frac{k+1}{k-1}}} = 0.686 \frac{A^* p_o}{\sqrt{RT_o}}}$$

Mass flow rate is varying linearly with area and density and inversly with the square root of temperature.

The important isentropic flow relations are plotted in Figure 16.9.

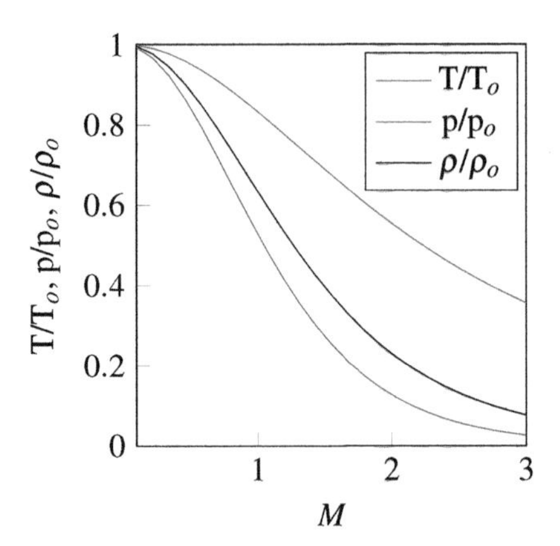

Figure 16.9 Isentropic flow relations.

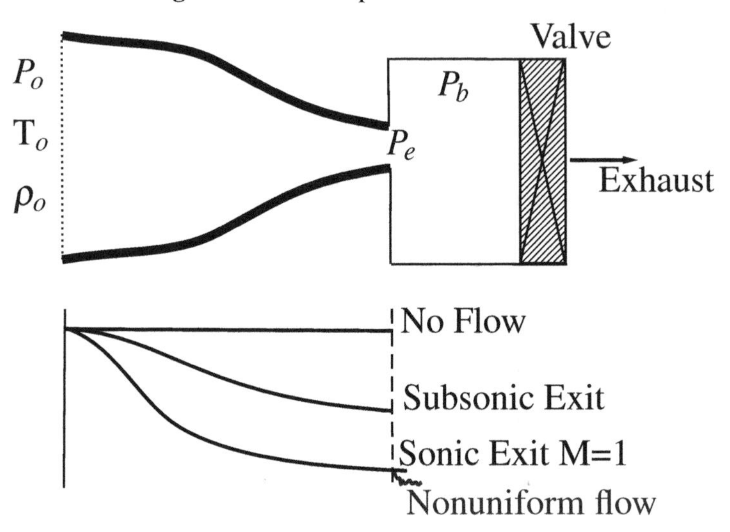

Figure 16.10 Flow configuration for converging nozzle.

16.6 CONVERGING NOZZLE

Converging nozzles are used in many industrial applications and their behaviors depend strongly on the region where gas is exhausting. This region is called the back pressure region. Figure 16.10 shows the distribution of pressure inside the converging nozzle as back pressure is reduced. The stagnation pressure is P_o, the exit pressure is P_e, the back pressure is P_b, and the pressure at the sonic condition is P^*. The Mach number at the exit is M_e.

Figure 16.11 shows the pressure ratios and their influence on mass flow rate.

(i) **No flow condition:** $P_b = P_o, P_b/P_o = 1; P_e = P_b$.
At exit plane $M_e = 0$.

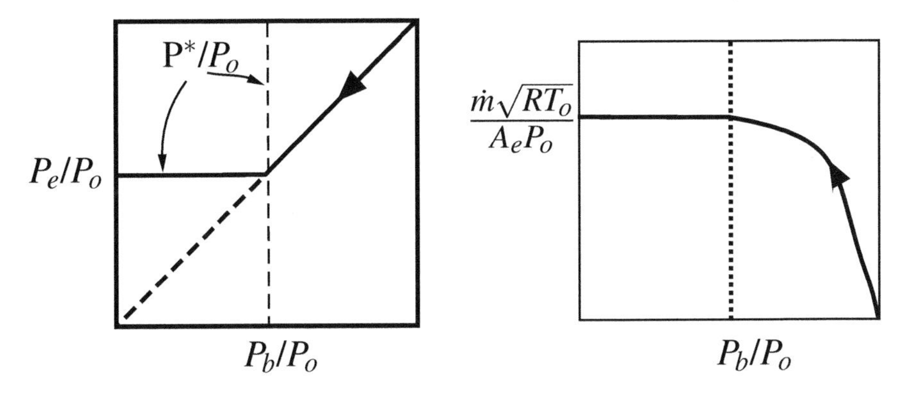

Figure 16.11 Pressure ratios and their influence on mass flow rate.

(ii) **Flow begins to increase as the back pressure is lowered:** $P_b > P^*$ or $P^*/P_b < P_b/P_o < 1$.
At exit plane $P_e = P_b$, $M_e < 1$.

(iii) **Flow increases to the choked flow limit as the back pressure is lowered to the critical pressure:** $P_b = P^*$ or $P^*/P_o = P_b/P_o < 1$.
At exit plane $P_e = P_b$, $M_e = 1$.

(iv) **Flow rate does not increase as the back pressure is lowered below the critical pressure:** $P_b < P^*$ or $P_b/P_o < P^*/P_o < 1$.
Pressure drop from P_e to P_b occurs outside the nozzle.

(v) **At exit plane:** $P_e = P^*, M_e = l$.

16.7 NORMAL SHOCKWAVES

A shockwave is a very thin region in a flow where a supersonic flow is decelerated to subsonic flow. The process is adiabatic but non-isentropic.

Mass flux:

$$G = \frac{\dot{m}}{A} = \rho_1 V_1 = \rho_2 V_2$$

$$\rho_1 A_1 V_1 = \rho_2 A_2 V_2$$

Momentum balance:

$$(p_1 - p_2)A = \dot{m}(V_2 - V_1)$$

$$(p_1 - p_2)\cancel{A} = \rho_1 \cancel{A} V_1 (V_2 - V_1)$$

$$p_1 - p_2 = \rho_1 V_1 (V_2 - V_1) = \rho_2 V_2 (V_2 - V_1)$$

$$(p_1 - p_2) = \rho_1 (V_1 V_2 - V_1^2)$$

$$\frac{(p_1 - p_2)}{\rho_2} = V_2^2 - V_1 V_2$$

$$(p_1 - p_2)\left[\frac{1}{\rho_1} + \frac{1}{\rho_2}\right] = V_2^2 - V_1^2$$

Energy balance: In case of high speed gas flow without heat and work interaction, we have

$$\frac{V_1^2}{2} + h_1 = \frac{V_2^2}{2} + h_2 = h_o$$

$$\boxed{h_o = \frac{V^2}{2} + \left(\frac{k}{k-1}\right)\frac{p}{\rho}} \qquad (16.10)$$

$$h_o = \frac{V_1^2}{2} + \left(\frac{k}{k-1}\right)\frac{p_1}{\rho_1} = \frac{V_2^2}{2} + \left(\frac{k}{k-1}\right)\frac{p_2}{\rho_2}$$

$$V_1^2 + \left(\frac{2k}{k-1}\right)\frac{p_1}{\rho_1} = V_2^2 + \left(\frac{2k}{k-1}\right)\frac{p_2}{\rho_2}$$

$$V_2^2 - V_1^2 = \left(\frac{2k}{k-1}\right)\left[\frac{p_1}{\rho_1} - \frac{p_2}{\rho_2}\right]$$

We substitute the expression for $V_2^2 - V_1^2$ into the above equation and we get:

$$(p_1 - p_2)\left[\frac{1}{\rho_1} + \frac{1}{\rho_2}\right] = \left(\frac{2k}{k-1}\right)\left[\frac{p_1}{\rho_1} - \frac{p_2}{\rho_2}\right]$$

$$\frac{\rho_2}{p_1}\left\{(p_1 - p_2)\left[\frac{1}{\rho_1} + \frac{1}{\rho_2}\right]\right\} = \frac{\rho_2}{p_1}\left\{\left(\frac{2k}{k-1}\right)\left[\frac{p_1}{\rho_1} - \frac{p_2}{\rho_2}\right]\right\}$$

$$\frac{\rho_2}{p_1}\left\{p_1(1 - \frac{p_2}{p_1})\left[\frac{1}{\rho_1} + \frac{1}{\rho_2}\right]\right\} = (1 - \frac{p_2}{p_1})\left[\frac{\rho_2 + \rho_1}{\rho_1 \cancel{\rho_2}}\right]\cancel{\rho_2}$$

$$\frac{\rho_2}{p_1}\left\{\left(\frac{2k}{k-1}\right)\left[\frac{p_1}{\rho_1} - \frac{p_2}{\rho_2}\right]\right\} = \left(\frac{2k}{k-1}\right)\left[\frac{\rho_2}{\rho_1} - \frac{p_2}{p_1}\right]$$

or this can also be written as

$$\frac{P_2}{P_1} = \frac{\left[\frac{\rho_2}{\rho_1}\left(\frac{\gamma+1}{\gamma-1}\right) - 1\right]}{\left[\left(\frac{\gamma+1}{\gamma-1}\right) - \frac{\rho_2}{\rho_1}\right]}$$

This equation is called a Rankine-Hugoniot equation and is plotted in Figure 16.12.

We can also rearrange the equation in the following form:

$$\frac{\rho_2}{\rho_1} = \frac{V_1}{V_2} = \frac{\left[\frac{P_2}{P_1}\left(\frac{\gamma+1}{\gamma-1}\right) + 1\right]}{\left[\left(\frac{\gamma+1}{\gamma-1}\right) - \frac{P_2}{P_1}\right]}$$

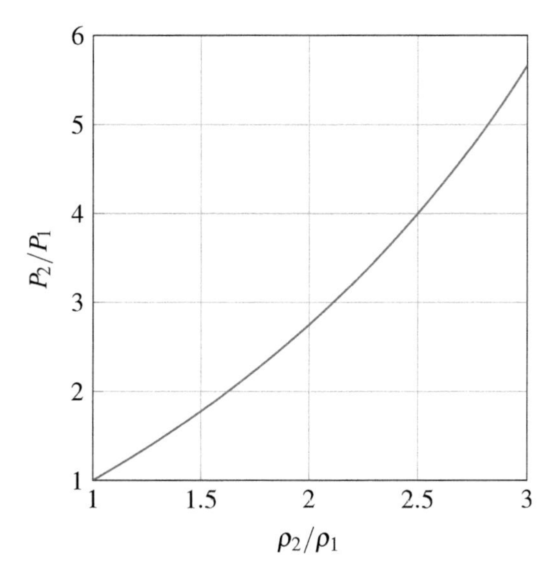

Figure 16.12 Plot of Rankine-Hugoniot equation for air.

From equation of state we have:

$$\frac{T_2}{T_1} = \frac{P_2}{P_1}\frac{\rho_1}{\rho_2}$$

$$\frac{T_2}{T_1} = \frac{\left[\left(\frac{\gamma+1}{\gamma-1}\right) + \frac{P_2}{P_1}\right]}{\left[\left(\frac{\gamma+1}{\gamma-1}\right) + \frac{P_1}{P_2}\right]}$$

The Rankine-Hugoniot equation gives us the **strength of shockwave**, defined as $\Delta p/p_1$.

Mach number after shockwave: So far we have seen that a shockwave would increase the pressure, density, and temperature of the fluid. We will now formulate an equation to estimate the Mach number after the shockwave.

Energy balance:

$$\frac{V_1^2}{2} + c_p T_1 = \frac{V_2^2}{2} + c_p T_2$$

$$\frac{V_1^2}{2} + c_p\left(\frac{a_1^2}{kR}\right) = \frac{V_2^2}{2} + c_p\left(\frac{a_2^2}{kR}\right)$$

$$\frac{k-1}{k} = \frac{R}{c_p}$$

$$\frac{V_1^2}{2} + \left(\frac{a_1^2}{k-1}\right) = \frac{V_2^2}{2} + \left(\frac{a_2^2}{k-1}\right)$$

leading to form

$$\left(\frac{T_2}{T_1}\right) = \left[\frac{1+\frac{(k-1)}{2}M_1^2}{1+\frac{(k-1)}{2}M_2^2}\right] \tag{16.11}$$

Momentum balance:

$$p_1A_1 + \rho_1 V_1^2 A_1 = p_2 A_2 + \rho_1 V_2^2 A_1$$

since $\rho V^2 = pkM^2$ we may write:

$$p_1(1+kM_1{}^2) = p_2(1+kM_2{}^2)$$

$$\frac{p_2}{p_1} = \frac{(1+kM_1{}^2)}{(1+kM_2{}^2)} \tag{16.12}$$

Mass balance:

$$\rho_1 V_1 = \rho_2 V_2$$

$$\left(\frac{p_1}{RT_1}\right) M_1\sqrt{kRT_1} = \left(\frac{p_2}{RT_2}\right) M_2\sqrt{kRT_2}$$

$$\frac{M_1\sqrt{kRT_1}}{RT_1} = \left(\frac{p_2}{p_1}\right)\frac{M_2\sqrt{kRT_2}}{RT_2}$$

$$\left(\frac{M_1}{M_2}\right) = \left(\frac{p_2}{p_1}\right)\left(\frac{T_1}{T_2}\right)\sqrt{\frac{T_2}{T_1}} \tag{16.13}$$

Substituting Eq. 16.13 and Eq. 16.12 into Eq. 16.13 we have:

$$\frac{M_1}{M_2} = \frac{(1+kM_1{}^2)\sqrt{\frac{2+kM_1{}^2-M_1{}^2}{2+kM_2{}^2-M_2{}^2}}\left(2+kM_2{}^2-M_2{}^2\right)}{(1+kM_2{}^2)\left(2+kM_1{}^2-M_1{}^2\right)}$$

$$(k-1)(M_2^4 - M_1^4) - 2kM_1^2M_2^2(M_2^2 - M_1^2) + 2(M_2^2 - M_1^2) = 0$$

Solution of this equation is:

$$\boxed{M_2 = \pm\sqrt{\frac{2+(k-1)M_1^2}{2kM_1^2-(k-1)}}}$$

Since Mach number cannot be negative, we consider only the positive solution. Using this equation, we can now easily calculate the ratios of pressure, density, and temperature. We can write it for air ($k = 1.4$) as:

$$M_2 = \sqrt{\frac{2+0.4M_1{}^2}{2.8M_1{}^2-0.4}}$$

Applying limit we get:

$$\lim_{M_1\to\infty} M_2 = \lim_{M_1\to\infty} \sqrt{\frac{2+0.4{M_1}^2}{2.8{M_1}^2-0.4}} = 0.3779$$

This shows that at very high Mach numbers the normal shock will make flow subsonic, and the minimum possible Mach number M_2 after normal shock is 0.377. However we should keep in mind that at extremely high Mach number the gas will no longer be perfect gas.

16.7.1 ENTROPY RISE ACROSS SHOCKWAVE

According to the second law of thermodynamics, a thermodynamic property called entropy would rise due to the abrupt and sudden temperature rise. Second law of thermodynamics is variously defined as follows.

Second Law of Thermodynamics and Entropy

Lord Kelvin Statement (1851)

> It is impossible, by means of inanimate material agency, to derive mechanical effect from any portion of matter by cooling it below the temperature of the coldest of the surrounding objects.
>
> –Lord Kelvin

Plank Statement (1897)

> It is impossible, to construct a periodically working (cyclic) machine which only do raising of a weight and a cooling of Heat reservoir[a].
>
> –M. Plank

Clausius Statement (1850)

> It is impossible to construct a device that operates in a thermodynamic cycle and produces no effect other than the transfer of heat from a cooler to a hotter body.
>
> –R. Clausius

Hatsopoulos and Keenan (1965)

> A system having specified allowed states and an upper bound in volume can reach from any given state a stable state and leave no net effect on the environment.
>
> –G. N. Hatsopoulos and J. H. Keenan

From a fluid mechanics perspective, these statements of the second law put forth the concept of **thermodynamic limitations**. In order to quantify the limitations of a system m Clausius introduced an abstract quantity called entropy.

[a] Es ist unmöglich, eine periodisch funktionierende Maschine zu konstruieren, die weiter nichts bewirkt als Hebung einer Last und Abkühlung eines Wärmereservoirs.

Entropy rise across a shockwave is very important to determining the strength of the shockwave. Thermodynamically, we can write entropy rise for ideal gas as:

$$s_2 - s_1 = c_p \ln \frac{T_2}{T_1} - R \ln \frac{p_2}{p_1}$$

$$\frac{s_2 - s_1}{R} = \frac{c_p}{R} \ln \frac{T_2}{T_1} - \ln \frac{p_2}{p_1}$$

$$\frac{s_2 - s_1}{R} = \ln \left[\left(\frac{T_2}{T_1} \right)^{\left(\frac{k}{k-1}\right)} \left(\frac{p_2}{p_1} \right) \right]$$

Substituting the temperature and pressure ratios and expanding the series we get the following expression:

$$\boxed{\frac{s_2 - s_1}{R} \equiv \frac{2k}{(k+1)^2} \frac{\left(M_1{}^2 - 1\right)^3}{3}}$$

Figure 16.13 shows that the flow ahead of a normal shockwave must be supersonic.

16.8 PITOT-TUBE CORRECTION FOR COMPRESSIBLE FLOWS

A pitot tube is used to measure the stagnation pressure, however, in the compressible flow the shockwave forms in front of the pitot tube (see Figure 16.14).

$$p_o - p \neq \frac{1}{2} \rho V^2$$

We now start from isentropic relation:

$$\frac{p_o}{p} = \left(\frac{T_o}{T} \right)^{\frac{k}{k-1}} = \left[1 + \left(\frac{k-1}{2} \right) M^2 \right]^{\frac{k}{k-1}}$$

Using binomial expansion, this equation can be written as:

$$\frac{p_o}{p} = 1 + \frac{k}{2} M^2 + \frac{k}{8} M^4 + \frac{k(2-k)}{48} M^6 + \ldots \tag{16.14}$$

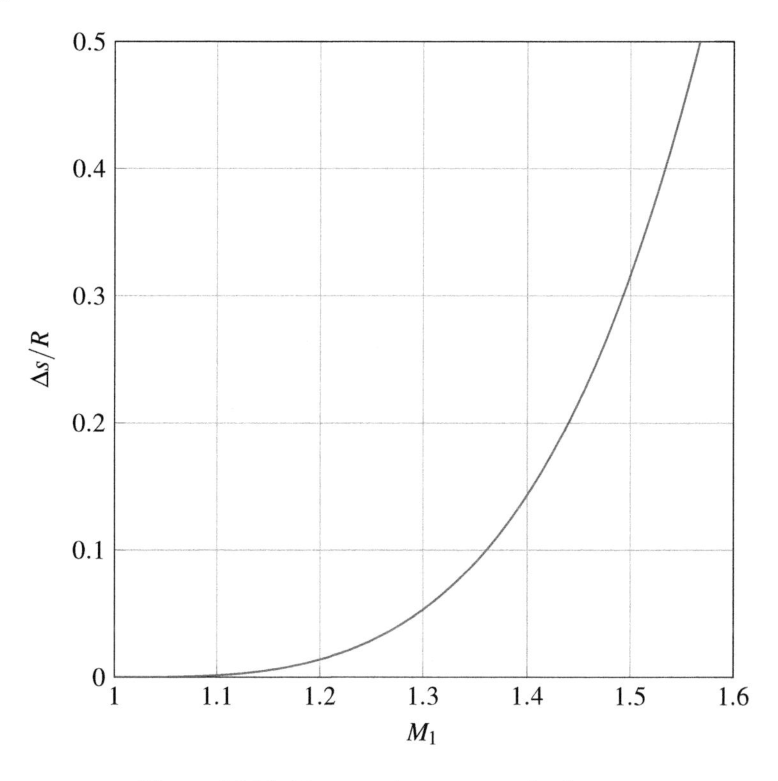

Figure 16.13 Entropy rise across a shockwave.

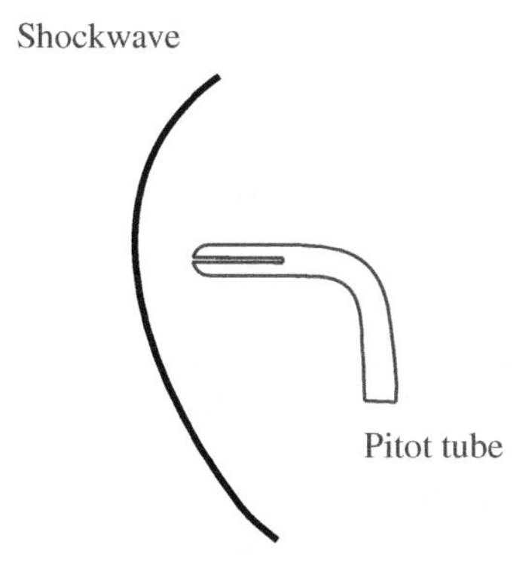

Figure 16.14 A pitot tube in supersonic flow will have a detached shock in front of it.

$$p_o - p = \frac{1}{2} pkM^2 \left[1 + \frac{M^2}{4} + \frac{(2-k)}{24} M^4 + ...\right]$$

$$\frac{p_o - p}{\frac{1}{2}\rho V^2} = \frac{p_o - p}{\frac{1}{2}pkM^2} = \left[1 + \frac{M^2}{4} + \frac{(2-k)}{24} M^4 + ...\right] = \alpha_c$$

Table 16.2
Correction Factor for Pitot Tube

M	0	0.1	0.2	0.3	0.4	0.5	0.6	0.7	0.8
α_c	1	1.003	1.01	1.023	1.041	1.064	1.093	1.129	1.17
Relative Error %	0	0.15	0.5	1.14	2.03	3.15	4.55	6.25	8.17

The term α_c is the correction factor that should be used when using the pitot tube in compressible flows.

The pressure rise across the shock is given by

$$\left(\frac{p_2}{p_1}\right) = \frac{(1+kM_1^2)}{(1+kM_2^2)}$$

However, it would be more convenient to have the ratio of stagnation pressure with static pressure

$$\frac{P_{o2}}{p_1} = \frac{P_{o2}}{p_2}\left(\frac{p_2}{p_1}\right) = \left[1+\left(\frac{k-1}{2}\right)M_2^2\right]\left[\frac{(1+kM_1^2)}{(1+kM_2^2)}\right]$$

$$\frac{P_{o2}}{p_1} = \left[\frac{(k+1)^{k+1}}{2kM_1^2-k+1}\left(\frac{M_1^2}{2}\right)^k\right]^{\frac{1}{k-1}}$$

This equation is called the Rayleigh formula. For $k = 1.4$ we have

$$\frac{P_{o2}}{p_1} = \frac{166.9M_1^7}{(7M_1^2-1)^{5/2}}$$

Example 16.3

Example A pitot tube is being used to measure the stagnation pressure of air moving at supersonic velocities. A shockwave has formed in front of the pitot tube. (i) Find the stagnation pressure after the shockwave, if the air static pressure is 150 kPa before the shockwave and Mach number is 4.
(ii) Find the Mach number, if the stagnation pressure after the shockwave is 3,000 kPa and air static pressure is 150 kPa before the shockwave.
Solution
(i) We can use the Rayleigh formula to measure the stagnation pressure after the shockwave. For $k = 1.4$ we have

$$\frac{P_{o2}}{p_1} = \frac{166.9M_1^7}{(7M_1^2-1)^{5/2}}$$

$$p_{o2} = \left(150 \cdot \left(\frac{166.9 \times 4^7}{(7 \times 4^2 - 1)^{\frac{5}{2}}}\right)\right) = 3159.80\ kPa$$

(ii) The stagnation pressure after the shockwave is 3,000 kPa, and air static pressure is 150 kPa before the shockwave:

$$\left(\frac{3000}{150} - \left(\frac{166.9 \cdot M^7}{(7 \cdot M^2 - 1)^{\frac{5}{2}}}\right)\right) = 0,$$

From trial and error we have $M = 3.895$.

16.9 FLOW THROUGH CONVERGING-DIVERGING NOZZLE

Swedish engineer Karl Gustaf Patrik de Laval, in 1890, developed the converging-diverging nozzle to increase the speed of a steam jet from sonic to supersonic speeds. This nozzle is now called the Laval nozzle, or converging-diverging nozzle, or simply CD nozzle. The converging-diverging nozzles are used in steam turbine power plants, missiles, and rockets (see Figure 16.15).

Figure 16.16 shows the pressure and Mach number distributions along a converging-diverging nozzle for different exit pressures. In curves *a* and *b*, the flow is subsonic throughout the nozzle. In curve *c*, the flow at minimum area or throat will be sonic, and therefore the pressure at the throat, and in converging section, cannot be decreased further. The further reduction in back pressure will not increase the flow rate, as the nozzle has already attained the maximum mass flow rate it can have and thus the "chocking" condition has been achieved. Between curve *c* and curve *j*, the shock will appear just after the throat, and as the pressure is reduced this shock wave will move in diverging section until it reaches the outlet. As the pressure is further lowered, the shock-wave will move ot of a nozzle and will appear as reflected shock inside the jet. Curve *j* is the "design" condition, i.e., the supersonic flow without any shockwaves. The condition "*k*" refers to the under expansion state when jet will expand as soon as it goes out in low pressure zone.

Figure 16.15 The rockets use converging-diverging nozzle.

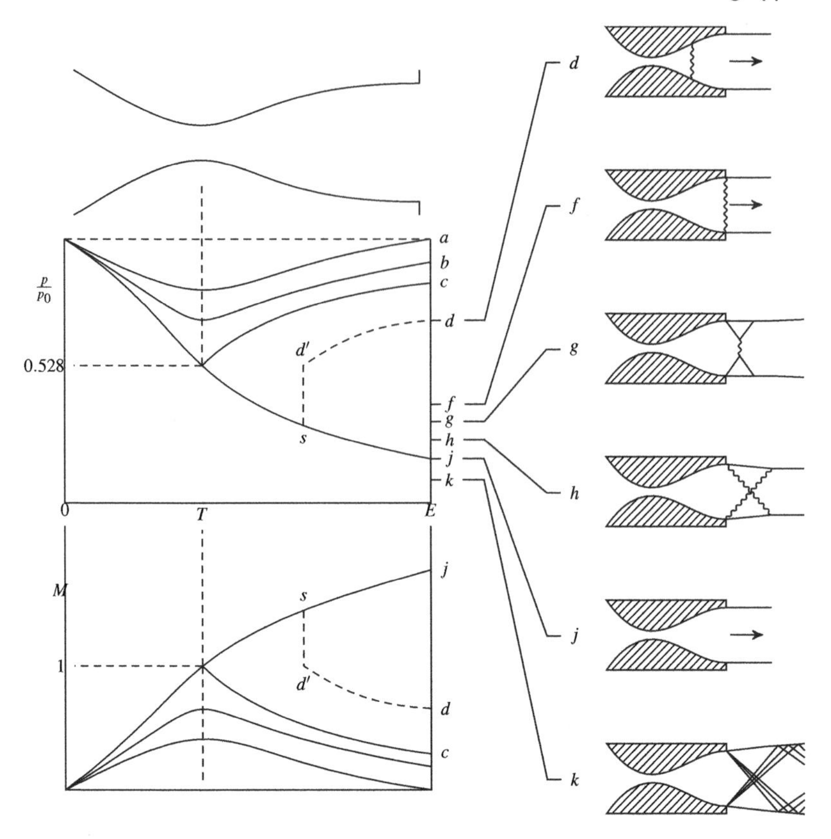

Figure 16.16 Laval nozzle and its flow regimes.

Example 16.4

Example A preliminary design of a wind tunnel to produce $M = 3$ at exit is desired. Mass flow rate is 1 kg/s and stagnation or reservoir conditions are $P_o = 90$ kPa, $T_o = 25°$C. Find the throat area, the outlet area, the velocity, the pressure, and the temperature at the outlet.

Solution We may use the maximum mass flow rate equation, as flow is already chocked

$$\dot{m}_{\max} = 0.686\frac{A^* p_o}{\sqrt{RT_o}} = 0.686\frac{A^* \times 90 \times 10^3 Pa}{\sqrt{287J/kgK \times (25+273)K}}$$

$$\dot{m}_{\max} = 1\frac{kg}{s} = 0.686\frac{A^* \times 90 \times 10^3 Pa}{\sqrt{287J/kgK \times (25+273)K}}$$

$$A^* = 0.00474m^2$$

Now using area ratio we have

$$\frac{A}{A^*} = 4.234$$

which gives exit area as $A = 0.02m^2$.

Using isentropic gas flow $k = 1.4$ tables we have:

$$\frac{p}{p_o} = 0.027, \frac{T}{T_o} = 0.357, \frac{\rho}{\rho_o} = 0.076$$

$$\rho_o = \frac{p_o}{RT_o} = \frac{90,000Pa}{287 \times 298K} = 1.0523kg/m^3$$

so the exit conditions are:

$$p = 2.43kPa,\ T = -166.6^oC,\ \rho = 0.08kg/m^3$$

16.10 OBLIQUE SHOCKWAVES

Figure 16.17 shows an oblique shockwave formed when a supersonic flow moved over a wedge or a ramp and flow is deflected by an angle θ. The resulting shockwave is inclined at an angle β as shown in Figure 16.18. The resolving velocities before and after shockwave indicate that shock relations developed for the normal shock are applicable for oblique shockwaves with M_1 replaced in them by $M_1\sin(\beta)$.

The relationship between θ and β is:

$$\tan\theta = \frac{2\cot\beta(M_1^2\sin^2\beta - 1)}{M_1^2\,[k + \cos(2\beta) + 2]}$$

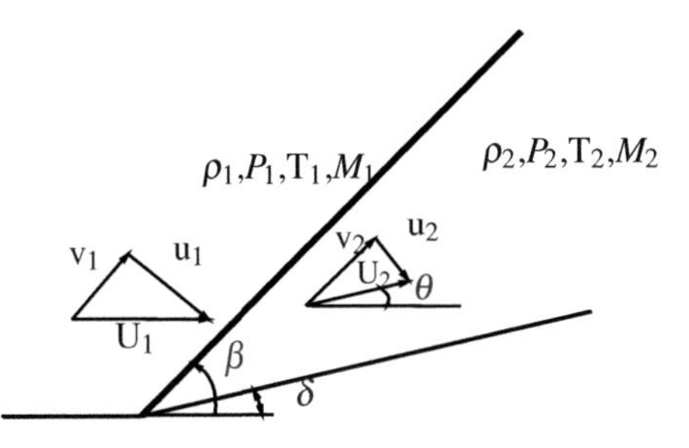

Figure 16.17 The results derived for normal shock are applicable to oblique shockwaves with M_1 replaced by $M_1\sin(\beta)$.

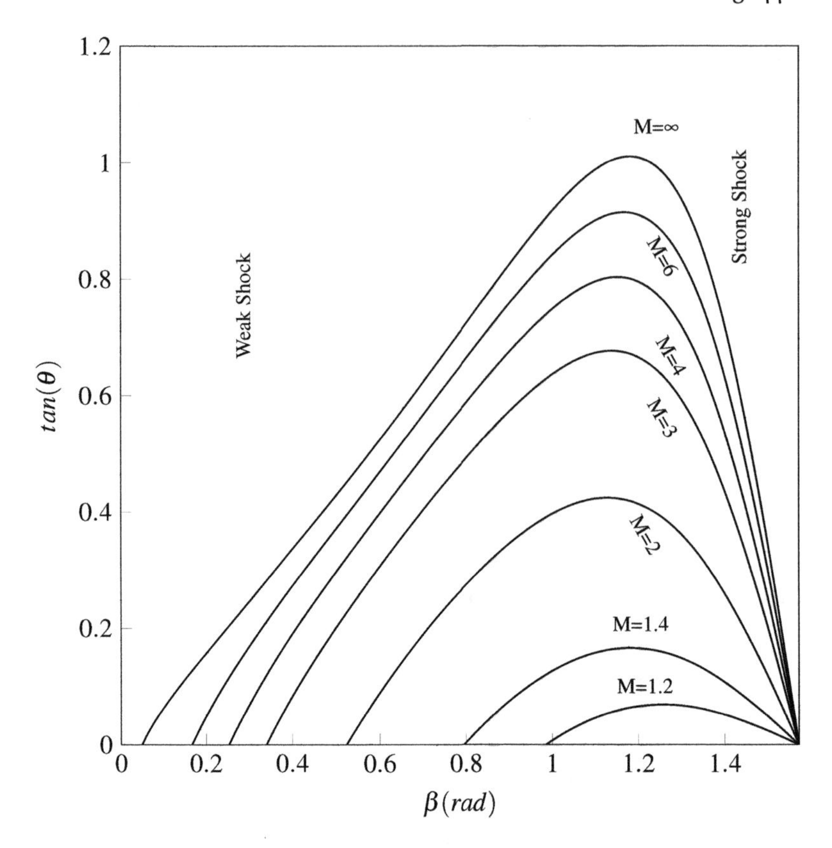

Figure 16.18 Oblique shockwave angle β plotted with flow deflection angle θ.

16.10.1 MACH CONE

Mach cone, which was previously discussed in this chapter, is actually an oblique shockwave.

$$M_1 \sin\beta \geq 1$$

The minimum β will occur when $M_1 \sin\beta \equiv 1$. If $M > 1$ then this is a shockwave. The rearrangement gives

$$\beta_{\min} = \sin^{-1}\left(\frac{1}{M_1}\right) = \beta_{Cone}$$

β_{Cone} is also called the Mach angle or Mach cone angle and it is the minimum possible shock angle.

16.10.2 DETACHED SHOCK

The shockwave will be attached to a wedge up to a certain angle. However, as the wedge angle increases, the shockwave will be detached from the wedge or cone

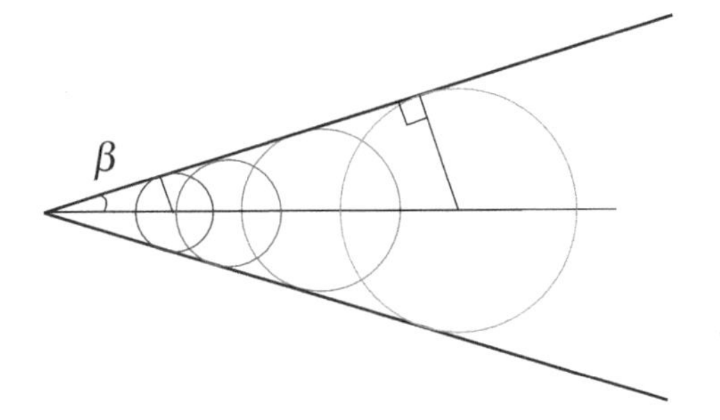

Figure 16.19 The movement of disturbance in a medium at $M > 1$. A Mach cone will form with angle β.

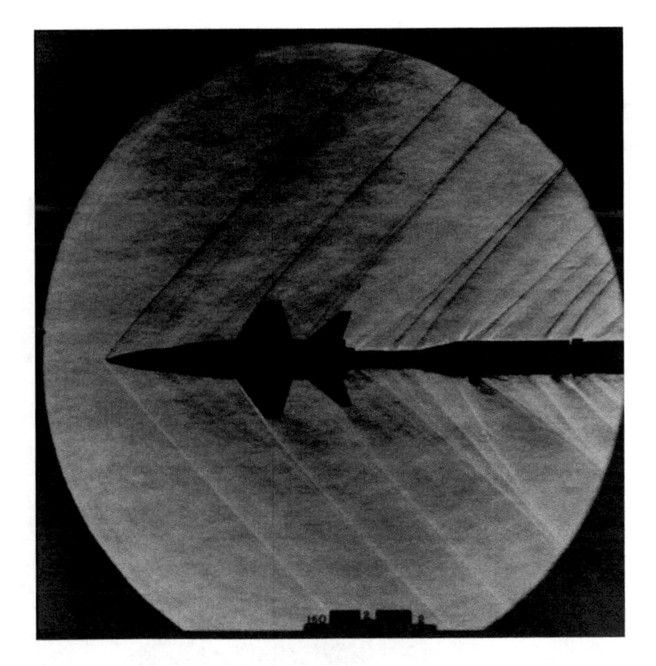

Figure 16.20 Oblique shockwaves in supersonic wind tunnel. (*Photograph published in Sixty Years of Aeronautical Research 1917–1977. By David A. Anderton (page 51) - a NASA publication. Picture taken from NASA Image Exchange, Courtesy of Dr. Marian Muste, IIHR - Hydroscience & Engineering, University of Iowa. Source: Intl. Assoc. for Hydro-Environment Engr. and Research*)

(see Figure 16.21). For analysis it is usually treated as normal shockwave for the estimation of flow quantities after the shockwave. Figure 16.22 shows the detached shockwave in a supersonic wind tunnel.

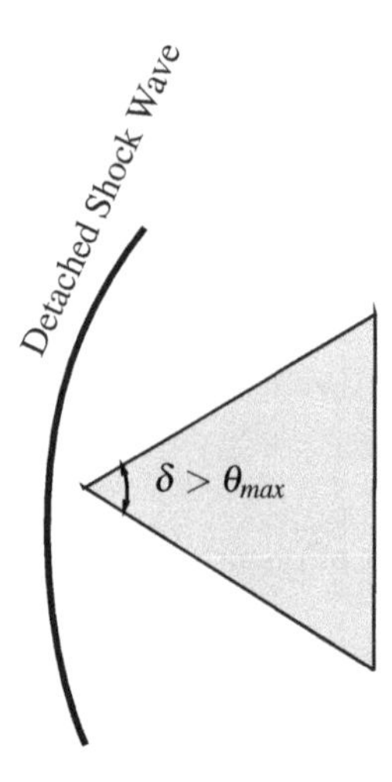

Figure 16.21 Detached shock wave will form if the wedge angle $\delta >$ shock deflection angle θ_{max}.

Figure 16.22 Detached shockwave in supersonic wind tunnel. (*Courtesy of Dr Marian Muste, IIHR - Hydroscience & Engineering, University of Iowa. Source: Int. Association for Hydro-Environment Engg. and Research*)

Example 16.5

Example A military aircraft traveling at a speed of 800 m/s in air at 4°C and 90 kPa. The aircraft's nose is like an asymmetric wedge. The flow deflection angle $\theta = 2$ degrees on the upper side of wedge, and 4 degrees on the lower

side. Find the type of shock experienced by the surface of the wedge.
Solution The acoustic speed is

$$a = \sqrt{kRT} = \sqrt{1.4 \times 287 \times 277} = 333.61 m/s$$

Using this, the Mach number is $V/a = 800/333.61 = 2.39$.
Since $\theta = 2°$, we solve the following equation by trial and error by assuming different

$$\tan\theta = \frac{2\cot\beta(M_1^2 \sin^2\beta - 1)}{M_1^2\,[k + \cos(2\beta) + 2]}$$

The β value of 26.15° is found. Now we calculate

$$M_1 \sin\beta = 1.056$$

This shows that shock on the upper side of wedge is a weak shockwave, as $M_1 \sin\beta$ is close to unity.

Repeating this procedure for the lower side gives $\beta = 27.7°$ for $\theta = 4°$. Also

$$M_1 \sin\beta = 1.11$$

shows that shock is also weak at the lower side.

Example 16.6

Example In a wind tunnel testing, it is important to match the Mach numbers if flow has become compressible. A prototype aircraft will fly at Mach number 0.7 at an altitude where pressure and temperature would be 7 kPa and −55.5°C. A 1/20th scale-down model of this aircraft is being tested in a wind tunnel. Calculate the total pressure of air in the tunnel that is necessary to achieve the dynamic similarity. The total temperature is 60°C. The dynamic viscosity is related to temperature as:

$$\mu = \mu_o \left(\frac{T}{T_o}\right)^{3/4}$$

where $\mu_o = 1.71 \times 10^{-5} Pa \cdot s$ and $T_o = 273K$
Solution
Prototype aircraft: The acoustic speed and speed of the aircraft are:

$$a = \sqrt{kRT} = \sqrt{1.4 \times 287 \times (-55.5 + 273)} = 295.62 m/s$$

$$u = M \cdot a = 0.7 \times 295.62 = 206.93 m/s$$

$$\rho = \frac{p}{RT} = \frac{7000}{287 \times 217.5} = 0.1121\ kg/m^3$$

$$\mu = \mu_o \left(\frac{T}{T_o}\right)^{3/4} = 1.71 \times 10^{-5} \left(\frac{217.5}{273}\right)^{3/4} = 0.144e-4$$

Model: Since Reynolds number on model and prototype should be the same, we can calculate the Reynolds number:

$$\frac{Re_p}{\ell_m} \equiv \frac{\rho_p u_p (\ell_p/\ell_m)}{\mu_p} = \frac{0.1121 \times 206.93 \times 20}{0.0000144} = 3.218 \times 10^7$$

$$\frac{T_o}{T} = 1 + \frac{(k-1)}{2} \cdot M^2 = 1.098$$

Now for $T_o = 333$ K for the model, the static temperature of air coming on the model is:

$$T_{model} = 303.27 \text{ K}$$

$$Re_m \equiv \frac{\rho_m u_m (\ell_m)}{\mu_m} = 3.218 \times 10^7 \ell_m$$

$$\mu_m = \mu_o \left(\frac{T}{T_o}\right)^{3/4} = 1.71 \times 10^{-5} \left(\frac{303.27}{273}\right)^{3/4} = 0.0000185$$

$$\frac{\rho_m u_m}{\mu_m} = 3.218 \times 10^7$$

$$\rho_m = \frac{3.218 \times 10^7 \times \mu_m}{u_m} = \frac{3.218 \times 10^7 \times \mu_m}{M \cdot \sqrt{kR(303.27)}} = 2.43$$

$$p_{o,m} = p \cdot \left(1 + \frac{(k-1)}{2} \cdot M^2\right)^{3.5} = 2.942 \times 10^5$$

It is necessary to have the total pressure available in the tunnel equal to this value, otherwise it would not be possible to achieve both the Mach and Reynolds numbers same on the model and the prototype.

16.11 FRICTIONAL FLOW IN A CONSTANT AREA DUCT

In this section, we consider the constant area duct with friction. The duct walls will pose friction to the flow, and entropy would rise until the flow become sonic. Figure 16.23 shows the representative control volume for the compressible flow through the duct with friction.

Continuity equation:

$$\dot{m} = \rho_1 A_1 V_1 = \rho_2 A_2 V_2 = constant$$

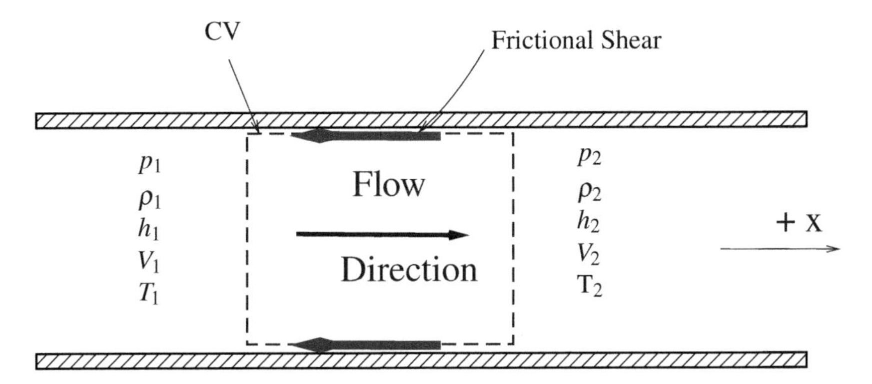

Figure 16.23 Control volume for steady one dimensional adiabatic flow.

$$G = \frac{\dot{m}}{A} = \rho_1 V_1 = \rho_2 V_2$$

Momentum equation:

$$-F_f + p_1 A_1 - p_2 A_2 = \rho_1 A_1 V_1^2 = \rho_2 A_2 V_2^2$$

First law of thermodynamics:

$$h_{o1} = h_1 + \frac{V_1^2}{2} = h_2 + \frac{V_2^2}{2} = h_{o2} = h + \frac{V^2}{2}$$

$$p = \rho RT$$

Second law of thermodynamics:

$$s_2 > s_1$$

$$Tds = dh - \frac{dp}{\rho}$$

$$Tds = du - RT\frac{d\rho}{\rho}$$

$$ds = c_p\frac{dT}{T} - R\frac{dp}{p} = c_v\frac{dT}{T} - R\frac{d\rho}{\rho}$$

$$s_2 - s_1 = c_p \ln\left(\frac{T_2}{T_1}\right) - R\ln\left(\frac{p_2}{p_1}\right) = c_v \ln\left(\frac{T_2}{T_1}\right) - R\ln\left(\frac{\rho_2}{\rho_1}\right)$$

The equation of Fanno line

$$s - s_1 = c_v \ln\left(\frac{T}{T_1}\right) - R\ln\left(\frac{\rho}{\rho_1}\right)$$

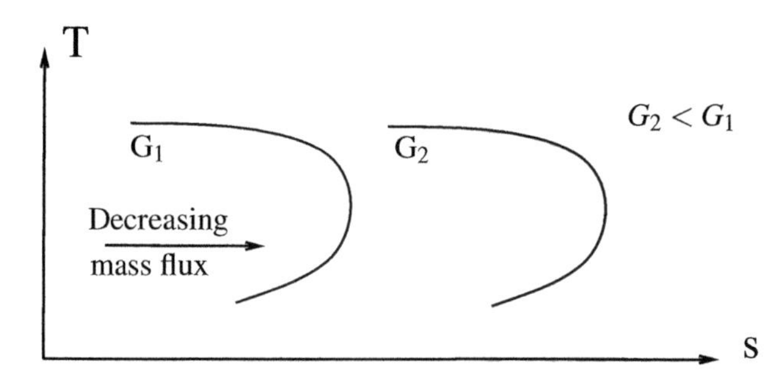

Figure 16.24 T-S plane for the compressible frictional flow. The decreasing mass flux would cause the increase in entropy.

$$s - s_1 = c_v \ln\left(\frac{T}{T_1}\right) - R\ln\left(\frac{V}{V_1}\right)$$

$$h = c_p T$$

$$V = \sqrt{2c_p(T_o - T)}$$

$$V_1 = \sqrt{2c_p(T_o - T_1)}$$

$$s - s_1 = c_v \ln\left(\frac{T}{T_1}\right) - \frac{R}{2}\ln\left(\frac{T_o - T}{T_o - T_1}\right)$$

$$s = c_v \ln(T) - c_v \ln(T_1) + c_v\left(\frac{k-1}{2}\right)\ln(T_o - T) - c_v\left(\frac{k-1}{2}\right)\ln(T_o - T_1) + s_1$$

$$\frac{ds}{dT} = c_v\left[\frac{1}{T} - \left(\frac{k-1}{2}\right)\left(\frac{1}{T_o - T}\right)\right] = 0$$

Figure 16.24 shows the shift in mass flux with the compressible flow through the duct with friction.

16.11.1 FANNO LINE FLOW EQUATIONS

The balance of forces on the control volume can be written as:

$$\left[pA - \left(p + \frac{dp}{dx}\delta x\right)A\right] - \tau_o(\pi \cdot D \cdot \delta x) = \rho AV\left[\left(V + \frac{dV}{dx}\delta x\right) - V\right]$$

Rearrangement gives

$$-\frac{dp}{dx}\delta x \cdot A - \tau_o(\pi \cdot D \cdot \delta x) = \rho V \frac{dV}{dx}\delta x \cdot A$$

$$dp + \tau_o\left(\frac{4\delta x}{D}\right) + \rho V dV = 0$$

We now use the Fanning and Darcy definitions of skin friction coefficients to replace shear stress:

$$C_f = \frac{\tau_o}{\frac{\rho}{2}V^2} = \frac{2\tau_o}{\rho V^2}$$

$$f = 4C_f = \frac{8\tau_o}{\rho V^2}$$

$$\tau_o = \left(\frac{f \cdot \rho V^2}{8}\right)$$

We now adjust the shear force term in the force balance as

$$\frac{dp}{p} + f\left(\frac{\rho V^2}{p}\right)\left(\frac{\delta x}{2D}\right) + \frac{\rho V dV}{p} = 0$$

We now formulate various substitutions:
From the definition of Mach number, we write velocity as

$$V^2 = M^2 k \cdot p/\rho = M^2 kRT$$
$$\frac{\rho V^2}{p} = kM^2$$
$$\frac{\rho V^2}{p}\left(\frac{dV}{V}\right) = kM^2\left(\frac{dV}{V}\right)$$

$$\frac{\rho V}{p}dV = kM^2\left(\frac{dV}{V}\right) \tag{16.15}$$

Using Mach number again, we can form an equation

$$V^2 = M^2 kRT$$
$$2VdV = (M^2 kR)dT + kRT(2MdM)$$
$$\frac{2VdV}{V^2} = \frac{()dT}{T} + \frac{(2MdM)}{M^2}$$
$$2\frac{dV}{V} = \frac{dT}{T} + 2\frac{dM}{M}$$

$$\frac{dV}{V} = \frac{1}{2}\left(\frac{dT}{T} + 2\frac{dM}{M}\right) \tag{16.16}$$

Using energy equation, we have

$$h_o = h + \frac{V^2}{2} = c_p T + \frac{V^2}{2}$$
$$d(h_o) = d\left(c_p T + \frac{V^2}{2}\right) = c_p dT + VdV$$
$$0 = c_p dT + VdV$$
$$\frac{c_p dT + VdV}{V^2} = 0$$

$$\frac{dT}{T} = -M^2(k-1)\frac{dV}{V} \tag{16.17}$$

We adjust dV/V terms as

$$\begin{aligned}
&\tfrac{dV}{V} = \tfrac{1}{2}\left(-M^2(k-1)\tfrac{dV}{V} + 2\tfrac{dM}{M}\right)\\
&\tfrac{dV}{V} + \tfrac{1}{2}M^2(k-1)\tfrac{dV}{V} = \tfrac{dM}{M}\\
&\left(1 + M^2\tfrac{(k-1)}{2}\right)\tfrac{dV}{V} = \tfrac{dM}{M}
\end{aligned}$$

We substitute the above result into equation:

$$\frac{dV}{V} = \frac{dM/M}{\left(1 + M^2\frac{(k-1)}{2}\right)} \tag{16.18}$$

Now using ideal gas equation, we have

$$\begin{aligned}
&p = \rho RT\\
&dp = R[Td\rho + \rho dT]
\end{aligned}$$

$$\frac{dp}{p} = R\left[\frac{Td\rho}{\rho RT} + \frac{\rho dT}{\rho RT}\right] = \left[\frac{d\rho}{\rho} + \frac{dT}{T}\right] \tag{16.19}$$

From continuity equation, we can form the equation

$$\frac{d\rho}{\rho} = \frac{-dV}{v} = \frac{-1}{2}\frac{d(V^2)}{V^2} = \frac{-1}{2}\frac{d(M^2)}{M^2}$$

After these substitutions and rearrangement we have

$$\frac{fL}{D} = \frac{2}{k}\frac{dM}{M^3} - \left(\frac{k+1}{k}\right)\frac{dM}{M\left(\frac{k-1}{2}M^2 + 1\right)}$$

which may be integrated directly. By using the limits $x = 0, M = M_o; x = L, M = M$:

$$\frac{fL}{D} = \frac{-1}{kM^2}\Bigg|_{M_o}^{M} - \left(\frac{k+1}{2k}\right)\ln\left[\frac{M^2}{1 + \left(\frac{k-1}{2}\right)M^2}\right]_{M_o}^{M}$$

$$\frac{fL}{D} = \frac{1}{k}\left[\frac{1}{M_o^2} - \frac{1}{M^2}\right] + \left(\frac{k+1}{2k}\right)\ln\left[\left(\frac{M_o}{M}\right)^2\left(\frac{2 + (k-1)M^2}{2 + (k-1){M_o}^2}\right)\right]$$

Whatever the Mach number, the second law of thermodynamics has indicated that the flow is going towards the maximum entropy state, and now from the above equation we can infer that if M_o is greater than 1, then M at the end of the duct cannot be less than 1, and if M_o is less than 1, then M cannot be greater then 1.

For the limiting condition $M = 1$ and $k = 1.4$, we get

$$\frac{fL_{\max}}{D} = \frac{5}{7}\left(\frac{1}{M_o^2} - 1\right) + \frac{6}{7}ln\left(\frac{6M_o^2}{5 + M_o^2}\right)$$

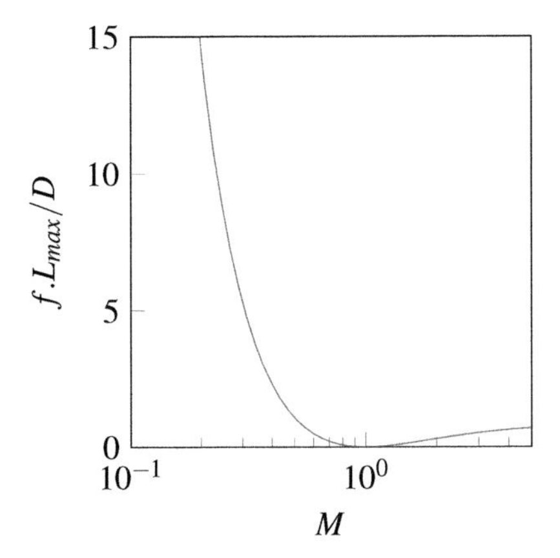

Figure 16.25 $f \cdot L_{max}/D$ vs. Mach number ($k = 1.4$).

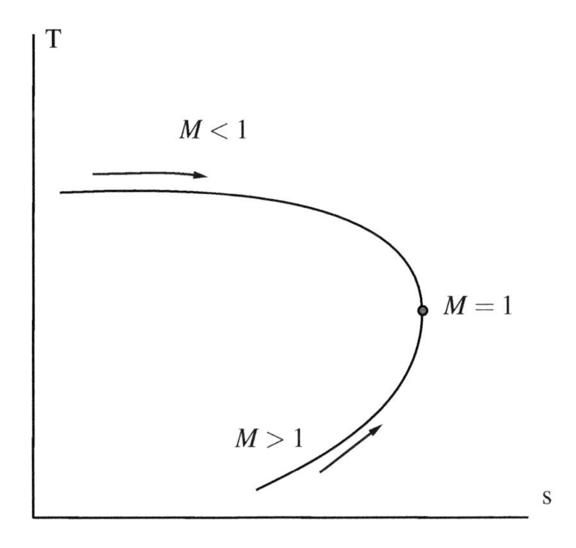

Figure 16.26 T-s diagram for frictional adiabatic flow in a constant-area duct.

A table of Fanno line relations has been provided in appendix[1] using Fanning's friction factor definition. Figure 16.25 shows $f \cdot L_{max}/D$ vs. Mach number ($k = 1.4$).

Figure 16.26 shows the schematic T-s diagram for frictional adiabatic flow in a constant-area duct.

Figure 16.27 shows the coordinates and nomenclature that will be used for the analysis of friction flow through duct. As in the derivation, we took the upper limit as $M = 1$, and we introduce the concept of maximum duct length that would lead to Mach number of unity. By subtracting the two lengths we can arrive at the real duct length under the given conditions.

[1] Appendix can be found online at `https://routledge.com/9781032324531`

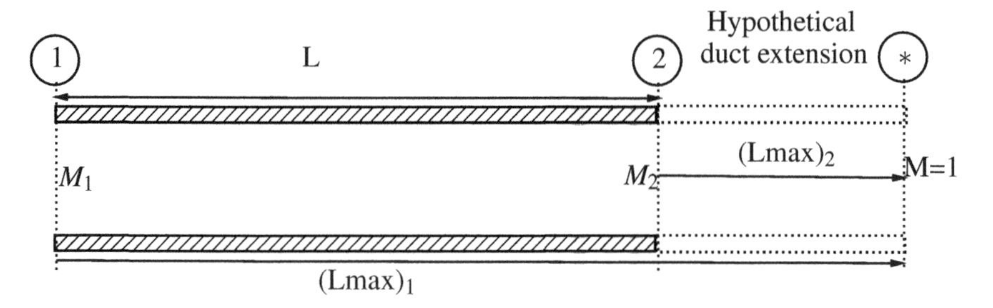

Figure 16.27 The concept of extended duct length and its nomenclature.

Table 16.3
Fanno Line Flow: Effect of Friction on Flow Properties

Property	Subsonic	Supersonic
Velocity	Increase	Decrease
Mach number	Increase	Decrease
Pressure	Decrease	Increase
Temperature	Decrease	Increase
Density	Decrease	Increase
Stagnation Enthalpy	Constant	Constant
Stagnation Pressure	Decrease	Decrease
Entropy	Increase	Increase

Using Darcy friction factor the procedure is

$$\frac{fL}{D} = \left(\frac{fL_{\max}}{D}\right)_1 - \left(\frac{fL_{\max}}{D}\right)_2$$

Using Fanning's friction factor the procedure is

$$\frac{4C_fL}{D} = \left(\frac{4C_fL_{\max}}{D}\right)_1 - \left(\frac{4C_fL_{\max}}{D}\right)_2$$

Example 16.7

Example Mass flow rate of 25 kg/s of air is flowing through a pipe of diameter of 0.3 m and length 1.4 m. The pipe walls are insulated and pipe has roughness of 0.002 m. The stagnation temperature is 90°C. Find the inlet Mach number.

Solution We assume $f = 0.02$ and estimate:

$$\frac{fL}{D} = \frac{0.02 \times 1.4}{0.3} = 0.0933$$

Using Fanno flow table we get Mach number

$$\frac{fL}{D} = \frac{(1-M^2)}{k \cdot M^2} + \frac{(k+1)}{(2 \cdot k)} \ln\left(\frac{(k+1) \cdot M^2}{2 \cdot \left(1 + \frac{(k-1)}{2} \cdot M^2\right)}\right)$$

$$M = 0.778$$

From table in appendix[a] we can approximate this Mach number as 0.78 and data is:

M	T_o/T	P_o/P	ρ_o/ρ	V*/V	$\frac{\dot{m}\sqrt{RT_o}}{P_oA}$	A/A*
0.78	1.122	1.495	1.333	1.611	0.654	1.047

Also from isentropic flow we have

$$\frac{T_o}{T} = 1.122$$

As $T_o = 90 + 273 = 363$ K
$T = 323$ K

$$M = 0.778 = \frac{V}{\sqrt{1.4 \times 287 \times 323}}$$

$V = 280.5 m/s$
Re $= 5.61 \times 10^6$ and $\varepsilon/D = 0.0066$ using Moody diagram we have $f =$ 0.032, therefore, we need to repeat this calculations

Using this new f we again calculate $fL/D = 0.1493$, which gives $M =$ 0.734.

M	T_o/T	P_o/P	ρ_o/ρ	V*/V	$\frac{\dot{m}\sqrt{RT_o}}{P_oA}$	A/A*
0.73	1.107	1.425	1.288	1.600	0.637	1.074

This gives $T = 327$ $V = 282$ m/s Re $= 5.6 \times 10^6$ and $f = 0.032$

This shows that the friction factor in the second trial is good enough and further calculations are not needed.

[a]Appendix can be found online at `https://routledge.com/9781032324531`

Example 16.8

Example A constant diameter duct of $d = 2$ cm is connected to a reservoir at pressure and temperature of 1,000 kPa and 600 K through a converging nozzle as shown in Figure 16.28. Determine the maximum mass flow rate through this system. Assume $f = 0.035$, $k = 1.4$, and $L = 20$ cm.

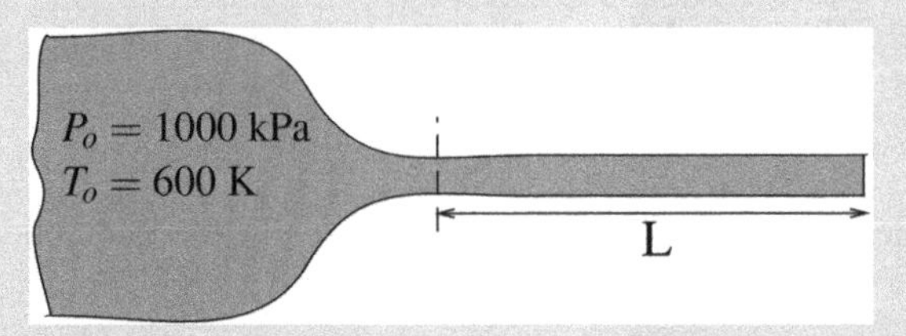

Figure 16.28 Schematic diagram of Example 16.8.

Solution We identify the convergent nozzle exit as station 1 and at the duct outlet as station 2. For maximum mass flow rate, the mach number at exit should be unity. Using the data given, we calculate fL/D, considering now $L = L_{max}$.

$$\frac{fL_{\text{max}}}{D} = 0.35$$

From Fanno line tables, or using following equation, we find the Mach number.

$$\frac{fL_{\text{max}}}{D} = \frac{5}{7}\left(\frac{1}{M_o^2} - 1\right) + \frac{6}{7}ln\left(\frac{6M_o^2}{5+M_o^2}\right)$$

$M_1 \approx 0.64$

$$\frac{p_o}{p} = \left(1 + \frac{\text{k}-1}{2}M_1{}^2\right)^{\frac{\text{k}}{k-1}} = 1.339$$

$$\frac{T_o}{T} = \left(1 + \frac{\text{k}-1}{2}M_1{}^2\right) = 1.087$$

This gives $p = 1,000\text{kPa}/1.339 = 7.46 \times 10^5$ Pa and $T = 600\text{K}/1.087 = 551.91\,\text{K}$.

$$\dot{m}_{\text{max}} = \frac{p_1}{R \cdot T_1} \cdot A \cdot M_1 \cdot \sqrt{k \cdot R \cdot T_1} = 0.459\, kg/s$$

16.12 FLOW IN CONSTANT AREA DUCT WITH HEAT TRANSFER

Consider a control volume for steady, one-dimensional, and frictionless flow of an ideal gas through the constant area duct with heat transfer, as illustrated in Figure 16.29. This is called Rayleigh flow or Rayleigh line flow. The flow is named after John William Strutt, 3rd Baron Rayleigh. Such flow represents the flow through aircraft engines where heat transfer cannot be ignored.

The continuity equation dictates that

$$\dot{m} = \rho_1 A_1 V_1 = \rho_2 A_2 V_2 = constant$$

As area is constant, we can also write the mass per unit area as

$$G = \dot{m}/A = \rho_1 V_1 = \rho_1 V_2 = \rho V$$

The momentum balance is

$$p_1 A_1 - p_2 A_2 = \rho_2 A_2 V_2^2 - \rho_1 A_1 V_1^2$$

$$A_1(p_1 + \rho_1 V_1^2) = A_2(p_2 + \rho_2 V_2^2) = A(p + \rho V^2) = constant$$

According to the first law of thermodynamics:

$$\dot{Q} = \dot{m}q = \dot{m}(h_{o2} - h_{o1}) = \dot{m}\left[\left(h + \frac{V^2}{2}\right)_2 - \left(h + \frac{V^2}{2}\right)_1\right]$$

From the second law of thermodynamics, we have

$$T.ds = dh - \frac{dp}{\rho}$$

$$T.ds = du - RT\frac{dp}{\rho}$$

The enthalpy, internal energy, and ideal gas relations are used to modify the equations

$$dh = c_p dT$$

$$du = c_v dT$$

$$p = \rho RT$$

$$ds = c_p \frac{dT}{T} - R\frac{dp}{p} = c_v \frac{dT}{T} - R\frac{d\rho}{\rho}$$

$$s_2 - s_1 = c_p \ln\left(\frac{T_2}{T_1}\right) - R\ln\left(\frac{p_2}{p_1}\right) = c_v \ln\left(\frac{T_2}{T_1}\right) - R\ln\left(\frac{\rho_2}{\rho_1}\right)$$

Introducing specific heats in terms of gas constant, we have

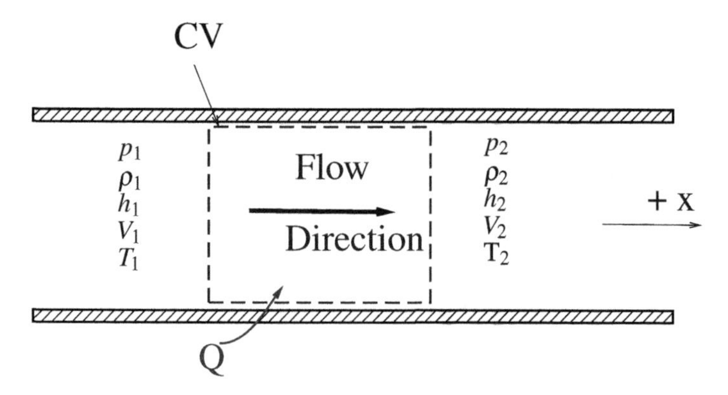

Figure 16.29 Control volume for steady, one-dimensional frictionless flow with heat transfer.

$$\left(\frac{s_2 - s_1}{R}\right) = \ln\left(\frac{(T_2/T_1)^{k/(k-1)}}{p_2/p_1}\right)$$

Introducing $V^2 = M^2kRT$ we have $p + \rho V^2 = p(1 + kM^2) = constant$

$$\frac{dp}{p} = -\frac{2kM}{1 + kM^2}dM$$

$$\rho V = \left(\frac{p}{RT}\right)\left(\frac{M}{\sqrt{kRT}}\right) = pM\sqrt{\left(\frac{k}{TR}\right)} = const$$

By logarithmic differentiation, we have

$$\frac{dT}{T} = 2\left(\frac{1 - kM^2}{M(1 + kM^2)}\right)dM$$

The static temperature variation will be

$$\frac{p_1}{p_2} = \frac{1 + kM_2{}^2}{1 + kM_1{}^2}$$

$$\frac{T_1}{T_2} = \left(\frac{M_1}{M_2}\right)^2 \frac{\left[1 + kM_2{}^2\right]^2}{\left[1 + kM_1{}^2\right]^2}$$

Density ratio will take the form:

$$\frac{\rho_1}{\rho_2} = \left(\frac{M_2}{M_1}\right)^2 \frac{\left[1 + kM_1{}^2\right]}{\left[1 + kM_2{}^2\right]}$$

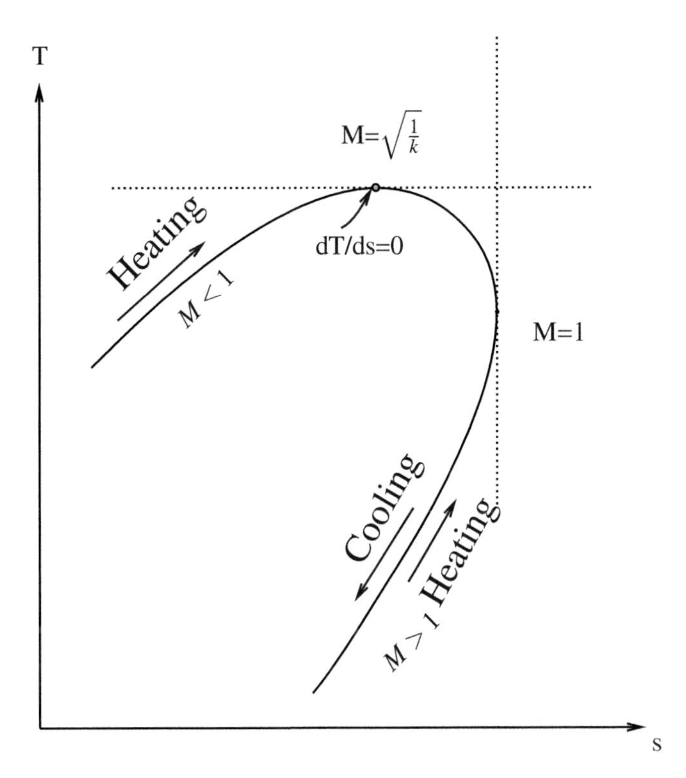

Figure 16.30 Rayleigh flow.

We can also formulate the entropy relation in terms of Mach number:

$$\left(\frac{s_2 - s_1}{R}\right) = \ln\left[\left(\frac{1+kM_1{}^2}{1+kM_2{}^2}\right)^{\frac{k+1}{k-1}}\left(\frac{M_2}{M_1}\right)^{\frac{2k}{k-1}}\right]$$

From isentropic gas flow relations we have

$$\frac{T_{o1}}{T_1} = 1 + \left(\frac{k-1}{2}\right)M_1^2$$

and

$$\frac{T_{o2}}{T_2} = 1 + \left(\frac{k-1}{2}\right)M_2^2$$

Taking the ratio of the above two equations gives

$$\frac{T_{o1}}{T_{o2}} = \left(\frac{T_1}{T_2}\right)\left(\frac{2+(k-1)M_1^2}{2+(k-1)M_2^2}\right)$$

Table 16.4
Trend in Flow Properties for Rayleigh Flow

Property	Heating $M > 1$	Heating $M < 1$	Cooling $M > 1$	Cooling $M < 1$
Static Pressure, p	Increases	Decreases	Decreases	Increases
Velocity, V	Decreases	Increases	Increases	Decreases
Stag. Temperature, T_o	Increases	Increases	Decreases	Decreases
Static Density, ρ	Increases	Decreases	Decreases	Increases
Static Temperature, T	Increases	Increases for $M < 1/\sqrt{(k)}$ Decreases for $M > 1/\sqrt{(k)}$	Decreases	Decreases for $M < 1/\sqrt{(k)}$ Increases for $M > 1/\sqrt{(k)}$

Inserting static temperature ratios and setting $M_2 = 1$ we can cast the relation

$$\frac{T_{o1}}{T_o^*} = \frac{M^2(k+1)}{(1+kM^2)^2}\left[2+(k-1)M^2\right]$$

Figure 16.30 shows the T-S plane for Rayleigh flow. In the case of heating of supersonic flow, the flow velocity would be reduced and it would reduce to sonic speed, whereas, in the case of subsonic flow and heating, the flow speed is increased and flow would reach Mach number $M = \sqrt{(1/k)}$. Table 16.4 lists the trend in flow properties for Rayleigh flow.

Example 16.9

Example Air is flowing through a 5-in. diameter pipe at inlet velocity $V_1 =$ 400 ft/s, $T_1 = 80°F$ and $p_1 = 50$ psia. How much heat transfer per unit mass is required to achieve the sonic condition at the exit of the duct? What will be the pressure, temperature, and velocity at the location where the Mach number is 0.6?

Solution The universal gas constant is

$$R_u = 1545\frac{ft \cdot lb_f}{lb_{mol} \cdot {}^oR}$$

The R value for air is

$$R = R_u/28.97 = 53.3$$

We can calculate the Mach number at the pipe inlet as:

$$M_1 = \frac{V_1}{\sqrt{k \cdot R \cdot T_1 \cdot 32.17}} = \frac{400}{\sqrt{1.4 \cdot 53.3.(80+460) \cdot 32.17}} = 0.351$$

If the Mach number at exit is unity, then the isentropic stagnation temperature at the exit is

$$T_o^* = \frac{T_{o1} \cdot \left(1 + k \cdot M_1^2\right)^2}{(k+1) \cdot M_1^2 \cdot \left(2 + (k-1) \cdot M_1^2\right)}$$

$$T_o^* = \frac{553.3 \times \left(1 + 1.4 \times (0.351)^2\right)^2}{(1.4+1) \times (0.351)^2 \times \left(2 + (1.4-1) \times (0.351)^2\right)} = 1253^o R$$

The heat transfer per unit slug of flowing fluid is

$$q = c_p(T_o^* - T_{o1}) \times 32.17 = 5407.14 Btu/slug$$

For the exit sonic conditions the static pressure is

$$\frac{p_2}{p_1} = \frac{1 + kM_1^2}{1 + kM_2^2}$$

$$\frac{p^*}{50} = \frac{1 + 1.4(0.351)^2}{1 + 1.4(1)^2}$$

$$p^* = 24.4 psia$$

For the exit sonic conditions the static temperature is

$$T^* = T_1 \left(\frac{1 + kM_1^2}{M_1(k+1)}\right)^2 = 1044.72^o R$$

and the flow exit velocity from pipe will be

$$a^* = \sqrt{k \cdot R \cdot T^* \cdot 32.17} = \sqrt{1.4 \times 53.3 \times 1044.72 \times 32.17} = 1583.62 ft/s$$

For the location where Mach number is 0.6 we have static pressure

$$p = \left(\frac{1+k}{1+kM^2}\right) p^* = \frac{24.4 psia \times 2.4}{1 + 1.4 \times 0.6^2} = 38.98 psia$$

Static temperature

$$T = T^* \left(\frac{M(1+k)}{1+kM^2}\right)^2 = 1044.72^o R \left(\frac{0.6(1+1.4)}{1 + 1.4 \times 0.6^2}\right)^2 = 957.7^o R$$

and velocity

$$V = M\sqrt{k \cdot R \cdot T \cdot 32.17} = 909.74 ft/s$$

PROBLEMS

16P-1 A jet flying at an altitude of 10,000 m has increased its speed from 200 m/s to a speed of 300 m/s, as shown in Figure 16.31. Find the Mach numbers.

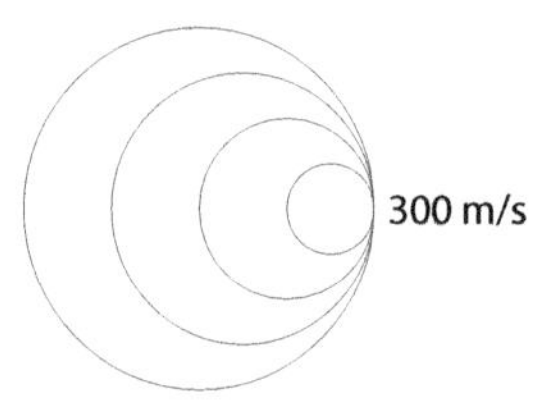

Figure 16.31 Figure of problem 16P-1. The movement of disturbance in a medium parallel to ground.

16P-2 The area of a converging nozzle is gradually reduced according to relation:

$$A = A_{inlet}(1 - 0.1x)$$

where x is the coordinate in the direction of the flow and the inlet area is 0.5 m^2. The length of nozzle is 9 m and Mach number varies according to relation:

$$M(x) = \tanh(0.3 \cdot x)$$

Find the temperature and pressure variation at the nozzle outlet if stagnation pressure is 200 kPa and the nozzle is open to atmosphere.

16P-3 Air at temperature 450K and pressure 2×10^5 N/m^2 moves in a converging duct from inlet to outlet. The speed of air is 200 m/s. Find the outlet conditions if the flow is chocked.

16P-4 Air is supplied from a reservoir where pressure is 1.4×10^5 N/m^2 and temperature is 300° K. The back pressure is maintained at 0.95×10^5 N/m^2. The area ratio (A/A*) at convergent nozzle exit is 1.25. Find the mass flow rate. The exit area is 0.00065 m^2.

16P-5 Air at pressure and temperature of 1.4×10^5 N/m^2 and temperature of 300° K respectively, and discharged through a convergent-divergent nozzle into a space at pressure of 0.95×10^5 N/m^2. The area ratio (A/A*) at Converging-Diverging nozzle exit is 1.25. The nozzle is following the design condition, and exit area is 0.0005 m^2.

16P-6 Air at stagnation pressure of 7×10^5 N/m^2 and temperature 300° K expanded through a frictionless convergent-divergent nozzle to an exhaust pressure of 5×10^5 N/m^2. The exit area ratio is 1.2. Find the Mach number at exit.

16P-7 A well-insulated spherical air tank having diameter of 1.7 m is used to blow-down air. The tank was initially charged to the absolute pressure of 2.54 Mpa and tempertaure of 470 K. The tank is blowndown for 40 s during which 40 kg of gas leaves the tank. Find the air temperature in the tank after the blowdown event and also estimate the nozzle throat diameter.

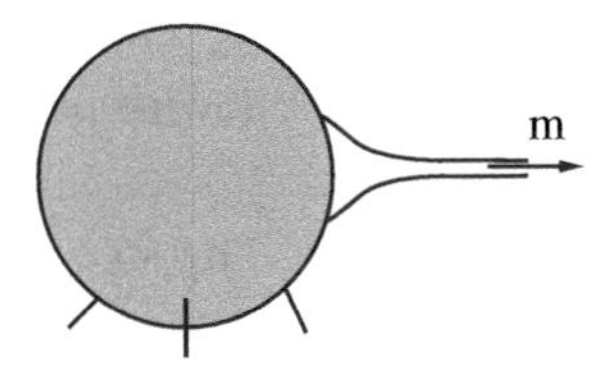

Figure 16.32 Figure of problem 16P7.

16P-8 An oblique shock has formed on a wedge, which caused a 15° deflection of flow. The stagnation pressure and temperature are 7×10^5 N/m^2 and 290° K. Find both strong and weak solutions, if M_1 is 3.

16P-9 A pitot tube is used to measure the speed in a supersonic wind tunnel. Find the upstream Mach number if the pressure recorded by pitot tube is eight times of the free-stream flow.

16P-10 A reentry vehicle is entering into Earth's atmosphere and a detached shock-wave has formed, causing heating and drag on the entry vehicle. Assuming perfect and ideal gas model for air flow around the vehicle ($k = 1.4$, $R = 287$ J/kg · K). Find, flow Mach number experienced by vehicle surface, the stagnation pressure, and velocity of flow between the shock and the vehicle. Ignore the heating effects, the ablation, the gas ionization/dissociation, and the friction on surface of the vehicle.

$$p_1 = 1000 \text{ Pa}, \quad T_1 = 214 \text{ K}, \quad T_{max} = 900 \text{ K}.$$

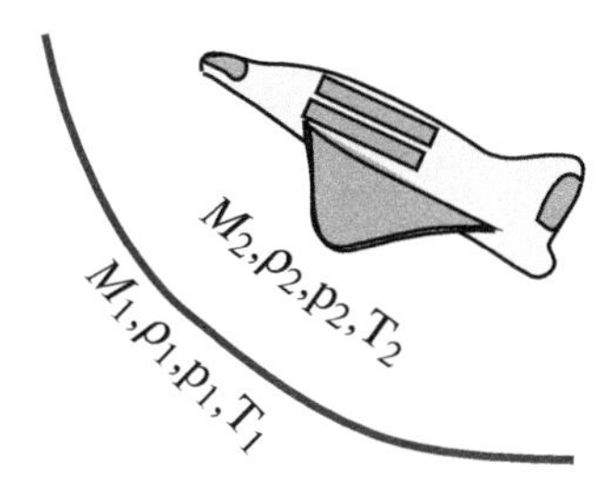

Figure 16.33 Figure of problem 16P10.

16P-11 The Van der Waal's model for gas is:

$$\left(p + \frac{\alpha}{\upsilon^2}\right)(\upsilon - \beta) = RT$$

where α and β are defined as:

$$\alpha = 3p_c v_c^2;\ \beta = \frac{1}{8}\left(\frac{RT_c}{p_c}\right);\ v_c = \frac{3}{8}\left(\frac{RT_c}{p_c}\right)$$

Prove that the acoustic speed of gas following this model is:

$$a = \sqrt{\left[\frac{RT}{1-\beta\rho} + \frac{\rho\beta RT}{(1-\beta\rho)^2} - 2\alpha\rho\right]}$$

16P-12 At inlet Mach number of 0.6, air enters a frictionless conduit having $C_f =$ 0.005. The inlet pressure and temperature are 170 kPa and 400° K. The conduit length is 0.43 m. Find the exit Mach number, pressure, and temperature.

REFERENCES

J. Bloomer, Practical fluid Mechanics for Engineering Applications, Marcel Dekker, USA, 2000.

F. Sharipov, Hypersonic flow of rarefied gas near the Brazilian satellite during its reentry into atmosphere, Brazilian J. Physics, vol. 33, no. 2, June, 2003.

17 Turbomachinery

Turbomachinery involves a collection of blades or buckets and flow passages that are arranged around an axis of rotation to form a rotor or an impeller. When fluid is used to rotate the rotor and shaft, power is produced; such a device is called a turbine. If electrical power is used to create the shaft work and subsequently energy is added to the fluid to raise its level, then such a device is a turbomachinery-based pump. In this chapter, the focus is placed on such rotating devices.

■ ■ ■ ■ ■ ■

Learning outcomes: After finishing this chapter, you should be able to:

- Formulate the Euler's turbomachinery equation.
- Know how to use specific speed in selection of turbines and pumps.
- Understand the cavitation phenomenon and its prevention.
- Understand the basic design principles for the energy extraction and energy adding devices.

Turbomachines are mechanical devices that either get energy from a fluid (turbine) or add energy to a liquid (pump)/gas (compressor) owing to dynamic interactions between the device and the fluid. While these devices' actual design and construction usually require significant understanding and effort, their primary operating principles are the same. Fans are one of the most common examples of turbomachinery devices, in which the air is pushed with a little pressure change across the blades. Cooling fans are not only used in industry, in aircraft, but also used quite commonly in household devices and even in desktop computers to cool the processors (see Figure 17.1).

Many turbomachines have some housing or casing surrounding the rotating blades or rotor, forming an internal flow passage through which the fluid flows. Not all the devices have enclosed the casings. Some, like windmills, wind turbines, and fans, have no casings around them. Some turbomachines include stationary blades or vanes besides rotor blades. These stationary vanes can be accelerated the flow and thus serve as nozzles. Or, these vanes can be set to diffuse the flow and act as diffusers.

17.1 DIMENSIONAL ANALYSIS AND RELEVANT PI-GROUPS

Important variables in turbomachines are: flow rate (Q), head (h), geometrical configuration, which can be represented by some characteristic diameter (D), other pertinent lengths, and surface roughness (ε), Pump shaft rotational speed (ω), fluid

DOI: 10.1201/9781003315117-17

Figure 17.1 (a) Cooling fan used in computers; (b) compressors are used in gas turbines.

viscosity (μ), and fluid density (ρ). The dependent variables are:

$$f(D, \ell, \varepsilon, Q, \omega, \mu, \rho)$$

Π groups are:

$$f\left(\frac{\ell}{D}, \frac{\varepsilon}{D}, \frac{Q}{\omega D^3}, \frac{\rho \omega D^2}{\mu}\right)$$

In turbomachinery literature these pi groups are identified by the following names:

$$\textbf{Head Coeff}\,(C_H) = \frac{g h_a}{\omega^2 D^2}$$

$$\textbf{Flow Coeff}\,(C_Q) = \frac{Q}{\omega D^3}$$

$$\textbf{Power Coeff}\,(C_P) = \frac{\dot{W}_{shaft}}{\rho \omega^3 D^5}$$

Experiments have revealed the following relationship between the pi-groups:

$$C_H = f(C_Q), C_P = f(C_Q), \eta = f(C_Q)$$

A useful term can be obtained by eliminating diameter D between the flow coefficient and the head rise coefficient. This is accomplished by raising the flow coefficient to an appropriate exponent and dividing this result by the head coefficient raised to another appropriate exponent:

$$N_s = \frac{\sqrt{C_Q}}{C_H^{3/4}} \tag{17.1}$$

For any pump it is customary to specify a value of specific speed at the flow coefficient corresponding to peak efficiency only. For pumps with low Q and high

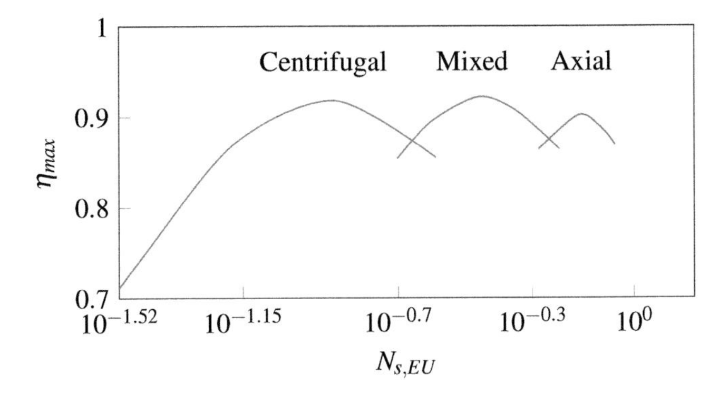

Figure 17.2 Maximum efficiency as a function of pump specific speed.

heads the specific speed is low compared to a pump with high Q and low heads. Centrifugal pumps typically are low-capacity, high head pumps, and therefore have low specific speeds.

In European countries, rotations per second (Hz) instead of rad/s units are used and specific speed is defined as:

$$N_{s,EU} = \frac{\omega_{Hz}\sqrt{Q_{m^3/s}}}{(gH_m)^{3/4}}$$

The subscripts in the above formula are the units.

The units used in the USA are a mixture of different systems, and specific speed is usually reported in the following units:

$$N_{s,US} = \frac{N_{rpm}\sqrt{Q_{gpm}}}{H_{ft}^{3/4}}$$

The subscripts in the above formula are the units; gpm is gallons per minute and rpm is revolutions per minute. Note that the value of gravitational acceleration is dropped in this formula. This shows that it is necessary for the buyer to check the units of specific speed from the pump manufacturers to ensure that the correct pump is selected.

17.2 CLASSIFICATION

Turbomachines are classified as axial-flow, mixed-flow, or radial-flow machines, depending on the predominant direction of the fluid motion relative to the rotor's axis as the fluid passes the blades.

1. For an axial-flow device, the fluid maintains a significant axial-flow direction component from the inlet to the outlet of the rotor. For a radial-flow machine,

Figure 17.3 (a) Windmill is one of an earliest type of axial flow turbine. (b) The modern wind turbines.

the flow across the blades involves a substantial radial-flow component at the rotor inlet, exit, or both. The windmill and wind turbines are axial flow turbomachinery devices (see Figure 17.3).

2. In mixed-flow machines, there may be significant radial- and axial-flow velocity components for the flow through the rotor row.
3. In radial-flow machines the fluid will be ejected radially outward.

17.3 EULER'S TURBOMACHINERY EQUATION

Starting from Reynolds transport theorem, we can develop the balance of moment of momentum as:

$$\frac{dN}{dt} = \int_{CS} \eta \rho \mathbf{V} \cdot dA + \frac{\partial}{\partial t} \int_{CV} \eta \rho d\vartheta \tag{17.2}$$

Now as $\eta = $ N/mass we have $\eta = 1$ (in case of mass).

$$\frac{dm}{dt} = \dot{m} = \int_{CS} \rho \mathbf{V} \cdot dA + \frac{\partial}{\partial t} \int_{CV} \rho d\vartheta \tag{17.3}$$

for steady incompressible flows:

$$\dot{m} = \int_{CS} \rho \mathbf{V} \cdot dA \tag{17.4}$$

The moment of momentum (**H**) is defined as:

$$(\mathbf{V} \times \mathbf{r}) \rho d\vartheta$$

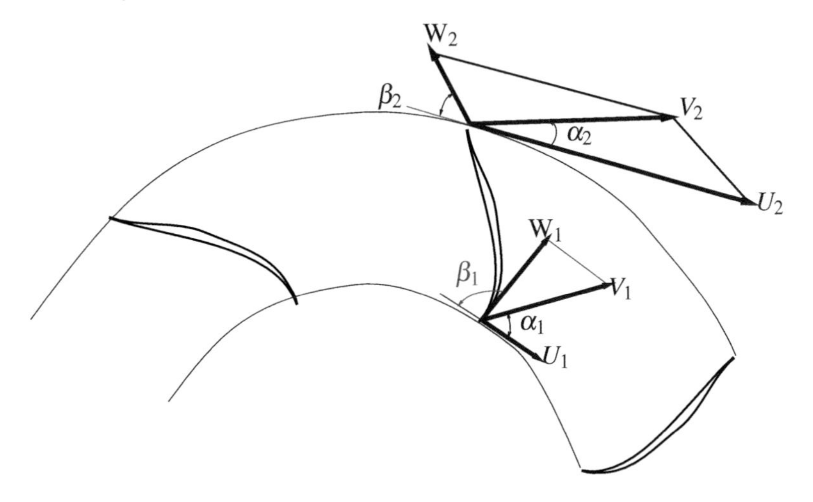

Figure 17.4 Velocity vector diagrams at inlet and outlet of an impeller.

where $\eta = H/m = \mathbf{V} \times \mathbf{r}$

$$\frac{dH}{dt} = \int_{CS} (\mathbf{V} \times \mathbf{r}) \rho \mathbf{V} \cdot dA + \frac{\partial}{\partial t} \int_{CV} (\mathbf{V} \times \mathbf{r}) \rho d\vartheta \tag{17.5}$$

for turbomachine rotors with steady-in-the-mean or steady (on average cyclical flow):

$$\frac{dH}{dt} = \int_{CS} (\mathbf{V} \times \mathbf{r}) \rho \mathbf{V} \cdot dA + \cancel{\frac{\partial}{\partial t} \int_{CV} (\mathbf{V} \times \mathbf{r}) \rho d\vartheta} \tag{17.6}$$

Also as $\mathbf{H}$ = Moment of Momentum and (momentum) $\times$ $\mathbf{r}$, we can rewrite the functionality as:

$$\frac{dH}{dt} = \mathbf{F} \times \mathbf{r}$$

$$\mathbf{F} \times \mathbf{r} = \int_{CS} (\mathbf{V} \times \mathbf{r}) \rho \mathbf{V} \cdot dA \tag{17.7}$$

The left-hand side of this equation represents the sum of the external torques acting on CV, and the right-hand side is the net rate of flow of moment-of-momentum (angular momentum) through CV.

Referring to Figure 17.4, we can formulated the one-dimensional simplified model for the flow through a turbomachine rotor with section 1 as inlet and 2 as outlet. Taking the sign convention that (+) is associated with mass flow rate into the control volume and the (−) as sign associated with the outflow from the control volume. In Figure 17.4, U is the blade velocity, V is the absolute fluid velocity that is seen by an observer sitting stationary and watching the blade movement. Further, W is the relative velocity that is observed by an observer who is riding on the blade.

$$T_{shaft} = -\dot{m}_1 (r_1 V_{\theta 1}) + \dot{m}_2 (r_2 V_{\theta 2}) \tag{17.8}$$

T_{shaft} is the shaft torque applied to the contents of the control volume. If V_θ and U are in the same direction, then V_θ is positive. If V_θ and U are in the opposite direction, then V_θ is negative. If T_{shaft} is in the same direction as rotation then it is positive, otherwise it is negative.

As shaft power $\dot{W}_{shaft}$ is related to shaft torque and angular velocity, one can write:

$$\dot{W}_{shaft} = -\dot{m}_1(U_1V_{\theta 1}) + \dot{m}_2(U_2V_{\theta 2}) \tag{17.9}$$

In terms of work per unit mass, $\dot{W}_{shaft}/\dot{m} = \dot{w}_{shaft}$:

$$\dot{w}_{shaft} = -(U_1V_{\theta 1}) + (U_2V_{\theta 2}) \tag{17.10}$$

These are the basic turbomachine equations whether the machines are radial, mixed, or axial-flow devices and for compressible and incompressible flows.

17.4 THE CENTRIFUGAL PUMP

One of the most widely used radial-flow turbomachines is the centrifugal pump. This type of pump has two major components: an impeller connected to a spinning shaft and a stationary casing, housing, or volute enclosing the impeller. The impeller is composed of several blades, frequently called vanes, organized in a regular pattern around the shaft. As the impeller rotates, flow is sucked in through the eye of the impeller and proceeds radially outward. Energy is added to the liquid by the rotating blades, and both pressure and absolute velocity are enhanced as the fluid progresses from the eye to the periphery of the blades. For the plainest type of centrifugal pump, the fluid flows straight into a volute-shaped casing. The casing shape reduces the velocity as the fluid leaves the impeller, and this reduction in kinetic energy is transformed into a gain in pressure.

The ideal or maximum head rise possible, h_i is found from:

$$\dot{w}_{shaft} = \rho g Q h_i$$

$$h_i = \frac{1}{g}(U_2V_{\theta 2} - U_1V_{\theta 1}) \tag{17.11}$$

h_i is the amount of energy per unit weight of fluid added to the fluid by the pump. The actual head rise realized by the fluid is less than the ideal amount by the head loss suffered. Often the fluid has no tangential component of velocity or swirl, as it enters the impeller; i.e., the angle between the absolute velocity and the tangential direction is 90°. In such a case the equation reduces to:

$$h_i = \frac{1}{g}(U_2V_{\theta 2}) \tag{17.12}$$

$$cot\beta_2 = \frac{U_2 - V_{\theta 2}}{V_{r2}}$$

$$h_i = \frac{1}{g}(U_2^2 - U_2V_{r2}cot\beta_2) \tag{17.13}$$

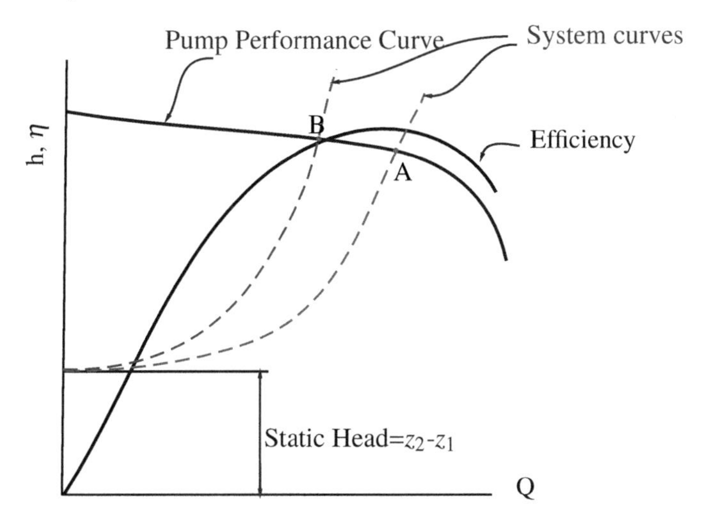

Figure 17.5 The pump characteristics curves.

$$Q = 2\pi r_2 b_2 V_{r2}$$

where b_2 is the impeller blade height at the radius r_2.

$$h_i = \frac{1}{g}\left(U_2^2 - U_2(\frac{Q}{2\pi r_2 b_2})cot\beta_2\right)$$

In actual pumps, angle β_1 lies in range: $15^\circ - 50^\circ$
In actual pumps, angle β_2 lies in range: $15^\circ - 35^\circ$ (nominally $20^\circ - 25^\circ$)
Blades with $\beta_1 < 90^\circ$ are called Backward curved. Blades with $\beta_1 > 90^\circ$ are called Forward curved. (usually not designed)

Eq. 17.13 can be cast into form

$$h_i = C_1 - C_2 Q$$

where $C_1 = U_2^2/g$ and C_2 is defined as:

$$C_2 = \frac{1}{g}\left[\frac{U_2}{2\pi r_2 b_2}cot\beta_2\right]$$

The constant C_1 is often called the shutoff head, as it the ideal head developed by the pump for zero flow rate.

17.5 PUMP CHARACTERISTIC CURVE

Figure 17.5 shows pump characteristic curve. The actual head rise h_a, gained by fluid flowing through a pump can be determined with an experimental arrangement (Pump Test Rig).

$$h_a = \frac{(p_2 - p_1)}{\gamma} + (z_2 - z_1) + \frac{(V_2^2 - V_1^2)}{2g}$$

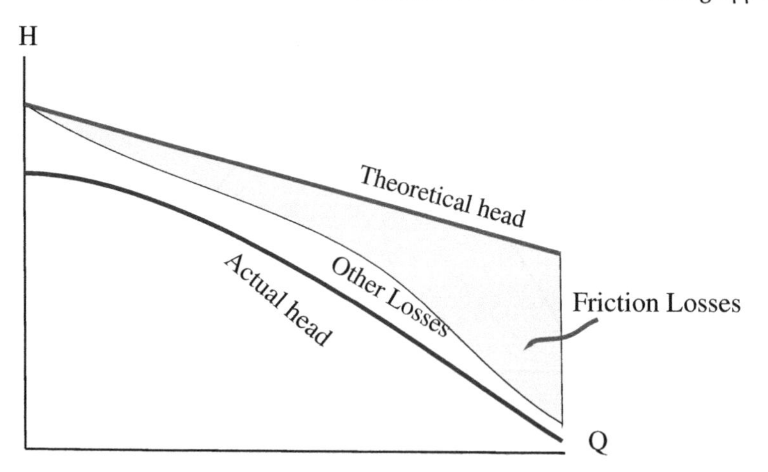

Figure 17.6 The actual head curve of pump will be different from the ideal head curve due to losses inside pump.

$$\text{Power gained by fluid} = \gamma Q h_a$$

Mechanical efficiency of a pump is defined as the ratio of useful mechanical power transferred to the flow by the pump to the shaft work, required to drive the pump.

$$\text{Efficiency} = \eta_p = \frac{\gamma Q h_a}{\dot{W}_{shaft}}$$

The overall pump efficiency is affected by the hydraulic losses in the pump, as previously discussed, and in addition, by the mechanical losses in the bearings and seals. There may also be some power loss due to leakage of the fluid between the back surface of the impeller hub plate and the casing, or through other pump components. This leakage contribution to the overall efficiency is called the volumetric loss. Thus, the overall efficiency arises from three sources, the hydraulic efficiency, $\eta = \eta_h \eta_m \eta_v$.

Mechanical losse arise due to friction in the bearings and the mechanical sealing.

A part of fluid will be entrapped between the housing and the impeller and would give rise to significant shear force which is classified as **Disc friction losses**.

There are volumetric losses due to leakage from the gaps between the impeller and the housing.

Pump-motor efficiency is defined as the ratio of useful mechanical power transmitted to the fluid over the electrical motor power required to drive the pump.

Figure 17.6 shows graphically the influence of losses on the pump performance.

Example 17.1

Example A centrifugal pump with no tangential velocity component at inlet is designed to deliver water at 80 cfm and has the following dimensions

Parameter	Inlet	Outlet
Radius, r [in]	3.7	11
Blade width, b [in]	0.46	0.3
Blade angle, β [degree]	20	45

Determine the design speed, the outlet absolute flow angle (α_2), the theoretical head developed (H) by the pump, and the minimum mechanical power delivered to the pump in horsepower.

Solution

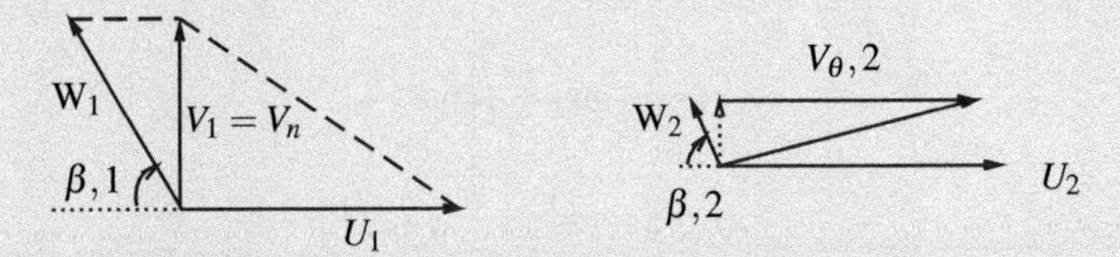

Figure 17.7 The inlet and out velocity triangles at impeller.

From Euler's turbomachinery equation we have

$$\dot{W}_{shaft} = (U_2V_{\theta,2} - U_1V_{\theta,1})\dot{m}$$

$$h = \frac{\dot{W}_{shaft}}{\dot{m}g} = \frac{1}{g}(U_2V_{\theta,2} - U_1V_{\theta,1})$$

From continuity we have

$$V_n = \frac{Q}{2\pi rb} = w\sin\beta$$

$$w = \frac{V_n}{\sin\beta}$$

$$\frac{V_{n,1}}{V_{n,2}} = \frac{A_2}{A_1} = \frac{r_2b_2}{r_1b_1}$$

The tangential velocity is

$$V_\theta = U - V_{rb}\cos\beta = U - \frac{V_n}{\sin\beta}\cos\beta$$

$$V_\theta = U - \frac{Q}{2\pi rb}\cot\beta$$

Rearranging and solving for ω we get

$$U_1 - \frac{Q}{2\pi r_1b_1}\cot\beta_1 = 0$$

$$\omega = \frac{Q}{2\pi r_1^2 b_1} \cot\beta_1$$

Using the data we have

$$\omega = 160.15 \frac{rad}{s}$$

The absolute velocities are

$$U_1 = \omega r_1 = 49.3 \frac{ft}{s}$$

$$U_2 = \omega r_2 = 146.8 \frac{ft}{s}$$

$$V_{n,2} = \frac{Q}{2\pi r_2 b_2} = 9.26 \frac{ft}{s}$$

$$V_{\theta,2} = U_2 - \frac{Q}{2\pi r_2 b_2} \cot\beta_2 = 137.53 \frac{ft}{s}$$

$$\alpha_2 = \tan^{-1} \frac{V_{\theta,2}}{V_{n,2}} = 1.503 rad = 86.189 (degrees)$$

The shaft work is

$$W_{shaft} = U_2 V_{\theta,2} \rho Q = 5.22 \times 10^4 \frac{ft.lb_f}{s} = 94 hp$$

The ideal head developed is

$$h = \frac{W_{shaft}}{\rho Q g} = 627 ft$$

17.6 REAL HEAD CURVE

17.6.1 SHIFT IN PUMP OPERATION CURVE

The system equation is $h_p = (z_2 - z_1) + KQ^2$. System equation shows how the actual head gained by the fluid from the pump is related to the system parameters. In this case, the parameters include the change in elevation head and the losses due to friction. To select a pump for a particular application, it is necessary to utilize both the system curve, as determined by the system equation, and the pump performance curve. If both curves are plotted on the same graph, their intersection point represents the operating point for the system. That is, this point gives the head and flow rate that satisfies both the system equation and the pump equation. We want the operating point to be near the best efficiency point for the pump. For a given pump, it is clear that as the system equation changes, the operating point will shift (fouling).

The pump efficiency can be calculated as:

$$\eta_{pump} = \frac{Q\rho g h_a}{Q\rho g h_i + \omega T_{friction}}$$

where $T_{friction}$ is the torque to overcome bearing, seal, and spin losses often assumed around 10% of the ideal torque in normal circumstances.

Example 17.2

Example In a water pump, the pump casing acts as a diffuser, which converts 65% of the absolute velocity head at the impeller outlet to static pressure rise. Assume the torque to overcome bearing, seal, and spin losses is 9% of the ideal torque at $Q = 0.073\ \mathrm{m^3/s}$. The head loss through the pump suction and discharge channel is 0.70 times the radial component of velocity head leaving the impeller. Find the pump efficiency if $\omega = 173$ rad/s and ideal head developed is 12.37 m. Take $V_n = 7.74$ m/s and $V_\theta = 9.35$ m/s at impeller exit.

Solution We can calculate the absolute velocity at the impeller outlet V_2 as:

$$V_2 = \sqrt{V_{n2}^2 + V_{\theta,2}^2}$$

$$V_2 = 12.14 m/s$$

The actual head is

$$h_a = \frac{65}{100} \cdot \frac{V_2^2}{2 \cdot g} - \frac{0.7 \cdot V_{n2}{}^2}{2 \cdot g} = 2.21m$$

$$Torque_{friction} = \frac{Q \cdot \rho \cdot g \cdot h_i}{\omega} \cdot \%loss = \frac{9}{100}\left(\frac{0.073 \times 1000 \times 9.81 \times 12.37}{173}\right)$$

$$= 4.61\ N \cdot m$$

$$\eta_{pump} = \frac{Q\rho g h_a}{Q\rho g h_i + \omega.Torque_{friction}} = 16.44\%$$

17.7 PUMP AFFINITY LAWS

With pump scaling/affinity laws it is possible to experimentally determine the performance characteristics of one pump in the laboratory and then use these data to predict the corresponding characteristics for other pumps within the family under different operating conditions.

How a change in the operating speed, for a given pump, affects pump characteristics?
Same flow coefficient gives $\frac{Q_1}{Q_2} = \frac{\omega_1}{\omega_2}$ For a given pump operating at a given flow coefficient, the flow varies directly with speed, the head varies as the speed squared, and the power varies as the speed cubed. Useful in estimating the effect of changing pump speed when data are available from a pump test obtained by operating the pump at a particular speed.

How a change in the impeller diameter, D, of a geometrically similar family of pumps, operating at a given speed, affects pump characteristics?
For the same flow coefficient with $\omega_1 = \omega_2 : \frac{Q_1}{Q_2} = \frac{D_1^3}{D_2^3}$ For a family of geometrically similar pumps operating at a given speed and the same flow coefficient, the flow varies as the diameter cubed, the head varies as the diameter squared, and the power varies as the diameter raised to the fifth power.

Example 17.3

Example The characteristic curve for a centrifugal pump as given by a manufacturer can be written as:

$$H = 30 - 150 \cdot Q - 3200\,Q^2$$

Find the new equation for characteristic curve if the pump is being operated at 72% of it's design speed.

Solution The same pump is used in both the cases so impeller diameter is same. Also $\omega = 2\pi\,N$, so we cast the dimensionless parameters into N instead of ω. Since the speed of the impeller is reduced to new value of 72%, the flow rate and head will be changed.
We equate the head coefficient:

$$C_H = \frac{gH}{N^2 D^2}$$

$$\frac{gH_1}{{N_1}^2 {D_1}^2} = \frac{gH_2}{{N_2}^2 {D_2}^2}$$

$$H_2 = \left(\frac{N_2 D_2}{N_1 D_1}\right)^2 H_1$$

$$H_2 = \left(\frac{N_2}{N_1}\right)^2 H_1$$

$$H_2 = \left(\frac{0.72 N_1}{N_1}\right)^2 H_1 = 0.5184 H_1$$

$$H_1 = \frac{H_2}{0.5184}$$

From flow coefficient we have:

$$C_Q = \frac{Q}{ND^3}$$

$$\frac{Q_1}{N_1 D_1{}^3} = \frac{Q_2}{N_2 D_2{}^3}$$

$$Q_1 = \frac{Q_2}{0.72}$$

We substitute the new head and flow rate into the equation and we get the new characteristic curve equation as:

$$H_2 = 15.55 - 108Q_2 - 3200Q_2{}^2$$

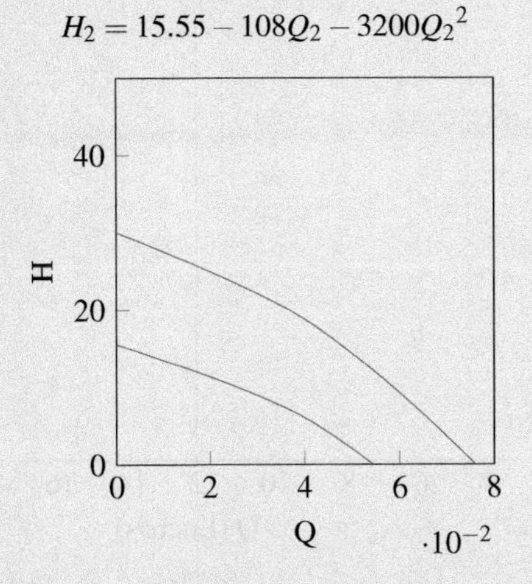

Figure 17.8 The pump's new characteristic curve is shown as bottom curve.

Pump characteristics chart from the manufacturers contains more information then what has been presented in the previous plots. A typical pump characteristics chart for two stage pump is shown in Figure 17.9. The manufacturer may combine all the impeller diameters which they are manufacturing for a given casing. Also for ready reference the lines of power and same efficiency will be made available, in case working conditions are changed. Thus, several important lines are incorporated into one single graph as shown in Figure 17.9.

17.8 THE PHENOMENON OF CAVITATION

Cavitation: The local boiling in liquids due to a sudden drop in pressure is called the cavitation.

Cavitation in a liquid can happens if the pressure is lowered at an approximately isothermal conditions. The phenomenon can be depicted by a process line in the thermodynamic phase diagram (see Figure 17.10).

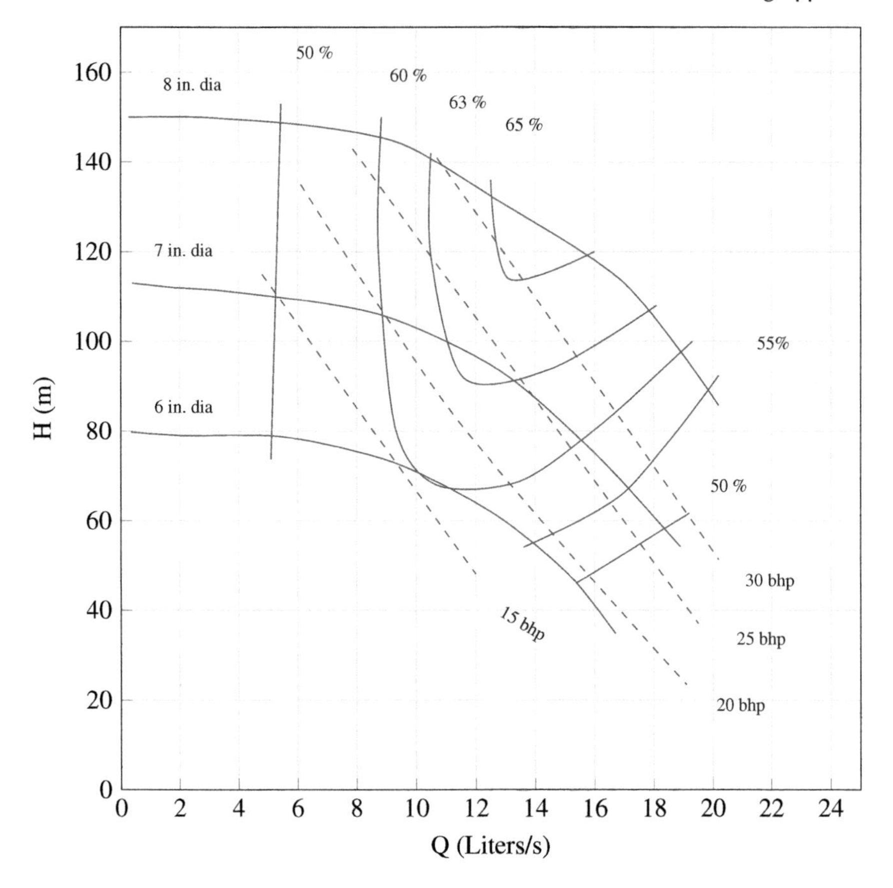

Figure 17.9 Performance curves for a two-stage centrifugal pump.

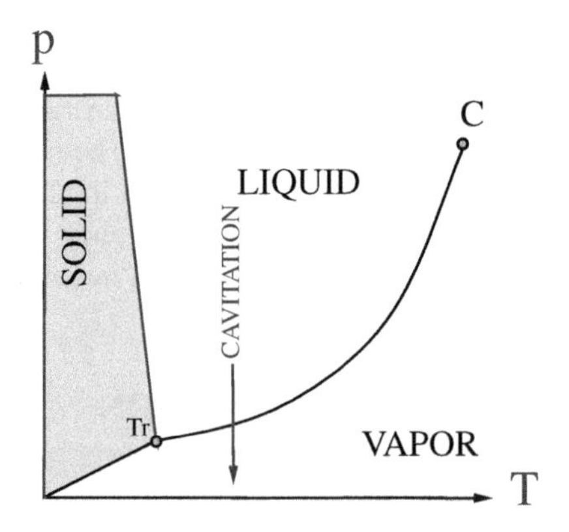

Figure 17.10 The phase diagram for understanding the cavitation phenomenon.

The cavitation phenomenon may happen at the pumping sequence of cryogenic liquids in rocket engines, flows through venturi nozzles, hydraulic valves, at the propeller blades of submarines and torpedo, etc. Cavitation as a phenomenon was first identified by Torricelli, and later by Leonard Euler. Sir Charles Parsons, in 1893, developed the fist cavitation tunnel and investigated the undesirable effect of cavitation on the performance of a ship's propellers.

Cavitation is an undesirable phenomenon in some cases, as when the bubbles formed reached a region of increased pressure, they violently collapse and thus create an impact on the blades and fluid devices that may even cause erosion of material. Thus cavitation may cause erosion, noise, and vibrations in the system simultaneously and should be prevented. The over pressures due to the implosion of bubbles can reach several thousand bars. The duration of collapse of a 1-cm radius spherical vapor bubble in water under an external pressure of one bar is approximately one millisecond (Franc and Michel, 2005).

In certain other applications, cavitation is desired and the phenomenon is exploited to be used for the cleaning of surfaces by ultrasonics. The cavitating jets are also used for the dispersion of particles in a liquid medium, for the production of emulsions and the electrolytic deposition.

An interesting phenomenon is acoustic cavitation in which the cavitation can be initiated due to oscillating pressure field applied over the free surface of a liquid.

Note that to the naked eye, the cavitation appears like a fuzzy white cloud.

Figure 17.11 shows the cavitation that happened over the propeller blades. The region downstream propeller shows a murky region as droplets filled the area. However, the stroboscopic photograph reveal the inception of bubbles starts from the blades.

Between 1923 and 1925, German engineer Dietrich Thoma (1881–1943) introduced the dimensionless parameter called the cavitation number. The cavitation number or cavitation index is defined using dynamical parameters as:

$$\sigma_v = \frac{P_{inlet} - P_v(T)}{\rho V_o^2}$$

where V_o is the velocity at the periphery of the runner or impeller, $P_v(T)$ is the vapor pressure which is a function of temperature.

For submerged jet the cavitation number is defined as:

$$\sigma_{v,jet} = \frac{P_e - P_v(T)}{\rho V_e^2}$$

where P_e and V_e are pressure and velocity inside the jet's core.

For pumps and turbines, the cavitation number is:

$$\sigma_{v,pump} = \frac{P_a - P_v(T) - \gamma H_s}{\rho V_o^2} = \frac{P_a - P_v(T) - \gamma H_s}{\gamma H}$$

The lesser the value of H_s, the higher the value of the $\sigma_{v,pump}$ would be. This parameter is also called cavitation parameter, or Thoma cavitation number. It is sometimes defined as follows and called the cavitation index:

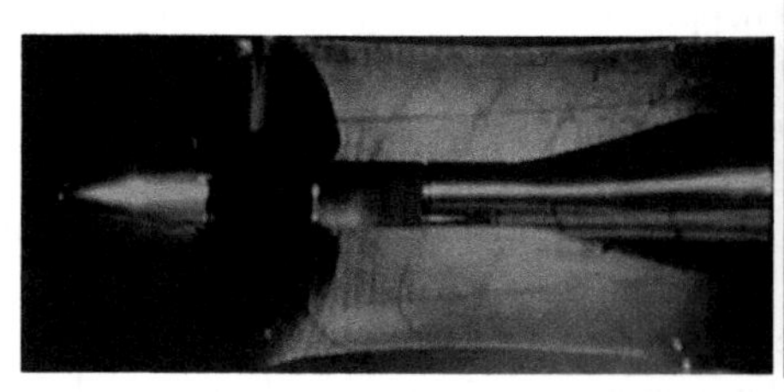
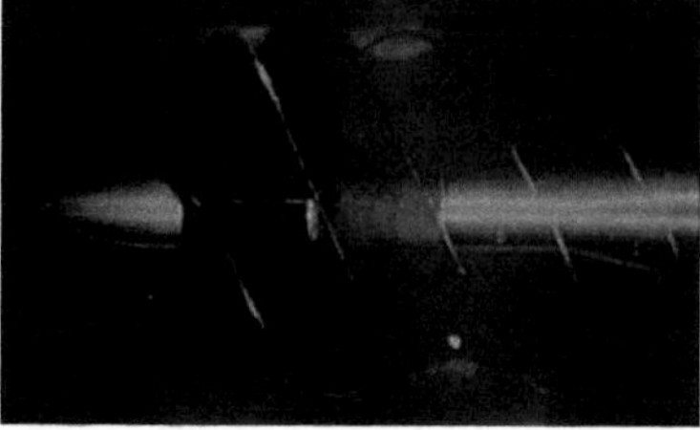

Figure 17.11 (Left) Cavitation phenomenon behind the propeller model as water passed over it at constant speed. (Right) Cavitation as visualized by stroboscopic photography. (*Picture from Ascher Shapiro's National Committee for Fluid Mechanics Films (NCFMF); Education Development Center Inc.; Author: Phillip Eisenberg; Source: Int. Asso. for Hydro-Environment Engg. and Research*)

$$\sigma_c = \frac{P_{ref} - P_v(T)}{\rho V_o^2/2}$$

where P_{ref} is the reference pressure.

Figure 17.12 schematically shows the different bubble sizes encountered in cavitation and corresponding cavitation index. Following cavitation regimes are based on the cavitation index (see Figure 17.12):

$\sigma_c = 1.8$ *Incipient Cavitation*
$0.3 < \sigma_c < 1.8$ *Developed Cavitation*
$\sigma_c < 0.3$ *Supercavitation*

Example 17.4

Example Water with density 997 kg/m^3 is flowing at speed of 30 m/s over an object. At a certain section the temperature is 10°C with gauge pressure 9.8 kPa and atmospheric pressure was 101 kPa. Calculate the cavitation index.

Solution At the temperature of 10°C the saturation vapor pressure $P_g = P_v =$ 1.228 kPa

$$\sigma_c = \frac{P_{ref} - P_v(T)}{\rho V_o^2/2}$$

$$\sigma_c = \frac{1000 \times [(101 + 9.8) - 1.223]}{\frac{997(30)^2}{2}} = 0.244$$

Note that we have used the absolute thermodynamic pressure value for the reference pressure.

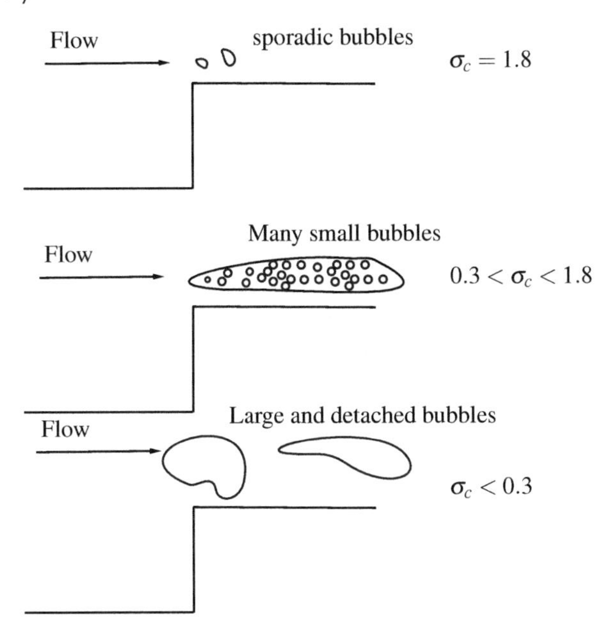

Figure 17.12 Cavitation index and relative bubble sizes for cases of no cavitation, incipient cavitation, and super cavitation.

Example 17.5

Example Tests on a pump model through pump test stand indicate a σ_v for pump is 0.10. A homologous unit is to be installed at a location where absolute pressure will be 13 psi and the water vapor pressure can be taken as 0.50 psi. The pump is expected to create water head of 80 ft. What is the maximum permissible suction head at the real location?

Solution

$$\sigma_{v,pump} = \frac{P_a - P_v(T) - \gamma H_s}{\rho V_o^2} = \frac{P_a - P_v(T) - \gamma H_s}{\gamma H}$$

$$H_s = \frac{P_a - P_v(T)}{\gamma} - H\sigma_{v,pmp}$$

$$H_s = \frac{(13 - 0.5)psi \times 144}{63lb_f/ft^3} - 0.1(80ft) = 20.57ft$$

In order to avoid cavitation, the real suction head at site must be less than 20.57 ft. For cavitation-less performance, the impeller must be fixed in a way that the suction setting for H_s results in a cavitation index greater than that of $\sigma_c = 0.1$.

17.9 NET POSITIVE SUCTION HEAD (NPSH)

In order to avoid cavitation, typically a threshold value of pressure is set in fluid systems.

Low pressures are commonly encountered on the suction side of a pump. Cavitation occurs when the liquid pressure at a location is reduced to the vapor pressure of the liquid. Cavitation phenomenon can cause a loss in efficiency and structural damage to the pump. There are actually two values of $NPSH$ of interest. $NPSH_r$ (required, that must be maintained, or exceeded, so that cavitation will not occur.) and $NPSH_a$ (available, which represents the head that actually occurs for the particular flow system, can be determined experimentally, or calculated if the system parameters are known). Pumps are tested to determine the value for $NPSH_r$, by either directly detecting cavitation, or by observing a change in the head-flow-rate curve (provided by manufacturer). To characterize the potential for cavitation, the difference between the total head on the suction side, near the pump impeller inlet:

$$NPSH_a = H_{sp} \pm H_s - h_f - H_{vp}$$

where

$NPSH_a$ Net Positive Suction Head,

H_{sp} Static pressure head (absolute) applied to the fluid,

H_s Elevation difference from the level of fluid in the reservoir to the pump inlet. It is positive if pump is below reservoir, otherwise it is set as negative,

h_f Head loss due to friction and minor losses,

H_{vp} Vapor pressure in terms of head.

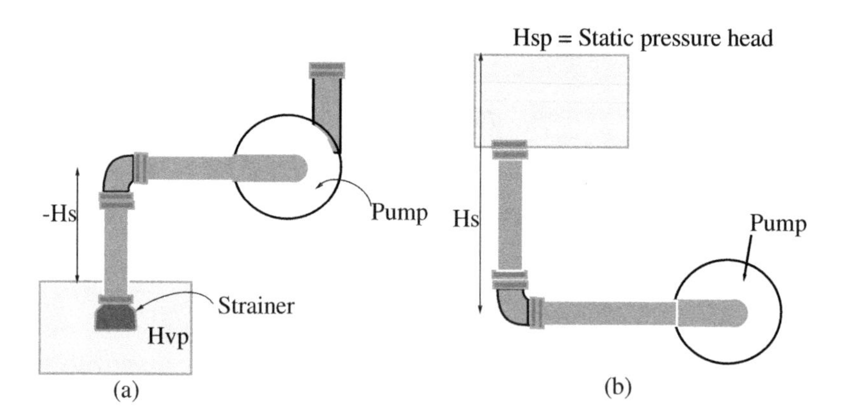

Figure 17.13 Pump suction line details if (a) tank is below pump or (b) tank is above pump.

For this calculation, absolute pressures are normally used since the vapor pressure is usually specified as an absolute pressure. For proper pump operation it is necessary that

$$NPSH_a \geq NPSH_r$$

Net Positive Suction Head required ($NPSH_r$) is supplied by the pump manufacturer.

Note that the negative *NPSH* mean that the fluid pressure has dropped below it's vapor pressure. In such a case, the piping layout and pressure conditions could cause the fluid boiling, and it is advisable to redesign such a system.

Figure Example 17.6

Example Determine the available *NPSH* for the system shown in the figure below handling flow rate of 33.6 L/min and comment on the viability of this system. The water ($\nu = 4.11\text{E-}07\ \text{m}^2/\text{s}$) level in the tank is 3 m above the pump inlet. The pipe is a commercial steel light rust ($\varepsilon = 0.1$ mm) Schedule 80 pipe, ID 1 in. (2.43 cm) with a total length of 15.0 m. The elbow is standard and the valve is a fully open globe valve. The fluid reservoir is a closed tank with a pressure of -30 kPa above water at 60°C. The atmospheric pressure is 101 kPa. Assume sharp-edged entrance into the reservoir. According to pump manufacturer, *NPSH* required for this particular pump is 3 m.

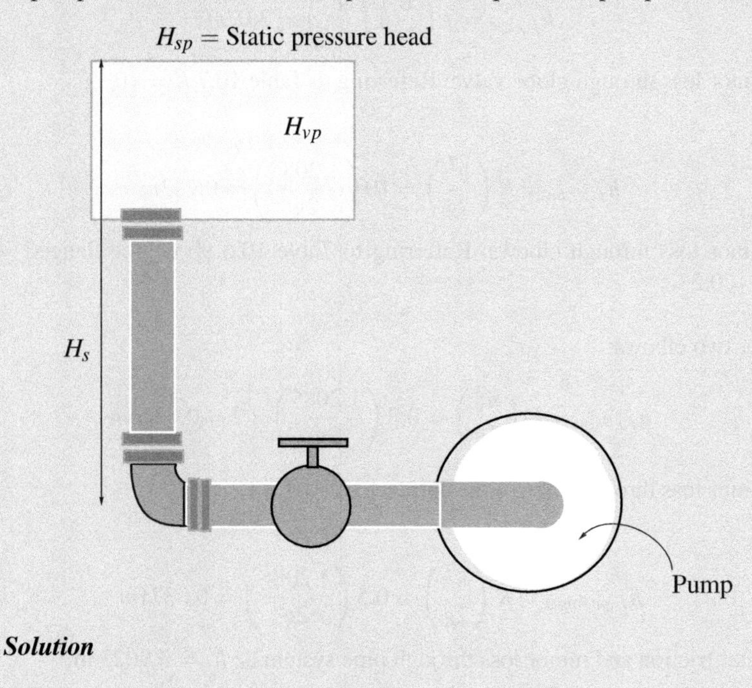

Solution

$$NPSH_a = H_{sp} \pm H_s - h_f - H_{vp}$$

$$H_{sp} = \frac{P_{abs}}{\gamma} = \frac{P_{atm} - P_{tank}}{\gamma} = \frac{(101-30)10^3}{9810} = 7.23m$$

$$H_s = +3m$$

$$H_{vp} = \frac{P_{sat@70C}}{\gamma} = \frac{31202}{9810} = 3.18m$$

$$h_f = h_{f,pipe} + \sum h_{minor}$$

$$\overline{u} = \frac{Q}{A} = \frac{0.00056}{\frac{\pi}{4}D^2} = 1.208m/s$$

$$\frac{\varepsilon}{D} \approx \frac{0.1}{2.43} = 0.0411$$

$$\mathrm{Re} = \frac{\overline{u}D}{\nu} = 7.114 \times 10^4$$

The Moody diagram gives the friction factor $f = 0.067$.

$$h_{f,pipe} = f\left(\frac{L}{D}\right)\left(\frac{\overline{u}^2}{2g}\right) = 3.07m$$

Minor loss through globe valve: Referring to Table 10.7 $K = 10$

$$h_{f,globe} = K\left(\frac{\overline{u}^2}{2g}\right) = 10\left(\frac{1.208^2}{2g}\right) = 0.743m$$

Minor loss through elbows: Referring to Table 10.6 90 degree flanged $K = 0.3$

For two elbows:

$$h_{f,elbow} = K\left(\frac{\overline{u}^2}{2g}\right) = 0.3\left(\frac{1.208^2}{2g}\right) \times 2 = 0.0446m$$

Minor loss through sharp edge entrance: $K = 0.5$

$$h_{f,entrance} = K\left(\frac{\overline{u}^2}{2g}\right) = 0.5\left(\frac{1.208^2}{2g}\right) = 0.0371m$$

Total friction and minor loss through pipe system is: $h_f = 3.9023$ m.

$$NPSH_a = H_{sp} \pm H_s - h_f - H_{vp}$$

$$NPSH_a = 7.23 + 3 - 3.9023 - 3.18 = 3.1476m$$

As *NPSH* available is positive and greater than *NPSH* required of 3 m, this is a viable system design.

The specific speed of the pumps is also a handy tool for checking the possibilities of cavitation:

$$N_s = \frac{{C_Q}^{1/2}}{{C_H}^{3/4}} = \frac{N\sqrt{Q}}{(gH)^{3/4}}$$

$$N_s < 0.12 \ \textit{No cavitation}$$

$$0.4 < N_s < 0.7 \ \textit{incipient caviation}$$

$$N_s > 0.7 \ \textit{cavitation possible}$$

Example 17.7

Example The total head at the best efficiency point for the pump handling 60 m^3/hr is 7 m. The impeller speed is 3,000 rpm. Check the possibilities of cavitation.

Solution The specific speed is defined:

$$N_s = \frac{{C_Q}^{1/2}}{{C_H}^{3/4}} = \frac{N\sqrt{Q}}{(gH)^{3/4}} = 0.27$$

Since $0.4 < N_s < 0.7$ *incipient caviation*, this is a case of incipient cavitation.

17.10 HYDRAULIC TURBINES

We can classify the turbines as either impulse or reaction turbines.

The specific speed is defined as:

$$N_{s,turbine} = \frac{N_{rpm}\sqrt{P_{kW}}}{(H_m)^{5/4}}$$

where *N* is speed in rpm, *P* is power in kW and *H* is head in meter. Table 17.1 lists some well-known water turbines and Figure 17.14 shows plot of turbine efficiencies versus the ratio of Q/Q_{max}, where Q_{max} is the maximum mass flow rate that can be handled by the turbine.

Table 17.1
Fact Sheet on Some Well-Known Water Turbines

Turbine	Ns	Efficiency (%)	Flow rate (m^3/s)	Vanes/buckets	Head (m)
Pelton		85-95	Low	20-40	300-800
Single jet	10 - 35				
2 Jets Pelton	10 - 45				
3 Jets Pelton	10 - 55				
4 Jets Pelton	10 - 70				
6 Jets Pelton	10 - 80				
Turgo	20 -80				
Francis	70 -500	Above 90	Medium	16-24	100-300
Propeller	600 - 900				
Kaplan	350 - 1100	90 - 93	Large	4-8	Below 100
Banki	20 - 80				

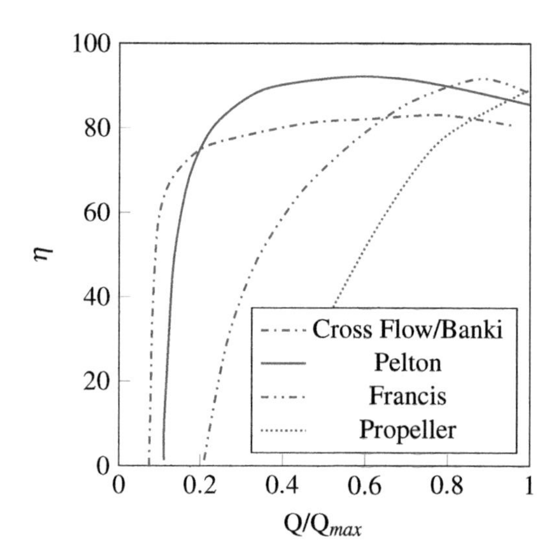

Figure 17.14 Turbine efficiencies.

Example 17.8

Example A proposed site for hydro-power development has a net head of 250 ft and design discharge of water flow of 580 cusecs. A 14 poles, 91%

efficient generator with 60 Hz frequency is planned to be used. What type of turbine would you recommend?

Solution We calculate the speed of the generator:

$$N = \frac{120f}{poles} = \frac{120(60Hz)}{14} = 514rpm$$

Power in hp is

$$P = \frac{\gamma QH\eta}{550} = \frac{62.43 \times 580 \times 250 \times 0.91}{550} = 14,970hp$$

converting it in SI units, it is 11.16 MW

$$N_{s,turbine} = \frac{N_{rpm}\sqrt{P_{hp}}}{(H_{ft})^{5/4}} = 63.2$$

and in the units of rpm, kW, m we have

$$N_{s,turbine} = \frac{N_{rpm}\sqrt{P_{kW}}}{(H_m)^{5/4}} = 76.5$$

Table 17.1 recommends that for such flow and head conditions the Francis turbine is suitable.

17.10.1 IMPULSE TURBINES

The Pelton's wheel is an example of pure impulse turbine in which all the head available is converted into an impulse via a nozzle and the jet of air is directed at the wheel of buckets. From Euler's turbomachinery equation:

$$T_{shaft} = \dot{m}(r_2V_2 - r_1V_1)$$

Approximating radius at inlet and outlet of bucket is same:

$$T_{shaft} = r\dot{m}(V_2 - V_1)$$

where V_2 is velocity at exit of the bucket and V_1 is the velocity at inlet of the bucket.

The velocity entering the control volume is V_j, and the exit velocity is

$$V_2 = (\omega r) - (V_j - \omega r)cos(\zeta)$$

where $\zeta = 180 - \beta$.

$$T_{shaft} = r\dot{m}\left[(\omega r) - (V_j - \omega r)cos\zeta - V_j\right]$$

$$T_{shaft} = -r\rho Q(1 + \cos\zeta)\left[V_j - \omega r\right]$$

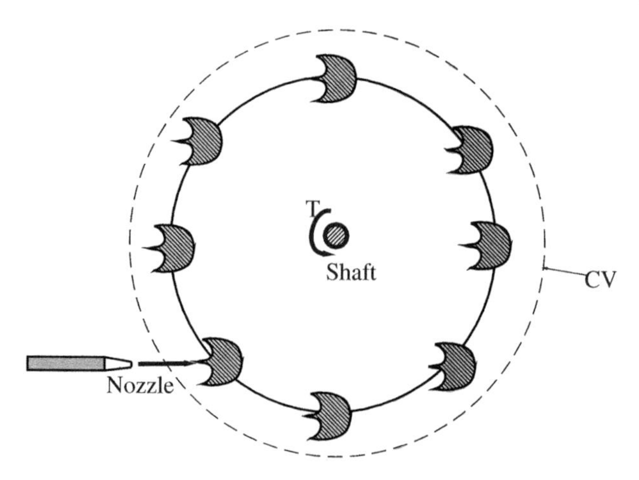

Figure 17.15 Control volume around Pelton wheel for moment of momentum derivation.

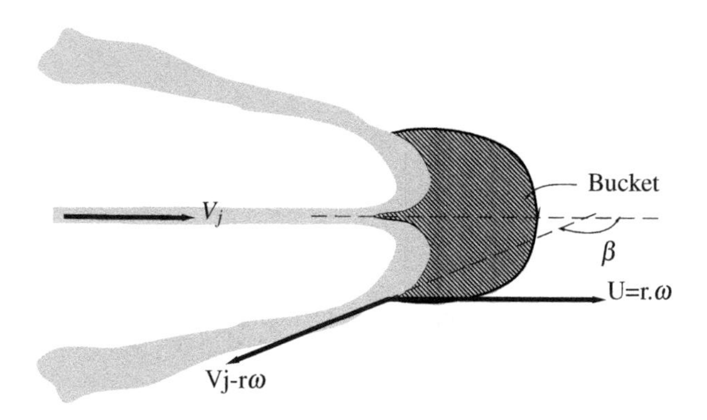

Figure 17.16 Velocity diagram of flow out of a Pelton wheel's bucket.

Since

$$\cos(\pi - \beta) = -\cos(\beta)$$

we can cast the equation as

$$T_{shaft} = -r\rho Q(1 - \cos\beta)\left[V_j - \omega r\right]$$

This is the torque developed by the wheel on CV. The opposite sign will give us the torque developed by the water on the wheel and thus

$$T_{shaft} = +r\rho Q(1 - \cos\beta)\left[V_j - \omega r\right]$$

We now introduce k, a coefficient accounting for loss of velocity moving across vanes or buckets.

$$T_{shaft} = +r\rho Q(1 - k\cos\beta)\left[V_j - \omega r\right]$$

Table 17.2
φ Factor Suggested for Different Specific Speeds

Ns (ft,hp,rpm)	2	3	4	5	6	7
φ	0.47	0.46	0.45	0.44	0.433	0.425

If bucket angle 180 degrees is selected then though theoretically it gives maximum power for a single bucket, with many buckets the water exiting from one bucket would strike the back side of neighboring buckets and thus the bucket angle β varies in range of 160 to 176 degrees. There are losses in the bucket and the most economical speed is suggested to be somewhat less than half of a jet's speed. In practice, a needle valve is used that controls the jet discharge by changing its area. The jet's velocity is calculated as:

$$V_j = C_v\sqrt{2gH_a}$$

where C_v is the discharge coefficient. The peripheral speed of the Pelton wheel can be calculated as:

$$U = \frac{D}{2}\omega$$

where D is the diameter of the wheel and ω is the rotation rate of the wheel.

The optimum peripheral speed also depends on the specific speed of the turbine and is expressed as:

$$U_{opt} = \varphi\sqrt{2gH_a}$$

where φ is a factor depending on specific speed of the turbine, tabulated in Table 17.2, and H_a is the head available.

If the diameter of the jet is d and the diameter of the Pelton wheel is D, from the center of the wheel to the center circle of the buckets it has been found through practice and experience that for the maximum efficiency the diameter ratio D/d should be about 54/N*s*. The specific speed is defined as:

$$N_s = \frac{N\sqrt{P}}{H^{5/4}}$$

where P is in horsepower, H is in feet, and N is in revolutions per minute.

It is found that for a single nozzle arrangement it is most efficient if the specific speed range is from 2 to 6.

Size of wheel: Doland (1954) gives the following equation for determining the size of Pelton wheels in inches:

$$D = 830\left(\frac{\sqrt{H_{ft}}}{N_{rpm}}\right)$$

Number of nozzles: If we increase the number of buckets on the wheel then it will be a huge structure and would create many structural challenges. Therefore, the flow rates are usually divided over the wheel by using many jets. The diameter of nozzle for this case is:

$$d_{nozzle} = \frac{0.54\sqrt{Q/n_{nozzle}}}{H^{5/4}}$$

where n_{nozzle} is the number of nozzles.

For horizontal axis runner the maximum jets used are two. However, for vertical axis wheel up to six jets can be used.

Number of buckets: A simple formula for number of buckets is:

$$N_{buckets} = \frac{1}{2}\left(\frac{D}{d}\right) + 15$$

where D is the wheel diameter, d is the nozzle diameter.

Bucket width: The bucket width should be at least three times the jet diameter.

Example 17.9

Example A Pelton wheel is installed to drive a generator for 60-cycle power. The head is 90 m (295 ft), and the discharge 0.040 m^3/s. Determine the diameter of the wheel and the speed of the wheel. The nozzle discharge coefficient is 0.9 and wheel efficiency is 0.75.

Solution The power in watts is:

$$P = \gamma Q H \eta = 9810 \times 0.04 \times 800 \times 0.75 = 26487 watts$$

converting it into horsepower power is 35.5 hp.
The head is 90 m (295 ft).
We assume that specific speed is 3.

$$N_s = \frac{N\sqrt{P}}{Q^{5/4}} = 3$$

from this we have roughly $N = 628$ rpm.
For synchronous coupling with generator the speed can be

$$N = \frac{120f}{poles}$$

in practice mostly even number of poles are used. So assuming **12** poles then $N = 600$ rpm. Using this as N we recalculate specific speed which is 2.86.
The peripheral speed is

$$U_{opt} = \varphi\sqrt{2gH} = 0.46\sqrt{2gH} = 19.32 m/s$$

$$\omega = \frac{2\pi N}{60} = \frac{2\pi(600)}{60} = 62.8 rad/s$$

$$U = \left(\frac{D}{2}\right)\omega$$

$$D = 2U/\omega = 0.615m$$

$$V_j = C_v\sqrt{2gH} = 37.8m/s$$

Since

$$A = \frac{Q}{V_j} = \frac{\pi}{4}d^2$$

we can write

$$d = \sqrt{\frac{4A}{\pi}} = \sqrt{\frac{4Q}{\pi V_j}} = 0.0367m$$

This gives $D/d = 16.77$. For best efficiency $D/d = 54/N_s$ is recommended. The data of this problem gives $D/d = 54/N_s = 18.85$
As the two values are very close, we consider this a solution.
Note: Using Doland formula:

$$D = 830\left(\frac{\sqrt{H_{ft}}}{N_{rpm}}\right)$$

$D = 22.6$ (inch) $= 0.57$ m is also close to this calculation. The bucket formula recommended 23 buckets on this wheel.

17.10.2 REACTION TURBINES

The large potential head is not always available and it is advisable to design turbines that can utilize the small heads. Such turbines in which the velocity changes within the blades are classifed as reaction turbines. Kaplan is one of the most widely used pure reaction turbine, shown in Figure 17.17, whereas, Francis turbine is a mixed type of turbine, shown in Figure 17.18. A reaction turbine is one in which flow takes place in a closed space under pressure. Generally, there are two types of reaction turbines: the Francis turbine and the axial-flow (or propeller) turbine. The flow through a reaction turbine may be radially inward, axial, or mixed (partially radial and partially axial).

To operate effectively, reaction turbines must have a submerged discharge. After passing through the runner, the water enters the draft tube, which directs the water to the point of release. The draft tube serves two functions: first, it enables the turbine to be set above tailwater level without losing any head. This helps produce reduced pressure at the upper end of the draft tube, limiting the height above tailwater at which the turbine runner can be set. Second, it helps reduce the head loss at

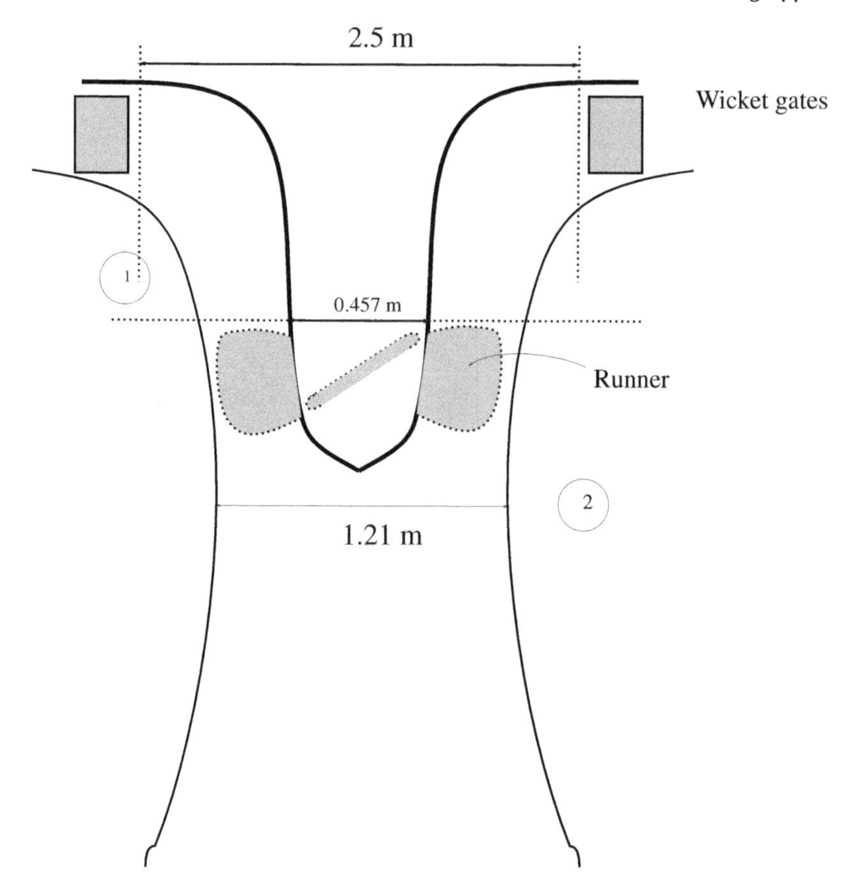

Figure 17.17 The Kaplan turbine with draft tube.

submerged discharge to increase the net head available to the turbine runner, which is accomplished by using a gradually diverging draft tube.

The Francis turbine is named after James B. Francis (1815–1892) was a famous American hydraulic engineer who, in 1849, designed and tested the first efficient inward-flow turbine. Water enters the scroll case in the typical Francis turbine and moves into the runner through a series of guide vanes with contracting passages that convert pressure head to velocity head. The vanes are called the wicket gates, which can be adjusted from time to time to control the amount and direction of the incoming liquid. They are usually operated by moving a shifting ring to which each gate is attached. The constant rotative speed of the runner under varying loads is attained by using governor that actuates a mechanism that adjusts the openings of gate. A relief valve or a surge tank is most of time necessary to avoid undesireable water hammering issues.

R_o is the radius till guide vanes from the axis of turbines

α_1 is the angle of the guide vanes at inlet

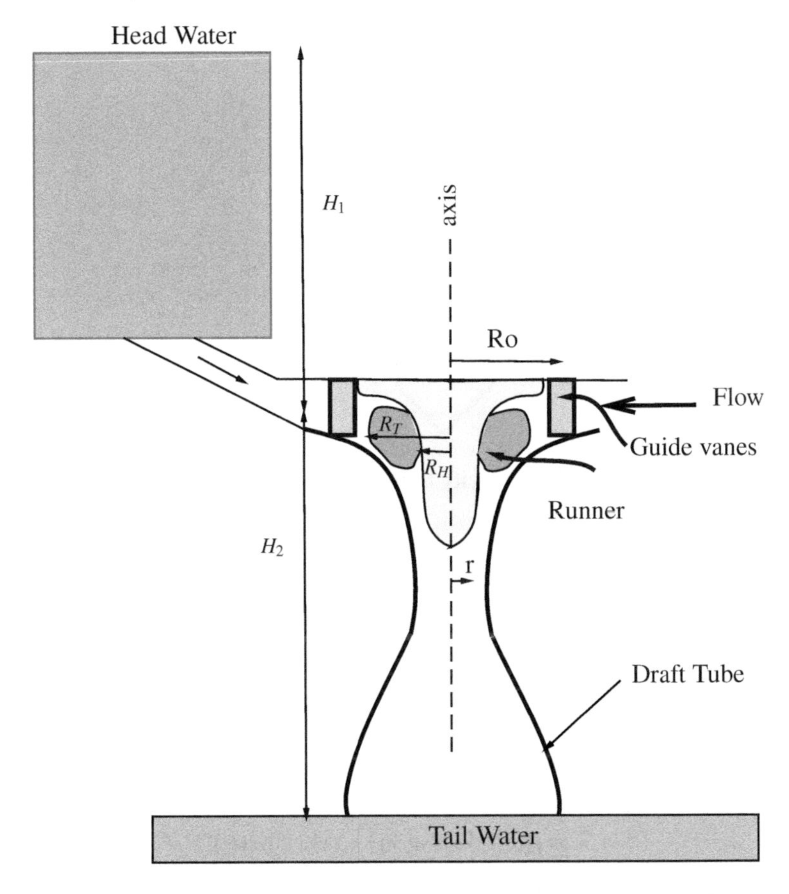

Figure 17.18 The Francis turbine.

α_2 is the angle of the guide vanes at outlet

V_o is the inlet speed of water entering through guide vanes

R_H is the radius of runner from axis till hub

R_T is the radius from axis till tip of the blade

β_1 angle at inlet with relative velocity

β_2 angle at outlet with relative velocity

V_a is the downward axial velocity

Figure 17.19 shows the inlet and outlet velocity triangles for the typical reaction turbine.

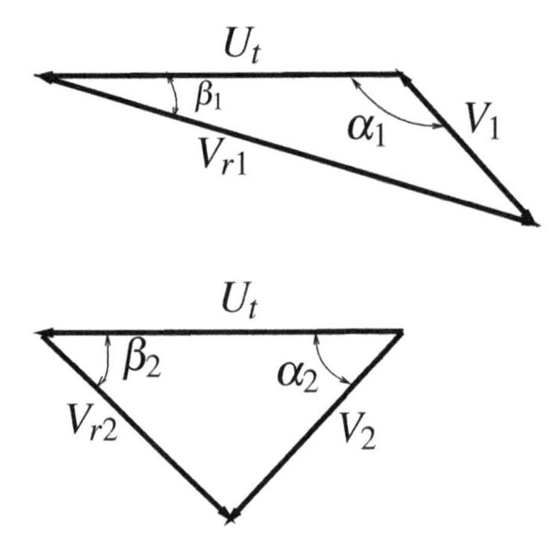

Figure 17.19 Velocity triangles at the inlet and outlet.

We can write the expression for the torque on the water:

$$T = \oint rV_\theta(\rho V \cdot dA)$$

assuming steady flow:

$$T = r_1 V_1 \cos(\alpha_1)(-\rho Q) + r_2 V_2 \cos(\alpha_2)(\rho Q)$$

$$T = \rho Q\left[r_2 V_2 \cos(\alpha_2) - r_1 V_1 \cos(\alpha_1)\right]$$

From velocity triangle gives

$$V_2 \cos \alpha_2 = U_{t2} - V_{r2} \cos \beta_2$$

The torque on turbine blades is

$$T_{turbine} = -\rho Q\left[r_2(U_{t2} - V_{r2}\cos\beta_2) - r_1 V_1 \cos(\alpha_1)\right]$$

$$T_{turbine} = \rho Q\left[r_2(V_{r2}\cos\beta_2 - U_{t2}) + r_1 V_1 \cos(\alpha_1)\right]$$

The power from turbine is

$$P = 2\pi NT = 2\pi N\rho Q\left[r_2(V_{r2}\cos\beta_2 - U_{t2}) + r_1 V_1 \cos(\alpha_1)\right]$$

$$P = N\rho Q\left[2\pi r_1 V_1 \cos(\alpha_1) - 2\pi r_2 V_2 \cos(\alpha_2)\right]$$

We can identify that the bracketed terms are the circulation along a closed path

$$\Gamma_1 = \oint (V \cdot d\ell) = 2\pi r_1 V_1 \cos(\alpha_1)$$

$$\Gamma_2 = \oint (V \cdot d\ell) = 2\pi r_2 V_2 \cos(\alpha_2)$$

Therefore the power can also be cast into the form

$$P = N\rho Q[\Gamma_1 - \Gamma_2]$$

In designing these turbines often the free vortex flow is assumed and considering the moment of momentum about the axis we can form an equation

$$V_t \cdot r = V_o R_o \cos\alpha$$

The flow rate of the water going through the blades can be computed as

$$Q = A.V = \pi(R_T^2 - R_H^2)V_a$$

This flow rate should be same as that entered through guide vanes therefore we can also write

$$Q = V_o \sin\alpha.A = (2\pi R_o b)V_o \sin\alpha$$

where b is the height of the guide vanes.

Empirical expression for runner design: deSiervo and deLeva (1976) gave the following equation for the Francis turbine runner:

$$D_{runner,Francis} = \frac{\sqrt{H}}{N}(26.2 + 0.211N_s)$$

where N is in rps, head (H) in meters, and N_s is in SI units.

$$D_{runner,Francis} = \frac{\sqrt{H_{ft}}}{N_{rpm}}\left(569.5 + 17.4N_s|_{hp,rpm,ft}\right)$$

17.11 CAVITATION IN HYDRAULIC TURBINES

The specific speed (Ns) for turbine is defined as:

$$Ns = \frac{N\sqrt{P}}{H^{5/4}}$$

where P is the shaft power, N is the rpm, and H is the available head for the turbine.

The cavitation number for the turbine is defined as:

$$\sigma_{v,turbine} = \frac{H_a - H_v - H_s}{H}$$

where

H is the working head of turbine (difference between head race and tail race levels in meters)

H_a is the atmospheric pressure head in meters of water,

H_s is the suction pressure head (or height of turbine inlet above tail race level) in meters,

H_v is the vapor pressure in meters of water corresponding to the water temperature.

The value of critical factor depends upon specific speed of the turbine.

To avoid cavitation in turbine, Thoma cavitation factor should be greater than critical cavitation factor which can be estimated by the following formulae:

Francis Turbine:

$$\sigma_{v,Francis} = 0.044\left(\frac{Ns}{100}\right)^2$$

Propeller Turbine:

$$\sigma_{v,propeller} = 0.0032\left(\frac{Ns}{100}\right)^{2.73} + 0.3$$

Kaplan Turbine:

$$\sigma_{v,Kaplan} = 1.1(\sigma_{v,propeller})$$

Example 17.10

Example An axial flow reaction turbine has the guide vane angled at $\alpha = 15°$. The volume flow rate of water is 30 m^3/s and head from head till tail water is 80 m. At the radial position of $r = 0.7$ m find the blade angle β if $R_H = 300$ mm, $R_T = 1.45$, $R_o = 2$ m and $b = 1.4$ m.

Solution We apply the first law of thermodynamics on the turbine system, assuming no heat transfer, no change in kinetic energy and enthalpy. The Euler's turbomahinery equation then can be written as:

$$\dot{W}_{shaft} = \dot{m}\left[U_2V_{\theta 2} - U_1V_{\theta 1}\right] = \rho Q g \Delta H$$

$\rho Q g \Delta H$ represents the rate of work due to change in potential head only.

$$g\Delta H = \left[U_2V_{\theta 2} - U_1V_{\theta 1}\right]$$

Also, we assume that tangential velocity is small at the outlet of axial flow turbine runner:

$$g\Delta H = \left[\cancel{U_2V_{\theta 2}} - U_1V_{\theta 1}\right]$$

$$g\Delta H = -U_1V_{\theta 1}$$

$$g\Delta H = -(r_1\omega)V_{\theta 1}$$

$$V_{\theta 1} = \frac{g\Delta H}{-(r_1\omega)} = \frac{9.81\times(-80m)}{-(0.7\times\omega)} = \frac{1121}{\omega}$$

Also from the free vortex design considerations

$$V_{\theta 1} = \frac{V_o R_o \cos\alpha}{r_1} = 18.20$$

This gives

$$\omega = 61.59 rad/s$$

Since we can calculate flow rate at inlet of guide vanes as well as in turbine runner, we have

$$Q = 2\pi b R_o V_o \sin\alpha$$

$$V_o = 6.595 m/s$$

Also

$$Q = V_a \pi (R_T^2 - R_H^2)$$

$$V_a = \frac{Q}{\pi(R_T^2 - R_H^2)} = 4.747 m/s$$

$$\beta_1 = \tan^{-1}\left(\frac{V_a}{r_1\omega - V_{\theta 1}}\right) = 10.79^\circ$$

Note that for the power from entire turbine we have to repeat this procedure at different radial positions on the blade which would give us variation in blade angle. The entire power would then be the summation of all the powers at different radial positions.

17.11.1 DRAFT TUBE

In the reaction turbine, the net head H shall be the difference between the energy levels upstream of the turbine and that of the tailrace. This can be written as

$$H_{net} = H_B - H_c = \left[z_B + \frac{p_B}{\gamma} + \frac{V_B^2}{2g}\right] - \left(\frac{V_C^2}{2g}\right)$$

The effective head available to act on the runner of a reaction turbine is

$$H_{eff} = H_{net} - \underbrace{K\frac{(V_1 - V_2)^2}{2g}}_{head\ loss\ due\ to\ draft\ tube} - \underbrace{\left[\frac{V_2^2}{2g} - \frac{V_C^2}{2g}\right]}_{head\ loss\ at\ discharge}$$

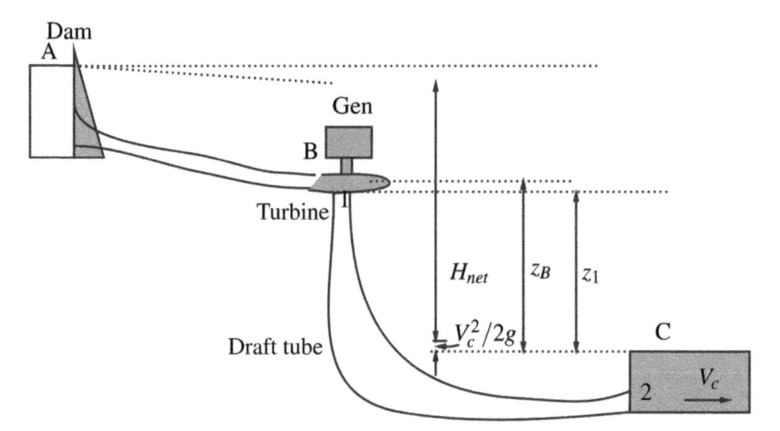

Figure 17.20 Net head in reaction turbine z_B is the draft head and V_c is the velocity in the tailrace.

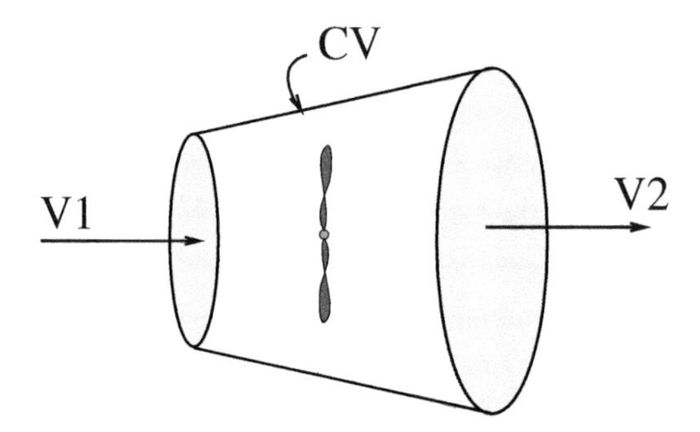

Figure 17.21 Control volume surrounding a wind turbine.

17.12 AXIAL FLOW WIND TURBINES

The horizontal axis wind turbines are a type of axial flow turbines. For analysis of such turbines we consider a control volume enclosing the wind turbine as shown in Figure 17.21. The incoming velocity is identified as V_1 and outflow velocity is V_2.

Figure 17.21 shows the control volume around a horizontal axis wind turbine. The flow is assumed to be steady and incompressible. We assume that the average wind speed through the rotor area of wind turbine is the average of the undisturbed wind speed before and after the wind turbine:

We start with estimating the mass flow rate through this control volume assuming that its a unidirectional flow with no mass leaving from the peripheral walls of the control volume.

$$\dot{m} = \rho V_m A = \rho A \left(\frac{V_1 + V_2}{2} \right)$$

where V_m is the mean velocity. We assume that this mean velocity is related with both inflow and outflow as

$$V_m = \frac{V_1 + V_2}{2}$$

The above formula is also known as the Froude's theorem (1883).

$$\dot{m} = \rho V_m A = \rho A \left(\frac{V_1 + V_2}{2} \right)$$

The rate of change of momentum through the control volume is

$$\Delta \dot{M} = \dot{m} V_2 - \dot{m} V_1 = \dot{m}(V_2 - V_1)$$

$$\Delta \dot{M} = \rho A \left(\frac{V_1 + V_2}{2} \right) (V_2 - V_1)$$

simplified as

$$\Delta \dot{M} = \frac{\rho A}{2}(V_2^2 - V_1^2)$$

We know that rate of change of momentum is force so

$$F = \frac{\rho A}{2}(V_2^2 - V_1^2)$$

The negative of this will be the reactive force of the turbine.

$$F = -\frac{\rho A}{2}(V_2^2 - V_1^2)$$

We now write the power developed by turbine rotor as

$$P = F \cdot V_m = -\frac{\rho A}{2}(V_2^2 - V_1^2) \left(\frac{V_1 + V_2}{2} \right)$$

$$P = \frac{-\rho A}{4}(V_2^2 - V_1^2)(V_1 + V_2)$$

Also, in case of no rotor the power of air would be

$$P_o = F \cdot V = \frac{\rho A}{2} V^2 \cdot V = \frac{\rho A}{2} V_1{}^3$$

We take the ratio of the two powers and simplification would lead to the following result:

$$C_p = \frac{P}{P_o} = \frac{1}{2} \left[(1 - (V_2/V_1)^2)((V_2/V_1) + 1) \right]$$

The parameter $\frac{P}{P_o}$ is called the power coefficient C_p in the literature. The plot of above equation is shown in Figure 17.22. As can be seen, the power is maximum

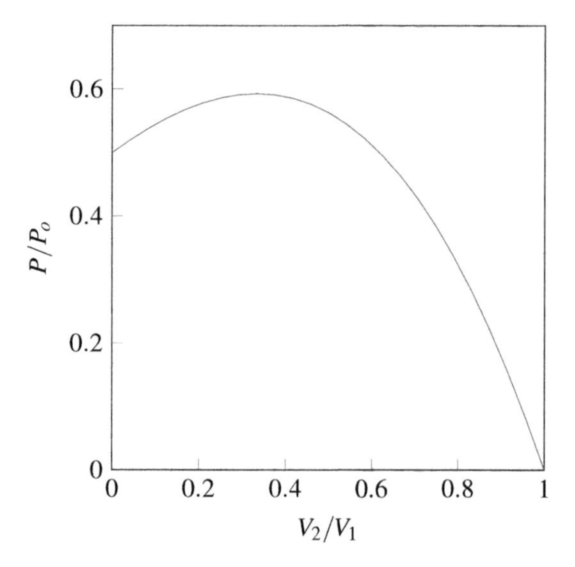

Figure 17.22 The power coefficient for an axial flow wind turbine vs. the speed ratios.

Table 17.3
λ and η_{max} Values of Some Modern Wind Turbines

Classification	Type	λ	η_{max}
Horizontal axis	Classic Windmill	1.3–2.5	30
	Modern two blade rotor	5.4–9.5	47
	Modern three blade rotor	2.3–5.3	46
	Loopwing turbines	1.5–4	40
Vertical axis	Savonius	<1	25–35
	Darrieus/Gorlov hybrid	14	25–35

when V_2/V_1 is 1/3, and the maximum value for the power extracted from the wind is 0.592 of the total power in the incoming wind. Betz's limit thus an added limit on power extraction from wind turbines along with the second law of thermodynamics (see Figure 17.22).

The turbines were analyzed over a variety of tip speed ratios defined as

$$\lambda = \frac{r \cdot \omega}{U_{in}}$$

where U_{in} is the incoming velocity.

The parameter is often plotted with turbine efficiency or power coefficient.

PROBLEMS

17P-1 A centrifugal water pump handling $Q = 0.065\frac{m^3}{s}$ has 17 cm diameter impeller and axial inlet flow is driven at 180 radians/s. The impeller vanes are backward-curved ($\beta_2 = 63$ degrees) and have axial width $b_2 = 3$ cm. Estimate the volume flow rate, head rise, power input and pump efficiency at the maximum efficiency point, if the torque has to overcome spin, bearing, and seal losses of 8% of the ideal torque. The pump casing acts as a diffuser, which converts 60% of the absolute velocity head at the impeller outlet to static pressure rise. Also, and the head loss through the pump suction and discharge channel is 0.75 times the radial component of velocity head leaving the impeller.

17P-2 A pump with $D = 600\ mm$ delivers $Q = 0.830\frac{m^3}{}s$ of water at $H = 15\ m$ at its best efficiency point. If the specific speed of the pump is 1.72 and the required motor input is 100 kW, find the shutoff head, H_o, and best efficiency, η. If the pump is required to run at another speed of 950 rpm, by scaling the performance curve, estimate the new flow rate, head, shutoff head and required power. The specific speed is defined in units of radians per second, meter and cubic meter per second.

17P-3 A centrifugal water pump operates at 1840 rpm; the impeller has backward-curved vanes with $\beta_2 = 60$ degree and $b_2 = 1.27\ cm$. At a flow rate of $0.03\frac{m^3}{s}$, the radial outlet velocity is $V_{n,2} = 3.34\frac{m}{s}$. Estimate the head this pump could deliver at 1350 rpm.

17P-4 A pump with $D = 525\ mm$ delivers $Q = 0.657\ m^3/s$ of water at $H = 13\ m$ at its best efficiency point. If the specific speed of the pump is 1.73 and the required input is 90 kW, determine the shutoff head, H_0, and best efficiency, η. It is needed that pump now run at 950 rpm. Using affinity laws find the new flow rate and head developed. The specific speed is defined in units of radians per second, meter and cubic meter per second. The density of water is 980 kg/m^3.

REFERENCES

Jean-Pierre Franc and Jean-Marie Michel, Fundamentals of Cavitation, Kluwer Academic Publishers, Netherlands, 2005.

J. J. Doland, Hydropower Engineering. New York: The Ronald Press Company, 1954.

F. de Siervo and F. de Leva, Modern Trends in Selecting and Designing Francis Turbines, Water Power and Dam Construction, Vol. 28, No. 3, 1976.

Index

9781032324531